Southern Hemisphere Paleo- and Neoclimates

Springer-Verlag Berlin Heidelberg GmbH

Peter Smolka • Wolfgang Volkheimer (Eds.)

Southern Hemisphere Paleo- and Neoclimates

Key Sites, Methods, Data and Models

With 133 Figures and 27 Tables

 Springer

EDITORS:

Dr. Peter Smolka
University of Münster
Geological Institute
Corrensstr. 24
48149 Münster
Germany
E-mail: smolka@uni-muenster.de

Prof. Dr. Wolfgang Volkheimer
Instituto Argentino de Nivología,
Glaciología y Ciencias Ambientales
Casilla de Correo 330
5500 Mendoza
Argentina
E-mail: volkheim@lab.cricyt.edu.ar

ISBN 978-3-642-64089-6   ISBN 978-3-642-59694-0 (eBook)
DOI 10.1007/978-3-642-59694-0

Additional material to this book can be downloaded from http://extras.springer.com
Library of Congress Cataloging-in-Publication Data
Smolka, Peter, 1960-Southern Hemisphere paleo- and neoclimates: key sites, methods, data and
models / Peter Smolka, Wolfgang Volkheimer (eds.). p. cm. Includes bibliographical references.
ISBN 978-3-642-64089-6   1.Paleoclimatology--Southern Hemisphere. 2. Southern Hemisphere--Climate.
I.Volkheimer, Wolfgang.II.Title QC884 .S66 2000 551.69181'4--dc21 00-028525

Cover Design: *design & production*, Heidelberg
Typesetting: Camera-ready by the editors

SPIN: 10680501   30/3136/xz – 5 4 3 2 1 0 – Printed on acid free paper

"This book is dedicated to

Dr. Vladislav Babuska

who,
as the
Scientific Secretary of the International Geological Correlation Program,

has helpfully guided and stimulated, through difficult times, many projects like this. The project members acknowledge gratefully his support and that of IGCP/IUGS/UNESCO."

Acknowledgement:
A project like IGCP-341 covers many aspects and involves a large number of people and working groups. While the direct results of several groups appear in this volume, others contributed significantly to the progress of the project through their cooperation, training courses, advice, data testing and helpful suggestions. In this context, special thanks are directed to Dr. Bruce Malamud (Cornell University, USA) who supported the project not only with enlighting training courses on the application of the theory of fractals to the interpretation of geological, paleoclimatological and paleooceanographical time-series, but also through dedicated efforts in the phase of final proofreading.

Contents

Chapter 3: Quaternary Climates

Chapter 4: Prequaternary Climates

Chapter 5: Modeling

List of Contributors

Anhuf, D.
Dr., Geographical Institute, University of Mannheim, L 9, 1-2, D-68131 Mannheim, Germany
anhuf@rumms.uni-mannheim.de

Barros, V.R.
Prof. Dr., Head of the Department of Atmoshperic Sciences, University of Buenos Aires, Pabellon II, 2nd Piso, Ciudad Universitaria, 1427 Buenos Aires, Argentina
barros@at1.fcen.uba.ar
barros@cw.at.fcen.edu.ar

Beckmann, G.
Dr., HUELS AG (Condea), Paul Baumann Str. 1, Postfach 1320, D-45772 Marl, Germany

Borromei, A.M.
Universidad Nacional del Sur, Departamento de Geologia, Catedra de Geologia Historica, Laboratorio de Palinologia. 8000 Bahia Blanca, Argentina
mquattro@criba.edu.ar

Calvo, A.
Prof. Dr., Instituto de Física Rosario (IFIR), UNR–CONICET, 27 de Febrero 210 bis, 2000 Rosario, Argentina
Rafa@sedal.usyd.edu.au
Rafa+@cs.cmu.edu

Castañeda, M.E.
Dr., Department of Atmoshperic Sciences, University of Buenos Aires, Pabellon II, 2nd Piso, Ciudad Universitaria, 1427 Buenos Aires, Argentina
Eliza@at1.fcen.uba.ar

Ceccatto, H.A
Professor, Instituto de Física Rosario (IFIR), UNR–CONICET, 27 de Febrero 210 bis, 2000 Rosario, Argentina
Cecatto@ifir.ifir.edu.ar

Compagnucci, R.H.
Dr., Department of Atmoshperic Sciences, University of Buenos Aires, Pabellon II, 2nd Piso, Ciudad Universitaria, 1427 Buenos Aires, Argentina
Rhc@at1.fcen.uba.ar

Corbella, H.
Dr., Museo Argentino Ciencias Naturales "B. Rivadavia", y Instituto Nacional de Investigacion de las Ciencias Naturales
Av. Angel Gallardo 470, CP. 1405, Buenos Aires, Argentina
Hcorbella@ibm.net

Doyle, M.
Dr., Department of Atmoshperic Sciences, University of Buenos Aires, Pabellon II, 2nd Piso, Ciudad Universitaria, 1427 Buenos Aires, Argentina
Doyle@at1.fcen.uba.ar

Dussel, P.
Dr., Instituto Argentino de Nivologia, Glaciologia y Ciencias Ambientales, Unidad de Historia Ambiental, Departamento de Dendrocronologia e Historia Ambiental, IANIGLA–CRICYT, Casilla de Correo 330, 5500 Mendoza, Argentina
mprieto@lab.cricyt.edu.ar

Espizua, L.
Dr., Instituto Argentino de Nivología, Glaciología y Ciencias Ambientales (IANIGLA–CRICYT), Casilla de Correo 330, 5500 Mendoza, Argentina
lespizua@lab.cricyt.edu.ar

Hämmerle H.,
Dr., NMI, University Tübingen, Gustav Werner Str. 3, D-72762 Reutlingen, Germany

Herrera, R.R.,
Dr., Instituto Argentino de Nivologia, Glaciologia y Ciencias Ambientales, Unidad de Historia Ambiental, Departamento de Dendrocronologia e Historia Ambiental, IANIGLA–CRICYT, 5500 Mendoza, Argentina
mprieto@lab.cricyt.edu.ar

Inacker, I.
Dr., NMI, University Tübingen, Gustav Werner Str. 3, D-72762 Reutlingen, Germany

Jacovkis, P.M.,
Prof. Dr., Instituto de Cálculo and Departamento de Computación, Facultad de Ciencias Exactas y Naturales, Universidad de Buenos Aires, Ciudad Universitaria, 1427 Buenos Aires, Argentina
jacovkis@dc.uba.ar
jacovkis@decanato.de.fcen.uba.ar

Klopries, B.
Dr., HUELS AG (Condea), Paul Baumann Str. 1, Postfach 1320, D-45772 Marl, Germany
burkhard.klopries@contensio.de

Latrubesse, E.M.
Prof. Dr., Universidade Federal de Goias-IESA, Campus Sanambaia, 74001-970, Goiana, GO, Brazil
Latrubes@virtualhouse.com.br
Latrubes@iesa.ufg.br

Leiva, J.C.
Dr., Instituto Argentino de Nivología, Glaciología y Ciencias Ambientales (IANIGLA–CRICYT), Casilla de Correo 330, 5500 Mendoza, Argentina
jcleiva@lab.cricyt.edu.ar

Lirio, J.M.
Dr., Instituto Antartico Argentino
Cerrito 1248, 1010 Buenos Aires, Argentina
Liriojm@yahoo.com

Llorens, R.
Lic., Instituto Argentino de Nivología, Glaciología y Ciencias Ambientales (IANIGLA–CRICYT), Casilla de Correo 330, 5500 Mendoza, Argentina

Markgraf, V.
Prof. Dr., Institute of Arctic and Alpine Research, University of Colorado, Boulder Colorado 80309, USA and
PAGES International Project Office, Baerenplatz 2, CH-3011 Bern, Switzerland
Markgraf@spot.colorado.edu
Markgraf@pages.unibe.ch

Nabel, P.
Dr., Museo Argentino Ciencias Naturales "B. Rivadavia", y Instituto Nacional de Investigacion de las Ciencias Naturales
Av. Angel Gallardo 470, CP 1405, Buenos Aires, Argentina
Pnabel@gecuat.gov.ar
Penabel@mail.retina.ar

Navone, H.D.
Professor, Instituto de Física Rosario (IFIR), UNR–CONICET, 27 de Febrero 210 bis, 2000 Rosario, Argentina
h.navone@elec.uq.edu.au

Norte, F.
Dr., Instituto Argentino de Nivología, Glaciología y Ciencias Ambientales (IANIGLA-CRICYT), Casilla de Correo 330, 5500 Mendoza, Argentina
Fnorte@lab.cricyt.edu.ar

Nuñez, H.J.
Dr., Instituto Antartico Argentino
Cerrito 1248, 1010 Buenos Aires, Argentina

Peschel, G.
Prof. Dr., Ernst-Moritz Arndt Universität, Greifswald, Institut für Geologische Wissenschaften, Germany

Petersen, N.
Prof. Dr., Head of Biomagnetics Group, Institut für Allgemeine und Angewandte Geophysik, Ludwig Maximilians Universität München, Theresienstr. 41, D-80333 München, Germany
Petersen@magbact.geophysik.uni-muenchen.de

Prieto, M.R.
Prof. Dr., Instituto Argentino de Nivologia, Glaciologia y Ciencias Ambientales, Unidad de Historia Ambiental, Departamento de Dendrocronologia e Historia Ambiental, IANIGLA–CRICYT, Casilla de Correo 330, 5500 Mendoza, Argentina
mprieto@lab.cricyt.edu.ar

Quattrocchio, M. E.
Prof. Dr., Universidad Nacional del Sur, Departemento de Geologia, Catedra de Geologia Historica, Laboratorio de Palinologia. 8000 Bahia Blanca, Argentina
mquattro@criba.edu.ar

Rinaldi, C.A.
Dr., Instituto Antartico Argentino
Cerrito 1248, 1010 Buenos Aires, Argentina

Runge, J.
Dr., University of Paderborn, Department of Physical Geography, D-33095 Paderborn, Germany
arung1@hrz.uni-paderborn.de

Smolka, P.
Dr., Geological Institute, University Muenster, Corrensstr. 24, D-48149 Muenster, Germany
Smolka@uni-muenster.de,

Tatur, A.
Dr., Institute of Ecology, Polish Academy of Sciences
Dziekanow Lesny k., Warszawy, 05-092 Lomianky, Poland

del Valle, R.A.
Dr., Instituto Antartico Argentino
Cerrito 1248, 1010 Buenos Aires, Argentina

Villalba, R.
Dr., Laboratorio de Dendrocronologia, IANIGLA–CRICYT, Casilla de Correo 330, 5500 Mendoza, Argentina and
Tree Ring Laboratory, Lamont-Doherty Earth Observatory, Columbia University, Palisades, NY 10964, USA
Ricardo@lab.cricyt.edu.ar

Volkheimer, W.
Prof. Dr., Director of the Instituto Argentino de Nivología, Glaciología y Ciencias Ambientales (IANIGLA–CRICYT), Casilla de Correo 330, 5500 Mendoza, Argentina
Volkheim@lab.cricyt.edu.ar

Wingenroth, M.C.
Dr., Instituto Argentino de Nivología, Glaciología y Ciencias Ambientales (IANIGLA–CRICYT), Casilla de Correo 330, 5500 Mendoza, Argentina
Wingenro@lab.cricyt.edu.ar

Introduction

P. Smolka, W. Volkheimer

Climate change and the impact of climate change will, if it happens, affect societies all over the world, not only in the Southern Hemisphere or at shallow coasts, but worldwide. This includes direct changes because of temperature and precipitation changes, as well as indirect changes such as vegetation belts. In addition, generally overlooked phenomena such as microbiological changes may occur, as CO2 is an important factor for several enzyme reactions, not only in the higher biota, but also in bacteria and viruses.

Growing insight arising from climate models (for example coupled ocean–atmosphere models) has shown that climate change may not be uniform worldwide (that is, describable through an increase of the *average* temperature by a certain amount). Climate change, if it happens, will be of a *differential* nature. This means that areas of increased temperature will be accompanied by areas of decreased temperature; areas with increased precipitation will be bordered by areas of drought. This is (or was) one of the reasons why climate change is so difficult to detect, as any parameter must be considered and tested locally (regionally) and not averaged globally. Furthermore, important key processes such as the removal of CO2 from the atmosphere through deep-water formation depend on subtle density equilibriums, difficult to reconstruct and even more difficult to predict.

In addition, *geological* studies carried out in totally different research programs, show, that climate change has been rapid, differential *and* of considerable magnitude (see for example, Bryson 1992; Dodson et al. 1993; McLean 1980; and Dowsett et al. 1994). Examples of programs include the ODP (Ocean Drilling Program), the IGBP (International Geosphere Biosphere Program), and the IGCP (International Geological Correlation Program). Evidence comes from pollen diagrams, stable isotopes, ice-core data, faunal and floral census, and transfer algorithms/functions based on these parameters. Many of these studies address both historical periods and the Quaternary, for example situations 18 thousand year ago, the Eem interglacial, or pronounced temperature-fluctuations that occurred during the last deglaciation event(s).

Although such "ice-house studies" contribute to knowledge of the mechanisms, timescales and especially magnitudes of climate change, they provide only limited insight into situations we might face in the future, namely "greenhouse scenarios." Fortunately, greenhouse scenarios (or better) "climates with temperatures warmer than the present at certain locations" did exist. They may have existed in historical times (the climate optimum around the 12th century), the Eem interglacial, and of course several times in the Neogene. Looking back into the past may be a key for assessing the future. Of course, not in the sense of prediction, but in the sense of checklists (maps and data) that show *how* times with *known* overall conditions (such as average warming) may have looked, including vegetation distributions, ocean currents, rain-fall patterns and other important economic parameters.

Three examples show the importance of this approach as it may, even for civil engineering and land-use planning, have a sometimes overlooked potential.

The first example is that the water supply of many cities depends directly or indirectly on glacier-runoff. Although it is often suspected that climate change may endanger these reservoirs (which is often true) the differential nature of climate change may increase precipitation in the form of snow in some areas, while in others more pessimistic scenarios apply. Consequently, study of historical and pre-historical glacier lines, especially when focusing on warmer times (12th century, Eem) may provide useful analogs for long-range reservoir planning.

Similar analogs exist for land-use planning: Here, changes of vegetation-belts may, even on a local and regional scale, be, depending on the type of land-use, be either favorable or non-favorable (or even both) for economy. The differential nature of expected climate change means however that the results of one study area cannot be transferred to another area. It requires that experiences and methodologies may be transferred, but in each area of the world, the relevant studies must be executed again. For this complex set of questions, IGCP-341 presents several case studies from the Mendoza area in central South America. Here studies on glacier advances (and retreats) are included as well as examples of studies on historical and prehistorical vegetation changes.

On a larger scale, climate change affects whole regions. Case studies focusing on natural deforestation and reforestation of Amazonia, and related phenomena in Central Africa, show the wide range of both natural variability and environments man has to face if he conducts experiments in non-linear systems, as shown by three key-studies from these areas. Consequently, a further key-study dedicated to the statistical assessment of already ongoing climate change in South America is included. This shows the differential nature of climate change as well as a decrease of precipitation in an area adjacent to Southern Amazonia.

The phenomena discussed until now are more or less regional and short-term fluctuations. An important question however is: Would or could climate change affect the ocean-circulation and if yes: Are the fluctuations of regional importance or could (did) the circulation pattern change even qualitatively? This question is of paramount importance, since the removal of CO_2 from the atmosphere depends crucially on the operation of the "global conveyor belt."

Consequently, for various Neogene time slices, faunal and floral communities, mainly from DSDP and ODP drillsites, have been studied to establish time series of temperatures well back into the Neogene. The synoptic study of hemisphere and worldwide-distributed time series means however that a worldwide uniform stratigraphic database exists. Therefore, within IGCP-341, a worldwide uniform stratigraphic standard with data (age ranges) for the Neogene and Paleogene was established. Stratigraphic age-ranges for about 80 000 fossils expressed in million years are now available.

Many atmospheric parameters such as precipitation and wind fields cannot be reconstructed directly. Therefore, the reconstructed oceans, which represent an equilibrium of the interior ocean dynamics, are coupled with an atmospheric general circulation model (ccm3.6 from NCAR). For reasons of convenience and easy access, this model is adapted from Cray environments to Windows NT (see accompanying CD).

After having studied ice-house scenarios thoroughly, the scientific community has focused more and more on green-house scenarios, whether these represent Neogene (IGCP-341), Paleocene/Eocene, or Cretaceous equilibria. As the purpose of IGCP-341 has been to integrate data, reconstructions and modeling it has sometimes been thought of as the "CLIMAP of the warmer times." This is only partially correct, although worldwide maps, databases and models are provided. What IGCP-341 however has done and what this book provides for the reader, are methods and case studies that show new pathways when addressing climate change.

The reader will find in this book a variety of methods, ranging from conventional ones already well established in the literature, to others that are not as commonly used, such as transfer algorithms for paleotemperature assessment (both in the terrestrial and marine realm) and neural networks.

It is the intent of IGCP-341 that this book be applicable to a wide variety of disciplines and levels, including geologists, land-use planners, meteorologists testing models in non-analogue situations, paleoclimatologists, university professors teaching "paleoclimatology beyond isotopes" or even the reader who is "just interested in the Earth." Readers might find that our Earth is even more fascinating (and worth to protect) than they thought.

For the project members

Peter Smolka   Wolfgang Volkheimer

References

Bryson RA (1992) A Macrophysical Model of the Holocene Intertropical Convergence and Jetstream Positions and Rainfall for the Saharan Region. Meteor. and Atmos. Phys. 47:247–258

Dodson JR, Fullagar R, Head L (1993) Dynamics of environment and people in forested crescents of temperate Australia. In: Dodson JR (ed): The Naive Lands: prehistory and environmental change in Australia and the southwest Pacific. Longman, Melbourne

Dowsett H, Thomson R, Barron J, Cronin T, Fleming F, Ishman S, Poore R, Willard D, Holtz Jr. T (1994) Joint investigations of the Middle Pliocene climate I: PRISM paleoenvironmental reconstructions. Global and Planetary Change, 9(3-4):169–196

McLean R (1980) The land-sea interface of small tropical islands: Morphodynamics and man. In: Brookfield HC (ed) Population-environment relations in tropical islands: The case of eastern Fiji. UNESCO/UNFPA technical notes 13:149–175

Chapter 1: Methods

This first chapter is on methods for the analysis of paleoclimates and neoclimates. The methods presented in these six works include neural networks, modeling fluid flow, dendroclimatology, atmospheric modeling, paleomagnetism and the creation of pole–equator transects in reconstructing environmental change.

Among the most challenging questions of global change is prediction of the future. Thus, the first contribution in this chapter (Calvo et al.) is dedicated to neural networks. They show how to forecast the Indian Monsoon based on past monsoon data. Actual data and examples can be found on the accompanying CD.

Predicting the development of river networks is of high importance for civil engineering. Knowledge of preferred future river paths, which follow precipitation changes (see also Barros et al., chapter 2), permits not only the optimization of maintenance works but supports optimum land-use planning. Furthermore "prediction" of past river networks supports geological reconstructions as the combination of reconstructions AND modeling may fill important gaps in knowledge. Applications of the work by Jacovkis lie not only in paleoclimatology, but also in the oil industry (reservoir sandstones).

The work of Villalba shows the high potential of dendroclimatology, especially if applied on a hemisphere-wide scale. This includes the assessment of ENSO anomalies (El Nino Southern Oscillation) as well as the reconstruction of past teleconnections, temperature and precipitation gradients on a hemisphere-wide scale.

Climate change not only affects rocks, sediments and isotopes. The constituents of the atmosphere, namely O_2 and CO_2 take part in important enzymatic reactions. As microbiota have CO_2 levels that are comparable to the atmospheric level of CO_2, a change of the CO_2 concentration by some ppm is a major environmental stress for such biota. It could be expected that environmental stress causes organisms to adapt accordingly. Beckman et al. used a common approach in chemistry and microbiology: They consider the geosystem as a linked chemical reactor and study self organizing processes that governed the chemical history of the atmosphere. In addition, chemical aspects of climate change, such as changes of the pH of the oceans, are considered with microbiological aspects, such as the potential for new pests and diseases.

Nabel and Petersen used paleomagnetic studies to approximate past climate changes in Pampean soils. While in the "cold" Chinese loess plateau this method is well established, the generalization to other environmental conditions, especially the transfer to warm and arid environments, needs some modifications.

Finally, in this chapter, Markgraf presents a synopsis of current knowledge of environmental changes that have been observed along one of the Pole-Equator-Pole transects.

Neural Network Analysis of Time Series: Applications to Climatic Data

R.A. Calvo[1], H.D. Navone[2], H.A. Ceccatto[3]
Instituto de Física Rosario (IFIR), UNR-CONICET, 27 de Febrero 210 bis, 2000
Rosario, Argentina
(1) rafa@sedal.usyd.edu.au, rafa+@cs.cmu.edu
(2) h.navone@elec.uq.edu.au
(3) cecatto@ifir.ifir.edu.ar, neurus@ifir.ifir.edu.ar

Abstract: Artificial neural networks are parallel computational algorithms that can simulate very efficiently complex dynamical systems. In this work we show, by means of real-world applications, that this technique can be a useful tool in the analysis of time-series data related to climate. We study two time series: the first one characterizes the solar activity as measured by the annual mean value of the sunspot number (Wolf number); the second one is the record of summer monsoon rainfall over India. Both records are often used in the literature as benchmarks for testing new statistical techniques. From these studies we conclude that artificial neural networks can advantageously substitute conventional methods of time series analysis. Moreover, they reveal themselves as a promising way of making predictions of climatic phenomena.

1.1. Introduction

One of the basic tenets of modern meteorology is making predictions. With this purpose in mind, there are at least three well defined ways of proceeding, depending on the degree of understanding and complexity of the phenomenon under consideration. The first, most satisfactory approach, consists of developing comprehensive physical/numerical models, of which the so-called general circulation models for climate simulation (WMO 1986) are a prime example. When this "first principles" approach can not be pursued, but still some understanding of the phenomenon guides the intuition of investigators, making predictions can still be possible by developing empirical approaches. In meteorology, diagnostic studies of historical datasets, and the selection of suitable predictor parameters by an objective search, have been traditionally very useful. Finally, when there is not even a reasonable qualitative understanding of the phenomenon at hand, or when it is the result of complex interactions involving many independent and irreducible degrees of freedom, it is always possible to blindly resort to more or less sophisticated statistical techniques. For instance, linear methods of analyzing time series from weather and climatic processes have had some success, although their predictive power is limited when confronted with systems whose irregular behaviour is a result of low-dimensional chaos (Farmer and Sidorowich 1988). In this case, advances in the last decade in the theory of dynamical systems, and the development of nonlinear methods (Tong 1983), offer some hope in weather forecasting (Elsner and Tsonis 1992).

As the most important objective of the World Climate Research Programme, climate prediction has received increased attention in recent years (Hastenrath 1985). On the other hand, the problems of weather and climate forecasting offer a unique area for testing and developing nonlinear algorithms. In this sense, artificial neural networks (ANN) have been recently established as a reliable tool for time-series analysis (Weigend et al. 1990; Elsner and Tsonis 1992).

The purpose of this work is to show that these computational architectures can be advantageously used as a replacement of standard linear statistical techniques. Instead of working on computer-generated data, this will be exemplified with two problems of practical interest: The first one is the study of the solar activity as measured by the annual mean value of the sunspot number (Wolf number); the second is the record of summer monsoon rainfall over India. Both records are often used in the literature as benchmarks for testing new statistical methods. The Wolf sunspot number gives an idea of the amount of radiation and charged particles emited by the sun. Variations in solar activity are of interest since they could produce changes in the global environment and, perhaps, influence the climate. The second series corresponds to the mean rainfall produced by the summer (June–September) monsoon in India. Predictions of this index are of paramount importance for agricultural planning in this country. On the other hand, the global implications of this phenomenon, and its connections with similar ones in other parts of the world, have created a widespread interest in its prediction.

First, we give a brief introduction to ANN and describe their training process. Second, we apply them to the analysis of the above mentioned time series. Finally, we compare the results obtained with those produced by standard statistical techniques, allowing us to establish the reliability of our approach. From these studies we conclude that ANN are a new and promising way of making predictions of climatic phenomena.

1.2 Artificial Neural Networks: A Capsule Introduction

Artificial neural networks are parallel computational structures of highly interconnected simple processors, called neurons, which simulate to some extent the structure and functioning of the brain (Rumelhart and McClelland 1986). In particular, in this work we will use the so called "feed-forward" ANN (Fig. 1.1). These architectures have a group of neurons, the input layer, which are fed by external stimuli (I_k). The input units send these stimuli to hidden neurons (not connected to the environment), grouped in one or more internal layers. These hidden units

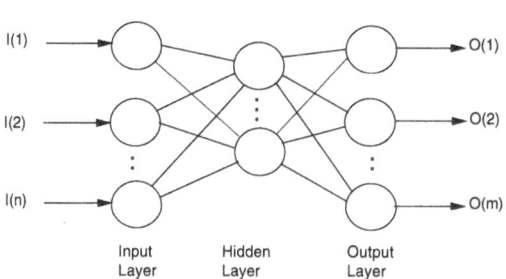

Fig. 1.1. Feed-forward artificial neural network.

process the information they receive, and pass their results to the last group of neurons, the output layer. Neurons in the output layer produce the final response ($\mathbf{O}_k$) to the external stimuli. The units are connected through information channels, whose strengths ("weights") have to be determined in order to properly relate inputs to desired outputs. The computations carried out inside each neuron amount to: (i) performing the weighted average of its impinging inputs; (ii) sending this average through a (biased) sigmoid function; (iii) forwarding the sigmoid function output to the next layer of neurons (Fig. 1.2). These calculations are generally performed synchronously by all neurons in a given layer, so that the stimulus-response delay depends on the number of layers. The process of adjusting weights and biases is known as network training, and the algorithm which performs this task is the so called "backpropagation rule" (Rumelhart et al. 1986). It essentially consists of a gradient-descent algorithm to reduce the error between actual and desired network outputs, which modifies weights and biases going backward from the output-layer to the input-layer connections. Despite the simplicity of the calculations involved, a feed-forward ANN with a sufficient number of hidden units is capable of performing any arbitrary mapping of n-dimensional inputs to m-dimensional outputs (Cybenko 1988, 1989). Although in general the optimal network architecture is not known, the nature of the problem often gives hints for selecting the proper number of input and output neurons. In the following, we indicate an ANN with n input units, h hidden units and m output units by $n{:}h{:}m$.

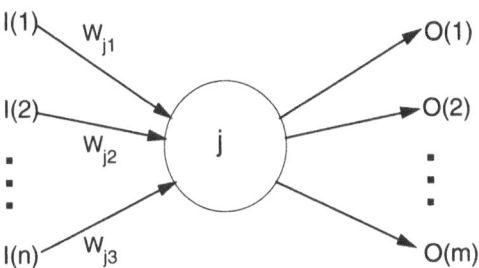

Fig. 1.2. The jth neuron with weights $\mathbf{W}_j$.

Typically, solving a problem by ANN requires the following: (a) choosing a suitable network architecture (number of layers and number of neurons in each layer); (b) selection of a large, representative set of training patterns (input–output pairs carrying enough information on the data set), (c) training the network to relate the inputs to the corresponding outputs, by modifying weights and biases according to the backpropagation rule. As mentioned above, this corresponds to minimizing an error function

$$Err = \sum_{k=1}^{N} (\mathbf{T}_k - \mathbf{O}_k)^2,$$

where $\mathbf{T}_k$ is the desired output ("target"), $\mathbf{O}_k$ the actual output produced by the network as a response to input $\mathbf{I}_k$, and N the total number of training patterns. In general every pair ($\mathbf{I}_k$, $\mathbf{T}_k$) has to be presented many times to the network in order for it to (approximately) learn the mapping. The length of the training process is usually measured in terms of single presentations of the whole training set ("epochs"). A frequent pitfall in this part of the process is getting trapped in high local minima of the error function, which can be partially avoided by judiciously modifying the parameters that control the gradient-descent algorithm ("learning rate" and "momentum"). A successful training experiment generally allows the network to capture the essential relationships among inputs and outputs. In such cases, the trained net shows remarkable generalization capabilities, being able to relate input–output pairs not included in the training set. However, the architecture used is crucial: "smaller-than-needed" nets do not learn the examples, while "bigger-than-required" nets in most cases overfit the data, learning undesirable (noisy) features which degrade their generalization performance. In the latter case, a way to cope with the problem is keeping a small number of input–output examples without being presented to the network, to monitor its generalization performance on this "cross-validation" set while learning the rest of the training data. The generalization performance is usually appraised by following the time evolution of the Normalized Mean Square Error, $NMSE = (Err) / (\sigma^2 M)$, where M is the number of patterns in the cross-validation set and σ is the standard deviation of the validation targets (Note that $NMSE = 1$ for a network that only learns to predict the target average). In this work we will use this method to stop the training processes before overfitting the noisy time series of solar activity and monsoon rainfall. Other useful quantities to characterize the network performance are the correlation coefficient between predicted and actual targets (CC) and the explained variance (EV) (ratio of variances of predicted and desired outputs). Good reference books on ANN are, for example, Hertz et al. (1991) and Bishop (1995).

1.3. Prediction of Solar Activity

The solar activity is characterized, among other quantities, by means of the relative Wolf sunspot number. In this section we will concentrate our studies on this index, since it has the largest data set compared to other activity measurements. In particular, we will consider the predictability of its annual mean value. The time series of the Wolf number shows a frequency-spectrum with a large maximum at around 11.0 years, the usually cited (short) period, and also other important periods between 9.5 and 13.7 years. Even longer periodicities from 58 to 200 years have been recently determined. This complex behavior, and the superimposed noisy structure due to measurement errors, require sophisticated methods in order to reconstruct the intrinsic dynamics of the solar activity.

The ANN approach is described in the previous section; here we present the application of this method to the prediction of the annual mean sunspot number. First, we will discuss the performance of two standard forecasting techniques that are considered particularly good for this type of application (Cerrito 1992). For instance, using observations for the years 1770–1869, Box and Jenkins (1976) showed that, within their technique of time-series analysis, a third-order autoregressive model (AR(3)) is optimal for these data. However, they also stated that the model is not a particularly good fit of the data. Cerrito (1992) proposed a new technique using the nonparametric kernel density estimator. The use of the density estimator requires the assumption that the data are strictly stationary, completely excluding the possibility of periodicity. This means that in the strict sense it can not be applied to the sunspot data. The performances of these two methods in forecasting the years 1870–1889 can be seen in Table 1.1.

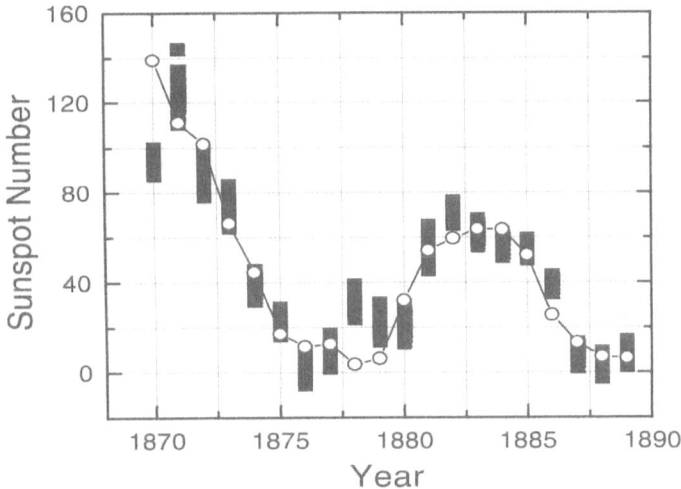

Fig. 1.3. Prediction of solar activity in the cross-validation interval 1870–1889. Solid squares correspond to the results of 100 different networks; open circles are the observed values.

In our case, we used the sunspot values for the years 1770–1869 as the training set (data set in the statistical methods) and 1870–1889 for cross-validation (prediction set). The results of 100 training experiments with different initialization of weights and biases is shown in Fig. 1.3. The statistical distribution of the *RMSE* for the cross-validation interval is given in Fig. 1.4. In all cases, we considered 12:3:1 networks. These results were obtained using a learning rate (the parameter that controls the step size at which the gradient-descent algorithm seeks for the minimum of the error function) equal to 0.5 during the first 5000 epochs, and then reducing it to 0.1 for the rest of the training process. The best-trained network has a $RMSE_{1870-1889} = 13.7$, although slightly better results can be obtained by monitoring the training process and suitably changing the learning parameters along it (Calvo et al. 1995). As can be seen from Table 1.1, the ANN fits the validation targets much better than the other methods.

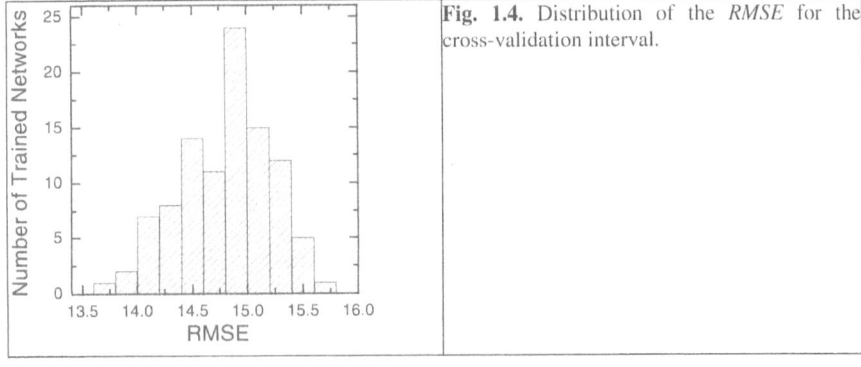

Fig. 1.4. Distribution of the *RMSE* for the cross-validation interval.

Table 1.1. Performance of methods for solar-activity prediction (1870–1899).

Method	RMSE	NMSE	CC	EV
AR(3)	31.2	0.68	0.92	0.12
NPKDE	26.5	0.49	0.87	0.31
ANN	13.7	0.13	0.93	0.89

Prediction methods: AR(3): Autoregressive order 3 (Box and Jenkins 1976); NPKDE: Non Parametric Kernel Density Estimator (Cerrito 1990); ANN: Artificial Neural Network Statistics: *RMSE*: Root Mean Square Error; *NMSE*: Normalized Mean Square Error; *CC*: Correlation Coefficient (Pearson): Best: 1.0, Worst: 0.0; *EV*: Explained Variance: Best: 1.0, Worst: 0.0

Notice however, that the whole procedure is not a genuine prediction of the cross-validation interval, since we used the information in this interval to stop the network training. Notwithstanding this, the numbers in Table 1.1 do characterize the network's ability to capture the intrinsic dynamics of solar activity. Furthermore, Fig. 1.4 shows that this property does not rely crucially on a fortunate training process. Actual forecasts for the next complete cycle of solar activity by the ANN method can be found in Calvo et al. (1995). This work also shows that the prediction power of the ANN does not deteriorate sensibly for long-term forecasting (a whole cycle in this case).

1.4. Prediction of Summer Monsoon Rainfall over India

The prediction of summer monsoon rainfall (SMR) over India has been the subject of several recent papers (Wu 1985; Das 1987; Shukla and Mooley 1987; Hastenrath 1987, 1988; Mooley and Paolino 1988; Basu and Andharia 1992) because of its paramount importance for agricultural planning in this country. These studies, mainly directed to find suitable predictor parameters, indicate that a substantial portion of the interannual rainfall variance can be predicted from antecedent departures in the large-scale circulation setting. The predictors belong to the "atmosphere-ocean-land anomaly complex", and can be grouped (Hastenrath 1988) into three families of preseason indicators: (i) upper-air flow over India; (ii) life cycle of the Southern Oscillation; (iii) heat low development over southern Asia and establishment of meridional pressure gradient including cross-equatorial flow over the Indian ocean. All the above mentioned studies used limited data sets and treated the SMR time series essentially as a stochastic process (Pandit and Yu 1983), producing forecast equations which are regression relations with one or more predictor variables. In particular, Shukla and Mooley (1987) (SM) concluded that two circulation features, namely (i) life cycle of the Southern Oscillation and (ii) seasonal transition of the midtropospheric circulation over India, show the most significant relationship with monsoon rainfall. Using the data given in SM for these two predictors, we obtained (Navone and Ceccatto 1994) the results shown in Table 1.2 (labeled as SM). Keeping with the standard practice in the literature, we fit the regression coefficients using the first 30 years of the record (1939–1968) and predicted the period 1969–1984.

Basu and Andharia (1992) (BA) have recently presented an alternative approach based on chaos theory (Moon 1987), which treats the SMR time series as deterministic but, possibly, chaotic. The method they followed essentially reconstructs the assumed deterministic dynamics of monsoon rainfall (Henderson and Wells 1988). First, for proper reconstruction of the attractor, they used standard methods to estimate an embedding dimension $M = 7$; second, they approximated the dynamics by $R_{k+M+1} = F(R_{k+1}, ..., R_{k+M})$ with the prediction function F taken as a general second-degree polynomial with seven variables. In this equation R_{k+M+1} represents the estimated rainfall at year $(k+M+1)$ as obtained dynamically from the records at the previous $M = 7$ years. Finally, they determined the 36 coefficients of the proposed polynomial by a standard least-squares fit to the 119-year (1871–1989) time series of SMR over India. In order to compare with SM results, we have repeated (Navone and Ceccatto 1994) these calculations, fitting the polynomial coefficients with rainfall data for the years 1871–1968. Then, this polynomial was used to again forecast the rainfall for the years 1969–1984. The results obtained are presented in Table 1.2 as BA. Learning the SMR dynamics by ANN is a much more difficult task than the previous example on solar activity. This is due to multiple reasons, which include, among others, a shorter time series record, lack of apparent periodicities, stronger external perturbations and, probably, some nonstationary behavior of the intrinsic dynamics. We considered networks with a 7:4:1 architecture, which corresponds to 37 adjustable weights and biases. This is nearly the same number of free coefficients used in BA's polynomial fitting. To gain some understanding of the problem, we first trained 20 of these networks with the whole record 1878–1994, stopping the training at *NMSE* values near 0.7. Shown in Fig. 1.5 is the mean error produced by these 20 networks at each point (we plot these errors in units of the largest value corresponding to 1961).

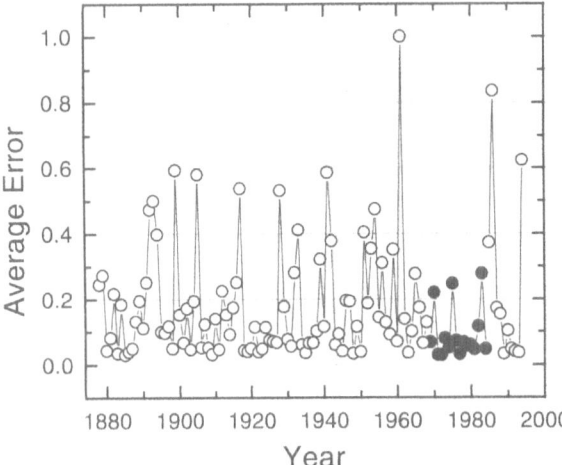

Fig. 1.5. Average error (in units of the largest value corresponding to 1961) produced by 20 networks trained to fit the whole record 1878–1994 of summer monsoon rainfall. Black dots correspond to the cross-validation interval 1969–1984.

Some points are well fit by most networks while others are relatively poorly approximated, and we checked that this pattern is only weakly dependent on the stopping *NMSE* value. Low errors correspond to periods of high predictability of the monsoon rainfall, which is the case for the interval 1969–1984 (black dots in Fig. 1.5), usually considered for testing new forecasting methods. This point has been already stressed in the literature (Hastenrath 1993). In the following we will use the interval 1969–1984 for cross-validation and will produce genuine predictions for the last 8 years of the record (1987–1994). We have trained approximately 5000 networks with random initialization of their free parameters. In this case we used first the gradient-descent algorithm with a small learning rate of 0.005 to avoid rapid overfitting, and continued the most promising training experiments with the conjugate-gradient method to refine the search. Only less than 1% of these training experiments were successful, with the $NMSE_{1969-1984}$ in the cross-validation interval having a low and shallow minimum. The final predictions of the best-trained networks (corresponding to $NMSE_{1969-1984}$ less than 0.4) are shown in Fig. 1.6. In particular, the best model has the following performance on the prediction interval: $RMSE_{1987-1994}=56.9$, $NMSE_{1987-1994}=0.35$, $CC_{1987-1994}=0.82$, and $EV_{1987-1994}=0.48$. By using ANN, it is also possible to combine BA's deterministic and SM's stochastic models of the SMR, producing a hybrid approach (Navone and Ceccatto 1994). First, we trained a 2:2:1 network to correlate predictors to rainfall data as in SM's work (here again the years 1871–1968 were used for training and the interval 1969–1984 for cross-validation). The results obtained with this network are given in Table 1.2 as 2:2:1. Secondly, we trained a 7:4:1 network to learn the dynamics of the time series, which performed on the cross-validation interval as indicated in the same Table. Finally, we linked both networks by connecting their output units to a new neuron.

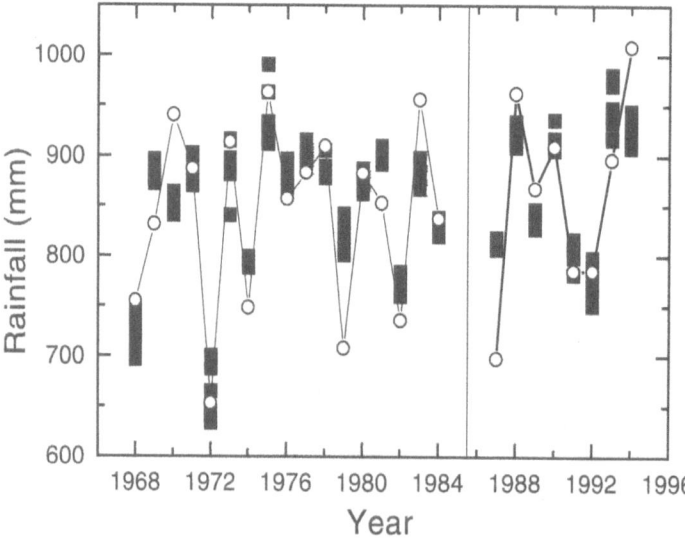

Fig. 1.6. Predictions of the best-trained networks (solid squares) and actual observations (open circles) for the cross-validation (1969–1984) and forecasting (1987–1994) intervals.

This produced a hierarchical (2+7):(2+4):(1+1):1 architecture with three additional parameters (2 weights and 1 bias). These parameters were determined by retraining the network (modifying only the new weights and bias) with the 1939–1968 part of the record. As above, we followed the evolution of the prediction capabilities of this system by monitoring its performance on the cross-validation interval. The final results, given in Table 1.2 as Hybrid, show the extremely good fitting capabilities of the ANN. In particular, the most severe 1972 drought is reproduced with nearly 2% of error (Navone and Ceccatto 1994).

Table 1.2. Performances of different methods for monsoon rainfall prediction in the period 1969–1984 (Navone and Ceccatto 1994).

Method	RMSE	NMSE	CC	EV
SM	41.5	0.24	0.89	0.56
2:2:1	33.6	0.16	0.94	0.56
BA	66.7	0.61	0.68	0.79
7:4:1	41.7	0.25	0.87	0.76
Hybrid	26.7	0.10	0.95	0.85

Prediction methods: SM: Shukla and Mooley (1987); BA: Basu and Andharia (1992). For discussion of each method, see text.
Statistics: *RMSE*: Root Mean Square Error; *NMSE*: Normalized Mean Square Error; *CC*: Correlation Coefficient (Pearson): Best: 1.0, Worst: 0.0; *EV*: Explained Variance: Best: 1.0, Worst: 0.0

1.6. Conclusions

In this work we have presented an exploratory use of ANN as sophisticated fitting functions, either for correlating data instead of the standard linear regression methods, or to reconstruct complex dynamics from time-series records. This was exemplified by considering two problems of practical interest, namely the predictions of solar activity and all-Indian summer monsoon rainfall. As a general outcome of this exercise, we found that ANN can be advantageously used in this context, showing comparable or better performance than conventional methods. This is particularly well demonstrated by the remarkable results of Tables 1.1 and 1.2. We conclude, in agreement with previous works (Weigend et al. 1990; Elsner and Tsonis 1992; Navone and Ceccatto 1994; Calvo et al. 1995), that neural networks are a reliable tool for time series analysis, providing a new and promising paradigm in the study of complex climatic phenomena.

Acknowledgements: One of the authors (HDN) wishes to thank the kind hospitality of the Intelligent Systems Research Group, Department of Computer Science and Electrical Engineering, University of Queensland (Australia), where part of this work was carried out. RAC and HDN acknowledge partial financial support from FOMEC and Universidad Nacional de Rosario.

References

Basu S, Andharia HI (1992) The chaotic time series of Indian monsoon rainfall and its prediction. Proc Indian Acad Sci (Earth Planet Sci) 101:27–34

Bishop CM (1995) Neural Networks for Pattern Recognition. Oxford University Press

Box G, Jenkins G (1976) Time Series Analysis, reviewed. Holden, Oakland

Calvo RA, Ceccatto HA, Piacentini RD (1995) Neural network prediction of solar activity. The Astrophysical Journal 444: 916–921

Cerrito PB (1992) Predicting Wolf's sunspot numbers with and without the assumption of periodicity. The Astrophysical Journal 393:795

Cybenko G (1988) Continuous valued neural network with two hidden layers are sufficient. Report, Department of Computer Science, Tufts University, Medford

Cybenko G (1989) Approximations by superpositions of a sigmoidal function. Math Control Signals Systems 2:303–314

Das PK (1987) Short- and long-range monsoon prediction in India. In: Fein JS, Stephens PL (eds) Monsoons. John Wiley, New York, pp 549–578

Elsner JB, Tsonis AA (1992) Nonlinear Prediction, Chaos, and Noise. Bull Am Met Soc 73:49–60

Farmer JD, Sidorowich JJ (1988) Exploiting chaos to predict the future and reduce noise. Theoretical Division and Center for Nonlinear Studies. Los Alamos National Laboratory, LA-UR-88–901

Hastenrath S (1985) Climate and Circulation of the Tropics. Reidel, pp 330–352

Hastenrath S (1987) On the prediction of Indian monsoon rainfall anomalies. J Climate Appl Meteor 26:847–857

Hastenrath S (1988) Prediction of Indian Monsoon Rainfall: Further Exploration. J Climate 1: 298–304

Hastenrath S, Greischar L (1993) Changing predictability of Indian monsoon rainfall anomalies. Proc Indian Acad Sci (Earth Planet Sci) 102:35–47

Henderson HW, Wells R (1988) Obtaining attractor dimensions from meteorological data. Adv Geophys 30:205–237

Hertz J, Krogh A, Palmer RG (1991) Introduction to the theory of neural computation. Santa Fe Institute studies in the sciences of complexity. Lectures notes: v 1 (Computation and neural systems series). Addison-Wesley

Moon F (1987) Chaotic vibrations: An introduction for applied scientists and engineers. Wiley, New York

Mooley DA, Paolino DA (1988) A predictive monsoon signal in the surface level thermal field over India. Mon Wea Rev 116:256–264

Navone HD, Ceccatto HA (1994) Predicting Indian monsoon rainfall: a neural network approach Climate Dynamics 10:305–312

Pandit SM, Yu SM (1983) Time Series and System Analysis with Applications. Wiley

Rumelhart DE, McClelland JL (1986) Parallel distributed processing. MIT Press, London

Rumelhart DE, Hinton GE, Williams RJ (1986) Learning representations by backpropagating errors. Nature 323:533–536

Shukla M, Mooley DA (1987) Empirical Prediction of the Summer Monsoon Rainfall over India. Mon Wea Rev 115:695–703

Tong H (1983) Threshold Models in Nonlinear Time Series Analysis. Springer-Verlag

Weigend AS, Huberman BA, Rumelhart D (1990) Predicting the Future: A Connectionist Approach. Intl J of Neural Sys 1:193–209

World Meteorological Organization (1986) Program on long-range forecasting research. LRF Research Rep. Series No. 6 (WMO/TD-No. 87), pp 824

Wu MC (1985) On the prediction of the Indian summer monsoon. Pap Meteor Res (Taipeh) 8(1):34–44

One-Dimensional Hydrodynamic Flow in Complex Networks: State of the Art, Some Applications and Generalizations

Pablo M. Jacovkis
Instituto de Cálculo and Departamento de Computación, Facultad de Ciencias Exactas y Naturales, Universidad de Buenos Aires, Ciudad Universitaria, 1427 Buenos Aires, Argentina
jacovkis@dc.uba.ar, jacovkis@decanato.de.fcen.uba.ar

Abstract: Climates differing from the present may result in other run-off systems, resulting in a change on any scale of river networks. While this is only of minor importance in the case of small creeks, large rivers adjusting to new equilibria may result in significant changes in erosion, sedimentation, and water discharge. Knowing more about such changes is therefore essential. Fluvial deltas and basins may be simulated by means of one-dimensional mathematical models that evolve in time and take into account the partial differential equations that govern the flow in each reach and the special equations needed at the junction points of the network. In this work, some theoretical and numerical analysis of this type of modeling is described, and some implications and generalizations are indicated.

2.1. Introduction

Past or future climates differing from the present may cause river networks to adjust to new equilibria, as erosion, deposition and water-discharge patterns may change. Especially in the case of larger rivers (for instance, the Paraná, Amazonas, Ganges) the knowledge about such potential future equilibria can be of great economic importance. For example, maintenance works of dams can focus on sites of potential future erosion and the sites of industrial settlements could be planned with respects to potential areas of accumulation. In this contribution, methods to treat the development of river-networks quantitatively are presented.

2.2. The One-Dimensional Hydrodynamic Equation

The one-dimensional gradually varied unsteady free-surface flow of a fluvial reach is governed by a system of two quasilinear hyperbolic partial differential equations, originally formulated by the great physicist and engineer Barré de Saint-Venant (1871); they may be seen in Stoker (1957), where they are derived from physical assumptions. These equations are

$$\partial(Q/S)/\partial t + \partial(Q^2/S^2)/2\partial x + g\partial Z/\partial t + gQ|Q|/D^2 = 0, \tag{2.1}$$

$$\partial S/\partial t + \partial Q/\partial x = 0, \tag{2.2}$$

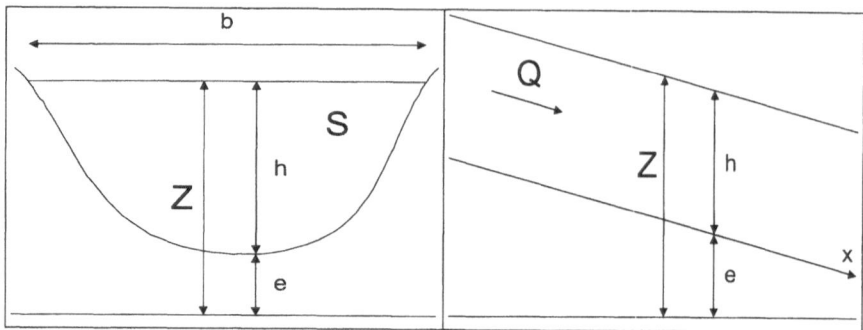

Fig. 2.1. Lateral (left) and longitudinal (right) section of a river. Further explanations in the text.

and they represent conservation of momentum and conservation of mass, respectively. Symbols in these two equations are the discharge $Q = Q(x, t)$, the stage $Z = Z(x, t)$, the wetted cross-section $S = S(Z(x, t), x)$, the acceleration due to gravity g, the conveyance $D = D(Z(x, t), x)$ (conveniently related to the shear stress acting on the bed), time t, and the spatial coordinate through the longitudinal axis of the channel or river x (see Fig. 2.1). The discharge may be replaced by the velocity and the stage by the vertical distance from the bed level to the free surface. Initial conditions $Q(x, t) = Q_0(x)$ and $Z(x, t) = Z_0(x)$ must be provided at initial time t. In addition, as each reach has a finite length, boundary conditions at one or both extreme points of the reach must be provided. The system of equations is then a mixed initial-boundary value problem. Possible boundary conditions are discharge, stage, or a one-to-one relationship between both.

With the simplifying assumptions (a) that the cross-sections of the channel or river are rectangular and prismatic and (b) resistance is neglected, the Saint-Venant equations (2.1) and (2.2) are usually called shallow-water equations; they may be studied as conservation laws. They can be written:

$$\partial u/\partial t + u\partial u/\partial x + g\partial h/\partial t = 0 \qquad (2.3)$$

$$\partial h/\partial t + \partial(uh)/\partial x = 0 \qquad (2.4)$$

where now $u = u(x, t) = Q/S = Q/(hb)$ is the mean velocity of the fluid, $h = h(x, t)$ is the height from the river bed to the free surface, and b is the cross-sectional width. It should be noted that $h = Z - e$, where e is the bed elevation measured from a fixed plane of reference; for rivers with a fixed bed, $e = e(x)$. Taking the vector form of Eqs. (2.3) and (2.4),

$$\partial \mathbf{w}/\partial t + \partial A\mathbf{w}/\partial x = 0, \qquad (2.5)$$

with $\mathbf{w} = (u, h)'$ and $A = \begin{bmatrix} u & g \\ h & u \end{bmatrix}$. We can write Eq. (2.5) as a conservation law.

The theory of conservation laws has developed spectacularly since the 1950s, due to the research of a number of applied mathematicians, particularly Lax (see for instance, Lax 1973; Oleinik 1957). The theory was developed in order to apply it to gas-dynamic equations, but it can equally well be applied to shallow water equations due to a remarkable analogy between both types of equations, observed for the first time by Riabouchinsky (1932). It is now known that under certain conditions, namely the so-called entropy condition and some conditions on the initial data, a weak solution exists and is unique. The weak solution coincides with a smooth solution except on the curves of discontinuity, where an integral equation must be satisfied for a sufficiently large class of functions. The generalization of the proof of existence for a scalar conservation law to systems of conservation laws is mainly the work of Glimm (1965).

2.3. The Numerical Approach

As with most partial differential equations, the Saint-Venant equations can usually not be solved analytically; therefore, numerical methods must be employed. This means that the differential equations are replaced by algebraic equations that approximate the true (and unknown) solution. Many numerical methods adopted for modeling fluvial reaches have been proposed by hydraulic engineers. Abbott (1979) outlines in detail several numerical methods. The most often used numerical methods in fluvial hydraulics are finite-difference methods (see for instance, Richtmyer and Morton 1967; Hall and Porsching 1990). They belong to two different classes: (a) explicit methods, that yield simpler programs and computations, but often require very small time steps, so that many iterations are necessary to simulate long periods of time, and (b) implicit methods, that permit larger time steps but require the solution of an algebraic system of equations at each time step.

A finite-difference numerical method for solving a system of partial differential equations over a finite spatial reach $x_0 < x < x_L$ for times t with $t^0 < t < t^N$ consists in constructing a finite grid of discretization points (x_i, t^n) over the region of interest $(x_0 < x_i < x_L, t^0 < t^n < t^N)$, where $x_i = x_0 + i\Delta x$, $t^n = t^0 + n\Delta t$, and Δx and Δt are the spatial and time steps. Consequently, the derivatives are replaced by quotients of differences, which consist of values of the functions and their arguments at the discretization points. If now v^n is the vector formed by the unknown approximate values of Q and Z (or u and h, if we are using Eqs. (2.3) and (2.4) at the discretization points) at points $x_0, x_1,...,x_L$ and time t^n, we obtain for each time step an algebraic system

$$B_1 v^{n+1} = B_0 v^n + f \qquad (2.6)$$

where $B_1 = B_1(\Delta x, \Delta t)$ and $B_0 = B_0(\Delta x, \Delta t)$ are functions depending on the space and time steps and on the numerical method used. Vector f takes the boundary conditions into account and, in the general case, some non-homogeneous terms. For a linear numerical method, B_1 and B_0 are matrices. If $B_1 = I$, the identity matrix (i.e., $Iv = v$ for all vectors and matrices v), the numerical method is explicit, else it is implicit. Of course, it is not necessary for the region of interest to be a rectangle in the time-space plane; more complex geometries are also admissible. Furthermore, space and time intervals need not be constant, they may vary depending on the discretization points of the grid. If we know the value of v^0 (the approximate solution at the initial time that is constructed using the known initial values) then we can solve Eq. (2.6) at each time step. This means we have an iterative way to obtain our solution. Powerful theorems in numerical analysis allow us to adopt or reject numerical methods, depending on their capability to obtain the approximate solutions. Other criteria are the difficulties in the computer implementation, the computer memory and time required, and the order of approximation, that is, how fast the approximate values converge to the true values when space and time steps Δx and Δt approximate zero.

If the matrix B_1 of the system in an implicit method is banded, that is, entries $b_{i,j}$ of matrix B_1 are zero if $|i-j| > k$ for certain (small) k or, more intuitively, if the only nonzero entries are not too far from the diagonal, enormous computational efforts are saved. This is because special methods are available for which the amount of computer memory and time necessary required for solving system (2.6) are (roughly) proportional to n (the order of matrix B_1), instead of being (roughly) proportional to n^2 and n^3, respectively. For simple reaches, most implicit numerical methods yield banded matrices. For instance, the popular Preissmann method (Ligget and Cunge 1975) discretizes a given function f and its derivatives as follows:

$$f_i^n \sim \theta(f_i^{n+1} + f_{i+1}^{n+1})/2 + (1 - \theta)(f_i^n + f_{i+1}^n)/2$$

$$\partial f/\partial t = (f_i^{n+1} - f_i^n + f_{i+1}^{n+1} - f_{i+1}^n)/(2\Delta t)$$

$$\partial f/\partial x = \theta(f_{i+1}^{n+1} - f_i^{n+1})/\Delta x + (1 - \theta)(f_{i+1}^n - f_i^n)\Delta x$$

where it is necessary that $\frac{1}{2} < \theta < 1$ for the scheme to be stable. To apply the Preissmann method to the Saint-Venant equations, we expand their non-linear terms in a Taylor series up to the first order term in Δt and neglect the remainder. If one boundary condition is given at each extreme point (that is the case in the subcritical case) the B_1 matrix obtained has the form

$$\begin{bmatrix}
x & x & x & & & & & & & & & & \\
x & x & x & & & & & & & & & & \\
 & & x & x & x & x & & & & & & & \\
 & & x & x & x & x & & & & & & & \\
 & & & & x & x & x & x & & & & & \\
 & & & & x & x & x & x & & & & & \\
 & & & & & & & \cdot & & & & & \\
 & & & & & & & & \cdot & & & & \\
 & & & & & & & & x & x & x & x & \\
 & & & & & & & & x & x & x & x & \\
 & & & & & & & & & & x & x & x & x \\
 & & & & & & & & & & x & x & x & x \\
 & & & & & & & & & & & & x & x & x \\
 & & & & & & & & & & & & x & x & x
\end{bmatrix}$$

where each x means a nonzero entry in the matrix. It is easy to see that matrix B_1 is banded, and may be further reduced to a tridiagonal matrix.

2.4. Calibration

In addition to (a) the theoretical approach to conservation laws, (b) the numerical methods for solving hyperbolic partial differential equations, and (c) the computer implementation of the mathematical model, another problem must be taken into account: the calibration of the model. It is usually possible to record or infer actual initial and boundary conditions, and to store geometric data (i.e., form of the cross-sections at different spatial positions). However, it is in generally very difficult to get accurate data of conveyances (the conveyances indicate the influence of the shear stress acted on the bed), due to the fact that they depend on many factors: cross-sectional geometries, type of bed, bed vegetation, etc. Therefore, in order to run models that approximate real situations, a calibration of conveyances must be performed, that is, conveyances must be found for which the model reproduces with reasonable accuracy real situations. In theory, this could be done through a theoretical mathematical approach considering the calibration as an inverse problem. In practice, on the one hand, inverse problems tend to be ill-conditioned, so that important technical difficulties appear when focusing the calibration from a too theoretical standpoint, and, on the other hand, the amount of data to be calibrated is so large that the task is almost prohibitive. For calibration, therefore, an empirical approach is appropriate, where the guidance of an experienced hydraulic engineering is fundamental. After calibration, that is, after the model reproduces as exactly as possible real situations, a validation is needed, before numerical experiments can be performed: historical data not used for calibration are compared with the results of the model and, in order for the mathematical model to be well calibrated, the difference between computed and historical data must be not greater than the difference obtained during the calibration process.

2.5. Fluvial Basins and Deltas

For fluvial basins and deltas, the situation is more complicated. Besides the Saint-Venant equations, we must introduce compatibility equations at the junctions of the network, that is, at points where two tributaries flow into a downstream reach or a reach bifurcates into two branches. They are, in first-order approximation

$$Z_i = Z_j = Z_k, \tag{2.7}$$

$$Q_i + Q_j = Q_k, \tag{2.8}$$

where i and j denote the downstream extreme points of the tributaries that flow into the reach whose upstream extreme point is denoted by k (see Figs. 2.2 and 2.3). Equation (2.8) holds also for a reach whose downstream extreme point k bifurcates into two branches whose upstream extreme points are i and j.

Stoker (1957) introduced the compatibility conditions (inner boundaries), Eqs. (2.7) and (2.8), at the junctions of the network, and modeled the junction of the Ohio and Mississippi rivers. After his pioneering work, other authors solved numerically various fluvial networks (Quinn and Wylie 1972; Fread 1973; Gradowczyk and Jacovkis 1974; Wood et al. 1975; Li et al. 1983; Joliffe 1984).

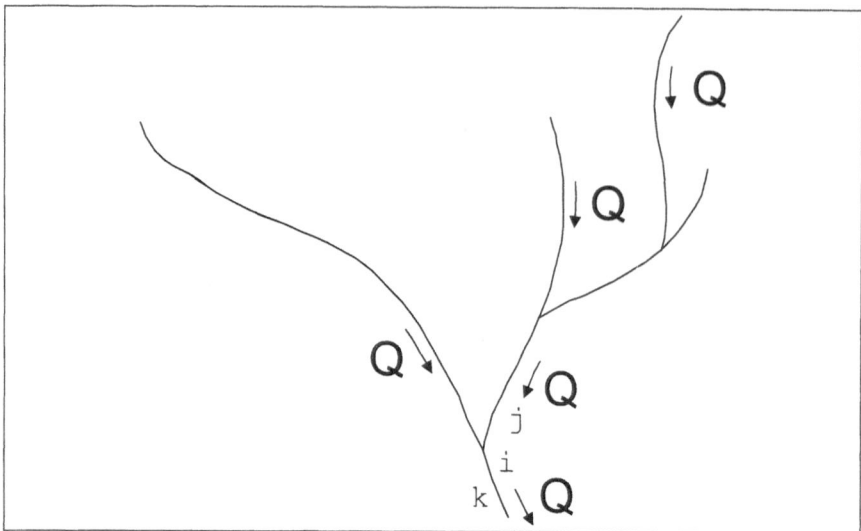

Fig. 2.2. An arborescent fluvial network.

In general, however, the system to be solved at each time step was not organized in a structurally convenient form (i.e., the matrix of the system was sparse but unstructured). We designed an extremely efficient algorithm for arborescent networks (originating in fluvial basins) that essentially consists in transversing the tree-like basin in what is called a *postorder* in graph-theory terminology. The method conserves the band structure of the matrix (Jacovkis 1989). It is impossible to construct a similar algorithm for a deltaic structure, but a block structure of the matrix was discovered that permits an efficient algorithm.

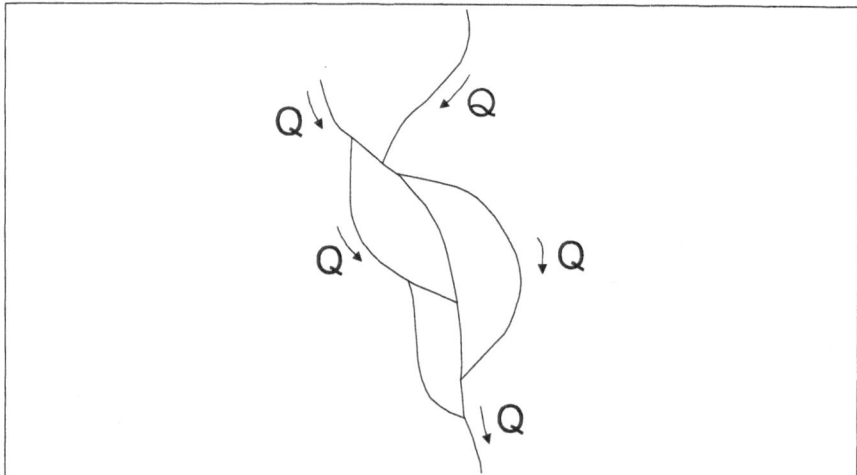

Fig. 2.3. A deltaic fluvial network.

(Jacovkis 1990). We call a computational algorithm "efficient" if the time necessary to solve a problem for a fixed simulation time, and the computer memory required, are proportional to the amount of data. The models of the author have been used for modeling several Argentine river basins and deltas, for instance the Limay-Neuquén- Negro river basin and the Paraná river delta.

From the theoretical (and also practical) point of view, there exists an interesting problem: where must the boundary conditions be given? This is a nontrivial problem, because we now may have a large number of open extreme points, depending on the complexity of the network. For a simple reach, there are only two extreme points, i.e., end points such as beginning and end (the river mouth). It has been proven that if the flow is subcritical (where "subcritical flow" has a meaning analogous to "subsonic flow" in gas dynamics) one boundary condition must be given at each extreme point, but if the flow is supercritical (that is, a wave generated by, say, a thrown pebble cannot propagate upward against the current, see supersonic gas dynamics) both boundaries must be given at the upstream extreme point. The example of the "thrown pebble" clearly shows that for a subcritical flow information goes downstream and upstream, but for a supercritical flow information propagates only downstream (condition $u > gh$). We generalized this idea to networks: for subcritical flow, one boundary condition must be given at each open extreme point; for supercritical flow, the situation is more complex, and a rather complicated theorem results (see Jacovkis 1991). Obviously, fluvial deltas and river basins may also be represented by a two-dimensional system of partial differential equations, but the advantages of the one-dimensional modeling are twofold: on the one hand, less computational efforts are required, and, on the other hand, less field data are necessary. This is particularly important when we try to apply the fluvial model for paleoenvironmental studies, as field data are of course a very scarce resource.

2.6. The Mobile-Bed Model

Suppose now that the flow is moving over an erodible bed, that is, the bed elevation e measured from a fixed plane of reference changes with time because bed particles slide and roll along the bed. For simple channels with prismatic and rectangular cross-sections, the shallow water equations become

$$\partial u/\partial t + u\partial u/\partial x + g\partial h/\partial x + g\partial e/\partial x + \mu\, u|u|/h = 0, \tag{2.9}$$

$$\partial h/\partial t + h\partial u/\partial x + u\partial h/\partial x = 0, \tag{2.10}$$

$$\partial e/\partial t + \partial G/\partial x = 0, \tag{2.11}$$

where $e = e(x, t)$ is the (movable) bed height, measured from a plane of reference, $G = G(u, h)$ is the solid discharge for unit width and μ is a dimensionless resistance coefficient conveniently related to the conveyance D. Several empirical formulae for μ and G have been proposed, because the bed movement is not yet sufficiently known to replace them by an exact equation. According to Gradowczyk (1968) for instance, we have

$$\mu = \Lambda\, (d_m/h)^n, \tag{2.12}$$

$$G(u, h) = \chi(|\tau| - \tau_0)^r \operatorname{sign} \tau \text{ if } |\tau| > \tau_0$$
$$= 0 \text{ if } |\tau| < \tau_0, \tag{2.13}$$

where τ, the shear stress transmitted by the fluid to the bed, is obtained as $\tau = \rho \mu u |u| = \rho \Lambda (d_m/h)^n (u|u|)$. Here Λ, χ, n and r are empirical constants and d_m is the mean diameter of the bed material. When $n = 1/3$ and $\Lambda = g/441$ (g in m/sec), Eq. (2.12) is the well-known Strickler's empirical formula, assumed to be valid when the bed configuration is flat. The threshold shear stress of the bed material is τ. When $\chi = 8/(g\rho^{1/2}\Delta\rho p)$, with p the porosity of the bed material, $\Delta\rho = \rho_s - \rho$, ρ_s the bed density, ρ the density of the water, and $r = 3/2$, Eq. (2.13) is the Meyer-Peter and Muller formula expressed in terms of volume of transported material. The variable G is given in m/sec if t is given in ton/m.

This model type allows very efficient simulations using data supposed to correspond to Paleo- and Neoclimates. Therefore, different hypotheses may be tested, and their feasibility analyzed. Unless the number of discretization points of the network is prohibitively large (a most unplausible assumption in the situations being treated), a personal computer suffices to do the computations. Besides, we are working in a very interesting generalization where we have studied (Jacovkis 1991 and 1995) the extended hydrodynamic system of Eqs. (2.9), (2.10) and (2.11) with mobile beds. We are currently integrating it with the network model to describe complex deltaic mobile-bed systems. The mobile-bed hypothesis is much more realistic than a fixed-bed hypothesis for simulations of long time periods. Particularly, this means that both warmer and colder analogs from the past, as well as potential future climatic equilibria, can be studied. This roughly corresponds to the changed precipitation trends (areas of increased precipitation vs areas of decreased precipitation) presented by Barros et al., in this volume.

Acknowledgements: The author gratefully acknowledges Professors Susana Bidner, Guillermo Marshall and Angel Menéndez for comments and suggestions and the Argentine National Council for Science and Technology and the University of Buenos Aires for financial support.

References

Abbott, M (1979) *Computational Hydraulics*, Pitman, London

Fread DC (1973) Techniques for implicit dynamic routing with tributaries, *Water Resources Research*, 8:918–926

Glimm J (1965) Solutions in the large for nonlinear hyperbolic systems of equations, *Comm. Pure Appl. Math.* 73:256–274

Gradowczyk MH (1968) Wave propagation and boundary instability in erodible-bed channels, *J. Fluid Mech*, 33:93–112

Gradowczyk MH, Jacovkis PM (1974) Un modelo matemático para regímenes impermanentes en redes fluviales complejas, Segundas Jornadas Latinoamericanas de Computación, UTN, Buenos Aires

Hall CA, Porsching, TA (1990) *Numerical analysis of partial differential equations*, Prentice–Hall, Englewood Cliffs, NJ

Jacovkis PM (1989) Modelos hidrodinámicos en cuencas fluviales, *Revista Internacional de Métodos Numéricos para Cálculo y Diseño en Ingeniería*, 5:295–320

Jacovkis PM (1990) Modelos numéricos hidrodinámicos en redes fluviales complejas, *Revista Internacional de Métodos Numéricos para Cálculo y Diseño en Ingeniería*, 6:543–572

Jacovkis PM (1991) Análisis de condiciones de contorno y cambio de régimen en modelos hidrodinámicos con fondo móvil, in: S. R. Idelsohn (ed) *Mecánica computacional*, 12:325–334, Santa Fe

Jacovkis PM (1991) One-dimensional hydrodynamic flow in complex networks and some generalizations, *SIAM J. Appl. Math*, 51:948–966

Jacovkis PM (1995) The hydrodynamic flow with a mobile bed: general and simplified approaches, in: D. Bainov (ed.), *Invited lectures and short communications delivered at the Sixth International Colloquium on Differential Equations*, Plovdiv, Bulgaria, 203–212

Joliffe JB (1984) Computation of dynamic waves in channel networks, *Journal of Hydraulic Engineering*, 110:1358–1370

Lax P (1973) *Hyperbolic systems of conservation laws and the mathematical theory of shock waves*, Society for Industrial and Applied Mathematics, Philadelphia

Li ZC, Zhan LJ, Wang, HL (1983) Difference methods of flow in branch channels, *Journal of Hydraulic Engineering*, 109:424–447

Liggett JA, Cunge JA (1975) Numerical methods of solutions of the unsteady flow equations.In: Mahmood K, Yevjevich V (eds) *Unsteady flow in open channels*, Vol. I, Water Research Publications, Fort Collins, 89–178

Oleinik O (1957) Discontinuous solutions of non-linear differential equations. *Usp. Mat. Nauk.* (*N. S.*), 12:3–73; English transl. in *Amer. Math. Soc. Transl. Ser.* 2, 26:95–172

Quinn F, Wylie B (1972) Transient analysis of the Detroit River by the implicit method, *Water Resources Research* 9:918–926

Riabouchinsky D (1932) Sur l'analogie hydraulique des mouvements d'un fluid compressible, *Institut de France, Académie des Sciences, Comptes Rendues*, 195–198

Richtmyer RD, Morton KW (1967) *Difference methods for initial value problems*, Interscience, New York

Saint-Venant, B (1871) Théorie du mouvement non-permanent des eaux avec application aux crues des rivières et à l'introduction des marées dans leur lit, *Comptes Rendues de l'Académie des Sciences de Paris*, 7:148–154 and 237–240

Stoker JJ (1957) *Water waves*, Interscience, New York

Wood FF, Harley BM, Perkins FG (1975) Transient flow routing in channel network. Research Report 75–1, International Institute for Applied Systems Analysis (IIASA), Laxemburg, Austria

Dendroclimatology: A Southern Hemisphere Perspective

Ricardo Villalba
Departamento de Dendrocronología e Historia Ambiental, IANIGLA-CRICYT,
Casilla de Correo 330, 5500 Mendoza, Argentina, and Tree-Ring Laboratory,
Lamont–Doherty Earth Observatory, Columbia University, Palisades, NY 10964,
USA.
ricardo@lab.cricyt.edu.ar

Abstract: It is difficult to isolate a distinctive dendroclimatology for the Southern Hemisphere. However, its geography raises some special issues of which dendroclimatologists in the Southern Hemisphere need to be aware. (1) Eighty percent of the Southern Hemisphere is covered by water. Heat transport by the Southern Oceans has special relevance as a major determinant of current climates and an influence on how climates were in the past and might be in the future. The potential of tree-rings to provide information on past ocean temperatures for the Southern Hemisphere has been noted, but quantitative estimates of past sea-surface temperature still need to be developed. (2) The Antarctic ice sheet contains more than 90% of the world's ice. Any climatic change in Antarctica would impact the global climate, but particularly that of the Southern Hemisphere. Tree-ring records at higher latitudes, in connection with ice-cores from Antarctica, should be used to understand the patterns of long-term atmospheric circulation at higher latitudes. (3) Although the influence of El Niño–Southern Oscillation (ENSO) is not exclusive to the Southern Hemisphere, there is an increasing level of confidence of ENSO phenomena as a useful predictor of climate variations in the Southern Hemisphere. Trees from Indonesia–Australia and South America, the geographical poles strongly impacted by ENSO, should be able to provide consistent information on past ENSO events. (4) Tree-ring records from the southern tip of South America should be employed to monitor any change in tree-growth related to abnormal levels of ultraviolet radiation related to the Antarctic ozone hole. (5) Pollution and acid rain are not widespread problems in the temperate ecosystems of the Southern Hemisphere. Consequently, tree-ring studies in the Southern Hemisphere offer the opportunity for properly assessing the relative importance of human interference (e.g. CO_2 fertilization) versus environmental changes on forests.

3.1. Introduction

The study of climate in the Southern Hemisphere contrasts with that in the Northern Hemisphere in several aspects. Not only are there important differences as a consequence of the different distribution of landmasses and oceans, but the smaller areas of land and the differences in historical and cultural development have delayed the processes leading to a better understanding of the Southern Hemisphere climates (Pittock et al. 1978; Graetz and Wilson 1996). Much has been written about climate variations in the Northern Hemisphere, in some cases as if it were the global story, but still relatively little is available for the Southern Hemisphere (Barry 1978).

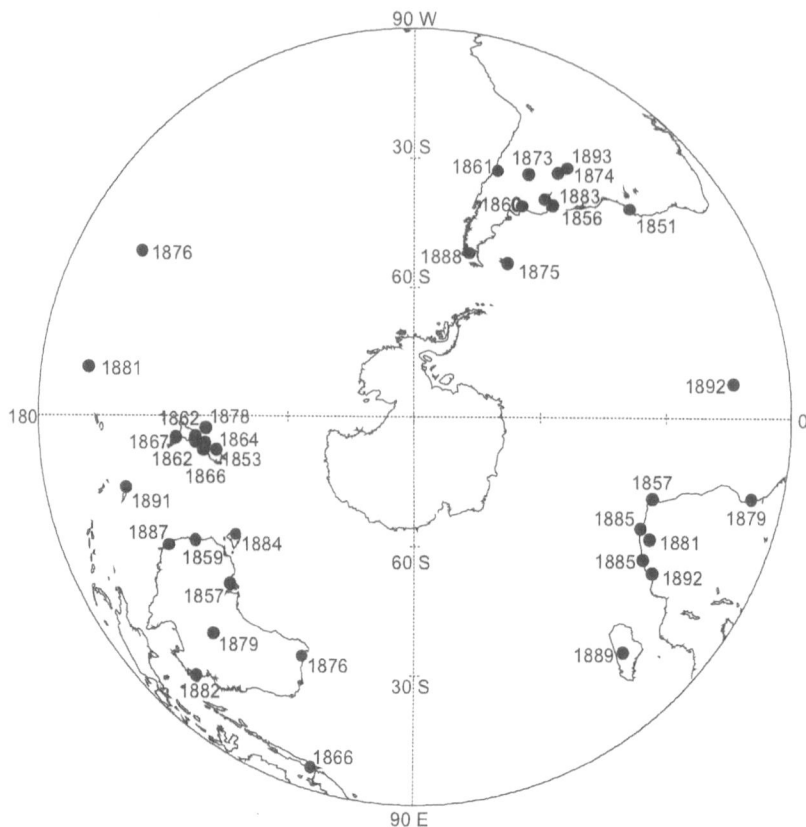

Fig. 3.1. Locations of the 36 stations with temperature records longer than 100 years in the Southern Hemisphere. For each station the first year of record is indicated.

Dendroclimatology, the use of tree-rings as proxy climate indicators, is not an exception to this Southern Hemisphere trend of lagging in climatic knowledge. Despite the excellence of much recent work on dendrochronology and dendroclimatology emanating from the Northern Hemisphere (Briffa et al. 1992a, 1992b, 1998; Cook et al. 1997, 1998; D'Arrigo and Jacoby 1993; Graumlich 1993; Hughes and Brown 1992; Jacoby and D'Arrigo 1989; Luckman et al. 1997; Stahle and Cleaveland 1992, 1993), the contribution of the Southern Hemisphere dendroclimatology in evaluating past climatic conditions has been more limited. Although geographical, historical, and cultural reasons for this difference are understandable, the consequences for a true understanding of the global climate are serious. Presently, we are facing a *global change*, which is not just a hemispheric concern. A complete understanding of the climatic behavior in one hemisphere is not possible without an understanding of that in the other (Pittock 1978). The two hemispheres are different in a number of important geographical features, which invalidate the global applicability of climatic models derived from only one hemisphere (Whetton et al. 1996).

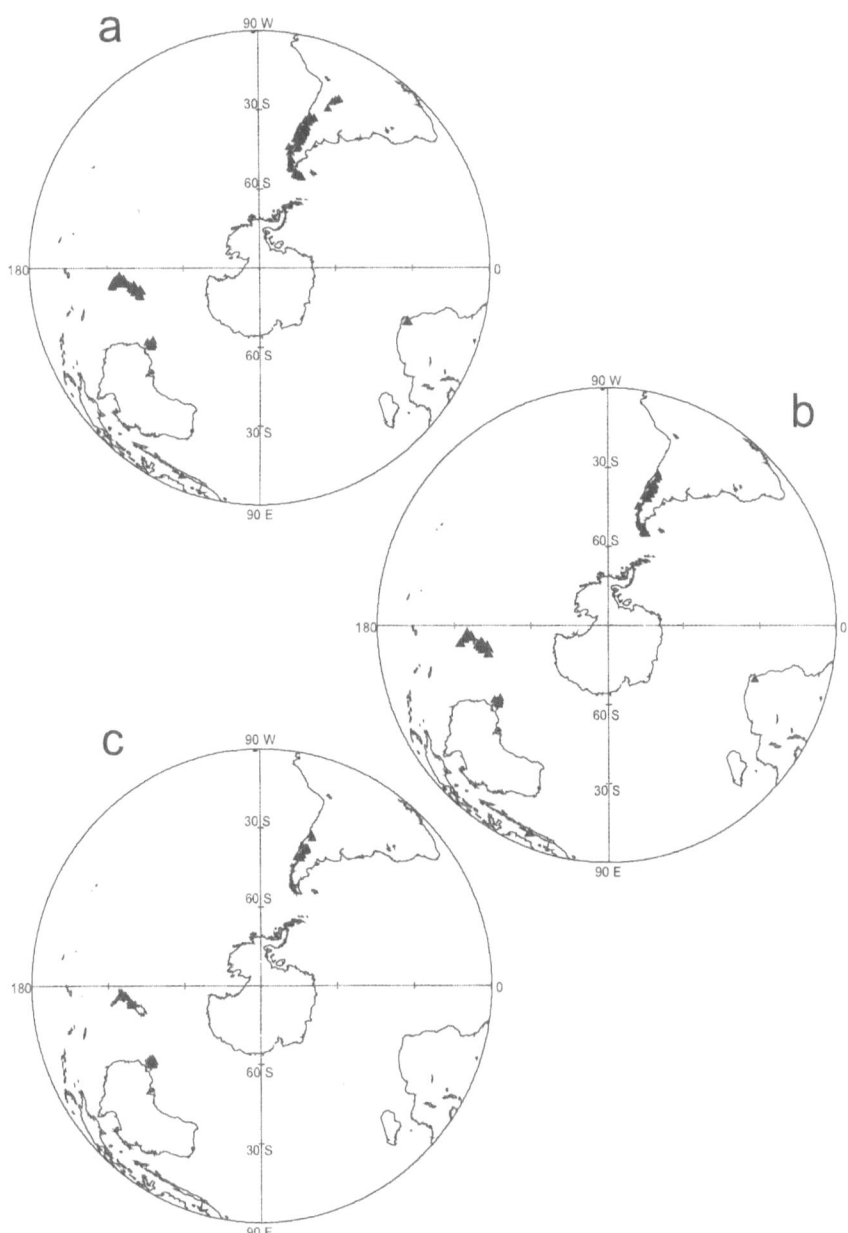

Fig. 3.2. Tree-ring chronologies (▲) in the Southern Hemisphere covering the last (from top to bottom): (**a**) 200, (**b**) 400 and (**c**) 600 years.

In the Southern Hemisphere most climatologists using weather station data work with meteorological records that rarely exceed 100 years. For the whole hemisphere, there are only 36 meteorological stations with reliable temperature records of more than 100 years (Fig. 3.1). Few of them date back to the 1850's (Jones et al. 1986a). Precipitation records longer than 100 years in the Southern Hemisphere are more abundant but unevenly distributed (Eischeid et al. 1991). Consequently, climatological studies in the Southern Hemisphere concentrate on short-term variabilities. In this short period of time, decade- to century-long trends are difficult to evaluate with any statistical certainty. Longer records of climatic variations are needed to validate any current climatic change. Annual- or seasonal-resolved climatic records from tree-rings, lasting for several centuries or a few millennia, are one of the most appropriate tools to evaluate long-term natural variabilities and any climatic change that differ substantially from these natural climatic variabilities.

Figures 3.2a, b, and c show the available coverage of tree-ring records in the Southern Hemisphere for the last 200, 400, and 600 years, respectively. For these intervals, tree-ring chronologies constitute the most numerous and wide spread proxy-record with annual or seasonal resolution in the Southern Hemisphere.

Analyses of precipitation data from the Southern Hemisphere show the existence of simultaneous anomalies over southern Africa, Australia, South America, and parts of the South Pacific (Wright 1977; Stoeckenius 1981; Pittock 1984). Analyses of surface pressure and upper air (Van Loon and Rogers 1981; Rogers and Van Loon 1982; Mo and White 1985; Karoly 1989; Karoly et al. 1996) also show that climatic variations on widely separated lands of the Southern Hemisphere are closely related in response to variations in the large-scale atmospheric circulation. It is clear that changes in temperature and precipitation are not uniform over the Southern Hemisphere, but rather contain considerable large-scale coherent structures with regions of similar or opposite signs. Consistent changes in the hemispheric patterns of temperature and precipitation in the Southern Hemisphere, as those indicated by Pittock (1984), provide the climatological basis for comparing dendrochronological records from diverse regions in the Southern Hemisphere. On the other hand, the long-term perspective resulting from tree-rings analysis is vital to evaluate the stability of modern climatic patterns, and to establish the linkages between changes in circulation and temperature, as well as circulation and precipitation in order to build up an understanding of long-term climate changes. Due to the short interval of the instrumental data set in the Southern Hemisphere, some apparent relationships in current climate may change under a long-term perspective.

Analyses of sea-level pressure, precipitation and temperature fluctuations in the Southern Hemisphere have revealed the importance to climatic variability of a few features of the large-scale hemispheric circulation, for example, the Southern Oscillation, the latitude of the subtropical high pressure belts, the strength of the westerlies and the eccentricity of the circumpolar vortex (Fig. 3.3; Trenberth 1976; Streten 1980; Pittock 1984; Aceituno 1988; Karoly et al. 1996). These few features of the general circulation are associated with a large percentage of the interannual variability of Southern Hemisphere climate. A long-term perspective of their variations is highly desirable.

Fig. 3.3. Map of the Southern Hemisphere showing the geographical setting for hemispheric circulation indices (SOI: Southern Oscillation Index; TPI: Trans–Polar Index) and regional circulation indices (L_{aus}: Latitude of the Subtropical High Pressure Belt along the eastern Australia coast; L_{sa}: Latitude of the Subtropical High Pressure Belt along the western South American coast), as discussed in this paper.

Several tree-ring chronologies in the Southern Hemisphere have been developed in regions highly impacted by relatively minor changes in the large-scale circulation. For instance, the Southern Oscillation is strongly correlated with climate variations and tree growth over central Chile as well as other regions in South America (Aceituno 1988; Villalba 1994). Tree-growth variations along the Pacific coast of southern South America have provided information about past changes in latitudinal positions of the subtropical anticyclone of the Chilean coast (Villalba 1990a). Some chronologies in southeastern Australasia could reveal past latitudinal fluctuations of the subtropical anticyclone east of Australia (Cook et al. 1995). Recent studies indicate that tree-ring chronologies from South America, New Zealand, and Tasmania could provide a long-term perspective of large-scale changes at higher latitudes in the Southern Hemisphere (Villalba et al. 1997a). Consequently, the potential of tree-ring chronologies for unraveling the history of the major centers of action in the Southern Hemisphere during the last millennium is clearly evident and should be emphasized in any long-term climatic study.

Boninsegna (1992) and Norton and Palmer (1992) have presented detailed reviews of the most important advances in the field of dendroclimatology in South America and Australasia for the 1970's and 1980's, respectively. More recently, Boninsegna and Villalba (1996) reviewed the development of dendrochronology and dendroclimatology in the Southern Hemisphere. Large part of the impetus for dendrochronology in the Southern Hemisphere has come from visiting experts or from collaborative works with scientists from the Northern Hemisphere. The Southern Hemisphere sampling program initiated by the Laboratory of Tree-Ring Research, University of Arizona, in the 1970's, was the common starting point for modern dendroclimatological research in different regions of the Southern Hemisphere. Not only did this laboratory develop the first properly cross-dated and well replicated chronologies for the Southern Hemisphere (LaMarche et al. 1979a, b, c, d, e), but they also were directly or indirectly involved in the formation of local dendrochronological groups. Today, active groups of dendrochronologists work in Argentina, Australia, Chile, New Zealand, and South Africa. In this review, I refer to the most recent dendroclimatological work in the Southern Hemisphere and its relationships with particular features of the large-scale atmospheric circulation. Some previous dendroclimatological results, relevant from a Southern Hemisphere perspective, are also briefly discussed. Readers interested in the standard methods used in dendroclimatology will find comprehensive reviews of them in Fritts (1976, 1991), Hughes et al. (1982), Schweingruber (1988), and Cook and Kairiukstis (1990).

3.2. The South American Sector

The number of well-replicated tree-ring chronologies in South America has exponentially increased during the last four decades. The first two chronologies in South America were presented in 1956 by Schulman using *Austrocedrus chilensis* and *Araucaria araucana*, two conifers growing in xeric forests of northern Patagonia. In the 1970's, further sampling of almost exclusively these two species, resulted in the development of 32 new chronologies in Argentina and Chile (LaMarche et al. 1979a, b; Holmes 1982). In the late 1980's, the number of chronologies in South America increased to more than 80 (Boninsegna et al. 1989; Boninsegna 1992; Roig and Boninsegna 1990; Villalba et al. 1987, 1992). A large part of these chronologies was developed from collections of broadleaf species from the genera *Nothofagus, Adesmia, Cedrela*, and *Juglans* (Fig. 3.4). Extensive sample collections were also conducted during the early 1990's. The new collections were mainly based on *Fitzroya cupressoides* (Lara and Villalba 1994; Roig 1996; Villalba et al. 1996), *Austrocedrus chilensis* (Aravena et al. 1997; Villalba and Veblen 1997) and *Pilgerodendron uviferum* (Roig 1991; Szeicz 1997). At the end of 1998 the number of chronologies developed in South America will be close to 200. This relatively large tree-ring data set has been used to estimate some atmospheric circulation indices (Villalba 1990a; Villalba et al. 1997a), local and regional changes in temperature and precipitation (Boninsegna 1988; Boninsegna et al. 1989; Lara and Villalba 1993; Roig 1996; Roig and Boninsegna 1990; Villalba 1990b; Villalba et al. 1989, 1992, 1996, 1997b) and changes in river discharges (Holmes et al. 1982; Cobos and Boninsegna 1983). Other climatic related studies include glacier advances and retreats (Rabassa et al. 1986; Villalba et al. 1990) and volcanic eruptions (Villalba and Boninsegna 1992).

Significant correlations were found by Pittock (1980a, b) between variations in precipitation and temperature in southern South America and fluctuations of four indices of the general circulation: (i) an index of the Southern Oscillation (SO), (ii) an index of the pressure differences across the Antarctic between Tasmania and the Malvinas (Falkland) Islands, (iii) the latitude of the high-pressure belt off the coast of Chile, (iv) an index of the pressure fluctuations in the tropical Atlantic and mid-latitude circulation. Similar to other regions of the Southern Hemisphere, in South America, these circulation indexes explain a large percentage of the regional climatic variability. Several attempts to reconstruct their past fluctuations have been performed. However, due to the complex nature of the large-scale circulation, reconstructions of some indexes, such as the SO Index, are problematic (Cook 1992).

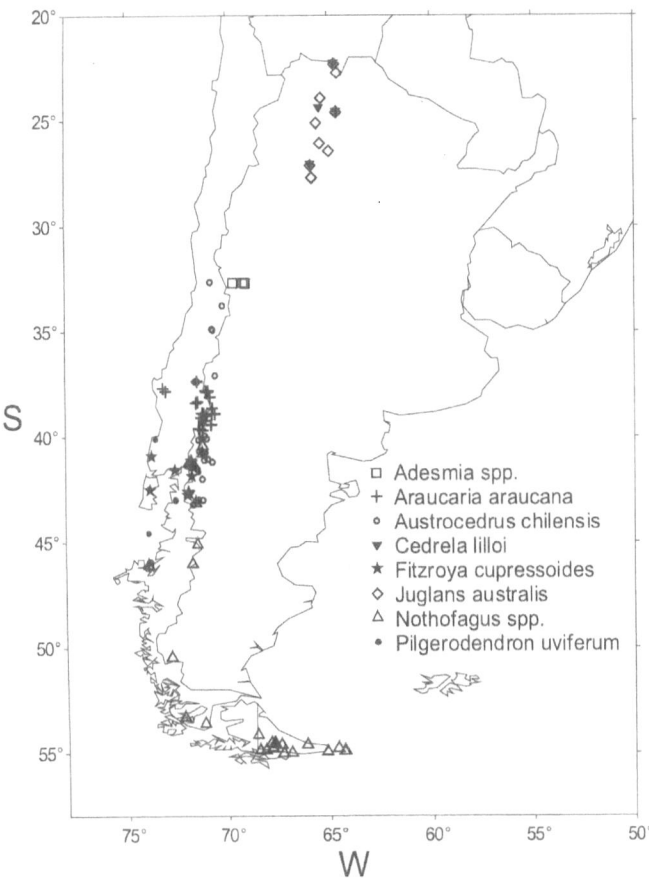

Fig. 3.4. Locations of tree-ring chronology sites in South America.

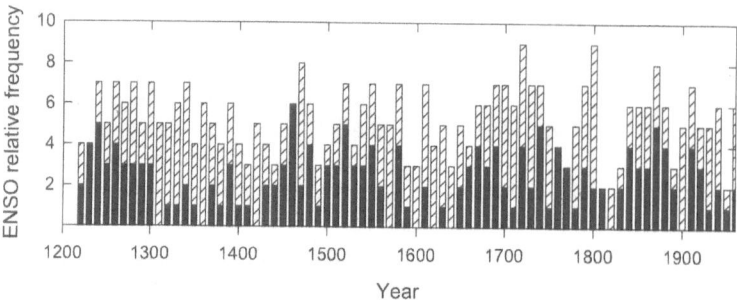

Fig. 3.5. A preliminary record of El Niño–Southern Oscillation (ENSO) events based on temperature and precipitation reconstructions from tree-rings. Number of years by decade with heavy winter rainfall (year 0) in central Chile concurrent with subsequent warm summers (year +1) in northern Patagonia (black bars), and low winter rainfall in central Chile concurrent with subsequent cool summer in northern Patagonia (dashed bars). Black and dashed bars are inferred to represent relative frequency of warm and cold ENSO events, respectively.

Several authors (Kiladis and Diaz 1989; Aceituno and Montecinos 1992; Lough 1991) have noted that the relationships between the SO and interannual variability in temperature and precipitation during the present century, vary significantly between events. In some events of the SO, opposite relationships to those expected were noted, particularly in extratropical regions. This event-to-event variability associated with the SO limits the successful reconstruction of the SO based on only one chronology or a set of chronologies from a relatively homogeneous climatic region. A first attempt to estimate decadal changes in the SO, based on the simultaneous use of tree-ring chronologies from regions in South America affected differentially by the SO, was conducted by Villalba (1994). Heavy winter rainfall in central Chile is associated with positive anomalies during the developing stage of warm events of the SO. Conversely, most cold events correspond to dry conditions (Rutllant and Fuenzalida 1991). On the other hand, in northern Patagonia positive departures of summer temperature follow warm events of the SO (Kiladis and Diaz 1989). Chronologies of Central Chile and Northern Patagonia were combined to derive a preliminary estimate of the SO since AD 1220 (Fig. 3.5). A reconstruction of winter precipitation in Central Chile (Boninsegna 1988) was simultaneously compared with a reconstruction of summer temperature in northern Patagonia (Villalba 1990b). For the 1877–1972 interval, based on a total of 22 years in which both series show simultaneously positive departures from the mean values, 17 of those years are related to warm events of the SO. A similar analysis for the interval 1886–1972 indicates that from a total of 25 years in which both series show negative departures from the mean, 18 cases can be associated with cold events of the SO (Villalba 1994). Based on the previous analysis, the occurrence of a warm event of the SO in years in which both reconstructions show positive departures (above average winter rainfall in central Chile, and above average summer temperature in northern Patagonia) is highly probable. Conversely, negative departures in both series could indicate years of cold events. Based on this analysis, the following periods were intervals of repeated recurrences of the El Niño–Southern Oscillation events (Fig. 3.5): AD 1240–1349, 1450–1489, 1510–1589, 1670–1759, 1780–1809 and 1840–1889. On the other hand, low event recurrence was recorded during the periods AD 1350–1449, 1590–1669 and 1810–1839. Quinn and Neal (1992) also recognize most of these periods of repeated incidences in historical records.

Some interesting features result from the simultaneous comparison between the reconstruction of the winter precipitation in Central Chile and the reconstruction of the summer temperature in northern Patagonia. In both reconstructions, the years 1468–69 represent the highest departures from their respective averages during the last 1000 years. A warm event of the SO, larger in magnitude than those recorded historically, is probably responsible for these extreme deviations in precipitation and temperature from these two regions that are separated by 1000 km. Another important warm event may also be associated with higher departures in the years 1395–96. Coincidentally, corrected radiocarbon dates from detrital wood contained in flood sediments on the northern coast of Peru, also indicate the occurrence of strong El Niño events (i.e. warm phase of the SO) in AD 1460±20 and 1380±40 (Wells 1987, 1990). Based on the Nile River flood data, Quinn (1992) also reported that 1468 was an El Niño year.

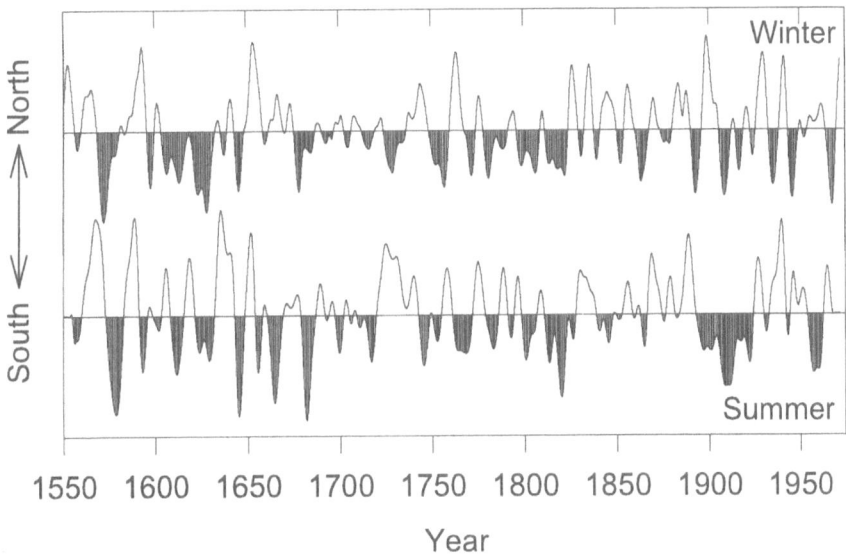

Fig. 3.6. Estimates of latitudinal positions of the Southeastern Pacific anticyclone in winter (upper) and summer (lower), inferred from tree-ring variations along the Chilean coast. Tree-ring estimates have been smoothed with a 7-year low-pass digital filter to emphasize low-frequency variance.

Changes in regional precipitation at middle latitudes are related to changes in geographical position and the intensity of subtropical anticyclones (Stewart 1975; Pittock 1980a, 1985).Strong relationships between the latitudes of the subtropical high pressure belt off the western coast of South America and precipitation in Central Chile have been reported by Mossman (1909), Rubin (1955), Pittock (1971, 1980a), and Minetti and Sierra (1989).The mean latitude of the subtropical high-pressure belt along the Chilean coast was determined from instrumental records of sea-level pressure by Pittock (1980a) for the 1941–1962 period. Eigenvector analysis of 17 tree-ring chronologies in Argentina and Chile was used to estimate the summer and winter past positions of the southeastern Pacific subtropical anticyclone (Villalba 1990a; Fig. 3.6). Analyses of both summer and winter estimates of the anticyclone positions indicate that low and high indexes are not randomly distributed through time. Regimes of unusually low or high indexes can persist for two or more years. Periods as long as nine years of low-index regime were recorded in the estimated records. The search for periodicity revealed that low or high indexes have a marked rhythm of 3.4–3.7 years in winter anticyclone positions and rhythms of 2, 3.2 and 5 years in summer latitudinal shifts. Frequencies of 3.2, 3.4–3.7, and 5 years for the Southeastern Pacific anticyclone positions reflect the dependence of the anticyclonic activity on El Niño–Southern Oscillation events. Eight of the ten most extreme northern locations from the summer estimates of the latitudinal positions of the southeastern Pacific anticyclone are associated with El Niño events proposed by Quinn and Neal (1992). In general, an association between a weaker southeastern Pacific anticyclone and El Niño event occurrence has been apparent for the last four centuries, although some anomalous cases are recognized (Villalba 1994).

A 3620-year reconstruction of summer temperatures has been developed for northern Patagonia (Lara and Villalba 1993). This record constitutes one of the longest annually resolved climatic reconstructions from tree rings. There is a potential for developing chronologies that extend 4000 to 5000 years using both living trees and sub-fossil material (dead snags, fallen logs and buried trunks; Lara et al. in preparation). Sub-fossil material from Pelluco, Chile, has been dated back to >45000 years B.P. using radiocarbon (Heusser 1981; Roig et al. 1998), demonstrating the potential of getting younger sub-fossil material that would be useful in building absolutely dated chronologies. For Tierra del Fuego (55° S), Roig et al. (1996) reported the development of a fragmented-floating chronology from subfossil *Nothofagus* woods. This chronology, which spans for the past 1400 years, show a persistent oscillatory mode centered at about 7 years, reflecting some similarities in tree growth over time.

3.3. The Australian Sector

Reviews by Ogden (1978, 1982), Dunwiddie and LaMarche (1980a), Norton (1990), and Norton and Palmer (1992) have outlined the development of the significant advances in dendrochronological research during the 1970's and 1980's involving Australian species. The first properly cross-dated and well replicated tree-ring width chronologies from Australia were developed by LaMarche et al. (1979d) as part of a major program in the 1970's on dendroclimatology of the Southern Hemisphere. Seventeen chronologies were developed from four conifers:

Arthrotaxis   cupressoides,   Arthrotaxis   selaginoides,   Callistris   robusta, Phyllocladus asplenifoliuos, and one broadleaf: *Nothofagus gunnii.* LaMarche and co-workers sampled 350 individuals from 38 sites in mainland Australia, but due to the difficulty in determining annual rings, only two short chronologies from *Callistris* were developed. From a total of 20 chronologies developed by LaMarche et al. (1979d) for Australia, 18 were from long-lived endemic conifers of Tasmania, thus showing the dissimilar spatial distribution of chronologies in the region (Fig. 3.7; Norton and Palmer 1992). The set of tree-ring chronologies developed by LaMarche et al. (1979d) was used to produce a preliminary reconstruction of temperature for Tasmania (LaMarche and Pittock 1982) and the stream-flow for eight rivers in western Tasmania (Campbell 1982). Additional details on the methods used in these reconstructions are given in Norton and Palmer (1992).

Most of the recent research effort in Australia dendrochronology has been directed towards developing long tree-ring chronologies. Cook et al. (1991) presented a 1089-year tree-ring chronology of *Lagarostrobos franklinii* from western Tasmania. Based on this precisely dated annual tree-ring chronology, a millennium temperature reconstruction for Tasmania was developed (Cook et al. 1992). Interestingly, the coldest and warmest reconstructed periods of the last 1000 years occurred over the past 100 years. An expansion of the Antarctic ice pack and the consequent irruption of large number of icebergs into lower latitudes (Burrows 1976) apparently preceded the temperature decline in the early 1900's. This cold interval also coincided with the occurrence of a cold anomaly of sea-surface temperature off the west coast of Tasmania that reduced air temperature over the island. The anomalous increase in tree growth since 1965 parallels the temperature trends in Tasmania during the last decades and the increased warming throughout much of the Southern Hemisphere (Jones et al. 1986b). This pattern of unusual cold followed by unusual warmth, suggests that the Southern Hemisphere, at least in the Tasmanian region, has suffered a significant reorganization of the ocean–atmosphere system during the 20th century.

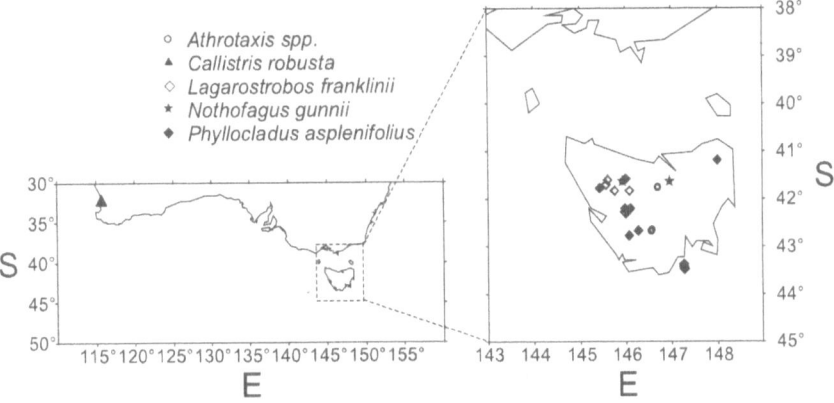

Fig. 3.7. Locations of tree-ring chronology sites in Australia.

More recently, the Tasmanian temperature reconstructions have been extended back to 1260 BC through inclusion of tree-rings series from several old stumps of *L. franklinii* found at the same original sampling site (Cook et al. 1996a, b). Spectral analysis of the most recent 2790-year period in the reconstruction indicates the existence of four oscillatory modes with periods of 31, 57, 77, and 200 years. Cook et al. (1995) suggested that the 31- and 55-year oscillations could be related to fluctuations in zonally averaged July sea-level pressure in the 40–50° S latitude zone (Enomoto 1991). Fluctuations in the 20–30 and 40–60 year periods were considered by Enomoto (1991) to be oscillations within waves number zero and one, which are related to the expansion–contraction and eccentricity of the circumpolar vortex, respectively. An expansion of the circumpolar vortex is associated with low sea-level pressure over Tasmania, which in turn produces a greater tendency for cool winds from the southwestern sector and lower temperatures over Tasmania. During the 1895–1920 interval, a clear link is observed between low sea-level pressure and below-average temperatures over Tasmania in summer (Cook et al. 1995).

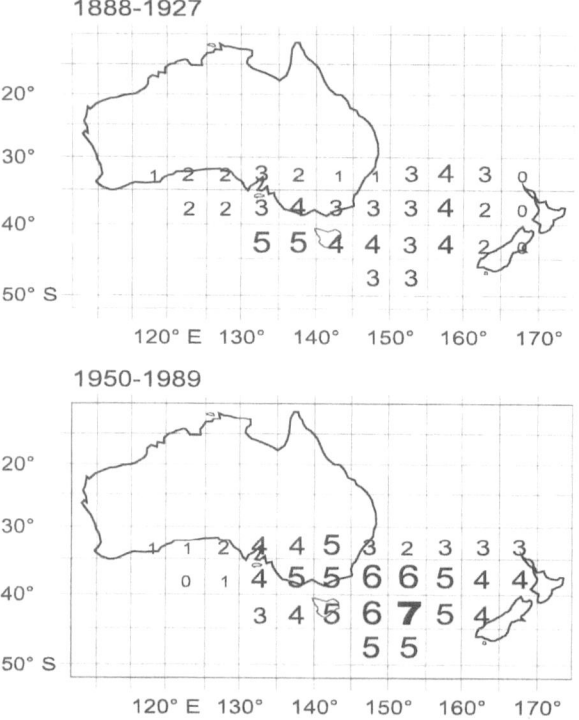

Fig. 3.8. Correlations between reconstructed summer temperature in Tasmania and sea surface temperature for two contrasting cold and warm intervals. The correlation field is weighted towards the ocean west of Tasmania during the cold interval (1888–1927), but east of Tasmania during the warm interval (1949–1988). Differences in atmospheric and oceanic circulation during the cold vs. warm intervals appear to be related to the changes in correlation (modified from Cook et al. 1995).

In contrast, both summer sea-level pressure and temperature have been above the instrumental record mean since 1960. Ocean circulation dynamics and external forcing caused by long-term solar could be related to the 77- and 200-year terms (Cook et al. 1995).The links between ocean temperatures in the Tasman Sea region and temperatures over Tasmania were also investigated by Cook et al. (1995, 1996a). For the 1888–1927 and 1949–1988 intervals of different climate regimes, sea-surface temperature data for 5 x 5 grid cells, covering the southern Australia–Tasmania–New Zealand sector, were correlated with the reconstructed temperature for Tasmania (Fig. 3.8). During the early period, the correlation field is weighted towards the west coast of Tasmania. In contrast, for the late period, the correlation field is weighted towards the Tasman Sea east of Tasmania. These results imply a change from a cold West Wind Drift displaced to the north during the first cold interval to a more southerly penetration of the warm East Australian current during the later warm interval. These changes in the ocean were coupled with a southward shift of the subtropical high pressure belt and an increase in the frequency of warm winds from the northeastern sector over Tasmania (Cook et al. 1995, 1996a). A network of *L. franklinii* chronologies has recently been developed in western Tasmania to explore changes in tree response to climate as function of elevation (Buckley et al. 1997). The vertical zonation of the west coast climate largely accounts for the differences in both the mean climatic regimes and the tree responses to climate between lowland and subalpine sites in Tasmania. When traditional dendroclimatic techniques are used, subalpine chronologies yield the most reliable information about past climatic variability in Tasmania (Buckley et al. 1997). An extensive sampling of cross-sections was taken by Schweingruber (1992) to evaluate the dendrochronological potential of approximately 670 species of shrubs and trees in southern Australia. Woody species growing in three different biogeographical zones, the montane/alpine forests, the savannah woodland and the desert, were considered. Clear growth zones, likely true annual rings, are formed in *Nothofagus cunninghamii* and some *Eucalyptus* species in the montane rain forest of southeastern Australia (Schweingruber 1992). These findings suggest the possibility of developing new tree-ring chronologies in southeastern Australia, a region strongly affected by changes in the subtropical high pressure cell off the Australian coast (Pittock 1975; Lough 1991; Fig. 3.3). In the Australian savannahs and deserts, some species are suitable for age determination; however, cross-dating is still problematic (Schweingruber 1992).

3.4. The New Zealand Sector

Ogden (1982), and Norton and Palmer (1992) gave general reviews of the use of dendroclimatology in the Australasia region, including New Zealand. Dunwiddie (1979) and Norton and Ogden (1987) presented reviews limited to New Zealand trees. Inconclusive results from the early dendrochronological attempts in New Zealand during the 1950's and 1960's have been attributed to cross-dating problems, use of short-lived species, and uncertainties about the annual nature of rings (Norton and Palmer 1992). Similar to other regions in the Southern Hemisphere, the initial stimulus for the most recent dendrochronological research in New Zealand came during the 1970's when LaMarche et al. (1979c) developed the first properly cross-dated and well-replicated chronologies.

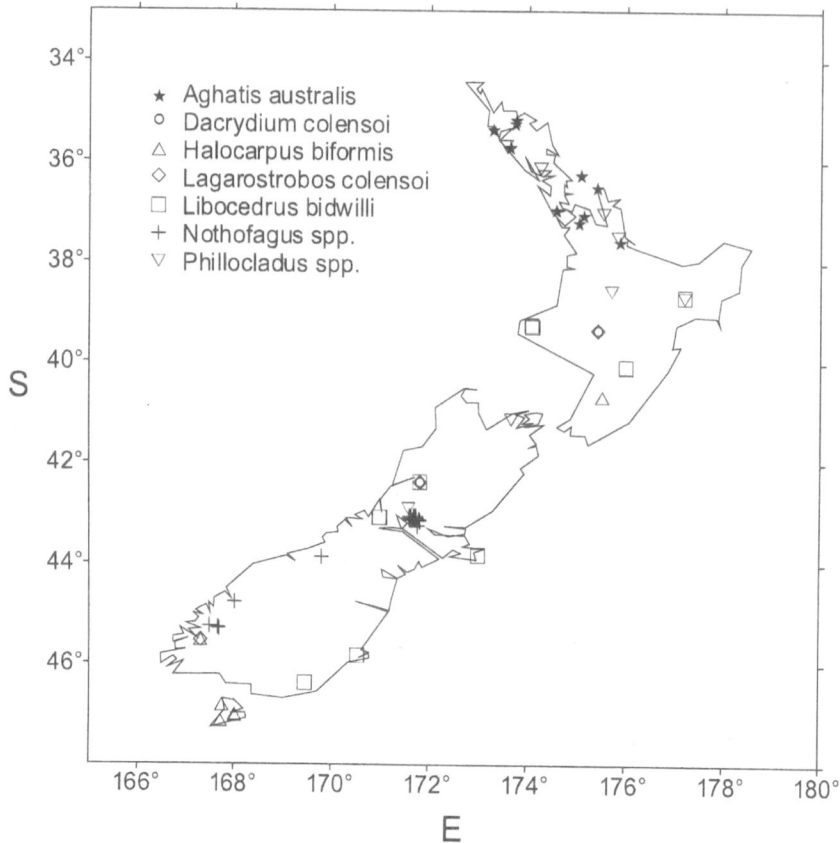

Fig. 3.9. Locations of tree-ring chronology sites in New Zealand.

Twenty-one chronologies were initially developed using seven conifers: Libocedrus bidwillii, Phyllocladus alpinus, P. trichomanoides, P. glaucus, Lagarostrobos colensoi, Halacarpus biformis, and Agathis australis. Since then, further sampling has established eight additional modern chronologies for Agathis australis (Ahmed and Ogden 1985), two for L. bidwillii, seven for Phyllocladus trichomanoides (Palmer 1989), two for P. glauca (Palmer 1989), 30 for Nothofagus solandri and N. menziesii from the South Island (Norton 1983a, b), and recently two chronologies for H. biformis from Stewart Island (D'Arrigo et al. 1995; Fig. 3.9). Subfossil material of A. australis was cross-dated by Bridge and Ogden (1985) and a floating chronology was finally developed. Based on radiocarbon dates, the subfossil chronology spanned the interval 3500–3000 years B.P. A second floating chronology from buried logs of Phyllocladus trichomanoides was developed by Palmer (1989) for the interval 2200–1800 years B.P.

Based on ten chronologies of *N. solandri* and *N. menziesii*, Norton et al. (1989) derived a reconstruction of New Zealand summer temperature to AD 1730. Other independent reconstructions of summer temperature for New Zealand have been presented by Palmer (1989), using a combination of *Phyllocladus trichomanoides* and *P. glauca* chronologies. Reconstructions of precipitation at the town of Canterbury and the stream-flow for the Hurunui River were developed by Norton (1987), employing four chronologies of *N. solandrii* growing on the east slope of the South Island. Finally, the impact of the Tambora eruption of April 1815 on endemic conifers of Australia and New Zealand has been investigated by Palmer and Ogden (1992), and on *Nothofagus* tree-ring series by Norton (1992).

Recently, reconstructions of temperature along with zonal and meridional flows over New Zealand for the period back to 1731 have been presented by Salinger et al. (1994). These reconstructions, which capture more low-frequency variability than the previous reconstructions, constitute the first attempt at using a combination of five endemic species in New Zealand for inferring past climatic variations. The two temperature reconstructions developed by Salinger et al. (1994) show the recent temperature increase in New Zealand to some extent. In contrast, the Stewart Island and Mangawhero chronologies show evidence that recent decades of warming in New Zealand were unusual, although not clearly unprecedented, relative to temperature conditions inferred from tree-ring records of prior centuries (D'Arrigo et al. 1995, 1998).

New Zealand is another region in the Southern Hemisphere influenced by El Niño–Southern Oscillation (ENSO) events (Gordon 1986). South and southwestern anomalous flow is observed over New Zealand during El Niño events. El Niño events are accompanied by temperature below normal over New Zealand in combination with drier conditions in the areas protected from southern winds. However wetter conditions prevail in the areas exposed to the southern winds. Tree-ring chronologies from New Zealand could be another important piece of information for revealing the long-term behavior of ENSO events. Besides, the presence of abundant subfossil wood, particularly in the North island of New Zealand, offer the opportunity for developing long-tree ring chronologies for the late Holocene (Norton and Palmer 1992).

3.5. The South African Sector

The application of dendrochronological techniques in South Africa has been more problematic than any other region in the Southern Hemisphere (LaMarche 1982). Virgin forests are extremely reduced in extent (less than 2% of South Africa's total area) and some of them have been intensively exploited since the early 20th century (Palmer and Pitman 1972). Besides the scarcity of useful forests, other common problems encountered in South African trees include the presence of ambiguous ring boundaries, multiple intra-annual bands, and severe ring wedging (Dyer 1982; LaMarche 1982). After a relative large sampling of cores and discs, two genera, *Podocarpus* and *Widdringtonia* were considered the most suitable (LaMarche et al. 1979e).

A precisely dated tree-ring chronology, 413 years old, has been developed using *Widdringtonia cedarbergensis* at a site near Cape Town (Dunwiddie and LaMarche 1980b). A climatic-response function shows significant relationships between *Widdringtonia* tree growth and local precipitation during spring and early summer.

According to Dunwiddie and LaMarche (1980b) there is potential for the development of much longer chronologies from *Widdringtonia cedarbergensis*. In the semi-arid environments in which *Widdringtonia* lives, the dead wood decay is slow, and wood from stumps, logs and remnants could be cross-dated with living trees. Frost rings are a consistent and well-replicated pattern in *Widdringtonia* tree rings (Dunwiddie and LaMarche 1980b). This gives the possibility of developing extensive frost-event records in South Africa during the growing season.

Three of the four *Podocarpus* species growing in South Africa were sampled by LaMarche et al. (1979). The wood of these species is characterized by severe ring wedging and extreme lobate growth (LaMarche 1982). Diffuse ring boundaries, intra-annual bands, and ring wedging have been partially attributed to the effects of the summer rainfall distribution (Dyer 1982). In South America, *Podocarpus* species present severe ring wedging under both summer-dominant or Mediterranean types of climate (Villalba et al. 1985; Villalba 1987, unpublished), which suggests for *Podocarpus* a genetic rather than an environmental control of ring-wedging formation.

A major point in South African dendroclimatology is the need for examining the large number of tropical and subtropical species to assess their dendrochronological potential (but see the pioneer work of Lilly 1977, and February 1996). Only during the last ten years have some tropical and subtropical species been successfully used in dendroclimatology (Jacoby 1989).

3.6. The Tropical Southern Hemisphere Sector

Indistinct tree-ring structures appear to be the major difficulty for applying dendrochronological methods in the tropics. However, many of the tropical species in regions with seasonal rainfall or flooding form distinct annual rings in their wood (Jacoby 1989). The pioneering works of Coster (1927, 1928) and Berlage (1931) in the tropics successfully identified annual rings in some tropical species. In 1931, Berlage presented the first tree-ring chronology from the tropics. Tree-ring variations in the 400-year chronology of *Tectona grandis* from central Java were correlated with precipitation. More recently, this chronology has been re-examined (Murphy and Whetton 1989; Jacoby and D'Arrigo 1990; D'Arrigo et al. 1994), showing important relationships with El Niño–Southern Oscillation events.

Several species showing identifiable annual rings were found in the inundated areas of tropical South America (Worbes 1985, 1989), in the Amazonian forests (Vetter and Botosso 1989), and in the tropical forests of northeastern Argentina (Boninsegna et al. 1989). No relationships between the growth of these tropical trees and climate have been established yet. Annual rings have also been reported in the conifer *Araucaria angustifolia* in southeastern Brazil (Sietz and Kanninen 1989).

In the subtropical forests of northwestern Argentina twelve chronologies were developed using the *Cedrela* and *Juglans* species, and correlations between climate variations and tree-ring widths were established (Villalba et al. 1985, 1987, 1992). This set of chronologies was used to examine the spatial patterns of radial growth and to reconstruct changes in seasonal and annual precipitation in subtropical Argentina since 1800. The dominant patterns of tree-growth variations reflect the direct effects of two of the most important types of rainfall variations in northwestern Argentina: One in which conditions vary uniformly across the region and another in which contrasting precipitation anomalies occur in eastern and western areas (Fig. 3.10).

The successful applicability of subtropical species in dendroclimatological studies is demonstrated by reconstructions of seasonal and annual precipitation in northwestern Argentina. On average, 60–80% of the variance in precipitation was explained using the subtropical tree-ring data set (Villalba et al. 1992). The major limitations of the subtropical reconstructions appear to be the shortness of the tree-ring series. Most of the chronologies in subtropical Argentina cover the last 150 years only, and few of them reach 300 years.

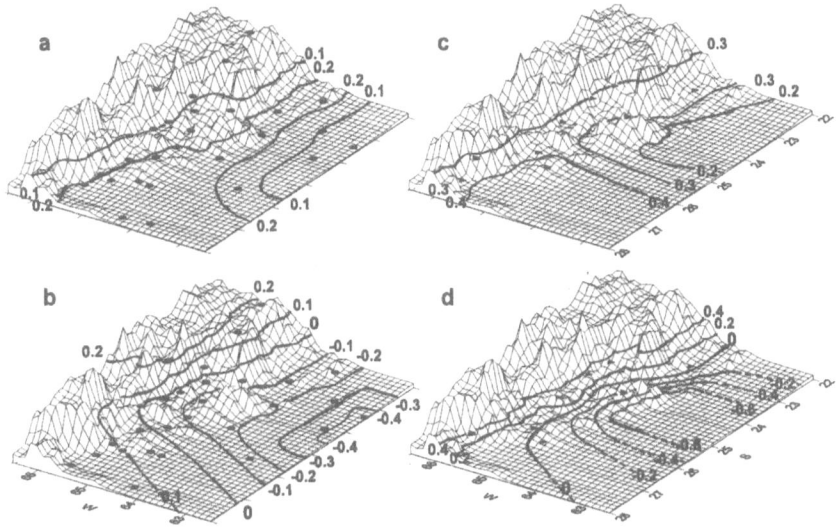

Fig. 3.10. Spatial patterns of climate and tree-growth anomalies in subtropical northwestern Argentina resulting from a principal component analysis of precipitation records and tree-ring chronologies. The first and second tree-ring patterns (**c** and **d**) clearly resemble the first and third precipitation pattern (**a** and **b**), respectively. Meteorological stations are indicated by dots. Tree-ring chronologies are indicated by triangles (modified from Villalba et al. 1992).

Concerning the Australian tropics, Ogden (1981) cites several species that form annual or quasi-annual rings. In open woodlands with summer rainfall distribution annual rings have been identified in *Callistris collumellaris*, *Diospyros ferrea* and some *Eucalyptus* sp. (Ogden 1982). Growth rings of some tropical conifers have also been carefully examined by Ash (1983a, b, 1985, 1986) in northeastern Australia and Fiji. Regular annual growth rings were observed in *Callistris macleayana*, and a significant correlation was found between annual ring widths and the extension of the wet seasons (Ash 1983b). Relationships between ring widths of other tropical conifers in Australia and precipitation have also been observed for *Araucaria cuninghamii* (Booth and Ryan 1985) and *Callistris collumellaris* (Perlinski 1983; cited in Norton 1990). A detailed study of the wood was necessary to discern the structure of the annual bands in *Pisonia grandis* growing in the coral cays near the coast of Queensland (Ogden 1981). Rings in *P. grandis* are wider when precipitation in the wet season is reduced, but higher precipitation during the dry season also favors tree growth.

Stahle et al. (1995, 1997) have begun to survey the incredibly diverse indigenous forests in tropical Africa for species suitable for dendroclimatology. Two tropical chronologies of *Pterocarpus angolensis* from Zimbabwe are strongly correlated with total rainfall amounts during the wet season. Both chronologies only date back to 1870 at present, but *P. angolensis* is the most important timber species in south tropical Africa, and old *P. angolensis* survives in buildings, furniture, canoes, and diverse implements. These sources of older wood may permit the eventual development of 200–300 year chronologies of *P. angolensis* in southeastern Africa (Stahle et al. 1997).

The above results emphasize the importance of dendroclimatological studies in the tropics. There is a need for continued work to develop tropical tree-ring chronologies of similar quality to those of temperate and subarctic zones. The use of different methods for age determination, such as induction of artificial boundaries, densitometry, radiocarbon dates, and correlation between periodical structure in the wood with phenological events, will be required to increase the number of tropical species useful in dendrochronology.

3.7. Particular Issues Confronting Dendroclimatology in the Southern Hemisphere

It is difficult to isolate a distinctive dendroclimatology for the Southern Hemisphere. There are no obvious reasons in the Southern Hemisphere to use completely different techniques or models for tree-ring based climatic reconstructions from those used in the Northern Hemisphere. However, the geography raises some special problems of which dendroclimatologists in the Southern Hemisphere need to be more conscious.

The Southern Hemisphere is the Ocean Hemisphere. Eighty percent of the Southern Hemisphere is covered by water. The mechanisms of heat transport by the Southern Oceans have special relevance not only as a major determinant of the patterns of current climates in the Southern Hemisphere, but also as an influence on how climates were in the past and might be in the future (Hamon and Godfrey 1978). Preliminary results indicate that information on past changes in ocean circulation features, mainly annual or seasonal temperature changes, could be retrieved from tree-rings. Tree-ring chronologies along the western coast of South America appear to contain a climatic signal strongly associated with the sea-surface temperatures of the surrounding oceans (Lara and Villalba 1993; Villalba 1990b). Regional anomalies in sea-surface temperature, such as the cooling trend recorded in the 1960's and 1970's around 40° S off the Chilean coast, have correctly been identified and documented using tree rings. In Tasmania, Cook et al. (1995) found trees that recorded contrasting cold-warm temperature patterns of the last 100 years associated with different regimes in the sea-surface temperature surrounding Tasmania. The next step for dendroclimatology in the Southern Hemisphere should be the use of tree-ring records to derive long-term series of physical oceanic parameters.

The Antarctic ice sheet is about 14 million km^2 in area and contains more than 90% of the world's ice. Any significant climatic change in Antarctica would affect the global climate, but particularly that of the Southern Hemisphere. Long tree-ring chronologies at higher latitudes in South America, the southern tip of Africa, Tasmania, and New Zealand, in connection with annually resolved ice-cores from Antarctica (Aristarain et al. 1990; Peel 1992), could provide the basis for a proper understanding of the patterns of long-term atmospheric circulation at higher latitudes in the Southern Hemisphere (Villalba et al. 1997). A simultaneous comparison of temperature reconstructions from Tasmania and northern Patagonia will be shown as an example of the potential of tree-ring records to provide insight into large-scale circulation at remote higher latitudes.

The Trans–Polar Index (TPI; Pittock 1980a) is defined as the surface pressure anomaly at Hobart, Tasmania (43° S, 147° E) minus that measured in Stanley, Malvinas (Falkland) Islands (52° S, 58° W). The TPI is a measure of eccentricity of the westerly vortex around the South Pole (wavenumber one), displaced either towards South America or eastern Australia. When the westerly vortex is displaced towards South America, precipitation increases at latitudes near 40° S over northern Patagonia, and is consistent with more westerly activity in the South American sector (Pittock 1980, 1984). The original TPI, calculated by Pittock (1980) for the interval 1942–1962, was recently updated to 1985 by Carleton (1989). In the 1951–1982 part of the TPI record, Carleton (1989) noted a decrease of the interannual variability, which is associated with a lack of correlation between the mean annual sea-level pressure between Stanley and Hobart. Temporal changes in the phase and/or amplitude of the Southern Hemisphere waves have been attributed to the changes recorded in TPI (Carleton 1989). Based on these relationships, during intervals dominated by wavenumber one features, tree-ring records from Tasmania are expected to be negatively related to tree-ring records in South America, but uncorrelated at other times. Effectively, when summer temperature reconstructions for the last 1000 years from Tasmania (Cook et al. 1992) and northern Patagonia (Villalba 1990a) are compared, significant negative correlations of 99.9% are found for some intervals (Fig. 3.11). During these intervals, centered around 1100–1200, 1550, and 1640–1760 (Fig. 3.11b), the eccentricity of the westerly vortex around the Pole (wavenumber one) appeared to be the dominant atmospheric feature forcing climatic variations at higher latitudes in the Southern Hemisphere. In contrast, the synchronism of temperature fluctuations in Tasmania and northern Patagonia points out a more central effect of the wavenumber zero. Simultaneous increases or decreases in temperature at Tasmania and northern Patagonia should be related to expansions or contractions rather than changes in the eccentricity of the circumpolar vortex. For instance, the lower temperatures recorded at the beginning of the 20th century in Tasmania and northern Patagonia (Fig. 3.11c) have been attributed to an expansion of the circumpolar vortex, in combination with iceberg irruptions at lower latitudes in the Southern Hemisphere (Lamb 1967; Burrows 1976; Cook et al. 1992).

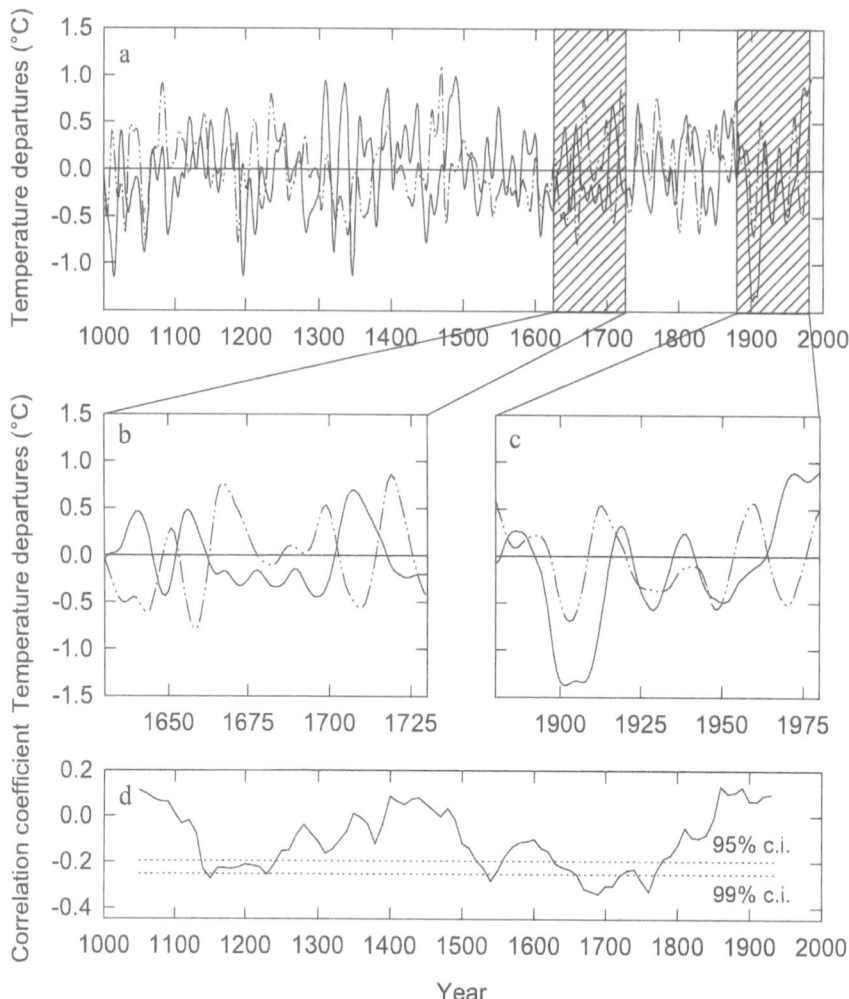

Fig. 3.11. Comparison of temperature variations during the last 1000 years in northern Patagonia (dash-dotted line) and Tasmania (solid line) reconstructed from tree-rings (**3.11a**). Periods of significant negative correlations between temperature reconstructions may reflect climatic variations at higher latitudes in the Southern Hemisphere, related to the circumpolar vortex eccentricity (wavenumber one, **3.11b**). In contrast, synchronous variations, such as that shown at the beginning of the 20th century, should be associated with the expansion–contraction of the circumpolar vortex (wavenumber zero, **3.11c**). Correlation coefficients between the two series were calculated for intervals of 100-year lengths, displaced 10 years from one another (**3.11d**).

Although the El Niño–Southern Oscillation (ENSO) influence is not exclusive to the Southern Hemisphere, there is an increasing level of confidence among climatologists regarding the capability of ENSO phenomena as a useful predictor of seasonal to interannual climate variations in South America, Australia, South Africa, and the tropical regions (Aceituno 1988; Nicholls 1988, 1992; Ropelewski and Halpert 1987). It was previously mentioned that chronologies along the Southern Hemisphere have a large potential for reconstructing past variations of ENSO. Trees from the Indonesian–Australian and South American sectors, which are the geographical extremities strongly impacted by ENSO events, should be considered in any attempt to reconstruct a high-resolution record of past ENSO events.the very strong El Niño event in 1396, which has recently been documented through the use of historical sources in Southern Asia (Grove 1998),provides a good example of the high potential of the Southern Hemisphere tree-ring records for reconstructing past ENSO events. For Central Chile, a region where the association between ENSO and abundant precipitation is well established (Rutlland and Fuenzalida 1991), the reconstructed precipitation for the year 1396 was +4.97 standard deviations (SDs) above the 1220–1972 long-term mean (Boninsegna 1988). In northern Patagonia, the estimated temperature during the summer of 1396 was +2.84 SDs above the tree-ring based mean for the last millenium (Villalba 1990b). Coincidentally, in the millennial Tasmania temperature reconstruction, the year 1396 is the warmest year (+2.12 SDs) in the 270-year interval 1160–1430 (Cook et al. 1992).

Uncertainty is a common feature in climatic model predictions, particularly in the Southern Hemisphere where large-scale circulation is not as well known as it is in the Northern Hemisphere (Pittock et al. 1978; Whetton et al. 1996). Long tree-ring reconstructions can also be used to validate regional predictions from General Circulation Models that are used to simulate future climates. For instance, physical reasoning and climatic-modeling results suggest a poleward shifts of the mid-latitude rainfall belt in the Southern Hemisphere, as related to a global warming of 2–4° C (Pittock and Salinger 1991; Whetton et al. 1996). Thus, an increase of summer precipitation should occur at the subtropical climatic zones of South America east of the Andes (Labraga 1997). These model predictions are supported by precipitation-sensitive tree-ring records from subtropical northwestern Argentina. These records show that the precipitation increase during the past three decades, characterized by a steady warming in subtropical South America, has been unprecedented over the past 200 years (Villalba et al. 1998).

Another challenge facing dendroclimatologists in the Southern Hemisphere is the so-called Antarctic ozone hole. Recent surveys indicate that the belt of depleted stratospheric ozone has spatially extended over Antarctica. In 1996, the estimated area covered by ozone total column values less than 200–220 m.atm.cm was close to 20 million km^2 (about twice the size of Europe), and persisted two months (WMO 1998). Tree-ring records from higher latitudes in the Southern Hemisphere, particularly from the southern tip of Argentina and Chile (50–55° S), should be employed to monitor any changes in tree-growth related to the abnormal levels of ultraviolet radiation.

Finally, we need to consider the more noticeable human impact on natural forests in the Northern Hemisphere compared to the Southern Hemisphere, especially in the temperate regions. Pollution and acid rain are not widespread problems in the Southern Hemisphere. Consequently, dendrochronological studies in the forests of the Southern Hemisphere offer the opportunity of assessing, more properly, the relative importance of human interference (e.g. CO_2 fertilization) versus environmental changes.

3.8. Final Remarks

The processes controlling climate are extremely complex. Therefore, some aspects are better interpreted from a paleo-record perspective since the shortness of the meteorological records severely limited the evaluation of some modern climate processes, which in turn may have recently been affected by anthropogenic influences. When climate changes recorded over the last 100 years are placed in the context of several centuries, some interesting implications emerge. In Tasmania, for instance, the coldest and warmest periods during the last 1000 years have occurred within a time-interval of 50 years in the 20th century. As pointed out by Cook et al. (1991), this atypical pattern of uncommon low followed by uncommon high temperatures makes it difficult to properly evaluate the recent trends in temperature in the Southern Hemisphere. At 40° S latitude in South America, neither meteorological nor dendroclimatological records provide evidence of the warming trend during the last decades of this century that can be related to anthropogenic causes (Lara and Villalba 1993). These local or regional complexities in climatic patterns emphasize the necessity of long, annually resolved records to properly evaluate recent climatic trends in the Southern Hemisphere.

Only 20% of the Southern Hemisphere is emergent land. Ice sheets and deserts represent a large component of the emergent lands. The paucity of potential dendroclimatological sites in the Southern Hemisphere emphasizes the need for climatic pattern analysis to determine the representation of particular sites and to facilitate the search for sites that are a more profitable use of the limited researchers and resources that exist. Hemispheric maps of the strength of regional temperature coherence, derived by Briffa and Jones (1993), show that the correlation decay length (a measure of the spatial coherence) in temperature is higher in the Southern than the Northern Hemisphere. These authors also indicate that individual proxies of temperature variations, from the Southern Hemisphere, have greater potential to represent annual changes, independent of their seasonal response. Consequently, dendroclimatology in the Southern Hemisphere should be directed towards reconstructions of middle- to large-scale climatic features of the atmospheric circulation, rather than spatial reconstructions of dense grids of climate variables.

The relationships between atmospheric circulation indexes of middle and high latitudes have been shown by Pittock (1984). The latitude of the subtropical high-pressure belt off the eastern coast of Australia(L_{aus}) and the Trans–Polar Index (TPI) are correlated at the 99.9% confidence level for the interval 1941–1960 ($r = 0.68$, $n = 20$; Fig. 3.3). Variations in the TPI and the latitude of the South Pacific high pressure cell along the Chilean (L_{sa}) coast appear unrelated on an annual basis. However, fluctuations of the TPI in summer are significantly (95% confidence level) correlated with L_{sa} variations($r = -0.45$, $n = 30$), which in turn are associated with the occurrence of El Niño–Southern Oscillation events (Aceituno 1988; Villalba 1994). These present interconnections between features of the large-scale circulation need to be evaluated for the recent past. Are these large-scale relationships stable over time? It has previously been shown that tree-ring series from opposite sites in the Southern Hemisphere (Tasmania and northern Patagonia) can provide useful information about the changes over time of some high latitude circulation features. An adequate set of high-quality proxy records (including tree-rings, ice cores and historical records), from key areas in the Southern Hemisphere, would be necessary to obtain a detailed view of past hemispheric large-scale climatic changes.

Future steps of dendroclimatological studies in the Southern Hemisphere should focus on the collection of high quality, new data from regions that are deficient in dendrochronological records. Another main priority is to extend the currently available dendrochronological records to the late Holocene, through additional samplings of living trees or sub-fossil materials. The lack of knowledge on the ecology and life-histories of the species was pointed out by Norton (1990) as being the most serious limitation of dendrochronological studies in the Southern Hemisphere. Recent ecological studies as those of Ogden (1985) on *Agathis* and *Nothofagus*, Veblen (1985a, b, 1989a, b) on *Nothofagus*, Burns (1991) on *Araucaria*, and Lara (1991) on *Fitzroya*, provide the basis for a better interpretation of dendrochronological studies on these species. Finally, the investigation of the basic processes responsible for year-to-year variations in tree-ring properties should also be considered. In the near future, some answers to these primary requirements will be provided by the current research on dendrochronology and dendroecology developed at local tree-ring laboratories in Argentina, Chile, Australia, New Zealand and South Africa, or in collaboration with laboratories from the Northern Hemisphere. Because of the similarities in composition and dynamics of the forests, cooperative research between laboratories in the Southern Hemisphere should also be encouraged.

Acknowledgements: Partial support for this review was provided by the Argentinean Research Council (CONICET), the NASA Global Change Program, and the National Geographic Society. For review and critical comments on the manuscript I thank José A. Boninsegna, Fidel A. Roig, and Wolfand Volkheimer (all at CONICET, Argentina), Peter Smolka (University of Münster, Germany), Mario Giovinetto (University of Calgary, Canada), John Ogden (University of Auckland, New Zealand) and Juan C. Aravena (University of Chile). E. Cook (Lamont–Doherty Earth Observatory, New York, USA) provided the *L. franklinii* tree-ring records from Tasmania and Fig. 3.8. J. Palmer (Lincoln University, New Zealand) supplied geographical and topographical information from the New Zealand chronologies. This article is Lamont–Doherty Earth Observatory Contribution No. 5893.

References

Aceituno, P (1988) On the functioning of the Southern Oscillation in the South American sector. Part I: Surface climate. Monthly Weather Review, 116:505–524

Aceituno P, Montecinos A (1992) Análisis de la estabilidad de la relación entre la Oscilación Sur y la precipitacion en America del Sur, Paleo ENSO Records Intern. Symp. Extended Abstracts. Ortlieb, L., and Machare, J. (eds), ORSTOM–CONCYTEC, Lima, p. 7–13

Ahmed M, Ogden J (1985) Modern New Zealand tree-ring chronologies 3. *Agathis australis* (Salisb.)-kauri. Tree Ring Bulletin 45:11–24

Aravena JC, Lequesne C, Jiménez H, Armesto JJ (1997) Sampling and data analysis in Chile. In: Jones PD, Briffa KR, Armesto JJ, Boninsegna JA (comp.), Final Report of the Project: Dendroclimatic reconstruction from southern South American temperate forests, C11–CT93–0336 (DG12 HSMU), 18 pp

Aristarain AJ, Jouzel J, Lorius C (1990) A 400 year isotope record of the Antarctic Peninsula climate. Geophysical Research Letters 17:2369–2372

Ash J (1983a) Growth rings in *Agathis robusta* and *Araucaria cunninghamii* from Tropical Australia. Australian Journal of Botany 31:269–275

Ash J (1983b) Tree rings in tropical *Callistris macleayana* F. Muell. Australian Journal of Botany 31: 277–281

Ash J (1985) Growth rings and longevity of *Agathis vitiensis* (Seemann) Beth and Hook. f. ex Drake in Fiji. Australian Journal of Botany 33:81–88

Ash J (1986) Growth rings, age and taxonomy of *Dacrydium* (Podocarpaceae) in Fiji. Australian Journal of Botany 34:197–205

Barry RG (1978) Climatic fluctuations during the periods of historical and instrumental record. In A.B. Pittock et al. (eds) Climatic change and variability. A Southern Perspective. Cambridge University Press, pp 150–166

Berlage HP Jr. (1931) Over het verband tusschen de dikte der jaarringen van djatiboomen (*Tectona grandis* L.F.) en den regenval op Java. Tectona 24:939–953

Boninsegna JA (1988) Santiago de Chile winter rainfall since 1220 as being reconstructed by tree rings. Quaternary of South America and Antarctic Peninsula 6:67–87

Boninsegna JA (1992) South American dendroclimatological records. In Bradley RS, Jones PD (eds) Climate since AD 1500. Routledge, London pp 446–462

Boninsegna JA, Villalba R (1996) Dendroclimatology in the Southern Hemisphere: Review and prospects. In Dean JS, Meko DM, Swetnam, TW (eds) Tree Rings, Environment and Humanity. Radiocarbon pp 127–141

Boninsegna JA, Keegan J, Jacoby GC, D'Arrigo RD, Holmes RL (1989) Dendrochronological studies in Tierra del Fuego, Argentina. Quaternary of South America and Antarctic Peninsula 7:315–326

Boninsegna JA, Villalba R, Amarilla L, Ocampo J (1989) Studies on tree rings, growth rates and age–size relationships of tropical trees in Misiones, Argentina. International Association of Wood Anatomists Bulletin n.s 10(2):161–169

Booth TH, Ryan P (1985) Climatic effects on the diameter growth of *Araucaria cunninghamii* Ait. ex. D. Don. Forest Ecology and Management 10:297–311

Bridge MC, Ogden J (1986) A sub-fossil kauri *(Agathis australis)* tree-ring chronology. Journal of the Royal Society of New Zealand 16:17–23

Briffa KR, Jones PD (1993) Global surface air temperature variations during the twentieth century: Part 2, implications for large-scale high-frequency palaeoclimatic studies. The Holocene 3:77–88

Briffa KR, Jones PD, Schweingruber FH (1992a) Tree-ring density reconstructions of summer temperature across western North America since AD 1600. Journal of Climate 5:735–754

Briffa KR, Jones PD, Bartholin TS, Eckstein D, Schweingruber FH, Karlén W, Zetterberg P, Eronen M (1992b) Fennoscandian summers from AD 500: Temperature changes on short and long timescales. Climatic Dynamics 7:111–119

Briffa KR, Jones PD, Schweingruber FH, Osborn TJ (1998) Influence of volcanic eruptions on Northern Hemisphere summer temperatures over the past 600 years. Nature 393:450–455

Buckley BM, Cook ER, Peterson MJ, Barbetti M (1997) A changing temperature response with elevation for *Lagarostrobos franklinii* in Tasmania, Australia. Climatic Change 36:477–498

Burns BR (1991) The regeneration dynamics of Araucaria araucana. Unpublished Ph.D. thesis, University of Colorado at Boulder

Burrows CJ (1976) Icebergs in the Southern Ocean. New Zealand Geographer 32:127–138

Campbell DA (1982) Preliminary estimates of summer streamflow for Tasmania. In: Hughes MK, Kelly PM, Pilcher PM, LaMarche VC Jr. (eds) Climate from Tree Rings. Cambridge University Press, Cambridge. pp 170–177

Carleton AM (1989) Antarctic sea–ice relationships with indices of the atmospheric circulation of the Southern Hemisphere. Climate Dynamics 2:207–220

Cobos DR, Boninsegna JA (1983) Fluctuations of some glaciers in the Upper Atuel river basin, Mendoza, Argentina. Quaternary of South America and Antarctic Peninsula 1:61–82

Cook ER (1992) Using tree rings to study past El Niño/Southern Oscillation influences on climate. In: Diaz H, Markgraf V (eds), El Niño. Historical and Paleoclimatic Aspects of the Southern Oscillation. Cambridge University Press, Cambridge, pp 203–214

Cook ER, Kairiukstis L (eds) (1990) Methods of Dendrochronology: Applications in the environmental Sciences. Kluwer, Dordrecht, The Netherlands

Cook ER, Bird T, Peterson M, Barbetti M, Buckley B, D'Arrigo, R, Francey R, Tans, P (1991) Climatic change in Tasmania inferred from a 1089-year tree-ring chronology of Huon Pine. Science 253:1266–1268

Cook ER, Bird T, Peterson M, Barbetti M, Buckley B, D'Arrigo R, Francey R (1992) Climatic change over the last millennium in Tasmania reconstructed from tree-rings. The Holocene 2:205–217

Cook ER, Buckley BM, D'Arrigo RD (1995) Interdecadal temperature oscillations in the Southern Hemisphere: Evidence from Tasmania tree rings since 300 BC In: National Research Council (ed), Natural Climate Variability on Decade–to–Century Time Scales. National Academy Press, Washington, D.C. pp 523–532

Cook ER, Francey RJ, Buckley BM, D'Arrigo RD (1996a) Recent increases in Tasmanian Huon pine ring widths from a subalpine stand: Natural climate variability, CO_2 fertilization, or greenhouse warming? Papers and Proceedings of the Royal Society of Tasmania 130:65–72

Cook ER, Buckley BM, D'Arrigo RD (1996b) Inter-decadal climate oscillations in the Tasmanian sector of the Southern Hemisphere: Evidence from tree rings over the past three millennia. In: Jones PD, Bradley RS, Jouzel J (eds), Climatic Variations and Forcing Mechanisms of the Last 2000 Years. NATO ASI Series I: Global Environmental Change, Vol. 41, pp 141–160

Cook ER, Meko DM, Stockton CW (1997) A new assessment of possible solar and lunar forcing of the bi-decadal drought rhythm in the western United States. Journal of Climate 10:1343–1356

Cook ER, Meko DM, Stahle DW, Cleaveland MK (1998) Drought reconstructions for the continental United States. Journal of Climate

Coster C (1927) Zur Anatomie und Physiologie der Zuwachszonen und Jahresringbildung in den Tropen. Ann. Jard. Bot. Buitenzorg 37:49–160

Coster C (1928) Zur Anatomie und Physiologie der Zuwachszonen und Jahresringbildung in den Tropen. Ann. Jard. Bot. Buitenzorg 38:1–114

D'Arrigo RD, Jacoby GC (1993) Secular trends in high northern latitude temperature reconstructions based on tree rings. Climatic Change 25:163–177

D'Arrigo RD, Jacoby GC, Krusic PJ (1994) Progress in Dendroclimatic Studies in Indonesia. TAO 5:349–363

D'Arrigo RD, Buckley BM, Cook ER, Wagner WS (1995) Temperature-sensitive tree-ring width chronologies of Pink Pine (*Halocarpus biformis*) from Stewart Island, New Zealand. Palaeogeogr, Palaeoclimatol, Palaeoecol 119:293–300

D'Arrigo RD, Cook ER, Salinger MJ, Palmer J, Krusic PJ, Buckley BM, Villalba R (1998) Tree-ring records from New Zealand: Long-term context for recent warming trend. Climate Dynamics 14:191–199

Dunwiddie PW (1979) Dendrochronological studies of indigenous New Zealand trees. New Zealand Journal of Botany 17:251–266

Dunwiddie PW, LaMarche VC (1980a) Dendrochronological characteristic of some native Australian trees. Australian Forestry 43:124–135

Dunwiddie PW, LaMarche VC (1980b) A climatically responsive tree-ring record for *Widdringtonia cedarbergensis*, Cape Province, South Africa. Nature 286:796–797

Dyer TGJ (1982) Southern Africa. In: Hughes MK, Kelly PM, Pilcher PM, LaMarche VC Jr. (eds) Climate from Tree Rings. Cambridge University Press, Cambridge, pp 82–83

Eischeid JK, Diaz HF, Bradley RS, Jones PD (1991) A comprehensive precipitation data set for global land areas. DOE Technical Report No. TR051. U.S. Dept. of Energy Carbon Dioxide Research Division, Washington. D.C. 81 pp

Enomoto H (1991) Fluctuations of snow accumulation in the Antarctic and sea level pressure in the Southern Hemisphere in the last 100 years. Climatic Change 18:67–87

February EC (1996) Plant xylem anatomy, dendrochronology and stable isotopes as tools in rainfall reconstruction in southern Africa. Ph.D. dissertation, The University of Cape Town, South Africa, 97 pp

Fritts H (1976) Tree-Rings and Climate. Academic Press, London

Fritts H (1991) Reconstructing Large-scale Climatic Patterns from Tree-Ring Data. The University of Arizona Press, Tucson

Gordon ND (1986) The Southern Oscillation and New Zealand Weather. Monthly Weather Review 114:371–387

Graetz RD, Wilson MA (1996) North-South: Where Is the divide? In: Giambelluca TW, Henderson–Sellers A (eds), Climate Change: Developing Southern Hemisphere Perspectives. John Wiley & Sons, Chichester, pp 3–33

Graumlich LJ (1993) A 1000-year record of temperature and precipitation in the Sierra Nevada. Quaternary Research 39:249–255

Grove RH (1998) Global impact of the 1789–93 El Niño. Nature 393: 318–319

Hamon BV, Godfrey JS (1978) The role of the oceans. In: Pittock, AB et al. (eds) Climatic change and variability. A Southern Perspective. Cambridge University Press, pp 31–52

Heusser C (1981) Palynology of the last interglacial-glacial cycle in midlatitudes of Southern Chile. Quaternary Research 16:293–321

Holmes RL (1982) Argentina and Chile. In: Hughes MK, Kelly PM, Pilcher PM, LaMarche VC Jr. (eds) Climate from Tree Rings. Cambridge University Press, Cambridge, pp 84–89

Holmes RL, Stockton CW, LaMarche VC (1982) Extension of riverflow records in Argentina. In: Hughes MK, Kelly PM, Pilcher PM, LaMarche VC Jr. (eds), Climate from Tree Rings. Cambridge University Press, Cambridge. pp 168–170

Hughes MK, Brown PM (1992). Drought frequency in central California since 101 BC recorded in giant sequoia tree-rings. Climate Dynamics 6:161–167

Hughes MK, Kelly PM, Pilcher PM, LaMarche VC Jr. (eds) (1982) Climate from Tree Rings. Cambridge University Press, Cambridge

Jacoby GC (1989) Overview of tree-ring analysis in tropical regions. IAWA Bulletin 10:99–100

Jacoby GC, D'Arrigo RD (1989) Reconstructed Northern Hemisphere annual temperature since 1671 based on high-latitude tree-ring data from North America. Climatic Change 14:39–59

Jacoby GC, D'Arrigo RD (1990) Teak (*Tectona grandis* L.F.), a tropical species of large-scale dendroclimatic potential. Dendrochronologia 8:83–98

Jones PD, Raper SCB, Goodess CM, Cherry BSG, Wigley TML (1986a) A grid point surface air temperature data set for the Southern Hemisphere. DOE Technical Report No. TR027. U.S. Dept. of Energy Carbon Dioxide Research Division, Washington. D.C. 73 pp

Jones, PD, Raper SCB, Wigley TML (1986b) Southern Hemisphere surface air temperature variations: 1851–1984. Journal of Climate and Applied Meteorology 25:1213–1230

Karoly DJ (1989) Southern Hemisphere circulation features associated with El Niño–Southern Oscillation events. Journal of Climate 2:1239–1252

Kiladis GN, Diaz HF (1989) Global climatic anomalies with extremes in the Southern Oscillation. Journal of Climate 2:1069–1090

Karoly DJ, Hope P, Jones PD (1996) Decadal variations of the Southern Hemisphere circulation. International Journal of Climatology 16:723–738

Labraga JC (1997) The climate change in South America due to doubling in the CO_2 concentration: intercomparison of general circulation model equilibrium experiments. International Journal of Climatology 17:377–398

LaMarche VC Jr. (1982) Comment. In: Hughes MK, Kelly PM, Pilcher PM, LaMarche VC Jr. (eds) Climate from Tree Rings. Cambridge University Press, Cambridge. pp 83–84

LaMarche, VC Jr., Pittock AB (1982) Preliminary temperature reconstructions for Tasmania. In: Hughes MK, Kelly PM, Pilcher PM, LaMarche VC Jr. (eds) Climate from Tree Rings. Cambridge University Press, Cambridge, pp 177–185

LaMarche VC Jr., Holmes RL, Dunwiddie PW, Drew LG (1979a) Tree-ring chronologies of the Southern Hemisphere. 1. Argentina. Chronology Series V. Laboratory of Tree-Ring Research, University of Arizona, Tucson

LaMarche VC Jr., Holmes RL, Dunwiddie PW, Drew LG (1979b) Tree-ring chronologies of the Southern Hemisphere. 2. Chile. Chronology Series V. Laboratory of Tree-Ring Research, University of Arizona, Tucson

LaMarche VC Jr., Holmes RL, Dunwiddie PW, Drew LG (1979c) Tree-ring chronologies of the Southern Hemisphere. 3. New Zealand. Chronology Series V. Laboratory of Tree-Ring Research, University of Arizona, Tucson

LaMarche VC Jr., Holmes RL, Dunwiddie PW, Drew LG (1979d) Tree-ring chronologies of the Southern Hemisphere. 4. Australia. Chronology Series V. Laboratory of Tree-Ring Research, University of Arizona, Tucson

LaMarche VC Jr., Holmes RL, Dunwiddie PW, Drew LG (1979e) Tree-ring chronologies of the Southern Hemisphere. 5. South Africa. Chronology Series V. Laboratory of Tree-Ring Research, University of Arizona, Tucson

Lamb HH (1967) On climatic variations affecting the Far South. In: Polar Meteorology, pp 428–453. W.M.O. Technical Report No 87, Geneva. (W.M.O. N 211, T.P. III)

Lara A (1991) The dynamics and disturbance regimes of Fitzroya cupressoides forest in the south-central Andes of Chile. Unpublished Ph. D. dissertation, University of Colorado at Boulder

Lara A, Villalba R (1993) A 3620-year temperature reconstruction from *Fitzroya cupressoides* tree rings in southern South America. Science 260:1104–1106

Lara A, Villalba R (1994) Potencialidad de *Fitzroya cupressoides* para reconstrucciones climáticas durante el Holoceno en Chile y Argentina. Revista Chilena de Historia Natural 67:443–451

Lilly MA (1977) An assessment of the dendrochronological potential of indigenous tree species in South Africa. Department of Geography and Environmental Studies, University of the Witwatersrand, Johannesburg, Occasional Paper No. 18

Lough JM (1991) Rainfall variations in Qeensland, Australia: 1891–1986. International Journal of Climatology 11:745–768

Luckman BH, Briffa KR, Jones PD, Schweingruber FH (1997) Tree-ring based reconstruction of summer temperatures at the Columbia Icefield, Alberta, Canada, AD 1073–1983. The Holocene 7:375–389

Minetti JL, Sierra EM (1989) The influence of general circulation patterns on humid and dry years in the Cuyo Andean region of Argentina. International Journal of Climatology 9:55–68

Mo KC, White GH (1985) Teleconnections in the Southern Hemisphere. Monthly Weather Review 113:22–37

Mossman RC (1909) The monsoons of the Chilean littoral (Preliminary note). Trans. Roy. Soc. Edinburg, XLVII, Part 1, No. 6, 137–141

Murphy JO, Whetton PH (1989) A re-analysis of a tree ring chronology from Java. Proceedings of the Koninklijke Nederlandse Akademie van Wetenschappen B92:241–257

Nicholls N (1988) El Niño–Southern Oscillation impact prediction. Bulletin of the American Meteorological Society 69:173–176

Nicholls N (1992) Historical El Niño/Southern Oscillation variability in the Australian region. In: Diaz H, Markgraf V (eds) El Niño. Historical and Paleoclimatic Aspects of the Southern Oscillation. Cambridge University Press, Cambridge, pp 151–174

Norton DA (1983a) Modern New Zealand tree-ring chronologies. I. *Nothofagus solandri*. Tree-Ring Bulletin 43:1–17

Norton DA (1983b) Modern New Zealand tree-ring chronologies. I. *Nothofagus menziesii*. Tree-Ring Bulletin 43:1–17

Norton DA (1987) Reconstructions of past river flow and precipitation in Canterbury, New Zealand, from analysis of tree-rings. Journal of Hydrology (New Zealand) 26:61–174

Norton DA (1990) Dendrochronology in the Southern Hemisphere. In: Cook E, Kairiukstis L (eds) Methods of Dendrochronology: Applications in the Environmental Sciences. Kluwer, Dordrecht, pp 17–21

Norton DA (1992) New Zealand temperatures, 1800–30. In: Harrington C (ed) The year without a summer? Climate in 1816. Syllogeus, National Museums of Canada. pp 516–520

Norton DA, Ogden J (1987) Dendrochronology: a review with emphasis on New Zealand applications. New Zealand Journal of Ecology 10:77–95

Norton DA, Palmer JG (1992) Dendroclimatic evidence from Australasia. In: Bradley RS, Jones PD (eds) Climate since AD 1500. Routledge, London pp 463–482

Norton DA, Briffa KR, Salinger MJ (1989) Reconstruction of New Zealand summer temperature to 1730 AD using dendroclimatic techniques. International Journal of Climatology 9:633–644

Ogden J (1978) On the dendrochronological potential of Australian trees. Australian Journal of Ecology 3:339–356

Ogden J (1981) Dendrochronological studies and the determination of trees ages in the Australian tropics. Journal of Biogeography 8:405–420

Ogden J (1982) Australasia. In: Hughes MK, Kelly PM, Pilcher PM, LaMarche VC Jr. (eds) Climate from Tree Rings. Cambridge University Press, Cambridge, pp 90–103

Ogden J (1985) An introduction to plant demography with special reference to New Zealand trees. New Zealand Journal of Botany 23:751–772

Palmer E, Pitman N (1972) Trees of South Africa. A.A. Balkema, Cape Town

Palmer JG (1989) A dendroclimatic study of *Phyllocladus trichromanoides* D.Don (tanekaha). Unpublished Ph.D. thesis, University of Auckland

Palmer JG, Ogden J (1992) Tree-ring chronologies from endemic Australian and New Zealand conifers. In: Harrington C (ed) The year without a summer? Climate in 1816. Syllogeus, National Museums of Canada, pp 510–515

Peel DA (1992) Ice core evidence from the Antarctic Peninsula region. In: Bradley RS, Jones PD (eds) Climate since AD 1500. Routledge, London, pp 549–571

Perlinski J (1983) Dendrochronology in a subtropical arid region. In: Pacific Science Association 15th Congress, Dunedin. Program, Abstracts, and Congress Information, Vol. 1

Pittock AB (1971) Rainfall and the general circulation. Proceedings of the International Conference on Weather Modification, Canberra, American Meteorological Society, pp 330–338

Pittock AB (1975) Climatic change and the patterns of variation in Australian rainfall. Search 6:498–504

Pittock AB (1978) An overview. In: Pittock AB et al. (eds) Climatic change and variability. A Southern Perspective. Cambridge University Press, pp 1–6

Pittock AB (1980a) Patterns of climatic variation in Argentina and Chile. I. Precipitation, 1931–1960. Monthly Weather Review 108:1347–1361

Pittock AB (1980b) Patterns of climatic variation in Argentina and Chile. II. Temperature, 1931–1960. Monthly Weather Review 108:1362–1369

Pittock AB (1983) Recent Climatic Change in Australia: Implications for a CO2-Warmed Earth. Climatic Change 5:321–340

Pittock AB (1984) On the reality, stability and usefulness of Southern Hemisphere teleconnections. Australian Meteorological Mag. 32:75–82

Pittock AB, Salinger MJ (1991) Southern Hemisphere Climate Scenarios. Climatic Change 18:205–222

Pittock AB, Frakes LA, Jensen DR, Peterson JA, Zillman JW (eds) (1978) Climatic change and variability. A Southern Perspective. Cambridge University Press, London

Quinn WH (1992) A study of Southern Oscillation-related climatic activity for AD 622–1990 incorporating Nile River data. In: Diaz H, Markgraf V (eds) El Niño. Historical and Paleoclimatic Aspects of the Southern Oscillation. Cambridge University Press, Cambridge, pp 119–150

Quinn WH, Neal VT (1992) The historical record of El Niño events. In: Bradley RS, Jones PD (eds) Climate since AD 1500. Routledge, London, pp 623–648

Rabassa J, Brandani A, Boninsegna JA, Cobos DR (1985) Glacier fluctuations during and since the Little Ice Age and forest colonization: Monte Tronador and Volcan Lanin, Northern Patagonian Andes. Proceedings of the International Symposium on Glacier Mass Balance, Fluctuations and Runoff, Alma Ata, October 1985, USSR

Rogers JC, Van Loon H (1982) Spatial variability of sea level pressure and 500 mb height anomalies over the Southern Hemisphere. Monthly Weather Review 110:1375–1392

Roig FA (1991) Dendrocronología y dendroclimatología del bosque de *Pilgerodendron uviferum* en su area norte de dispersión. Bolletin de la Sociedad Argentina de Botanica 27:217–234

Roig FA (1996) Dendroklimatologische Untersuchungen an *Fitzroya cupressoides* im Gebiet der Küstenkordillere und der Südlichen Anden. Ph.D. Thesis, Basel University, 158 pp

Roig FA, Boninsegna JA (1990) Environmental factors affecting growth of *Adesmia* communities as determined from tree rings. Dendrochronologia 8:39–66

Roig FA, Boninsegna JA (1992) Chiloe Island (Chile) summer precipitation reconstructed for 426 years from *Pilgerodendron uviferum* tree-ring chronologies. In: Bartholin T.S. et al. (eds) Tree Rings and Environment, Lundqua Report 34:277–280

Roig FA, Roig C, Rabassa J, Boninsegna JA (1996) Fuegian floating tree-ring chronology from subfossil *Nothofagus* wood. The Holocene 6: 469–476

Roig FA, LeQuesne C, Boninsegna JA, Lara A, Grudd H, Villagrán C, Jones PD (1998) A late Pleistocene tree-ring record from southern Chile: Implications for high-resolution paleoclimatic information. Abstracts, 1st IGBP PAGES Open Science Meeting, London, pp 110

Ropelewski CF, Halpert MS (1987) Global and regional scale precipitation patterns associated with the El Niño/Southern Oscillation. Monthly Weather Review 115:1606–1626

Rubin MJ (1955) An analysis of pressure anomalies in the Southern Hemisphere. Notos 4:11–16

Rutlland J, Fuenzalida H (1991) Synoptic aspects of the central Chile rainfall variability associated with the Southern Oscillation. International Journal of Climatology 11:63–76

Salinger MJ, Palmer JG, Jones PD, Briffa KR (1994) Reconstruction of New Zealand climate indices back to AD 1731 using dendroclimatic techniques: Some preliminary results. International Journal of Climatology 14:1135–1149

Schulman E (1956) Dendroclimatic Changes in Semiarid America. University of Arizona Press, Tucson

Schweingruber FH (1988) Tree rings, basics and applications of dendrochronology. D. Reidel, Boston

Schweingruber FH (1992) Annual growth rings and growth zones in woody plants in southern Australia. IAWA Bulletin 13:359–380

Seitz RA, Kanninen M (1989) Tree ring analysis of *Araucaria angustifolia* in southern Brazil: Preliminary results. IAWA Bulletin 10:170–174

Stahle DW, Cleaveland MK (1992) Reconstruction and analysis of spring rainfall over the southeastern U.S. for the past 1000 years. Bulletin of the American Meteorological Society 73:1947–1961

Stahle DW, Cleaveland MK (1993) Southern Oscillation extremes reconstructed from tree rings of the Sierra Madre Occidental and southern Great Plains. Journal of Climate 6:129–140

Stahle DW, Cleaveland MK, Maingi J, Munyao, J (1995) The dendroclimatology of *Vitex keniensis* in Kenya. EOS Supplement, Nov. 7, 1995

Stahle DW, Cleaveland MK, Haynes GA, Klinowicz J, Muskove P, Nyrvenya P, Nicholson (1997) Development of a rainfall-sensitive tree-ring chronology in Zimbabwe. Eight Symposium on Global Change Studies, Amer. Meteorol. Soc., Boston, pp 205–211

Stewart J (1975) Some changes in tropical rainfall regimes, Australia 1870–1970. In Proceedings of the WMO/IAMAP Symposium on Long-Term Climatic Fluctuations, Norwich. WMO-N 421, Geneva, pp 255–264

Stoeckenius T (1981) Interannual variations of tropical precipitation patterns. Monthly Weather Review 109:1233–1247

Streten NA (1980) Some synoptic indices of the Southern Hemisphere mean sea level circulation 1972–1977. Monthly Weather Review 108:18–36

Trenberth KE (1976) Fluctuations and trends in indices of the southern hemispheric circulation. Quarterly Journal of the Royal Meteorological Society 102:65–75

Van Loon H, Rogers JC (1981) The Southern Oscillation, Part II: Association with changes in the middle troposphere in the northern winter. Monthly Weather Review 109:1163–1168

Veblen TT (1985a) Stand dynamics in Chilean *Nothofagus* forest. In: Pickett STA, White PS (eds) Ecology of natural disturbance and patch dynamics. Academic Press, New York

Veblen TT (1985b) Forest development in tree-fall gaps in the temperate rain forest of Chile. National Geography Research 1:161–184

Veblen TT (1989a) *Nothofagus* regeneration in treefall gaps in northern Patagonia. Canadian Journal of Forest Research 19:365–371

Veblen TT (1989b) Tree regeneration responses to gaps along a trans-Andean gradient. Ecology 70:543–545

Vetter RE, Botosso PC (1989) Remarks on age and growth rate determination of Amazonian trees. IAWA Bulletin 10(2):133–145

Villalba R (1987) Tropical Dendroecology. Serie Cientifica 35:44–47

Villalba R (1990a) Latitude of the surface high-pressure belt over western South America during the last 500 years as inferred from tree-ring analysis. Quaternary of South America and Antarctic Peninsula 7:273–303

Villalba R (1990b) Climatic Fluctuations in Northern Patagonia During the Last 1000 Years as inferred from Tree-Ring Records. Quaternary Research 34(3):346–360

Villalba R (1994) Tree-ring and glacial evidence for the Medieval Warm Epoch and the Little Ice Age in southern South America. Climatic Change 26:183–197

Villalba R, Boninsegna JA (1992) Tree-Ring Changes in South America Following the Major Volcanic Eruptions between 1750 and 1970. In: Harrington C (ed) The year without a summer? Climate in 1816. Syllogeus, National Museums of Canada, pp 493–509

Villalba R, Veblen TT (1997) Spatial and temporal variations in tree growth along the forest-steppe ecotone in northern Patagonia. Canadian Journal of Forest Research 27:580–597

Villalba R, Boninsegna JA, Holmes RL (1985) *Cedrela angustifolia* and *Juglans australis*: Two New Tropical Species Useful in Dendrochronology. Tree-Ring Bulletin 45:25–36

Villalba R, Boninsegna JA, Ripalta A (1987). Climate, Site conditions and Tree-growth in Subtropical Northwestern Argentina. Canadian Journal of Forest Research 17(12):1527–1544

Villalba R, Boninsegna JA, Cobos DR (1989) A Tree-Ring Reconstruction of Summer Temperature Between AD 1500 and 1974 in Western Argentina. Third International Conference on Southern Hemisphere Meteorology and Oceanography, 196–197. Buenos Aires, Argentina

Villalba R, Holmes RL, Boninsegna JA (1992) Spatial patterns of climate and tree growth variations in subtropical northwestern Argentina. Journal of Biogeography 19:631-649

Villalba R, Leiva J, Rubulis S, Suarez JA Lenzano L (1990) Climate, tree-rings and glacial fluctuations in the Frias Valley, Rio Negro, Argentina. *Arctic and Alpine Research* 22:215–232

Villalba R, Boninsegna JA, Lara A, Veblen TT, Roig FA, Aravena JC, Ripalta A (1996) Interdecadal climatic variations in millennial temperature reconstructions from southern South America. In: Jones PD, Bradley RS, Jouzel J (eds), Climatic Variations and Forcing Mechanisms of the Last 2000 Years. NATO ASI Series I: Global Environmental Change, Vol. 41, pp 161–189

Villalba R, Cook ER, D'Arrigo RD, Jacoby GC, Jones PD, Salinger MJ, Palmer J (1997a) Sea-level pressure variability around Antarctica since AD 1750 inferred from subantarctic tree-ring records. Climate Dynamics 13: 375–390

Villalba R, Boninsegna JA, Veblen TT, Schmelter A, Rubulis S (1997b) Recent trends in tree-ring records from high elevation sites in the Andes of northern Patagonia. Climatic Change 36:425–454

Villalba R, Grau HR, Boninsegna JA, Jacoby GC, Ripalta A (1998) Tree-ring evidence for long-term precipitation changes in subtropical South America. International Journal of Climatology 18:1463–1478

Wells LE (1987) An alluvial record of el Niño events from northern coastal Peru. Journal of Geophysical Research 92:14463–14470

Wells LE (1990) Holocene history of the El Niño phenomenon as recorded in flood sediments of northern coastal Peru. Geology 18:1134–1137

Whetton P, Pittock AB, Labraga JC, Mullan, AB, Joubert A (1996) Southern Hemisphere climate: Comparing models with reality. In: Giambelluca TW, Henderson–Sellers A (eds), Climate Change: Developing Southern Hemisphere Perspectives. John Wiley & Sons, Chichester, pp 89–130

WMO (1998) The global climate system review, December 1993–May 1996. Nicholls JM (ed) World Climate and Monitoring Programme, WMO-No. 856, 95 pp

Worbes M (1985) Structural and other adaptations to long-term flooding by trees in Central Amazonia. Amazoniana 9:459–484

Worbes M (1989) Growth rings, increment and age of trees in inundation forests, savannahs and a mountain forest in the Neotropics. IAWA Bulletin 10:109–122

Wright PB (1977) The Southern Oscillation – Patterns and mechanisms of the teleconnections and the persistence. Hawaii Institute of Geophysics, Report HIG–77–12

The Chemical History of the Atmosphere: Self-Organizing Processes and Biological Consequences

G. Beckmann[1], B. Klopries[1], H. Hämmerle[2], O. Inacker[2], P. Smolka[3]

(1) HÜLS AG, Paul Baumann Str. 1, Postfach 1320, D-45772 Marl, Germany
(2) NMI, University Tübingen, Gustav Werner Str. 3, D-72762 Reutlingen, Germany
(3) University Münster, Geological Institute, D-48149 Münster, Germany

Abstract: During the Earth's history massive changes of the CO_2 content of the atmosphere have been observed. We describe the history of the atmosphere of the Earth with equations that consider both background from chemistry, engineering and geology. It could be demonstrated, that times of accelerated atmospheric change coincide with times of accelerated evolution, i.e. an increased rate of forming of new species and disappearance of old ones. Together with microbiological evidence, this suggests that the extremely high gradient of present atmospheric change should be regarded as considerably more dangerous than in the past. The set of equations is relatively simple. It considers only massive global changes and effects. Nevertheless it yields insight into the atmospheric history and geological processes. All biological systems are aqueous systems. In these systems CO_2 is soluble. Since CO_2 participates directly or indirectly in many important reactions in cells, the CO_2 pressure in the cell is of prime importance. In systems characterized by low CO_2 partial pressures, such as many microbiological systems, moderate changes of the atmospheric CO_2 partial pressure cause a massive impact on chemical equilibria. Thus for many biological systems, the change (not necessarily the absolute value) of CO_2 partial pressure means a major environmental stress, caused by the close connection between atmospheric partial pressure and the interior of the cells. This environmental stress forces all species to adapt genetically. Microbiota and viruses can adapt much faster to the new situation due to their short life span, resulting in a large number of new species. Statistically it has to be expected that among these new species there are at least some that are incompatible with other hitherto existing biota (including "macrobiota"), i.e. threatening them as pests and diseases. This aspect of environmental change has not yet been paid sufficient attention.

4.1. Introduction

During the Earth's history the chemical composition of the atmosphere changed considerably. The CO_2 content of the atmosphere has fallen, accompanied by an increase of the O_2 content since the Precambrian. This resulted in a more or less permanent environmental stress for evolving and adapting life. The present day man-made change of the CO_2 content is much faster than any other change of the CO_2 content that has been observed in the Earth's history.

This paper's discussion focuses on several aspects synchronously:

(1) *The effect of CO_2 increase on climate and sea level.* This is accomplished for several places of the world by both atmospheric modeling and geological

studies, concentrating on already realized warmer climatic equilibria. Although the direct physical greenhouse effect approaches an upper limit (Fig. 4.1) the potential existence of triggering effects for other processes of larger magnitude cannot be ruled out as the rapid climatic shifts are documented in both Neogene and Quaternary times series (see for example Frenzel 1991, Smolka 1991).

(2) *The impact of a change in the CO_2 content on biota.* This applies especially to those that (a) have a low CO_2 partial pressure in their cells and (b) permit a rapid intracellular adjustment to outer changes due to the lack of protecting tissue. These are mainly many bacteria, viruses, fungi and plants. A retrospective recapitulation of Earth history that, through geological and paleontological evidence, displays data that permit a test of chemical and microbiological theory by comparison of evolution and the history of the CO_2/O_2 ratio both through geological modeling and field studies.

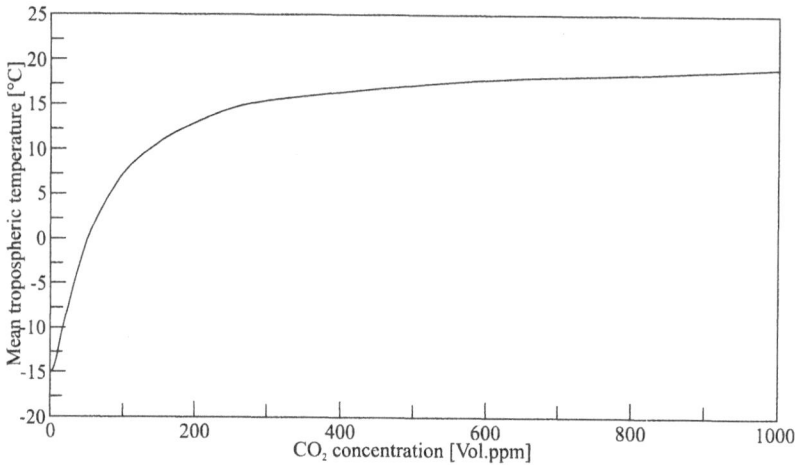

Fig. 4.1. Relation between CO_2 concentration (Vol. ppm) and temperature. Note the asymptotic nature of the curve.

4.2. A Self-Organizing Model of the Atmosphere: Boundary Conditions

Attempting to model the history of the Earth's atmosphere, including both the Precambrian and the Neogene, seems to be a hopeless enterprise, especially because of the number of processes involved. In addition several of these processes are complicated and, particularly with respect to a fully quantitative treatment, poorly understood (see for example Trabalka and Reichle 1986). These processes include the role of terrestrial and marine animals, weathering, soil-processes, rain, clouds, seasonal effects, photosynthesis, deep-water formation and ocean currents, to mention only a few.

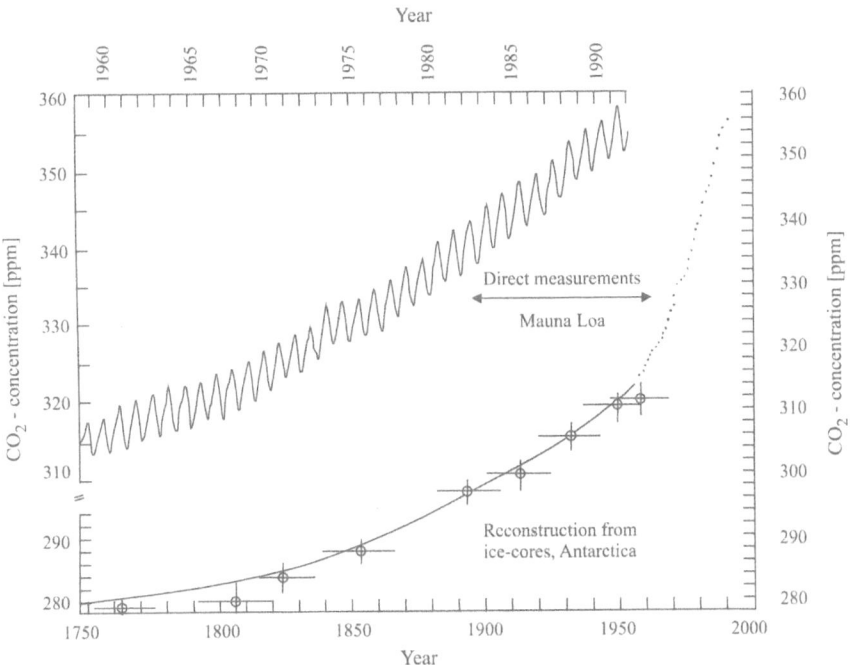

Fig. 4.2. Young history of atmospheric CO_2 concentration (after Schönwiese 1992).

A key for initiating a solution is offered by the problem itself (Keeling et al. 1989; Neftel et al. 1988, Schönwiese and Runge 1988). The upper graph of Fig. 4.2 shows CO_2 increase since 1958. The lower graph (reconstructions from ice-cores) since 1750. Oscillations in the upper curve are caused by seasonal change of Earth's vegetation coverage. During northern hemisphere summer, global vegetation coverage reaches its annual maximum due to large forested areas in Siberia, Canada and Northern Europe. A global maximum of photosynthesis causes an annual global minimum of CO_2. Average global photosynthesis reaches a minimum during southern hemisphere summer, because photosynthesis of smaller forested areas of the southern hemisphere cannot compensate for the CO_2 emissions during the northern hemisphere winter. These CO_2 emissions are (among other causes, including industrial activity) attributed to rotting of plants during the northern hemisphere winter. The difference between peaks amounts to about 6 ppm (parts per million), showing how promptly and sharply global CO_2 concentrations react to differences of local vegetation coverage.

This also includes indirect effects caused by changes in vegetation coverage, such as soil processes and microbiotic life. In addition, as it measures a *global* phenomenon, this curve integrates effects of annual fluctuations of marine photosynthesis and related processes. Consequently, despite the highly complex details of all subprocesses, including high local and regional variability, effective response of the global atmosphere is quite straightforward. This is confirmed by the lower curve of Fig. 4.2, the increase of CO_2 concentration since 1750.

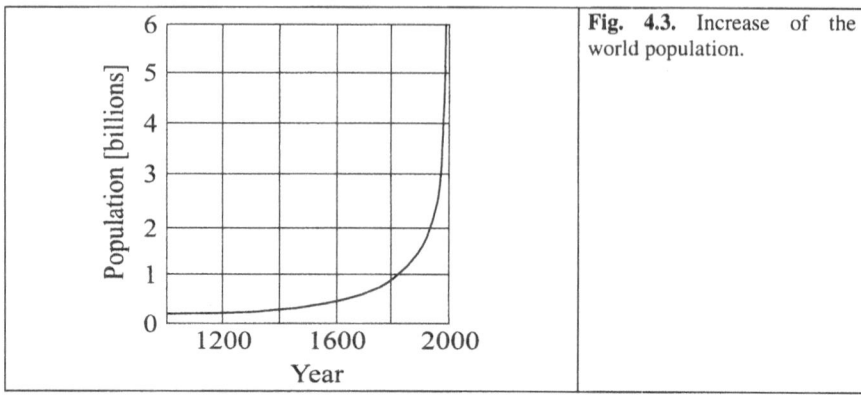

Fig. 4.3. Increase of the world population.

Despite the complexity of all the processes involved, including difficulties to realistically model the impact of social and economical processes, advances in technology, changes in behavior of consumption, economic crises, etc., the CO_2 increase correlates remarkably well with increase in world population (Fig. 4.3). This suggests that there is a chance for understanding processes on a global scale if it is possible to identify variables and processes that integrate several effects of spatial and temporal small- and medium-scale processes.

Therefore, to understand the *nature* of a process (not its regional variability) in many cases simple box-models may be very effective. In addition box-models permit the easy study of the nature of a process with other analytical techniques, such as the search for attractors, finding areas of stability, etc., (Bode 1992). In addition, there are several chemically important and geologically well-known processes that provide boundary conditions for quantitative studies: boundary conditions that, together with chemical equations, must not be violated. Such conditions are, for example, the existence of an atmosphere at all. This provides, together with a chemical process, the starting window of atmospheric equilibria about 4 by (billion years) ago. Another important boundary condition is the amount of Mesozoic carbonates, as their formation requires a considerable CO_2 depletion of the atmosphere. Thus this paper also demonstrates how the interaction of relatively complex processes can be understood by concentrating on the most important key variables and key equations. This is done through a straightforward and consequent application and integration of known data from different fields (that, considered alone, might have large error bars), and through known equations and laws, as well as boundary conditions that must not be violated.

Considering Earth as a large chemical reactor might be a straightforward experiment in the chemical industry yet it is uncommon in geology. Because in addition (a) used boundary conditions must be verified easily and (b) printing space needs to be kept short, the authors have attempted, wherever possible, to find references that are widespread and easily accessible. As all equations are outlined in the appendix, other boundary conditions can be included easily.

4.3. The Model

The scope of the model is a curve that describes the CO_2/O_2 ratio of the last 4 by. The box-model is designed such, that at any point, parameterizations can be substituted by other models of any complexity (for example, vegetation history). The applied method is comparable to the general procedure that is used to solve differential equations where fundamental chemical and physical equations are formulated (carbonate-equilibria, radiation balance, etc.). Geological boundary conditions are included.

These are, for example, the existence of an atmosphere, the increased appearance of carbonatic/phosphatic fossils around the Cambrian, the massive formation of carbonates in the Mesozoic, and the consequences for Mesozoic climate. The same applies to Cenozoic and recent (future?) times. A second and potentially third step is the parameterization of all equations included, in order to arrive at a self-organizing model. Such an algorithm will most probably never be able to run fully autonomously, because important boundary conditions changed through time (origin of life, terrestrial vegetation, paleogeography). A third step may be the study of the properties of this system in order to enhance insight into stability fields. This paper's geologic part focuses mainly on the first step. In its microbiological part, the focus is on the impact of CO_2 and CO_2 concentration on enzymatic reactions including evolutive aspects (CO_2 changes as environmental stress for microbiota, formation of new species including pests and diseases).

The average temperature of a planet consists (amongst many others) of two fundamental components: radiation effects (solar input, reflection and endogenic heat) and the effect of the atmosphere (natural and anthropogenic greenhouse effect). The natural greenhouse effect is closely linked to chemical equilibria in the oceans. This can be used for establishment of boundary-conditions.

4.3.1. Fundamental Concepts

The age of the Earth is set to 4.7 by. The solar constant changed (Junge 1981; Pollack 1979; and for an overview Kasting et al. 1988) according to astronomical data (Eq. 1). The temperature on the surface of the sun follows Eq. 2. After accretion of a planet, endogenic heat has to be accounted for (Eq. 3). The density of solar radiation a planet receives from the sun (Stefan–Boltzmann) is found in Eq. 4, and the radiation-budget on the surface of a planet (endogenic heat + solar radiation) accounting for the albedo (backscattering and radiation of heat) in Eq. 5. Using albedo and distance from the sun, the solution of Eq. 4 for temperature yields Eq. 6. The average temperature of a planet consists of the radiative component (black body and albedo yielding 259K for the recent Earth) and the natural greenhouse effect.

Using empirical data (ice-cores) for the relation between temperature and CO_2 (in vol. ppm) the natural greenhouse effect follows Eq. 7 (see the curve of the radiation-temperature and the intersection with the ordinate at ln 1 ppm CO_2). For the scope of the model this simple approximation is sufficient. Other models apply other approximations (see for example Heinze and Maier–Reimer 1992).

4.3.2. Initial Conditions

Before ocean-chemistry is considered, the concepts mentioned above, and the early history of Earth, Venus and Mars will be used to establish boundary conditions. The following sentences are not intended to be a thorough discussion of planetology concepts. The only intention is to sketch known processes and utilize them for defining a starting-window for the atmosphere of the Earth:

After accretion and subsequent heating, exhalating volatiles (CO_2, NO_x, NH_3, CH_4, H_2S) form a first atmosphere and a first ocean (crystal-water). Formation of the core and, potentially, separation of the moon (oldest lunar rocks about 4 by old, no distinct core, etc.) may also have been phenomena of this time (see also Crutzen and Müller 1989; and the overviews of Schneider 1989; Graedel and Crutzen 1989). Solar radiation (see equations) reached about 73.5% of the present-day value with a spectrum shifted towards the blue (Kasting et al. 1988). Other phenomena (emission of alpha-particles, neutrons, etc.) are not relevant for the model.

Venus reaches an average surface temperature of at least about 500° (distance to the sun about 108 million km, discharge of endogenic heat approximately 4000 W/m^2, a value very roughly estimated from industrial evidence provided by one of the authors (B.K.) the order of magnitude of a hot but not red radiating body. Consequently, formation of an ocean is not possible. The exhalating crystal-water remains gaseous. The water dissociates (UV radiation, ionizing particles) into hydrogen and hydroxylic radicals. The latter form hydrogen and various acids. The hydrogen migrates into space. Today the atmosphere of Venus is nearly free of water (<0.1%). The atmosphere roughly consists of 96% CO_2, 3.5% N_2 and about 0.5% acid gases. The pressure amounts to about 90 bars, the temperature about 470° C. The consequences are discussed below.

Mars is about 228 million km from the sun. The radius is about half the radius of Earth (6687 km); the volume about 0.15 times Earth's. Thus the resulting initial surface-radiation is roughly 600 W/m^2. Therefore an initial surface temperature of about 200° C is possible. Thus the existence of an ocean for about 2.5 by can be inferred. This is a time-span that, if terrestrial analogs are applicable (nature of the rocks, etc.), may have been sufficient for the formation of (micro)paleontological evidence for early life on Mars. The same applies to the potential for photosynthesis, free oxygen and ferric oxide. Again assuming comparable rocks (quantity of crystal-water) the maximum average ocean depth can be estimated to be roughly 1000 m. However photolysis and reduced gravitation (38% of the terrestrial value, escape velocity 5100 m/s compared to terrestrial 11 200 m/s) permits the diffusion of water vapor into higher atmospheric levels.

Whether the rather small number of craters may support a location of the ocean on the northern hemisphere of Mars is a question beyond the scope of this paper. After this work was finished the Mars pathfinder expedition of NASA yielded independently results that are consistent with the concepts of this paper (such as the former presence of water on Mars). Subsequent cooling reduced and finally (intersection with the 0° isotherm) stopped potential microbiological activity. Degradation of potentially available(?) biomass through free-oxygen frees CO_2 and subsequently reduces cooling, finally CO_2 freezes. Consequently (through observations) CO_2 covers frozen water on the poles.

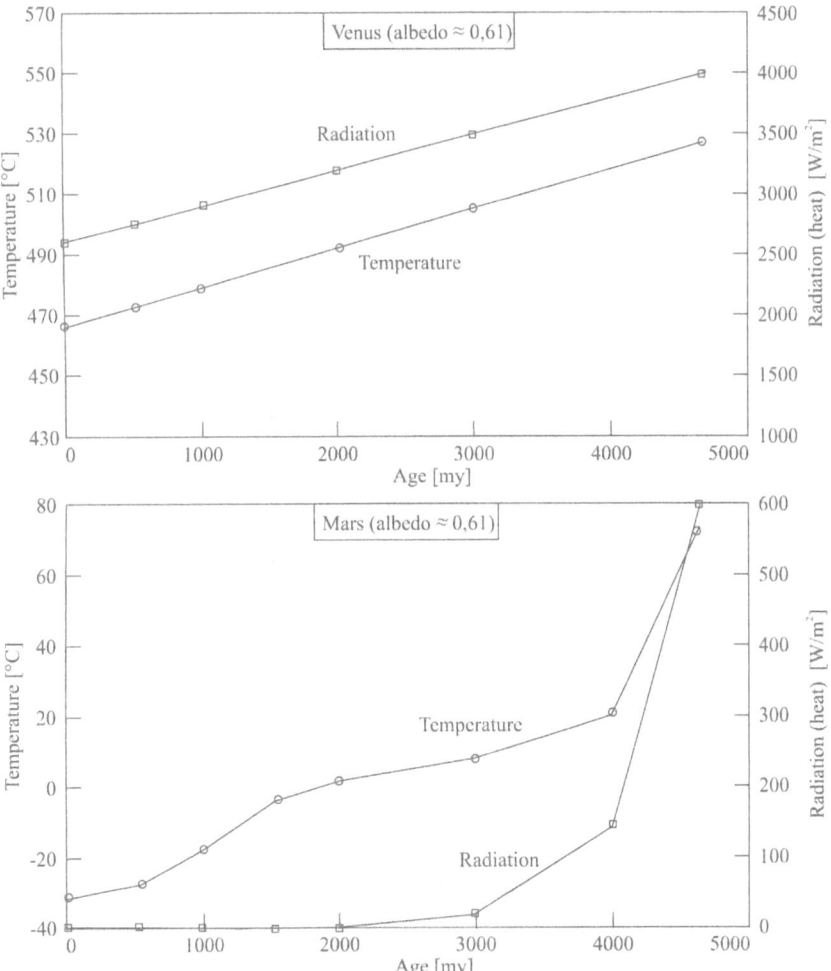

Fig. 4.4. Approximate history of the early atmospheres of Venus (top) and Mars (bottom).

Thus on Mars paleontological evidence may be expected. The same applies to (traces?) of organic carbon that should be expected in deeply buried sediments. Geophysical methods exist (as well as remote-operated drilling) that permit the prospecting for sediments with conductivities differing from the adjacent bedrock. Thus this concept can be tested with limited efforts.

Other workers may favor other concepts for the early history of the discussed planets. However, valid concepts must select conditions such that not only the properties of Earth, but also (roughly) those of Venus and Mars are approximated (Fig. 4.4). The reliability of the order of magnitude of the initial radiation of, for example Venus at 4000 W/m^2, may be disputed. Yet other values must be chosen such that corresponding values for Venus and Mars (diameter, distance, present atmosphere, frozen CO_2, etc.) are met. The same applies to Earth as well.

It was mentioned above that after accretion and subsequent heating, exhalating volatiles formed a first atmosphere and a first ocean (crystal-water). This means that the average surface-temperature never reached values high enough for lava-oceans. This is deduced from the following boundary conditions and fundamental chemical laws:

Above about 374° C water cannot be liquidized through pressure. Consequently, if such high temperatures had existed, the whole water would have been vaporized into the atmosphere (Eq. 9). The atmospheric pressure would have amounted to about 300 bar, with the partial pressure of water about 250 bar and that of CO_2 about 47 bar. The resulting natural greenhouse effect (Eq. 7) would have caused an additional rise of the temperature; the water (see Venus) would have been photolyzed and hydrogen diffusion into space would have occurred. The resulting present-day temperature would amount to about 300° C. Consequently, even in its very young history, the surface of the Earth was never so hot that it permitted complete coverage with lava.

The importance of this section for modeling is that the straightforward application of fundamental laws and the testing with as much boundary conditions as possible (atmosphere of Venus, ice on Mars, existence of an aqueous ocean on Earth and corresponding atmosphere, critical temperature of water, subsequent natural greenhouse effect, photolysis, etc.) may even for situations with very little data permit fairly precise assessments of geological conditions (Tstart < 374° C).

Applying an albedo of 0.61 (see Venus of today, Vogel 1977:404), a CO_2 partial-pressure of 2.4 bar (Eq. 9) and a radiation of about 4000 W/m^2 yields for the temperature of the Early Earth an initial starting value of 260° C. The corresponding H_2O partial pressure amounts to 48 bar (rounded up to an order of magnitude of about 50 bar, deduced from the *average* water column of about 700 m corresponding to the temperature of 260° C).

Assuming equilibrium between atmosphere and ocean (Eq. 11), the CO_2 concentration mentioned requires a pH of the ocean of about 6.2, e.g. less than the present-day value of 8.2 (see also Berger 1977, Shackleton 1977, and for an overview Broecker 1983:96).

This has vital consequences (also under Neoclimatic aspects) for the fossil record, especially with respect to the formation of carbonatic and phosphatic exoskeleton. This will be discussed in more detail below in together with observations for Mesozoic and Cenozoic times. In addition, the ocean contained NaCl, Ca- and Mg-Hydrogencarbonate and of course H_2CO_3. Consequently, the concentration of Ca-ions amounted to 1.0–1.2 g/l (explanation in the titled "Paleo- and Mesozoic boundary conditions").

4.3.3. The Precambrian

Life evolves between 4.5 and 3.5 by. In order to permit massive photosynthesis, the temperature (Nilsen et al. 1983) of the ocean must have fallen below about 50° C resulting in a corresponding H_2O partial pressure of 40 bar (and a corresponding rise in sea level of about 700 m, equivalent to about 250 million km^3 water). The main remaining constituents of the atmosphere are CO_2 and N_2. Volcanism resulted in acid rain while rivers have been alkaline (Kempe and Degens 1985; Vogel 1977; Strauch 1990). It is inferred that due to the precipitatation of the clouds as rain the albedo reduced to approximately 0.5, the figure for unvegetated land, covered with quartz sand and water. Thus the proxy starting value of 0.61 and the lower figure of 0.5 permits a sketch of the history of the early atmosphere as outlined in Figs. 4.5 and 4.6. During this time, the oxygen that was produced by photosynthesis was immediately consumed for oxidizing iron and other O_2 free matter (Eichler 1976; Fabian 1987; James and Trendall 1982; Trendall 1973). This means that microbiota did not need an efficient cell structure for the protection of, for example, DNS against free oxygen. The lack of free oxygen did not permit the formation of an ozone layer. Thus a parameterization of the photosynthetic activity of Precambrian life must be designed such, that up to about 1.3 by ago, all free oxygen is consumed by banded ironstones or other oxygen-free matter. For sensitivity studies, the known quantity of banded-ironstones however is only a lower limit, since, at least for short times, some free oxygen could be expected in the atmosphere (see below).

At the end of the Precambrian the oxic conditions have changed such that larger biota, still without exoskeleton, could be observed (Ediacara fauna). Because of the relatively large surface compared to the volume of the bodies, they were well adapted to absorb oxygen.

It might be inferred that an "acid" ocean with a quite low oxygen content was able to exist (Cloud and Glaessner 1982). The deprivation of CO_2 from the atmosphere through photosynthesis changed through the corresponding equilibrium the pH of the oceans (Eq. 11). In this context, from the paleontological point of view, special attention should be paid to the reduction of the acidity of the water. This is the physical precondition explaining why in the Cambrian, bones and calcareous shells could be formed in accordance with the pH-dependent solubilities of calcium hydroxylphosphate and calcium carbonate (pKs-values). These processes of precipitation, in which dissolved matter was converted to solids, are also influenced by the membrane action of cell walls, though this encapsulation effect is only limited. Thus through the rising pH the "carbonate formation capacity" improved so much, that not only cyanobacteria could benefit from the dissolved CO_3 and Ca ions. The conditions for anorganic carbonate-formation can be found in Eq. 12. Even accounting for some difference of the pH between the carbon-forming cell and seawater, it should be noted that the diffusion barriers are not effective enough to keep high gradients stable for a long time. This means that the equation characterizes the order of magnitude. The atmospheric equilibrium concentration for a pH of 6.9–7.0 is about 10% CO_2. In other words, the CO_2 content decreased in 2.8 by from about 55% down to 10%. Any parameterization of the photosynthetic activity must be formulated such that these equilibrium conditions are met.

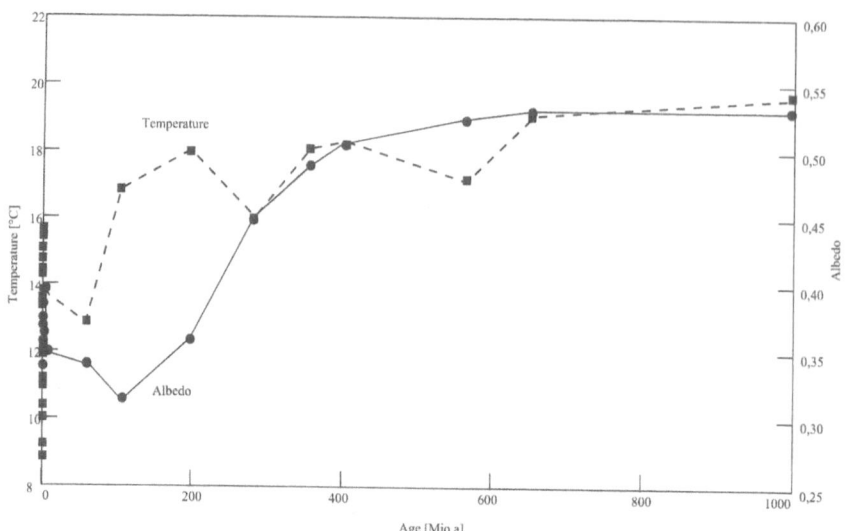

Fig. 4.5. Approximated average albedo of the Early Earth.

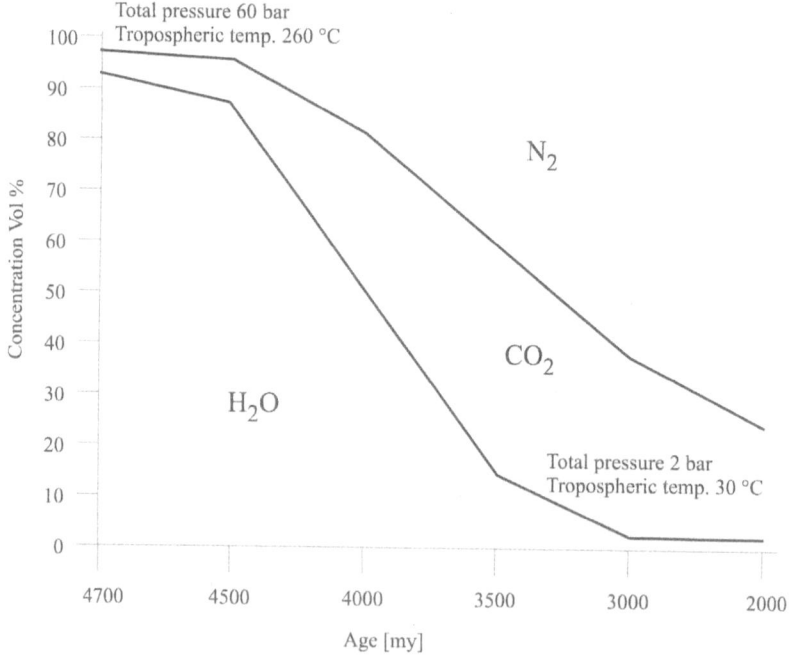

Fig. 4.6. Approximate composition of the Early Precambrian atmosphere of the Earth according to equations and concepts discussed in the text.

4.3.4. Paleo- and Mesozoic Boundary Conditions

From Cambrian times on, biota with calciumhydroxylphosphate (apatite) can be observed. Applying the above-mentioned principles, the corresponding pH amounts to 7.1–7.2 (CO_2/O_2 ratio and temperature accordingly). In this context the question arises whether a term for CO_2 production through animals should be introduced. The same applies to some compensation of an increase of the albedo through Paleozoic glaciations. For reasons of simplicity at this stage of model development such terms have been integrated into the planetary albedo.

From Devonian times on, the situation becomes complex and difficult to handle. The rate of O_2 production changes significantly through terrestrial vegetation. This causes a reduction of the albedo. Thus a self-organizing process originated with warming of the Earth through decrease of the effective albedo and, on the other hand, cooling through an increase of the O_2 content and subsequent increase of the effective albedo. Processes of this type normally perform oscillations. Thus a parameterization (or more elaborate models) of vegetation, including O_2 production, has to be designed such, that a total glaciation of the Earth is excluded. In addition, other boundary conditions have to be met:

Silurian and Lower Devonian plants had massive protective structures (cuticles, etc.) that (Remy and Remy 1977) permitted exposure to conditions that are comparable to present-day deserts (absence of protecting larger plants such as trees). The same applies to terrestrial animal life:

Massive protective structures, especially on the back (Ichtyostegales, Carrol 1988) and absence of animals exposing their skin directly to the sun (Strauch pers. comm.) can be observed. This picture changes dramatically towards Carbonian times. Consequently during Devonian/Carbonian times the quantity of free O_2 increased. The widespread formation of terrestrial redbeds (consumption of oxygen for oxidizing iron, see similar effects also in the Precambrian) is an additional indicator of free oxygen in the atmosphere.

Fossil evidence indicates indirectly that an amount of free oxygen that was sufficiently high to permit an ozone layer to form did not exist before Carbonian times. Devonian amphibians and other biota leaving the water (such as above-mentioned Ichtyostega always had massive dorsal shields that are impermeable for UV-radiation. The same applies to plants: all plants of Early Devonian times had cuticles protecting them efficiently against UV-radiation. Carbonian plants did not need such protectors any longer. Although this is only a qualitative indicator for the order of magnitude of the CO_2/O_2 ratio, the boundary conditions deduced from the amount of Mesozoic carbonates, that require a further shift of the CO_2/O_2 ratio towards O_2, are consistent with this interpretation. Other authors propose other curves of the CO_2 history (see works and data of Cloud and Glaessner 1982; Fabian 1987; overview of Schidlowski 1981; Schidlowski 1978; Lehmann and Hillmer 1980; Margulis et al. 1976; Schidlowski et al. 1975, including some interpretations that proposed present-day levels of O_2 or more in Carbonian times).

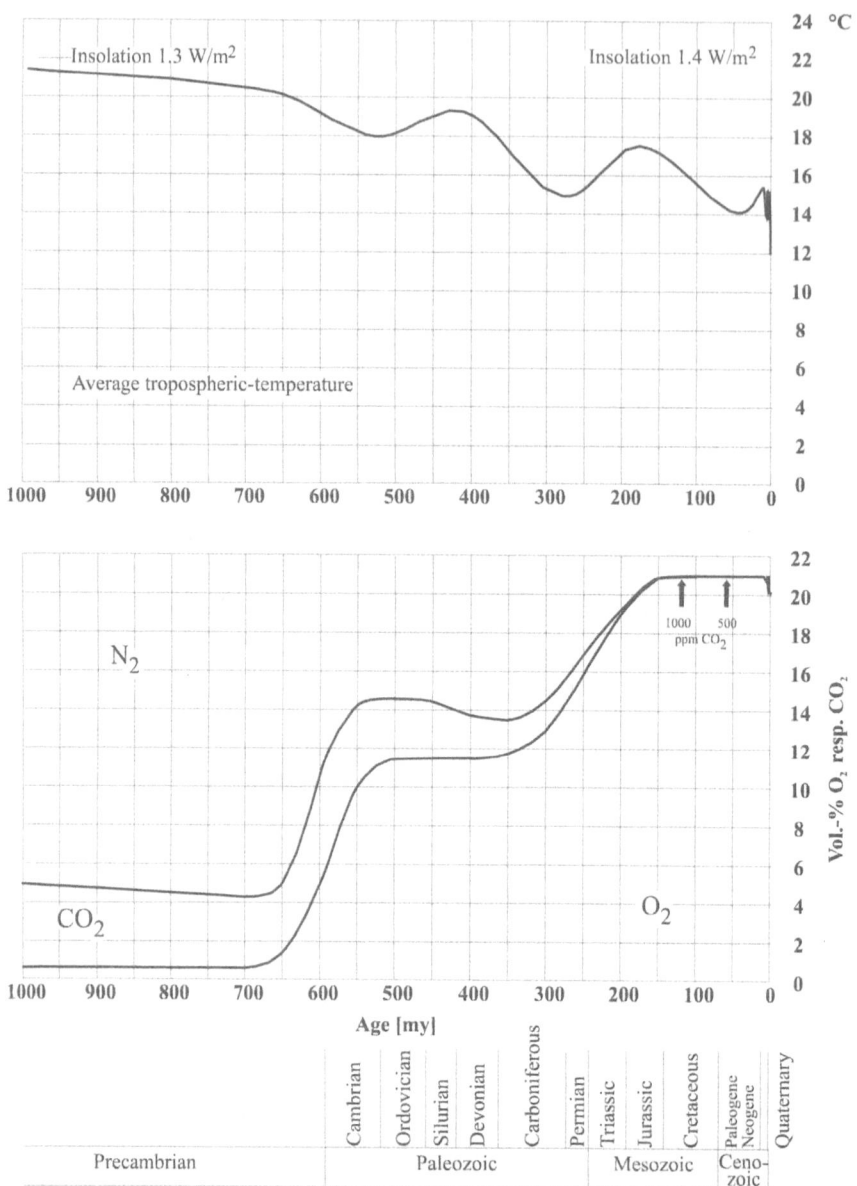

Fig. 4.7. Average tropospheric temperature (top) and average cumulative tropospheric composition (N_2, CO_2 and O_2) during the last 1000 My according to the equations and boundary conditions discussed in the text. Note the deflection of the curve in the Holocene.

Evidence that might be used against the concept of a relatively low O_2 content is the nature and quantity of vegetation as well as the considerable size of insects such as Meganeura ("dragon-fly"). At this point it may be argued that such large insects, which use tracheal respiration (inflow of air through apertures in the chitineous structure into tracheal tubes that pass it on to membrane-like tracheoles, which forward it directly through the point of consumption; the pumping is performed by movement of the body) would not be able to ensure the oxygen-supply at such low levels. It should be mentioned that nothing is known about the kinetics of the O_2 binding in the blood of these animals (hemoglobin, similar pigments, Fe[II] or Cu binding center, etc.). The same applies to physical parameters such as the membrane permeabilities that may vary numerically over a wide range. In addition the air in bronchioles of pulmonatory breathers exhibits a certain affinity to the air of the Carbonian (see also the chemical properties of blood and seawater). Since the tracheoles of Carbonian tracheates have comparable air input as the above mentioned pulmonatory breathers, they most probably have had even less difficulty ensuring oxygen supply, since an energy-intensive heating system of the mammals was not necessary at this time.

Another point is the size and the ability of an animal to fly. Under aspects of bionics (technological simulation of "inventions of nature") the air in the Carbonian had a higher viscosity and density than present-day air. Using aerodynamic principles (nonlinear), then under these conditions large objects glide and fly easier than small ones. These arguments provide some background that demonstrate how even qualitative and non-physical parameters could be integrated into the testing of hypotheses and formulation of boundary conditions.

In the Carbonian the CO_2/O_2 ratio changed due to the removal of organic carbon in the deposits of that time. This is parameterized as a sum-value. Either paleogeographic conditions (Gondwana glaciation, evolutive crisis, see also Stanley 1987), or the oscillatory nature of the system discussed, or both, caused a slight "counter-oscillation" of the CO_2/O_2 ratio in Upper Carbonian/Permian times (see Fig.4.7).

The following times are characterized by the decrease of CO_2, both through terrestrial and marine vegetation (algae). For the purpose of this paper, the ratio of both is insignificant. For subsequent studies, quantification may be interesting, since large Mesozoic forested areas can be excluded for paleogeographic reasons such as high sea level and the small area of land.

Permian and Mesozoic times are characterized by massive carbonate deposits (about $2.5*10^{15}$ t), a quantity that can be regarded as a lower estimate. Taking into account that the pH value was already near 7.3, a figure favorable for even anorganic carbonate formation, it becomes clear that even subtle changes towards the alkaline (for example inside the cells of marine microplankton) have significant effects. Consequently, from Permian times on, a self-supporting process commenced. This process consisted of initial pH, carbonate-formation through organisms, further shift of the pH towards the alkaline and better conditions for carbonate formation. From the overall quantity of carbonates the concentration of Ca-ions in seawater before these times can be roughly estimated: The order of magnitude was about 1.0–1.2 g/l. This explains the value that was mentioned above as concentration for Precambrian times. At the end of the Mesozoic the Ca concentration approached the present-day value of about 0.4 g/l.

This "de-acidification" of the sea (the shift towards the alkaline from a pH of about 7.3 to about 8) together with the corresponding CO_2 depletion of the atmosphere and the subsequent cooling can be regarded as permanent evolutionary stress which "life had to adopt to". Paleontologic evidence from the Mesozoic shows that life was able to keep up with this stress, as is demonstrated by the high species diversity and rapid change of life-forms in these times (not only microplankton but also larger species such as the ammonoidea). Thus the excellent stratigraphic conditions provided through Mesozoic biota may have their origin in simple chemical processes (as will be discussed with more detail below). The same applies to stress for terrestrial animals through the shift of both gas content and temperature of the atmosphere (for further details see the section on "biological impact").

4.3.5. The Cenozoic

Basically the same processes as mentioned above apply to the Cenozoic, namely: Occupation of terrestrial areas by plants, reduction of albedo, rise in temperature, accelerated growth of plants, CO_2 reduction down to the present-day ppm level, sometimes (paleogeographic conditions) formation of biogenic sedimentary deposits such as coal, rise of O_2 content, cooling (including glaciations at suitable places), subsequent oxidization of matter accessible to atmospheric O_2, rise of CO_2, etc. Times of vegetational optima (Eocene, Miocene) have been followed by times of cooler conditions (Oligocene cold-water formation off Antarctica, Middle–Miocene cold-waters in Labrador Sea/off-Greenland, etc.). Complications occur in Cenozoic times through the highly variable paleogeographic conditions (closing of the Isthmus of Panama, opening of the Drake-passage, variable exchange of water into the northern North Atlantic, uplift of large orogenes). The same process (see Figs. 4.2, 4.8, 4.9) may have operated during Pleistocene times, so the list of factors responsible for the ice-ages has to be extended by another one (see also Schwarzacher 1993; Imbrie and Berger 1984; Hays et al. 1976; and the overviews of Covey 1984 and Kuhle 1986). For reasons of readability, the curve of the atmospheric history is enlarged for Cenozoic times.

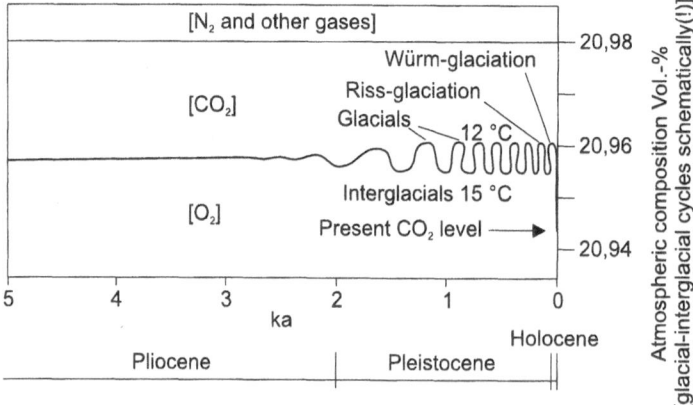

Fig. 4.8. Schematic sketch of average cumulative atmospheric composition (N_2, CO_2 and O_2) during the last 5 my. Note the deflection at the end of the Holocene that *is* drawn to scale.

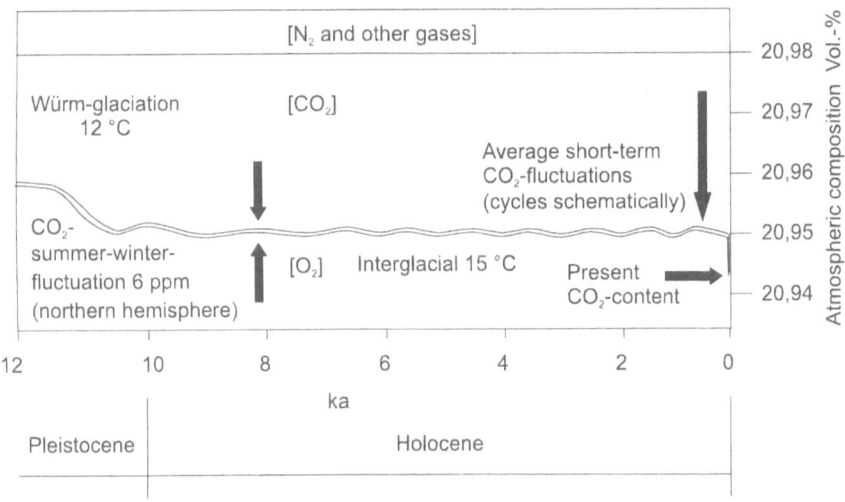

Fig. 4.9. Sketch of average cumulative atmospheric composition (N_2, CO_2 and O_2) during the last 12 ky. The distance of the two lines describes the 6 ppm difference between northern hemisphere summer and winter conditions. Note the deflection of the curve at the end of the 20th century.

The curve exhibits potential equilibria for various Neogene conditions (ice-ages, etc.) as well as (Figs. 4.8 and 4.9) an estimate for the Quaternary (order of magnitude). In this context the dramatic shift of the CO_2/O_2 ratio at the end of the Holocene must be noted. As mentioned above all technological systems consisting of at least an actor and a counter-actor tend to equilibrate through oscillations. The gradient of atmospheric change observed today (even if only the order of magnitude is considered) supercedes everything that has been observed or modeled for the geologic past. In other words, if boundary-conditions (geological data) and theory (chemical equations) are basically correct, then at the end of the Holocene an atmospheric event can be observed that, if applied to a self-organizing oscillating system, raises the question of stability. In addition it should be noted that an atmospheric change to 360, 500 and 1000 ppm is, regarding atmospheric conditions, a sudden set back to about 35 Ma (Oligocene), 60 Ma (Paleocene) and 120 Ma (Cretaceous), just to give an idea of the order of magnitude (see Fig. 4.7). In this context possible correlations between stomatal density and CO_2 content should be noted (van der Burgh et al. 1993).

Up to this point boundary conditions and fundamental equations governing the history of the CO_2/O_2 ratio have been discussed. Special attention has been paid to the integration of both theoretical concepts and geological data.

4.4. Biological Impact

CO_2 influences bioreactions: There are more than ten thousand known enzymes which catalyze different bioreactions. In approximately 5% of these reactions that occur in plants, larger animals and microorganisms, CO_2 is needed as a substrate or is produced as an end product. Thus changes of the CO_2 content of the atmosphere are not merely a geological phenomenon, but also of prime importance for past, present and future life.

Tissue, cells, organelles, for example mitochondria, which are referred to as biological subsystems, consist primarily of water, proteins and low molecular-weight substances such as salts and amino acids. In order to determine the solubility of CO_2 in biological systems, for a first approximation it is sufficient to consider the solubility of CO_2 in saline solutions. Dissolved CO_2 is present in physiological solutions in various forms: as physically dissolved CO_2, as carbonic acid and as bicarbonate. Depending on the pH of biological systems, the various forms of CO_2 represent different fractions of the total quantity of dissolved CO_2. In the physiological pH range of about 7.4, the quantity of physically dissolved CO_2 represents approximately 10% of the total CO_2. The remainder is essentially bound by bicarbonate ions. The importance of the pH for the activity of selected enzymes can be seen in Fig. 4.10. Above, the change of the pH of the oceans was discussed. In this context it is mentioned that solubility of CO_2 in biological systems is comparable to that of CO_2 in seawater. Thus, for a first approximation, there is no need to distinguish between terrestrial and marine organisms.

Cells and organelles, such as mitochondria and cell nuclei are surrounded by a biological membrane that basically consists of lipids (fats and fat-containing substances). Since CO_2 is more readily soluble in the fatty phase than in the aqueous phase, a biological membrane is not an effective barrier against exchange of CO_2. Thus an increase in the CO_2 concentration in the air will also have an effect on the cell fluid, the karylomorph. Consequently, under certain conditions a doubling of CO_2 concentration in the atmosphere will result in doubling the CO_2 concentration in many biological subsystems. This, as outlined further below, will affect the cell metabolism and thus be a major environmental stress to the cell. In aquatic systems, a rise of the CO_2 concentration causes a change of the pH. An increase of the CO_2 content of the air results in an increase of the physically dissolved CO_2 and of the concentration of carbonic acid. This results in an acidification of the intracellular medium with direct consequences for metabolic processes.

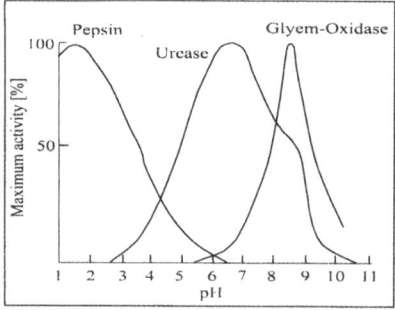 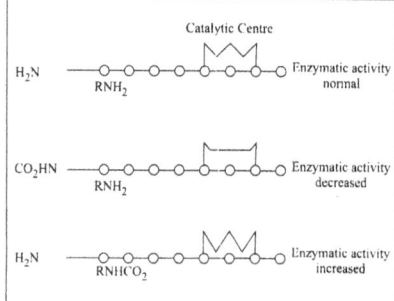

Fig. 4.10. Impact of pH changes on the activity of selected enzymes.

Fig. 4.11. Impact of CO_2 on enzymatic activity.

Figure 4.10 (from Lehninger 1977) shows that there are enzyme reactions that strongly depend on pH. Since a shift of pH results in a pronounced change of the metabolic conditions, the cellular systems have to counteract and compensate for the change of pH by buffering or outward transport of protons. The first case requires a change of the buffer base concentration, the latter, the outward transport of protons, requires energy. Both processes require an adaptation of the metabolic mechanisms to the changed environmental conditions.

Depending on the position of the optimum pH, biological systems react with varying sensitivities to a shift of the CO_2 concentration. A system with an optimum pH in slightly alkaline conditions will be more sensitive than a system having an optimum pH under neutral or even acid conditions.

Now the question arises whether the observed and predicted changes in atmospheric CO_2 content are sufficient enough to affect enzymatic reactions. Enzymatic reactions often follow the same course in different organisms, although the equilibrium state of bioreactions may vary. An increase of atmospheric CO_2 to, for example, 600 ppm can change the equilibrium state of a reaction differently in different organisms: In case of small, single-cell organisms or consumers of CO_2 (plants), there will be a strong relative change of CO_2 of about 100%. This causes a dramatic change of the stoichiometric conditions of the bioreactions considered.

In the cells of larger animals, for example man, the same absolute CO_2 change will cause relative CO_2 change of about only 0.5%. This is due to the already high CO_2 concentration in such biological systems. Human blood for example already contains 60 000 ppm CO_2. Thus in the latter case, changes of the atmospheric CO_2 content will have no direct effect on bioreactions.

From these considerations it can be concluded that even comparable bioreactions will react differently on an increase of atmospheric CO_2 content, depending on the partial pressure already prevailing in the system considered. Consequently, changes in the atmospheric CO_2 concentration will, from the microbiological point of view, have a much greater impact on plants and microorganisms than on higher animals.

CO_2 interacts with proteins in a number of important ways. Proteins are chain-molecules consisting of various amino acids. Direct interactions between CO_2 and these molecules are described in the literature. The best known are carbamino compounds, covalent compounds of CO_2 with free amino groups in the protein molecule. These reactions may take place at one end of the protein chain, but also in particular side-groups. Here the important point is, that these compounds can influence the three-dimensional structure of the proteins. However the correct structure of the proteins is a vital prerequisite for their performance. The formation of a carbamino compound can therefore lead to functional changes in various proteins (Fig. 4.11). In the case of hemoglobin, for example, the addition of CO_2 to the amino group at the end of the protein-chain can reduce the affinity of oxygen to the oxygen-binding center. Consequently oxygen can be released more easily to the respiratory active tissue, where considerable amounts of CO_2 (about 60 000 ppm) are present. In the lung, however, the reverse process takes place. Release of CO_2 reduces the CO_2 concentration in the blood to about 52 000 ppm. As a result, CO_2 is freed from the carbamino compounds. The affinity to haemoglobin for oxygen (heterotropic inhibition) increases correspondingly (see Kilmartin et al. 1973). In case of another protein, the enzyme ribulose diphosphate carboxylase (McFadden 1973), which is responsible for the primary formation of sugar from CO_2 in the assimilating parts of plants, activation of the enzyme is observed (homotropic activation). This process takes place at low CO_2 concentrations of 100–250 ppm. CO_2 also has other effects on proteins. The important fact is, that independent of its occurrence, either as substrate or as product of the reactions, CO_2 can affect the activity of proteins.

In addition, microbiota, plants and animals are affected in a different way. First of all there are two major groups regarding the exchange of CO_2 between complex biological systems, namely CO_2 producers (the respiratory active cells of microbes, plants and animals) and CO_2 consumers (photosynthetically active plants and some microorganisms). To reach the site of consumption (Fig. 4.12), after removal from the atmosphere CO_2 has to be transported through various cell membranes (vice-versa for CO_2 producers). The release or absorption of CO_2 may follow different paths. First, it can diffuse as physically dissolved CO_2 through aqueous phases (cytosol, etc.) and fatty phase (membranes, etc.). Second, it may be transported actively through cell membranes in bound form, as bicarbonate ion in aqueous solution. In the case of active transport such as enrichment of bicarbonate against a concentration gradient, energy has to be supplied. Thus, plants can obtain the necessary supply of CO_2 more easily when the CO_2 concentration in the atmosphere increases.

However, CO_2 is more difficult to release through nocturnal respiration processes, since the air outside the plant already contains more CO_2. Another important aspect is the passive transport of CO_2 in biological systems. This essentially depends on the geometry of the exchanger system, such as surface area and layer thickness. In addition it depends on the CO_2 concentration gradient.

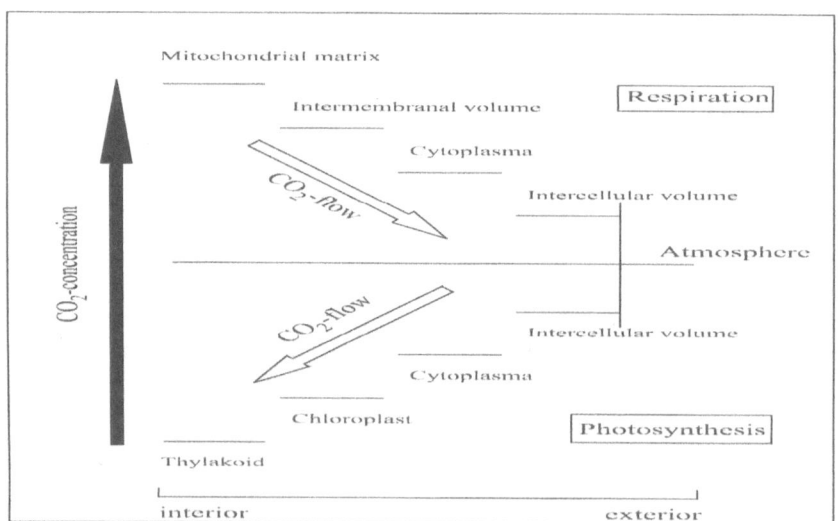

Fig. 4.12. Flow path of CO_2 in plants during photosynthesis and respiration.

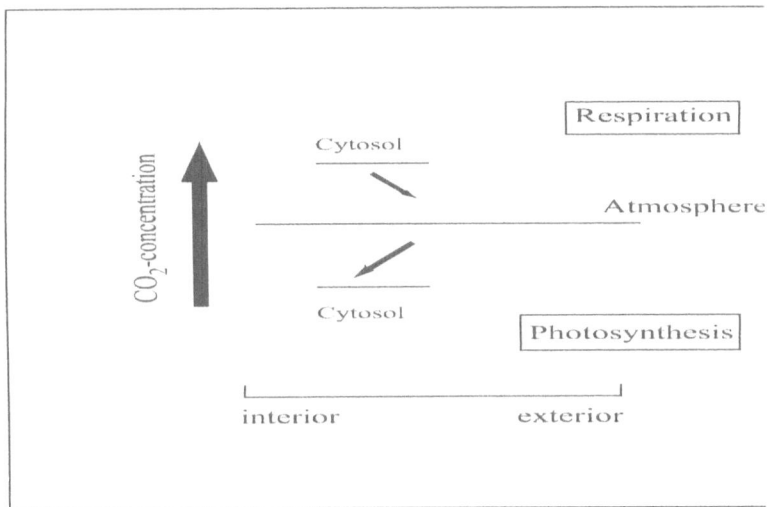

Fig. 4.13. Limited protective barriers of microorganisms against changes of atmospheric CO_2 content.

In order to ensure effective release of CO_2, the human lung has, for example, a surface area of about 70 m^2 and a layer thickness between the blood vessels and the alveoli of only 0.5 μm. The ratio between surface-area of the lung and the body-volume of man is a factor of 6000 smaller than the surface area of a bacterium and its volume. In other words, to achieve the same high relative exchange of surface area with the environment as a bacterium has, man would need a lung-surface area of about 0.5 km^2. Plants with leaves also possess large surface areas compared to their volume.

Consequently, even under the aspect of surface area, bacteria and microbiota in general are much more vulnerable to changes in the CO_2 concentration since they have no effective barriers against changes in the CO_2 content of air and water (Fig. 4.13). However, large CO_2 producers, such as man or animals, have both a high CO_2 level in the cells, and additional barriers which prolong the pathway of CO_2 exchange, such as respiratory organs and circulatory systems (Fig. 4.14).

Before returning back to evolution, paleontology and climate, the most important points of this microbiological section should summarized. The cell can be influenced by an increase of CO_2 in the atmosphere. In physically dissolved form, CO_2 can diffuse through cell walls. Organisms with large ratios between surface area and volume, such as microorganisms and plants, will be particularly affected.

In the interior of the cell, a change of the CO_2 concentration causes changes of the pH and the bicarbonate concentration. In order to ensure optimum functioning these values must be kept constant. In addition, CO_2 is both the substrate and product of many bioreactions. The equilibria of the discussed reactions depend on changes of the CO_2 content.

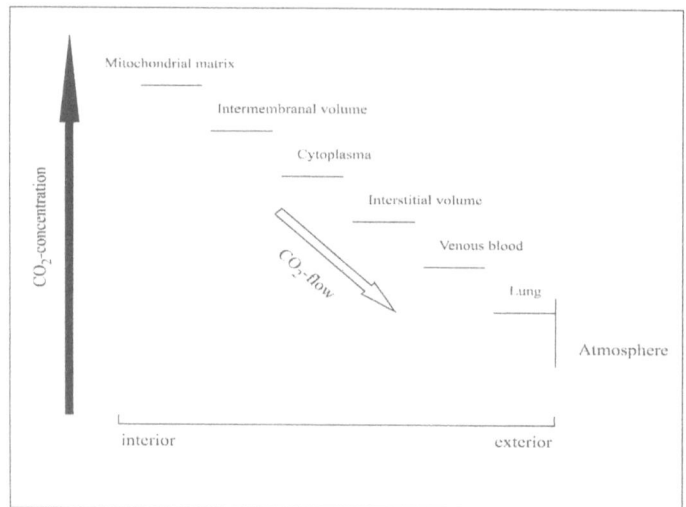

Fig. 4.14. Barriers CO_2 has to pass in large CO_2 producers.

In microorganisms and CO_2 consumers, such as plants, changes will be greater than in large CO_2 producers such as animals. An increase of atmospheric CO_2 to about 600 ppm means for man only a relative concentration change of 0.5%. For many microorganisms and plants the same absolute 600 ppm are a relative concentration change of about 100%. The changes in CO_2 concentration have a direct influence on the structure and function of proteins and enzymes caused by the direct binding of CO_2 into these macromolecules, therefore the impact of this environmental stress on evolution is discussed in the next section.

4.5. Evolutive Asymmetry

Stratigraphic evidence demonstrates two things:

(1) There are times of increasing diversity and of biological crises, for example the Permian or the K/T boundary. This statement does not deny the impact of proven extraterrestrial phenomena. It only means that large impacts had a modulating effect on population that was already subject to environmental stress and potentially, metaphorically spoken, "struggling for survival".

(2) There is a constant background of evolutive appearance and disappearance as demonstrated, for example, by some trilobites (Fig. 4.15). This figure also shows that the average lifespan of many species is remarkably short (approximately 1–3 My).

Stratigraphic ranges of some selected trilobite species

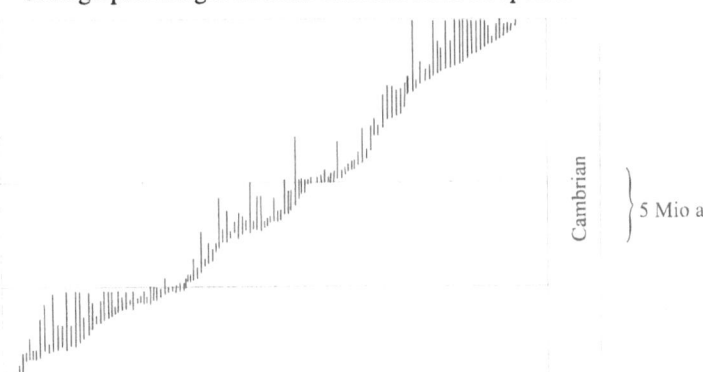

Fig. 4.15. Stratigraphic ranges of some selected trilobite species (according to Stanley 1987). Although many other potentially long-living species exist, this chart shows that there is a constant "coming and going" as "background phenomenon".

Regarding environmental stress as one major cause for evolutive adaption (e.g. a statistical distribution of mutations and subsequent survival of the fittest) the only major evolutive stresses of the early Paleozoic are the deacidification of the oceans and the increase of free oxygen. From Fig. 4.7 it can be deduced that at that time, within 5 My, the CO_2 concentration decreased by 1000 ppm, while the O_2 concentration increased by 2000 ppm. This means that most of the trilobites disappeared after a change of 400 ppm in the CO_2 concentration.

From the preceding section we know, that for larger animals, especially CO_2 producers, the shifts in the CO_2/O_2 ratio are well below any toxic limits. However, we also know that for microbiota (bacteria and related, such as viruses) even a gradual change in the CO_2/O_2 ratio means a massive environmental stress. In addition, generations of microbiota have a very-short life span (new generations within hours). Consequently, compared to macrobiota per unit of time, more new species occur through statistical mutation and more species are subject to a test for adaptation to a changing environment.

Thus the faster the change in atmospheric composition, the greater will be the pace of development of new microorganisms. For this phenomenon, the advance of microbiotic evolution compared to that of macrobiota, we introduce the term "evolutionary asymmetry".

If this observation is applied to evolution we can conclude that as a matter of statistics among new microorganisms there are, over many generations, also microorganisms that turn out to be incompatible to already existing biota. This means they are pathogenic pests and diseases.

In addition macrobiota often need for vital processes the assistance of bacteria and (potentially) viruses. This may not only apply to well-known examples as the digestion process in man, but also to marine environments, especially as the importance of marine bacteria and viruses was in the past not fully recognized. The extinction of certain bacteria and virus, both in the marine and terrestrial realm may, without being a pest, have contributed to the extinction of some biota while biota occupying the same niche but having different bacteriological requirements may have survived. Whether this process explains the survival of Nautiloids (while Ammonoidea have been extinct) at the end of the Cretaceous is a question to be tested against the fossil records. The same applies to several classes of microfossils.

Therefore times with a large gradient of atmospheric change could be expected to be times of fast evolution affecting (nearly) *all* groups of biota either directly (microbiota) or indirectly through evolutionary asymmetry. *Incompatibilities* (new pests, viruses and diseases) support the extinction of existing biota including macrobiota. The niches freed will be occupied by mutants of these macrobiota, even if they are *not* more efficient (in reproduction, etc.) compared to the "old" owner of the niche. In this case the criterion for "survival of the fittest" is the ability to endure (or utilize) the new situation caused by the changed environment (already known) and the microbiological challenge caused by evolutionary asymmetry.

4.6. Neoclimates

It is estimated that a significant genetic change that is sufficient to create a new species requires on average an order of magnitude of about ten thousand generations. Microorganisms reach this figure within ten years. Larger biota need more time, so, as outlined above, statistically the chance for the evolution of new microbiotic incompatibilities (pests and diseases) through evolutionary asymmetry increases in present times. See in this context the extremely steep gradient of atmospheric change at the end of the Holocene (Figs. 4.7–4.9).

In the sections above, the authors did not distinguish between microorganisms and viruses. The reason is, that from the point of view of the virus, the host-cell with its CO_2 concentration, pH, etc., is a changing habitat that requires adaptation. In addition, in its inanimate phase (during the passage from cell to cell, from one host to another) the virus can both be influenced by chemical reactions and is therefore also subject to selection.

At this point it may be argued that the present-day change in CO_2 concentration is too small to influence organisms and their future evolution. Here two points have to be kept in mind: (1) The relative change in the CO_2 concentration from, for example 270 ppm to 360 ppm means a change of about 30% of the CO_2/O_2 ratio. The chemical potential (the change in the free enthalpy of a system, related to the change in the number of moles) of all chemical reactions related to CO_2 changes significantly; the shift amounts about to 4%. The chemical potential is decisive for any physiological reaction. (2) Even the absolute change in the CO_2 relative to O_2 concentration is significant. It lies in the same range as the climatically effective changes of the absolute temperature.

Experiments and experiences in greenhouses incidentally show that an increase in the CO_2 concentration, combined with the "greenhouse effect", first leads to an increased growth rate (Nilsen et al. 1983), but at the same time results in a considerable increase in the threat from fungi and pests. These usually are controlled by means of crop-protection agents.

4.7. Impact on the Future of Mankind

What do the microbiologic and geological findings discussed above mean for the future? Following the preceding discussion, first of all the steep gradient of the change of the CO_2/O_2 ratio suggests an increased probability for new diseases (epidemics) and symptoms. Evidence, that shows that this process is already active, is given by numerous descriptions of new or modified pathogens, particularly from arboreal diseases. Furthermore, microbes seem to be increasingly aggressive, virulent and resistant. Here Malaria (although subject to other "environmental" stresses) is a well-known example; Penicillin is very different compared to CO_2 and if applied to fighting Malaria it is also a much more powerful bactericide. Nevertheless, this example highlights the "tactics" and timescales employed by microorganisms to cope with environmental change (e.g. to "adapt" in short and measurable times to the changing "environments" provided by the various Penicillins).

Therefore, the potential for environmental damage and threats to biological evolution caused by direct or indirect impact of change of the CO_2 concentration have to be discussed and compared with recent evidence. Several workers (Bazzaz and Reekie 1989; Bazzaz and Garbutt 1988; Fajer et al. 1989; Gates et al. 1983; McFadden 1973; Strain and Cure 1985; Dixon pers. comm.) are concerned with the consequences of these phenomena for plant growth and for communities of plants and animals. Examples of evidence include increased susceptibility to stress, increasing destruction by pests, a decrease in stomata with adverse effects on the water-balance of plants, changes in competitive behavior and disturbed system equilibria. These externally visible phenomena are monitored throughout the world. The causative background, as suggested by microbiological and chemical theory (chemical equilibria, etc.), the changed CO_2 concentration and its impact on cell functions, especially enzyme reactions, has to be studied with much more detail than in the past. At present, there is some understanding regarding the reactions of enzymes, which regulate photosynthesis, and night-respiration cycles of plants. Whether the enzyme that is responsible for photosynthesis (ribulose diphosphate carboxylase) is initially *stimulated* by the CO_2 increase and whether on the other hand, the enzyme, that is responsible for the respiration at night (ketuglutaric acid decarboxylase), initiating the opposite reaction, is inhibited by an increase of CO_2, is a question that is not yet finally solved. If this concept were valid this would imply for plants that an increase of the CO_2 content would cause a higher formation of glucose, the product of the photosynthesis reaction. All other substances that are produced during respiration at night would be produced to a lesser extent.

This draws attention to another important and not yet fully understood phenomenon: the atmospheric pollutants, SO_2, NO_x and ozone seem, in elevated concentrations, to modify the function of the above-mentioned two enzymes antagonistically to CO_2 (Guderian and Klumpp 1989). This means that the pollutants SO_2, NO_x and ozone seem to compensate the effect of another pollutant (CO_2). If this is substantiated through further studies, this suggests a new assessment of the widespread forest damage-syndrome; a limited increase in SO_2, NO_x and/or ozone concentrations in the air might compensate some of the effects of CO_2-induced problems, at least regarding forests. A pronounced increase, however, results in the typical SO_2, NO_x and ozone damage syndromes, while "typical"(?) CO_2-damage syndromes appear in atmospheres lacking the pollutants SO_2, NO_x and ozone.

4.8. Aspects of the Global Carbon Budget

IGCP-341 focuses not only on paleoclimates, but also on neoclimates, that is, it focuses on ongoing processes and the present-day action of man. As the details of the global carbon fluxes are difficult to quantify, there had been some discussions about uncertainties or even a "missing sink". Thus some thoughts need to be presented on the CO_2 balance:

The present concentration of CO_2 is about 360 ppm (Fig. 4.2). Furthermore, the atmosphere contains other gases containing carbon: CO, CH_4 and other large hydrocarbon molecules. Under the prevailing conditions in the atmosphere, they are all converted into CO_2 in short times (about 10 years), so their respective concentrations depend on their replenishment. Consequently this study focused on CO_2. The gas CO_2 (like O_2, N_2 and Ar) is one of the gases that disperses quickly in the troposphere; therefore no long-lasting lateral concentration gradients occur. The gas CO_2 is in addition a ubiquitous gas. This means that the discussed phenomena apply to any place on Earth.

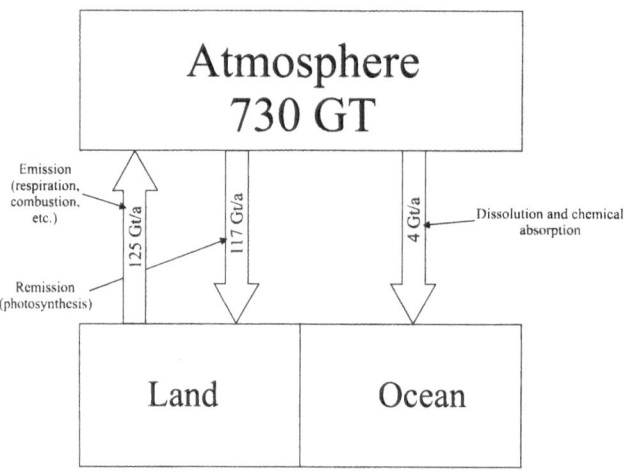

Fig. 4.16. Simplified carbon balance.

The fact that the atmosphere of the Earth contains less CO_2 than, for example, Venus or Mars is caused by a dynamic process: removal of CO_2 through photosynthesis and other biotic processes (and subsequent deposition in carbonates/fossil fuels) versus respiration, rotting and combustion. Figure 4.16 briefly sketches the carbon balance with values in Gt (gigatonnes) (Fond der Chemischen Industrie 1990; Kohlmaier et al. 1988). At present (1992) there are 730 Gt of carbon in the atmosphere (equivalent to 360 ppm CO_2).

Every year (upward arrow, Fig. 4.16) 125 Gt are emitted into the atmosphere and consist of a natural component of 119 Gt (rotting, natural combustion, respiration) and a man-made component of 6 Gt (mainly industrial combustion). The downward arrow (on left, Fig. 4.16) illustrates the effect of photosynthesis, the remission of 117 Gt. The downward arrow (on right, Fig. 4.16) represents the absorption of 4 Gt of carbon in the ocean as a result of physical solution followed by subsequent processes such as the reaction with calcium carbonate (calcium carbonate + CO_2 + H_2O = calcium bicarbonate). As a result of this CO_2 input, the acidity of the ocean increases and the lysoclines, such as the CCD rise.

According to this balance, the remaining 4 Gt remain in the atmosphere, which means an increase of the CO_2 concentration of 2 ppm per year. This figure also can be read off the Mauna Loa curve (measured data), so speculations about uncertainties are not necessary. In other words, the known figures about carbon fluxes are quite well approximated, a "missing sink" does not exist. It is the atmosphere.

4.9. Summary and Conclusions

The following ten points summarize the conclusion reached in this paper:

(1) A set of boundary conditions and fundamental chemical equations that must not be violated is used to approximate the history of the Earth's atmosphere from beginning to present. Application of these principles to the atmospheres of Venus and Mars yield results that are consistent with the data. In the early stages of planetary formation, the average maximum temperature was less than about 374° C (probably less than 260° C). Massive photosynthesis commenced between 4.5 and 3.5 Ga at temperatures below 50° C. While minor quantities of free oxygen (i.e. oxygen not bound in banded ironstones, etc.) might have been observable in Precambrian times as well, significant amounts of free oxygen were observed only from Eocambrian/Cambrian times on. After reaching a plateau of about 12% O_2 that persisted through most of the Paleozoic, during Mesozoic times the massive formation of carbonates together with chemical equilibria resulted in a decrease of CO_2 to the present-day value until at least the Early Cenozoic.

(2) Many processes with complex details (influence of soil, animal life, vegetation, ocean circulation) have been parameterized as "effective parameters". The concentration on key parameters and key processes provides "target windows" for more detailed models that otherwise (for example through a "bottom to top approach") are difficult to reach. Measured evidence (the "Mauna Loa curve", together with geological background information) demonstrates that the approach of integrating complex details into key processes is valid and consistent with the data.

(3) Times of changes of the CO_2/O_2 ratio are times of pronounced evolutionary change. Environmental stress, especially to microbiota, is caused by the *gradient* of the CO_2 content, not by its absolute value.

(4) The gas CO_2 affects bioreactions, especially in the microbiological realm. It is a catalyst, substrate and the product of many reactions (especially enzymatic reactions). Furthermore CO_2 is of paramount importance for the pH value of cells (microorganisms, plants) and the chemical activity, a key value for the optimum function of biochemical reactions. Since CO_2 is soluble in lipids (cell walls) approximately as well as in water, cell walls of microorganisms provide no effective barrier against changes of atmospheric CO_2 content. In addition, compared to CO_2 producing macroorganisms such as man, microorganisms have a low CO_2 concentration in the cell liquid. While for species like man (60 000 ppm CO_2 concentration in the blood) a change of some 100 ppm means only a relative change of 0.5%, for microorganisms a change of some 100 ppm (as has been observed in recent times and is expected for the future) means a relative change of 100%. This causes a major environmental stress. The same applies to other biota that is characterized by both a low CO_2 content in the cell liquid and a large surface compared to body volume, such as in many plants.

(5) Microorganisms have a fast reproduction rate. One generation reproduces in hours, whereas 10 000 generations, the estimated average time for a genetically new species, reproduce in years. While for macroorganisms the same average number of generations can be expected to form a new species, the time needed is much longer. Thus microorganisms are under an evolutive aspect in advance to macroorganisms when adapting to new environmental conditions. This phenomenon, which has not yet received enough attention, is called evolutionary asymmetry.

(6) Increased CO_2 contents in greenhouses first causes accelerated plant growth (for a short time only). On the other hand, the resistance against plant diseases decreases. Experiences with crop-protecting agents (herbicides, bactericides, etc.) show that an adaptation to new environmental conditions, both in greenhouses and in the open environment, occurs on a very short timescale. The important factor causing adaptation is the *change* of environmental conditions, such as the CO_2 gradient, not its absolute value.

(7) Statistically, among all new microbiological species that evolve, some could be expected that turn out to be incompatible to existing macroorganisms and result in new diseases and pests. Therefore, in addition to the known causes of extinctions, microbiological evolution is identified as an additional important evolutionary factor.

(8) In addition, due to the same process, bacteria and other microorganism that are vital for the survival of macroorganisms, for example participating in the digestion process, may become extinct. This explains why some and similar species of the same ecological niche become extinct while others with similar requirements survive. As this may occur both in the terrestrial and marine realm, and as the importance of marine bacteria and viruses become more recognized, another cause of mass extinctions needs attention.

(9) Therefore, free ecological niches of extinct macroorganisms can be occupied by other species with similar requirements. Such species need not necessarily be in advance of extinct species through better reproduction rates or other properties. They are just able to cope with microbiota incompatible to other macrobiota. The term "survival of the fittest" is extended by compatibility to new microbiological and enzymatic challenges. This process may be one cause for the sequence of so many "similar" species through time, species that are "technologically" not in advance of their precursors but are more compatible to new microbiological and enzymatic conditions than others.

(10) The CO_2 content of the atmosphere is a result of a process consisting of self regulating and thus oscillating activators and inhibitors. These are amongst others: increase of marine and terrestrial vegetation, reduction of O_2, cooling, reduction of vegetation, rotting and release of CO_2, increase of temperature, increase of vegetation. The natural oscillations of this process can be observed on any time scale, from the Mauna Loa curve (annual) through the ice-ages (potential additional factor), the Neogene oscillations and the Phanerozoic as well.
The sudden and significant deviation of the CO2-content from the window of self-regulation at the end of the Holocene through anthropogenic impact (Figs. 4.7–4.9) is thus a challenge not only under the aspect of the evolution of new pests and deseases but also under the aspect of stability at all. Thus the formulation of (new) stability conditions and *any* stability analysis, are important geological tasks for the future. The same applies to extension of the model formulated here by successive replacement of parameterized boundary conditions through model output of the various subsystems (ocean, atmosphere, vegetation, etc.).

The high gradient of climate change (in about 100 years a set back that compares to 35 (60) My of Earth history) paces microbiological evolution. This needs further study as well. The same applies to the study of the direct biochemical impact of changed CO_2 concentrations on enzymatic reactions, especially in conjunction with arboreal and animal (mankind?) diseases.
The principles discussed above also have technological and social aspects. Despite various political agreements the atmosphere's CO_2 content is expected to rise. The rise (2–3%?) is caused by both increased consumption (0–1% depending on technological progress) and growth of the world's population (about 2%). This means that the agreements of, for example, the Toronto Conference have no chance of being achieved, even if all countries fulfill their respective obligations. Another important factor is an increase in the massive use of fossil fuels, namely coal in several densely populated areas of the world, such as, for example, China.
In order to reach a significant effect, from the industrial viewpoint, new technologies have to be applied not only in northern hemisphere countries but also in all other countries. Financial constraints show this is unrealistic. Although any technology that reduces CO_2 emissions should be considered, more thought has to be spent on the right-hand arrow of Fig. 4.16, the remission through forestation. From the geological point of view (population dynamics) an increasing CO_2 emission per capita (e.g. rising standard of life with the technologies available),

even in the developing countries with older technologies, is not a problem if the population and thus utilization of the substratum (the atmosphere) is reduced.

Acknowledgement: This study financed by HÜLS AG, Germany. We gratefully acknowledge critical remarks of F. Strauch (paleontology), R. Guderian (biology) and G. Peschel (geophysics).

References

Bazzaz FA, Garbutt K (1988) The response of annuals in competive neighbourhoods: Effect of elevated CO_2. Ecology 69(4):937–946

Bazzaz FA, Reekie EG (1989) Competition and patterns of resource use among seddlings of five tropical trees grown at ambient and elevated CO_2. Oecologia 79:212–222

Berger WH (1977) Deep-Sea carbonate and the deglaciation spike in ptreopods and foraminifera. Nature 269:303–303

Bode, M. (1992) Beschreibung strukturbildender Prozesse in eindimensionalen Reaktions-Diffusions-Systemen durch Reduktion auf Amplitudengleichungen und Elementarstrukturen (Description of structur-forming processes in one-dimensional reaction-diffusion-systems through reduction on amplitude-equations and elementary structures).- Diss. Muenster

Broecker WS (1983) Der Ozean.– Spektrum der Wiss. 1983(11):96–107

Campbell JA (1975) Allgemeine Chemie. (Simultangleichgewichte, pp 536–538). Verlag Chemie, Weinheim

Carrol RL (1988) Vertebrate Paleontology and Evolution. Freeman

Cloud P, Glaessner MF (1982) The ediacaran period and system: Metazoa inherit the earth. Science 217:783–792

Covey C (1984) Erdbahn und Eiszeiten. Spektrum der Wissenschaft 4:84–93

Crutzen PJ, Müller M (1989) Das Ende des blauen Planeten? Der Klimakollaps – Gefahren und Auswege. CH Beck Verlag, München

Dixon RK: Project Leader "Forest Effects Project", EPA Global Change Research Programme, Pers. Comm.

Eichler J (1976) Handbook of stratabound deposits. In: Wolf KH (ed) Handbook of stratabound deposits Vol 7. Elsevier, Amsterdam pp 157

Fabian P 1987 Atmosphäre und Umwelt. Springer, Berlin

Fajer ED, Bowers DM, Bazzaz FA (1989) The effect of enriched carbon dioxide atmospheres on plant–insect herbivore interactions. Science 243(4895):1198–1200

Fond der Chemischen Industrie (1990) Folienserie des Fonds der Chemischen Industrie zur Förderung der Chemie und Biologischen Chemie im Verband der chemischen Industrie e.V., Frankfurt/M., 22, Umweltbereich Luft, 1987 (Series of transparencies of the Fund of the Chemical Industry for the support of Chemistry and Biological Chemistry in the Association of the Chemical Industry e.V., Vol. 22)

Frenzel B (1991) Klimageschichtliche Probleme der letzten 130000 Jahre (Paleoclimatic Problems of the last 130Ka). Akademie der Wissenschaften und der Literatur, Mainz

Gates DM, Strain BR, Weber JA (1983) Encyclopedia of Plant Physiology, New Series, Vol. 12D, p 503, Springer Verlag

Graedel, TE, Crutzen PJ (1989) Veränderungen der Atmosphäre. Spektrum der Wiss. 1989(11):58–68

Guderian R, Klumpp G (1989) Staub. Reinhaltung der Luft 49:255

Hays JD, Imbrie J, Shackleton NJ (1976) Variations in the earth's orbit: Pacemaker of the ice-ages. Science 194:1121–1132

Heinze C, Maier–Reimer E (1992) The Hamburg Oceanic Carbon Cycle Circulation Model (Cycle 1). Deutsches Klimarechenzentrum, Technical Report No. 5, Hamburg

Imbrie J, Berger A (1984) Milankovitch and climate change. Elsevier, Amsterdam

James, HL, Trendall AF (1982) Phys. Chem. Sci. Res. Rep. 3:199

Junge C (1981) Die Entwicklung der Erdatmosphäre. Naturwissenschaften 68:236–244

Kasting JF, Toon OB, Pollack JB (1988) Die Entwicklung des Klimas auf den erdähnlichen Planeten. Spektrum der Wiss., 4:45–53

Keeling CD, Bacastow RB, Carter AF, Piper SC, Whorf TP, Heimann M, Mook WG, Roeloffzen (1989) Geophys. Monographs 55:305 AGU, Washington

Kempe S, Degens ET (1985) An early soda ocean? Chemical Geology 53:95–108

Kilmartin JV, Fogg J, Luzzana M, Rossi–Bernardi L (1973) The role of α-Amino Groups of the α- and β Chains of Human Hemoglobin in Oxygen-linked Binding of Carbon Dioxide. J. Biol. Chem. 248(20):7039–7043

Kohlmaier GH, Janececk A, Kindermann J, Benderoth G, Klaudius A (1988) Zeitliche Entwicklung des globalen Kohlenstoff–Zyklus (History of the global Carbon Cycle). Workshop held in the Evang. Akademie Arnolshein/Taunus pp 9–24 (Institut für Physikalische und Theoretische Chemie der Johann–Wolfgang–Goethe Universität, Frankfurt/M)

Kuhle M (1986) Die Vergletscherung Tibets und die Entstehung von Eiszeiten. Spektrum der Wiss. 9:42

Lehmann U, Hillmer G (1980) Wirbellose Tiere der Vorzeit. Enke Verlag

Lehninger A (1977) Biochemie. Verlag Chemie, Weinheim

Margulis L, Walker JGC, Rambler H (1976) Reassessment of the roles of oxygen and ultraviolet light in Precambrian evolution. Nature 264:620–624

McFadden BA (1973) Autotrophic CO_2 assimilation and the evolution of Ribulose Disphosphate Carboxylase. Bact. Rev. 73(3), 289–319

Neftel A, Oeschger H Staffelbach T, Stauffer B (1988) CO_2 record in the Byrd ice-core 50,000–5,000 years BP. Nature 331:609–611

Nilsen S, Hovland K, Dons C, Sletten SP (1983) Effect of CO_2-enrichment on photosynthesis, growth and yield of tomato. Scientia Hortic. 20:1–14

Pollack B (1979) Climatic change on the terrestrial planets. Icarus 37(3):479–553

Remy W, Remy R (1977) Die Floren des Erdaltertums. Glückauf Verlag, Essen

Schidlowski M (1978) Origin of Life. In: Evolution of the Earth's atmosphere: Current State and Exploratory Concepts. Center of Scientific publications, Tokio, pp 3–20

Schidlowski M (1981) Die Geschichte der Erdatmosphäre. Spektrum der Wiss. 4:17

Schidlowski M, Eichmann R, Junge CE (1975) Precambrian sedimentary carbonates: Carbon and oxygen isotope geochemistry and implications for the terrestrial oxygen budget. Precambrian Research 1:1–69

Schneider SH (1989) Veränderungen des Klimas. Spektrum der Wiss. 1989(11):70–79

Schönwiese CD (1992) Vulkanismus und Klimageschichte (Volcanism and climatic history). Umweltwissenschaften und Schadstoff-Forschung 4(4), 239-245

Schönwiese CD, Runge K (1988) Der anthropogene Spurengaseinfluß auf das globale Klima. Berichte des Instituts für Meteorologie und Geophysik, Johann–Wolfgang–Goethe Universität, Frankfurt/M

Schwarzacher W (1993) Cyclostratigraphy and the Milankovitch Theory. Developments in Sedimentology 52 Elsevier, Amsterdam

Shackleton NJ (1977) Tropical rainforest history and the equatorial pacific carbonate dissolution cycles. In: Andersen NR and Malahoff A (eds) The fate of fossil fuel CO_2 in the oceans. Plenum Press

Smolka PP (1991) Neogene Ozeane. Die Geowissenschaften 9:347–351

Stanley SM (1987) Extinction. Scientific American Library / Freeman, New York

Strain BR, Cure JD (1985) Direct Effect of Increasing Carbon Dioxide on Vegetation. US Dept. of Energy

Strauch F (1990) Brinkmanns Abriß der Geologie, Vol. 2:39, Enke–Verlag, Stuttgart

Trabalka JR, Reichle DE (1986) The Changing Carbon Cycle. Springer Verlag, Berlin, Heidelberg, New York

Trendall AF (1973) Unesco Earth Sciences Ser. 9:257

van der Burgh J, Visscher H, Dilcher DL, Kürschner WM (1993) Paleoatmospheric Signatures in Neogene fossil leaves. Science 260(5115), 1788–1790

Vogel H (1977) Probleme aus der Physik. Springer–Verlag, Heidelberg

Equations and Data

> **(1)** *Solar Constant as function of time* t *and age of Earth* t_0
>
> $S_k = 10^{**}\{\log[1360 - (360^*t/t0)]\}$

> **(2)** *Temperature at the sun's surface*
>
> $T(t) = [S_k^*(a^2/R^2)\delta]^{(1/4)}$
>
> *Or in rough approximation:*
>
> $T_s(t) = \exp(\{[-\ln(T_h) - \ln(T_0)]/t_0\}^*t) + \ln(T_h)$
>
> *a:* Distance Planet-Sun (for Earth: 150 Mio km)
>
> *R:* Radius of the sun: 689 000 km
>
> T_h: Present temperature (about 6000 K)
>
> T_0: Temperature at 4.7 Ga (5557 K)

> **(3)** *"Endogenic heat" of a planet (roughly approximated)*
>
> $W_p = \exp(\{[\ln(W_0) - \ln(0.1)]/t_0\}^*t + \ln(0.1))$
>
> W_0: Endogenic heat after accretion, ca. 4000 W m^{-2}
>
> 0.1: Present endogenic heat

> **(4)** *Density of solar radiation at distance* a *from the sun for fast rotating planets and sufficient heat dissipation to the dark side, for example earth (Stefan–Boltzmann, black body).*
>
> $W_s = \sigma * T_s(t)^4 * R^2 / (2a)^2$
>
> σ: Stefan–Boltzmann constant 5.67^*10^{-8} W m^{-2} K^{-4}
>
> R: Radius of the sun (689 000 km)
>
> a: Distance of a planet from the sun (Earth: 150 million km)

> **(5)** *Radiation budget on the surface of a planet including the effect of the albedo*
>
> $W_g = W_s * (1 - \alpha) + W_p * \alpha$

> **(6)** *Solving Stefan–Boltzmann law for* T *yields "black–radiation" temperature of a planet*
>
> $T_b = (W_g / \sigma)^{1/4}$

> **(7)** *Total temperature (Tges) = black-radiation temperature + "greenhouse effect"*
>
> $Tges = T_b + [34.5^*(\ln CO_2)/6.5]$ (CO_2 in vol. ppm, Tges in Kelvin)

> **(8)** *Mass of the ocean:* 1.35^*10^{18} t water

(9) *Pressure Equilibria*

Temperature and partial pressure, energy content, phases of various gases (N_2, CO_2, H_2O, O_2), are derived from equilibria diagrams (IS-diagrams), available through standard chemistry texts (e.g. "CRS handbooks"). At first, CO_2 partial pressure can be assumed, yielding temperature. Adding CO_2 partial pressure and the equilibrium condition with the ocean, other partial pressures can be obtained. Considering chemical equilibria quantitatively (carbonate formation, etc.), leads to both consistencies and contradictions, both for equilibrium considered as well as for the sequence through time (quantities of sedimented carbonates, ocean deacidification, etc.). This iterative process can be performed both explicitly and algorithmically.

(10) One might arrive at initial estimates for O_2 production using banded ironstone amounts. Even if this approach were possible (stratigraphy, etc.) this would be a lower bound. For *budget* however, the absolute value of produced O_2 is not important, but rather the quantity of *free* oxygen. This free O_2 is of interest, because this is O_2 that was available for saturating reduced iron on land and forming the ozone layer. The curves discussed only deal with this *free* O_2, not with the total O_2 produced.

(11) *pH of the oceans as function of atmospheric CO_2*

$pH = -\log[pH_{val}*(Vol\ ppm\ CO_2 * patm)^{1/2}]$ with: $pH_{val} = 4.5*10^{-10}$ or

where $pHval = [(3.98*10^{-11})*(1/30)*(1.00*10^{-6})*0.157*(p_{atm}\ in\ bar)]^{1/2}$

$3.98*10^{-11} = (CO_3)^2 / (HCO_3) = 10^{-10.4}$ (Eq. 12)

$1/30$ = molar solubility of CO_2 in H_2O, $20°\,C$ (Law of Henry, Campbell 1975)

44 (approx. molar mass of air) / 28 (approx. molar mass of CO_2) = 1.57 (*0.1 to link between ppm and Pascal)

$(1.00*10^{-6})*0.157$: Factor to express vol ppm CO_2 as partial pressure. In order to derive partial pressure from concentration this factor is multiplied with total pressure of atmosphere (p_{atm} in bar). The present pressure is 1 bar, for the geological past higher pressures have to be accounted for (see figures).

(12) *Carbonate formation in equilibrium:*

Ca concentration recent: 0.4 g/l

Hydrogen-carbonate: 0.03 g/l

Solubility product $CaCO_3$ pL = 8.32

pH: 8.0

Boundary-equilibrium for molar fraction $CO_3/HCO_3 = 5.57*10^{-3}$ (from equilibrium $HCO_3 + H_2O = CO_3 + H$ and solubility product of $CaCO_3$), see Campbell 1975:535 "Simultaneous equilibria explained for mussel example"

$(CO_3)^2/(HCO_3) = 10^{-10.4}$

$CO_3 = 10^{-8.36}$ / Ca (recent) with 0.4 g/l

250 Ma (for example): Ca= 0.88 g/l = 0.022 mol/l

$CO_2aq = 10^2 (HCO_3)^2$

$HCO_3 = 5.74*10^{-5} * (PCO_2)^{1/2}$

Considering the ocean mass in equilibrium: 1 ppm increase of CO_2 yields $HCO_3(1) / HCO_3(2) = [PCO_2(1) / PCO_2(2)]^{1/2} = (540.00/538.46)^{1/2} = 1.001439$

At saturation concentration 545.460 weight ppm CO_2 amount to $4.23633*10^{-4}$ mol/l. For 547.000 ppm $4.242434*10^{-4}$ mol/l result.

The difference concentration amounts to $6.096*10^{-7}$ mol/l. This equivalates to $(6.096*10^{-7}$ mol CO_2 / l) * $(4.4*10^{-2}$ kg/mol) $= 2.682*10^{-8}$ kg CO_2 / l H_2O (with a water density about 1 this yields approx. $2.682*10^{-8}$ kg CO_2 / kg H_2O)

Total absorption of ocean then gives: $(1.35*10^{21}$ kg H_2O) * $(6.096*10^{-7}$ mol/l CO_2) * $(4.4*10^{-2}$ kg/mol) $= 3.62*10^{13}$ kg $= 36.2$ Gt CO_2 for the ocean.

A rise in atmosphere of 1 ppm CO_2 yields 7.85 Gt, resulting in $[7.85/(7.85+36.6)] = 0.177$, with distribution coefficient $(1.0 - 0.177) = 82\%$.

(Please note the definition of the distribution coefficient). The 82% applies for the equilibrium case. In reality only about 50% are reached permitting further solution of CO_2 in the ocean for a limited time (up to equilibrium).

The following pK–values have been applied:

$PCO_2/101300 * CO_2aq = 30$ p (in Pascal)

$CO_2aq + H_2O = H_2CO_3$, pK = 3.16

$H_2CO_3 = HCO_3^- + H^+$, pK = 3.30

$CO_2aq + 2H_2O = HCO_3^- + H^+$, pK = 6.46

$HCO^{3-} + H_2O = H^+ + CO_3^-$, pK = 10.40

Critical temperature of water: 374° C

Carbon balance:

720 Gt: Atmosphere

552 Gt: Forests

5000 Gt: Coal deposits

35 Gt: Petroleum

120 Gt: Gas

5 Gt/year: Fossil fuels (emission)

3 Gt/year: Forest fires (man-made)

Magnetic Signatures of Pampean Soils, A Case Study

Paulina Nabel[1] and Nikolai Petersen[2]
(1) Museo Argentino Ciencias Naturales "B. Rivadavia", y Instituto Nacional de Investigacion de las Ciencias Naturales, Av. Angel Gallardo 470, CP 1405, Buenos Aires, Argentina
Pnabel@gecuat.gov.ar, Penabel@mail.retina.ar
(2) Head of Biomagnetics Group, Institut fuer Allgemeine und Angewandte Geophysik, Ludwig–Maximilians Universität München, Theresienstr. 41, D-80333 München, Germany
Petersen@magbact.geophysik.uni-muenchen.de

Abstract: In several places in the world measurements of the magnetic susceptibility have been used to detect environmental changes. This study reassesses this problem with special emphasis on the validation of some general conclusions that are often found in the literature. This study links magnetic susceptibility measurements of Pampean soils with present environmental conditions. It could be shown that also in the Argentine Pampa, where such studies have just been started, susceptibility measurements can be used for paleoclimatic studies. In contrast to the results from the cold Chinese Loess plateau the calibration is however different.

5.1. Introduction

The areas of loess deposition, like the Pampa region, provide the best deposits to study the records of climatic changes in continental environments. Proxy climate data from loess have the potential to improve our knowledge of Pleistocene climate change (Beget et al. 1990).

"To understand global changes of the past, or to predict the changes expected in the future, one would like to know a detailed history of environmental changes throughout the full reach of the past, and for every region of the globe" (Eddy 1992).

The paleosols interbedded in Pampean loess are recorders of warmer environmental conditions that took place during the Quaternary in these regions. The alteration of loess sediments, that originated during dry and cold environmental conditions, with paleosols, has been used as an indicator of climatic changes that occurred during the Pleistocene (van Houten 1982).

Although distinguishing between loess (dryer conditions) and paleosols (moist conditions) is an important tool for deciphering continental climatic history, only in ideal cases is this analysis easy. Normally, the analysis of long field sections or even many drillholes for finding paleosols is a difficult task, sometimes requiring the application of different methods.

Studies from different researchers (Liu Tungsheng 1985a; Heller et al. 1987; Heller and Liu Tungsheng 1986; Kukla et al. 1988; Zhou et al. 1990; Maher and

Thompson 1991, 1992; Xiuming et al. 1992; Evans and Heller 1994) amongst others, and our own, have shown that by the measurement of the magnetic susceptibility, a quantitative indicator exists for the presence of paleo-soils in continental sedimentary sections. These serves as a proxy measure of climate. On account of this, magnetic susceptibility measurements have been used in paleoclimatic reconstructions for Quaternary sediments.

Since susceptibility measurements are a geophysical routine to be performed inexpensively and rapidly, they can be applied by other workers devoted to continental paleoclimatology. Studies have also been performed using susceptibility measurements with a high-time resolution for fully quantitative analysis, comparable to isotopic records (Heller and Liu Tungsheng 1986; Kukla et al. 1988, 1990; Bloemendal and deMenocal 1989; Beget et al. 1990) enhancing both insight into climate dynamics and allowing high-resolution regional correlation.

Although magnetic susceptibility measurements have been recommended to be used as a secondary climatic indicator (Liu Tungsheng et al. 1985b), our studies show that a regional calibration is necessary before they can be used as a paleoclimatic tool (Nabel et al. 1993). In particular, the susceptibility measurements carried out on the studied Quaternary paleosols of the Pampa region, have shown opposite behavior from those reported for European and Chinese paleosols. While in European and Chinese sediments the susceptibility values are commonly higher in the paleosols than in the interbedded loess in the Pampean region, most of the magnetic susceptibility measurements that have been recorded so far show lower values for paleosols than for loesses.

Even though, in the Pampean region the changes in magnetic susceptibility values along the profiles also reflect environmental changes, it is necessary to review their correlation. This is especially important because some of the mechanisms discussed may also apply to other areas of the world. Recently, a similar behavior has been reported about loess–paleosol sequences from Alaska (Beget and Hawkins 1989) and from Siberia (Rutter et al. 1995).

Before this study was carried out, the discussed Pampean Quaternary sequences had been dated. In addition the duration of the deposition of the sediments was determined. As a result of this chronological approach, we recognize the Hisisa Geosol from the Ensenada Formation which is located at the end of Matuyama Chron (Nabel et al. 1993). The accurate time setting of this paleosol suggests an environmental change in the Pampean region in the beginning of Brunhes Chron, 780 ka (thousand years ago). Loess deposits characterize the Brunhes onset suggesting a change to dryer and probably cooler environmental conditions (Nabel 1993).

Before susceptibility-values of paleosols can be used for paleoclimatic interpretation, a reliable and calibrated comparative model has to be established. Thus the first part of this contribution focuses on mechanisms of susceptibility behavior and precautions to be observed when applying this method routinely. In the second part, results from recent Pampean soils are presented.

5.2. Soil Magnetism

The susceptibility is a measure of the ease with which a material can be magnetized. A high susceptibility means "easy to magnetize", a low value, the opposite. Magnetic minerals (ferri- and antiferromagnetic) have the highest susceptibility values (about 5×10^{-10} m/kg), while diamagnetic ones present low and negative values (about -10^{-10} m/kg). Between them, paramagnetic minerals present low positive susceptibilities (about 10^{-10} m/kg). Table 5.1a lists specific susceptibilities for various minerals.

Magnetic susceptibility has been suggested to be an indicator of soil-forming processes (Mullins 1977). It depends not only on the concentration and type, but also on the shape and size of magnetic minerals in the soil (Oades and Townsed 1963; Dearing et al. 1985, 1986; Zheng et al. 1991). Furthermore, the magnetic mineral content and related magnetic susceptibility values vary with different environmental parameters that play a role in soil forming processes (Mullins 1977; Hilton 1987; Maher and Taylor 1988; Kukla et al. 1988).

Iron mobility and its sensitivity to ambient environmental conditions are conferred by its ability to change valence state, to form complexes with other soil components and to form a wide range of environmentally-specific oxide-phases (Maher 1986). The magnetic minerals are mostly iron oxides and hydroxides. Even at low concentrations, iron oxides have a high pigmenting power and determine the color of many soils. Soil structure and fabric are also influenced by variations in iron mineral forms and distributions. All these physical properties of the soils are helpful in explaining soil genesis and are important for naming and classifying soils (Schwertmann and Taylor 1977). The magnetic properties of a soil sample reflect the different magnetic properties of a wide range of soil-forming minerals. Table 5.1b lists the most common oxides and hydroxides, their magnetic status, the susceptibility values and the reported environmental associations.

Table 5.1a. Specific susceptibilities for various minerals (after Thompson and Oldfield 1986)

Remanence-carrying minerals (10^{-8} m^3 kg^{-1})		Other iron bearing minerals (10^{-9} m^3 kg^{-1})		Other minerals and materials (10^{-8} m^3 kg^{-1})	
Iron (αFe)	2×10^7	Olivines (Mg,Fe)$_2$SiO$_4$	1–130	Water (H$_2$O)	–0.9
Magnetite (Fe$_3$O$_4$)	5×10^4	Amphiboles (Mg,Fe,Al)	16–100	Halite (NaCl)	–0.6
Pyrhotite (Fe$_2$S$_3$)	5×10^3	Siderite (FeCO$_3$)	100	Quartz (SiO2)	–0.6
Ilmenite (FeTiO$_3$)	200	Pyroxenes (Mg,Fe)$_2$Si$_2$O$_6$	100	Calcite (CaCO$_3$)	–0.5
Lepidocrocite (γFeOOH)	70	Biotites (Mg,Fe,Al)	5–95	Feldspar (Ca,Na,K,Al,silicate)	–0.5
Hematite (Fe$_2$O$_3$)	60	Nontmita (Fe-rich clay)	~90	Kaolinite (clay min.)	~2
Goethite (αFeOOH)	70	Chamosite (Ox.chlorite)	~90	Montmorrillonite (clay)	~5
		Epidote (Ca,Fe,Al)	~30	Illite (clay min.)	~15
		Pyrite (FeS$_2$)	~30	Plastic (e.g. PVC)	~0.5
		Chalcopyrite (CuFeS$_2$)	~3		

Table 5.1b. Iron Oxides in Soils (after Schwertmann and Taylor 1977, Mullins 1977, Maher 1986)

Mineral	Magnetic Susceptibility $(10^{-8}\ m^3\ kg^{-1})$	Magnetic Status	Reported Environments
Hematite (αFe_2O_3)	27, 31 up to 63	Canted antiferromagnetic	Relatively dry, highly oxidized soils, usually in areas of elevated temperature
Goethite ($\alpha FeOOH$)	12.5, 35, 38, up to 126	Canted Antiferromagnetic	Moister soils, abundant in well-drained temperate areas
Maghematite (χFe_2O_3)	4.1×10^4, 4.4×10^4	Ferrimagnetic	Abundant in highly weathered tropical / subtropical soils
Lepidocrocite (YFfeOOH)	50–75, 69	Paramagnetic	Occurs in poorly drained soils
Ferrihydrite ($5Fe_2O_39H_2O$)	70	Paramagnetic	Poorly drained and podsolized soils
Magnetite (Fe_3O_4)	$5–10 \times 10^4$ $3.9–5.8 \times 10^4$	Ferrimagnetic	Primarily derived, from soil firing or from biogenesis

These minerals are either of detrital origin or they are authigenic. If they are primarily formed, weathering processes will be the most important factor. Iron could be liberated in the ionic form, and transported into the soil system. Subsequently released from the mineral lattice, this iron may be oxidized and precipitated in situ. Alternatively, it may be taken into the soil solution (as Fe^{+2}) and precipitated elsewhere in the profile, depending on the oxidative status of the micro-environment of the soil (Maher 1986).

As a note, reduction-oxidation cycles occurring under normal pedogenic conditions allow the conversion from weakly magnetic forms of iron oxide and hydroxide, to strongly magnetic forms such as maghemite or magnetite (Mullins 1977). Organic matter content and associated microorganisms, water regime (Petersen et al. 1992), as well as the chemical composition of solutions and burning, all play important roles in the formation or dissolution of iron minerals. Particularly, magnetite formation during soil development (in situ origin), has been reported to be both of inorganic (Maher and Thompson 1992) and organic origin (Petersen et al. 1986, 1993; Fassbinder et al. 1990).

5.3. Pampean Soils

A key area for the study of Pampean soils is the east and northeast of the province of Buenos Aires, Argentina (Fig. 5.1). Soil samples were taken from the same edaphic dominium (ED) of the Buenos Aires Province. Thus they formed under similar climatic conditions, but in different geomorphologic settings (Fig. 5.1).

Table 5.2. Explanation of the codes used for the description of the edaphic dominium (after CIRN 1989).

Edaphic Dominium					
	Asoc. Index			Textural Family	
Edaphic Dominium	Soil		Soil	Order	
11a	M17ac2+M8tc+F5btc		M17	ac	2
	D Na H			Main Group	Subgroup
	Limiting Agents				

If the numerator appears for only one soil (consociation), we suppose that the unit consists of more than 85% of this soil. If two soils appear, we estimate 60% and 40% respectively. In the case of three soils: 50%, 30% and 20%. When the soils are part of an association, their symbols are separated by a (–); if they are part of a complex, the symbols are separated by a (+). The "textural family" is used only for dominant soils.

Order	Maingroup	Subgroup		Family	Texture	Limiting Agents	
E	Entisol	ac	acuic	2	Fine	o	Np Limitation
E3	Haplacuent	ae	aeric	5	Fine silt	d	Drainage deficient
F	Alfisol	ag	taptoargic			h	Suscept. due to hydric erosion
F5	Natracualf	Tc	tipic			h'	Ongoing hydric erosion
M	Molisol	Hat	histic			Na	Sodium alkalinity at more than 50 cm depth
M7	Haplacuol					Na'	Sodium alkalinity at less than 50 cm depth
M8	Natracuol					Sa	Salinity
M17	Hapludol						

The Buenos Aires Province has a moderate climate. The mean annual temperature amounts to 16° C. Due to the proximity of the Atlantic Ocean, both the daily and the annual variations of the temperature are low. The precipitation amounts to about 900 mm per year. As a result of the South Atlantic anticyclone and low continental pressure, there is a relative increase of winds during spring and summer. In winter, a high-pressure center settles on the continent and W and SW winds are predominant (CIRN 1989).

The Buenos Aires province is characterized by a wide plane of Quaternary age that is underlain by eolian, fluvial and marine sediments. As no barriers to atmospheric circulation exist, rough climatic changes caused by the N–S movement of respective air-masses can frequently be observed. The Edaphic dominiums have been established from soil characterization and from the distribution of geomorphologic and lithological features. Sometimes they coincide with geomorphological subunits; sometime they form their subdivisions (Fig. 5.1).

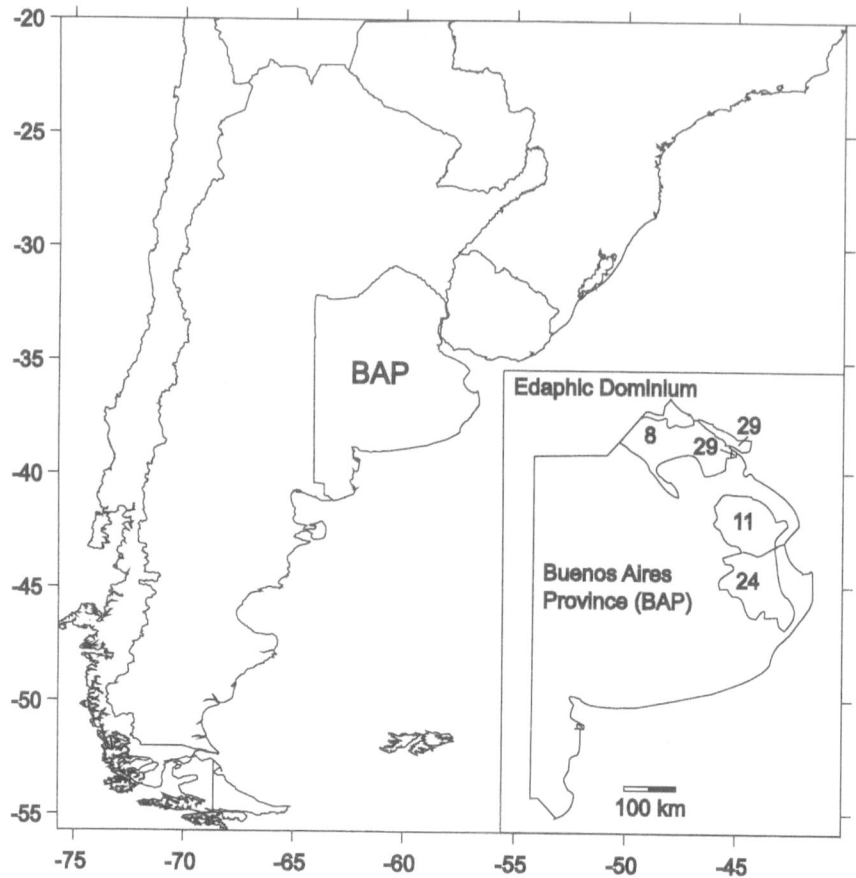

Fig. 5.1. Location of the study area. Numbers indicate the edaphic dominiums as explained in Table 5.2.

A and B horizons of soils from the edaphic dominium (ED) 8, 11, 24 and 29 were sampled (see Table 5.2 for the explanation of the classification). ED 8 has been developed over a wavy loess plain formed by (8a M17tc2 /O) soils in the upper parts and by (8c M8tc2+F5tc+M17ac /Na' D) in the depressed areas. This ED belongs to the "Pampa ondulada" region and is located in the north of the province.

ED 11 has been developed over a flat loessic (or loesslike) plain, characterized by (11a M17ac2+M8tc+F5tc /D Na' h) soils. The ED 24 is located in the "Pampa deprimida" region. It developed over a flat-concave plain formed by (24a M8tc2-M18:na-M18:ag/Na D) soils. The 29 ED belongs to the delta of the Parana river, characterized by (29a M7ht5+E3ae+M18ac /D Na Sa) soils.

5.4. Materials and Methods

To study the relationship between the magnetic behavior of soils from the Pampean region with their environment, different measurements were performed.

Samples from A and B horizons of soils were extracted using rigid plastic tubes of 10 cm diameter and subsampled at 2 cm intervals. All measurements were carried out on natural samples, room temperature dried, taking care to maintain their structure.

The samples were first subject to a range of magnetic measurements to characterize the directional parameters such as declination and inclination of the NRM (Natural Remanent Magnetization) together with their intensity. One aim was to study both their magnetic characteristics and the stability of the soil-magnetism. The measurements were carried out with a Cryogenic magnetometer. Low-field magnetic susceptibility (X) measurements were carried out with a "Bartington Instruments" single sample susceptibility sensor. This parameter is the ratio of magnetization induced to intensity of the magnetizing field measured within a low magnetic field (0.1 mT) with no induced remanence. It has been measured on a mass-specific basis and is roughly proportional to the concentration of ferrimagnetic minerals within a sample.

Frequency dependent susceptibilities (X FD) were measured with a "Bartington Instruments" dual frequency (1 and 10 kHz) susceptibility sensor (see also Dearing et al. 1985). The variation of the susceptibility with frequency indicates the presence of grains lying at the "stable single domain"/"superparamagnetic domain" boundary.

Identification of magnetic mineralogy has been carried out on magnetic phases, extracted and concentrated from an ultrasonically dispersed soil suspension, circulating around a continuous extraction system (Petersen et al. 1986). The extracts were used in strong-field thermomagnetic analysis to determine Curie points and for observation under transmission electron microscopy (TEM).

Magnetic bacteria analysis was carried out to check potential biogenic origin of in situ formation of magnetite that occurs under soil-forming conditions. Samples were stored undisturbed in the ambient geomagnetic field at room temperature, protected from direct light (Petersen et al. 1986; Vali et al. 1987; Spring et al. 1993).

5.5. Results

Soil sediments have been shown to possess a natural remanent magnetization (NRM) attributed to the statistical alignment of magnetic grains in the Earth's magnetic field. All 98 samples were subjected to directional analysis. Most of their NRM directions were well grouped, but inclination values are generally low. NRM directions of the upper horizons from 24a and some of the 8c soils do not show a good grouping in spite of the presence of higher inclination values. The water content of the soils does not seem to be a disturbing factor in the attainment of the magnetic directions (Verosub et al. 1979).

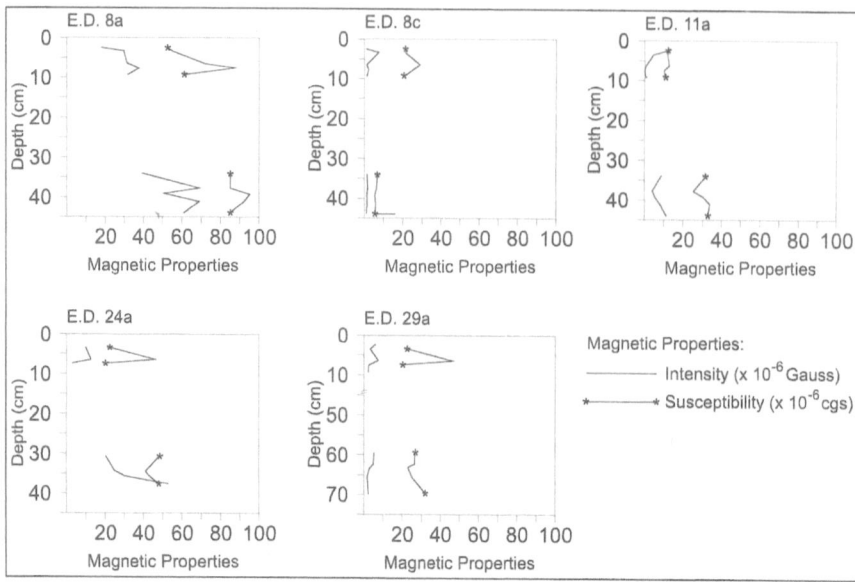

Fig. 5.2. Variation of magnetic properties in the discussed edaphic dominiums (ED) with depth. Further explanation can be found in the text.

To analyze the magnetic domain, state and magnetic grain size, high- and low-frequency susceptibility measurements were carried out. Our results of the frequency-dependent susceptibility (Xfd) show the same susceptibility values for high- and low-frequency measurements suggest that no frequency-dependent susceptibility was found; this is often common in very fine grained magnetite of around 0.02 um diameter (Maher 1988). Xfd is sensitive to the presence of sub-micron-sized grains, especially those spanning the "superparamagnetic"/"single domain" boundary characterized with grain-sizes of 0.01–0.03 um (Maher and Taylor 1988).

Fig. 5.2 illustrates the variation of susceptibility with depth of the sampled Pampean soils. B horizons from soils from ED (edaphic dominium) 8a, 11a and 24a show increased susceptibility values in relation to their A horizons, while soils from ED 29a and 8c show the opposite trend. The soil from ED 8a has the maximum values of susceptibility and shows an increase of X values with depth (Fig. 5.2a). These soils are located in the well-drained "Pampa ondulada" region. Soils belonging to ED 11a show susceptibility values about half of the ones above. They also show an increase with depth (Fig. 5.2c). These soils are located in a flat plain and at lower topographic levels. Soils from ED 24a show similar susceptibility values between A and B horizons. They are located in a flat-concave plain. Soils from ED 8c are waterlogged. They show the lowest susceptibility values (Fig. 5.2b) but a relative increase of susceptibility in the top of the soil. These soils are located in the depressed areas of the "Pampa ondulada" region. Soils from ED 29a belong to the delta region. The susceptibility values show also an increase towards the top of the soil (Fig. 5.2e).

Susceptibility and NRM intensity values show the same behavior indicating that both reflect the same sedimentary processes (Fig. 5.2).

A and B horizons from soils of ED 8a, 24a and 11a were sampled to test the existence of magnetite producing bacteria. The soil samples have been sealed in test tubes with nutrients for magnetic bacteria. The susceptibility increases with time. There is magnetite production only in the superficial samples, most in the A horizon from a soil of ED 24a. This sample clearly contains magnetite producing bacteria. In the B horizons there are no, or only few bacteria.

The fact that the soil belonging to ED 8a shows the highest susceptibility (and NRM intensity) values and shows relatively low content of magnetic bacteria, suggests that the present magnetic phase is of inorganic origin. Activity of magnetic bacteria in soils from ED 24a is probably responsible for the higher susceptibility of the top of the soil.

Strong-field thermomagnetic curves were performed with the aim to identify the mineralogical nature of the magnetic phase. The predominant magnetic value of Tc=570–600° C, suggests that magnetite (Fe_3O_4) is the main magnetic carrier.

5.6. Discussion

Different sources of magnetic minerals have been recognized in soils, they can arise from soil/rock erosion; volcanic and fly ash; eolian deposition of windblown soils; cosmic particles; source of pollution; authigenic formation and magnetic bacteria (Hilton 1987).

The soil from ED 8a has the maximum values of susceptibility and intensity (Fig. 5.2a) reflecting the highest magnetic mineral content. It shows an increase of X values with depth. There is no an increase of X values related to magnetic bacterial growth. No Xfd was recognized. Most of the soils in this ED are thin "Argiudol" without limiting elements. This behavior suggests that the magnetic signal is related to significant quantities of inherit magnetic iron oxides from its weathering parent material. The other soils from the sampled Pampean area have been developed over similar loess-like sediments and show a similar trend.

In soils from ED 11a the susceptibility and the intensities of the NRM decrease to a half of the samples from ED 8a (Fig. 5.2c). They show, however, the same trend, with an increase of X and I values with depth. This ED is formed by a complex consisting of Argiudol, Natracuol, and Natracualf soils (50%, 30%, 20%) with a deficient drainage, sodic alkalinity up to 50 cm depth. The susceptibility is caused by hydric erosion.

A similar behaviour is present in soils from ED 24a with an increase of magnetic bacteria in the topsoil. This ED is consists of an association of Natracuol Hapludol tapto-natrico and Hapludol tapto-argico, drainage deficient and with sodic alkalinity below 50 cm depth.

In delta soils from ED 29a, susceptibility values are similar to those of soils from ED 24a, but there is an increase of susceptibility towards the uppermost parts of the soil. The soil is formed by a complex of Haplacuol histico, Haplacuent albico and Hapludol acuico, limited by salinity, sodium alkalinity below 50 cm depth (drainage deficient).

In the waterlogged soil from ED 8c, susceptibility and NRM intensity diminish as a whole, but there are higher values in the top of the soil.

The soil is formed by a complex of typical Natracuol, typical Natracualf and Argiudol acuico with deficient drainage and sodium alkalinity up to 50 cm depth. The presence of sodium alkalinity in the first 50 cm of the soil profiles in ED 8c and 11a gives rise to the lowest susceptibility values, thus reducing the magnetic signal. In soils from ED 24a and 29a, sodium alkalinity occurs below 50 cm depth. This is probably the reason for the decrease of the susceptibility values in the B horizon of the ED 29a soils.

Due to the above, the main magnetic mineral input in this region seems to be the reworked (by water and wind) loess-like parent material with frequent contributions of volcanic ash. Moreover, as the parent materials are almost the same, and no lateral climatic changes exist between the studied soils, the difference in susceptibility values in our region seems to reflect the contribution of the following: topography (which controls the soil drainage, the erosion, and the transport by water); the sodium alkalinity (which dilutes the magnetic signal); and the presence of magnetic bacteria (which produces a magnetic input to susceptibility values of some of the soil levels).

Different correlations between soil features and magnetic parameters have been suggested. Statements that many well-drained soils throughout the world exhibit enhanced levels of secondary ferrimagnetic minerals (magnetite/maghemite) in their upper horizons can be found; these are detectable from magnetic susceptibility measurements (Le Borgne 1955; Mullins 1977; Dearing et al. 1986). The same applies to statements saying that waterlogged soils used to present the opposite behavior (Thompson and Oldfield 1986).

Nevertheless, our results show that these general statements have to be regionally witnessed. This study shows that local and regional characteristics related to sedimentological, hydrological geomorphological, climatic and biological features, as well as anthropogenic effects, result in specific local correlations between magnetic properties and environmental conditions.

5.7. Conclusions

The magnetic susceptibility can be used, after local calibration, to measure long time series of Quaternary loess-soil sequences. The calibration has to be performed mainly with respect to the sign, i.e. whether low susceptibilities values reflect paleosols (as in Pampa region) or loesses. The paleoclimatic and paleoenvironmental meaning of the susceptibility values have to be checked with present loess-soil units. For the Pampa region the main results can be summarized as follows:

1) The NRM (Natural Remanent Magnetization) formed simultaneously with the soil. The contribution of loess-like parent material seems to be the most important source of magnetic input in the studied area.
2) The topography controls the impact of erosion and transport by water. Susceptibility values diminish with the increase of water content in the soils.
3) The fact that soils with the highest susceptibility (and NRM intensity) values have a relative low content of magnetic bacteria suggests that most of the magnetization in this area is of inorganic origin.
4) The sodium alkalinity dilutes the magnetic signal.

In the Pampean region, low susceptibility values correspond to soils and high susceptibility values to loesses. Thus the measuring of susceptibilities along drill sections can be a tool to recognize sequences of paleosols and thus integrated environmental conditions.

Acknowledgements: The authors are especially grateful to Daniel Vargas for field work assistance and Monica Hanesch for carrying out the thermomagnetic measurements. We are also grateful to the Argentine Scientific Council (CONICET), to the Ludwig Maximilians Munich University and to the Argentine Museum of Natural Science for their institutional support.

References

Beget JE, Hawkins DB (1989) Influence of orbital parameters on Pleistocene loess deposition in Central Alaska. Nature 337: 151–153

Beget JE, Stone DB, Hawkins DB (1990) Paleoclimatic forcing of magnetic susceptibility variations in Alaska loess during the late Quaternary. Geology 18:40–43

Bloemendal J, deMenocal P (1989) Evidence for a change in the periodicity of tropical climate cycles at 2.4 Myr from whole-core magnetic susceptibility measurements. Nature 342:897–900

CIRN (1989) Mapa de suelos de la Provincia de Buenos Aires. Instituto Nacional de Tecnologia Agropecuaria. Proyecto PNUD ARG 85/019

Dearing JA, Maher BA, Oldfield F (1985) Geomorphological linkages between soils and sediments: the role of magnetic measurements. In: Richards KS, Arnett RR, Ellis, S (eds) Geomorphology and Soils. George Allen and Unwin, pp 441

Dearing J, Morton R, Price T, Foster I (1986) Tracing movements of topsoil by magnetic measurements: two case studies. Physics of the Earth and Planetary Interiors 42:93–104

Eddy J (1992) Past Global Changes (PAGES). Report 19. Global Change IGBP of ICSU

Evans ME, Heller F (1994) Magnetic enhancement and paleoclimate: study of a loess/palaeosol couplet across the loess plateau of China. Geophys. J. Int. 117:257–264

Fassbinder JW, Stanjek H, Vali H (1990) Occurrence of magnetic bacteria in soil. Nature 343:161–163

Heller F, Liu Tungsheng (1986) Paleoclimatic and sedimentary history from magnetic susceptibility of loess in China. Geophysical Research Letters 13(11):1169–1172

Heller F, Meili B, Wang J, Huamei L, Liu Tungsheng (1987) Magnetization and Sedimentation History of Loess in the Central Loess Plateau of China. In: Liu Tungsheng (ed) Aspects of Loess Research. China Ocean Press, pp 147–163

Hilton J (1987) A simple model for the interpretation of magnetic records in lacustrine and ocean sediments. Quaternary Research 27:160–166

Kukla G, Heller F, Liu Xiu Ming, Xu Tong Chun, Liu Tungsheng, An Zhi Sheng (1988) Pleistocene climates in China dated by magnetic susceptibility. Geology 16:811–814

Kukla G, An ZS, Melice JL, Gavin J, Xiao JL (1990) Magnetic susceptibility record of Chinese Loess. Trans. of the Royal Soc. of Edinburgh: Earth Sciences 81:263–288

Le Borgne E (1955) Susceptibilite magnetique anormale du sol superficiel. Ann.Geophys. 11:399–419

Liu Tungsheng (1985a) Loess and the environment. China Ocean Press Beijing:1–251

Liu Tungsheng, An Zhisheng, Yuan Baoyon and Han Jiamao (1985b) The Loess-Paleosol Sequence in China and Climatic History. Episodes 8(1):21–28

Maher B (1986) Characterization of soils by mineral magnetic measurement. Physics of the Earth and Planetary Interiors 42:76–92

Maher B, Taylor R (1988) Formation of ultrafine-grained magnetite in soils. Nature 336:368–370

Maher B, Thompson R (1991) Mineral magnetic record of Chinese loess and paleosols. Geology 19:3–6

Maher BA, Thompson R (1992) Paleoclimatic significance of the mineral Magnetic Record of the Chines Loess and Paleosols. Quaternary Research 37:155–170

Mullins C (1977) Magnetic susceptibility of the soil and its significance in Soil Science. Journal of Soil Science 28:223–246

Nabel P (1993) The Brunhes–Matuyama boundary in Pleistocene sediments of Buenos Aires Province, Argentina. Quaternary International 17:79–85

Nabel P, Camilión C, Machado G, Spiegelman A, Mormeneo L (1993) Magneto y litoestratigrafia de los sedimentos pampeanos en los alrededores de las Ciudad de Baradero Prov. de Buenos Aires. Rev. de la Asociación Geológica Argentina 48(3–4):193–206

Oades JM, Townsend WN (1963) The detection of ferromagnetic minerals in soils and clays. Journal of soil science 14:179–187

Petersen N, von Dobeneck T, Vali H (1986) Fossil bacterial magnetite in deep-sea sediments from the South Atlantic Ocean. Nature 320:611–615

Petersen N, Weiss D, Vali H (1992) Magnetic bacteria in lake sediments. In: Lowes FJ, Collinson D, Parry J, Runcorn S, Tozer D, Soward A (eds) Geomagnetism and Paleomagnetism. Kluwer Academic Publishers, pp 231–241

Petersen N, Schmidbauer E, Strattner M, Schueler D (1993) On the occurrence of bacterial magnetite in limnic sediments and soil. Ann. Geophys.11(Suppl. 1) C 90

Rutter NW, Chlachula J, Evans ME (1995) Magnetic susceptibility and remanence record of the kurtak loess, southern Siberia, Russia. In Terra Nostra. Abstracts XIII INQUA Congress, Berlin 3–10 August 1995, p. 235

Schwertmann U, Taylor R (1977) Iron oxides. In: Minerals in soil environment. Soil.Sci.Soc.Am. 145–180

Spring S, Amman R, Ludwig W, Schleifer K, van Gemerden H, Petersen N (1993) Dominating Role of an Unusual Magnetotactic Bacterium in the Microaerobic Zone of a Freshwater Sediment. Applied and Environmental Microbiology 59:2397–2403

Thompson R, Oldfield F (1986) Environmental Magnetism. Allen and Uwin, London, p. 219

Vali H, Forster O, Amarantidis G, Petersen N (1987) Magnetotactic bacteria and their magnetofossils in sediments. Earth and Planetary Science Letters 86:389–400

van Houten FB (1982) Ancient Soils and Ancient Climates. In: Climate in Earth History. Studies in Geophysics. National Academy Press. Washington D.C., pp 112–117

Verosub KL, Ensley RA, Ulrick JS (1979) The role of water content in the magnetization of sediments. Geophysical Research Letters. 6(4):226–228

Xiuming L, Shaw J, Liu T, Heller F, Yuan B (1992) Magnetic mineralogy of Chinese loess and its significance. Geophys. J. Int. 108:301–308

Zheng H, Oldfield F, Yu L, Shaw J, An Z (1991) The magnetic properties of particle-sized samples from the Luo Chuan loess section: evidence for pedogenesis. Phys. Earth Planet. Inter. 68:250–258

Zhou LP, Oldfield F, Wintle AG, Robinson SG, Wang JT (1990) Partly pedogenic origin of magnetic variations in Chinese loess. Nature 346, 737–739

Pole-Equator-Pole Paleoclimates of the Americas: PEP 1 A Review *

V. Markgraf
Institute of Arctic and Alpine Research, University of Colorado, Boulder Colorado
80309-0450, USA and
PAGES International Project Office, Baerenplatz 2, CH-3011 Bern, Switzerland
markgraf@spot.colorado.edu, markgraf@pages.unibe.ch

Abstract: This contribution reviews the approach proposed under the IGBP-PAGES mandate, to further our understanding of global climate change and its forcing, which calls for an inter-hemispheric analysis of present and past climate change in the Americas. Modern climate and Late Quaternary paleoclimate records from the Americas are reviewed to provide the background for addressing the task of inter-hemispheric climate correlation.

6.1. Introduction

Conditions are optimal for an interhemispheric comparative study of paleoclimate change in the Americas because of the symmetry of the North and South American land areas. Along this transect the land areas have similar poleward extent, similar distribution of mountains and lowlands, and similar symmetry of atmospheric and surface oceanic circulation. Two meetings held in Boulder, Colorado in 1991, and in Panama in 1993, were convened to formulate primary questions and design a plan for implementation of an interhemispheric paleoclimate agenda. The approach chosen called for the study of linkages between terrestrial and marine paleoclimate proxy records along two intersecting transects, a north to south transect along the west coast of the Americas from Alaska to Tierra del Fuego/Antarctica complemented by an equatorial trans-Pacific transect. Contributions at a third meeting held in 1998 in Mérida, Venezuela, showed that this inter-hemispheric approach of linking climate and paleoclimate records from all respective disciplines, provides a meaningful new understanding of climate variability, present and past (Markgraf 1998).

This approach takes into account the primary forcing the tropical Pacific ocean plays on marine and terrestrial environments of the Americas. It optimizes the retrieval of information on the coupling of the ocean-atmosphere-biosphere-cryosphere at frequencies from 10^0 to 10^4 years. A different focus is required for each frequency band to determine the specific classes of interaction between the climate-system processes and environmental responses. A high-resolution focus of PEP 1 examines changes in the signature of decadal modulation of interannual variability along a pole-equator-pole transect. Differences in amplitude and frequency of regional contrasts provide insights into the mechanisms and teleconnections that control decadal modulation of interannual variability. Working back from the present using instrumental records with the same temporal resolution as the geological records, the goal is to identify if and how modes of climate variability have changed in the recent past, such as during the Little Ice Age and the Medieval Warm Period (Bradley and Jones 1993; Hughes and Diaz 1994).

The highest frequency of interannual and interdecadal climate variance in the Americas is primarily related to the El Niño/Southern Oscillation system (ENSO). Understanding changes in the behavioral patterns of ENSO across the tropical Pacific and its extra-tropical climate teleconnection patterns during the last 2000 years is the focus of study at this spectral band (Glantz et al. 1991; Aceituno, 1988; Diaz and Kiladis 1992; Meehl and Branstator 1992).

At centennial to millennial frequencies of climate change, the following differences in boundary conditions are the first-order causes of climate variability in the Americas. Changes in the seasonal distribution of insolation, such as e.g. the low (high) seasonality contrast in the early Holocene summer in the southern (northern) hemisphere would explain the dampened (intensified) monsoonal activity, recognized in numerous records (Markgraf 1993; R.S. Thompson et al. 1993). Apparent changes in North Atlantic deep-water (thermohaline) circulation (Charles and Fairbanks 1992; Bond et al. 1992; Lehman et al. 1994), did influence atmospheric concentration of CO_2 and CH_4 apparently on a global scale (White et al. 1994), which in turn could synchronize inter-hemispheric climate change. Changes in sea-level, related to continental ice build-up, in sea-surface temperatures (SST), and in sea-ice extent have been suggested to determine intensity and latitudinal location of atmospheric circulation in response to changes in the latitudinal temperature gradient (e.g. Markgraf et al. 1992; Cook et al. 1992; Rind 1998).

Interhemispheric paleoclimate correlations addresses the transequatorial extent of the respective climate forcing by identifying changes in latitudinal and elevational temperature gradients through time; changes in intensity and location of upwelling and its relation to shifts in atmospheric circulation; changes in the surface circulation of the Pacific Ocean versus the North Atlantic ocean dynamics and their respective effects on tropical and extratropical climate teleconnections in the Americas.

PEP 1 addresses these issues at selected frequency bands to:

1) Develop a latitudinal transect of paleoclimate records for the last 150 ka to analyze climate variability at resolutions that vary from 10^0 to 10^4 years.
2) Determine the latitudinal and elevational extent of climate-change patterns, including rates, frequency, amplitude, and leads and lags, with special focus on high latitude (polar) amplification of climate changes.
3) Develop multi-proxy paleoclimate data sets that link marine and terrestrial records. These studies will focus on the recent evidence of global change and current warming, the instrumental record, the historical record, the mid-Holocene, the late-glacial Holocene transition, the last full-glacial, and the last interglacial.

4) Determine the mechanisms and forcings of climate change. Identify processes that transmit changes and if any phase relationships (spatial and temporal) occur in regions sensitive to various modes of atmospheric circulation.

Atmospheric and ocean circulation analyses have identified specific regions where sensitivity to change is high. These are regions of modern discontinuities, areas where gradients are steep and seasonal contrasts are large. Although more pronounced for the Arctic than the Antarctic, polar high latitudes have steep and seasonally changing atmospheric temperature and pressure gradients and are key areas to monitor the atmospheric circumpolar vortex (polar jet stream), which in turn influences intensity and location of the mid-latitude westerlies, north and south of the equator. Also ocean temperature gradients are steep in the circumpolar regions with far-reaching consequences for lower latitude marine circulation and upwelling. Polar latitudes are also areas of deep-water formation playing a crucial role for global thermohaline circulation. Steep and seasonally shifting atmospheric temperature and pressure gradients are also characteristic for subtropical latitudes, influencing intensity and direction of the trade-wind circulation. This in turn and in addition to basin-wide thermal dynamics influences location, intensity, and the physical and biological characteristics of upwelling zones that characterize these latitudes in both North and South America (C. Peterson et al. 1993). Superposed on these meridional discontinuities in the Americas and the adjacent oceans is the ENSO climate anomaly linked to changes in the tropical Pacific Ocean heat budget. Studies of atmospheric and marine teleconnection patterns show the influence of the tropical Pacific Ocean to extend to mid- and high-latitudes of both hemispheres (Glantz et al. 1991; Diaz and Markgraf 1992; Barnston 1993). Regionally and seasonally opposing precipitation and temperature anomaly patterns are the result, such as between southern Alaska, the Northwest Pacific, the North American Southwest, and the North American Southeast; between northern South America and the Caribbean, subtropical South America, and southernmost South America. Persistence of these regional patterns will help determine the past expression of ENSO.

6.2. Modern Climate and Climate Variability in the Western Hemisphere

Satellite images offer a synoptic view of climate processes on a global scale. For the climates of the Americas, an image for February 1 1994 (Fig. 6.1) helps to illustrate the active interaction between the tropics and extra-tropics that takes place on a day-to-day basis. This figure shows large upper tropospheric cloud flares emanating from the Intertropical Convergence Zone (ITCZ) in the tropical Pacific towards southern California, and southern Chile, respectively. From eastern Brazil upper tropospheric cloud streams are directed towards the North Atlantic and Europe. These cloud flares are visual markers for the continuous upper-level export of moisture and energy from the tropics to mid-latitudes. The locations where this export occurs can shift with time as a result of day-to-day changes in the synoptic weather patterns. Figure 6.1 also shows two examples of the mid-latitudes influencing tropical weather. There are two separate outbreaks of surface cold air that can be seen in this figure moving from the mid-latitudes of both hemispheres toward the tropics.

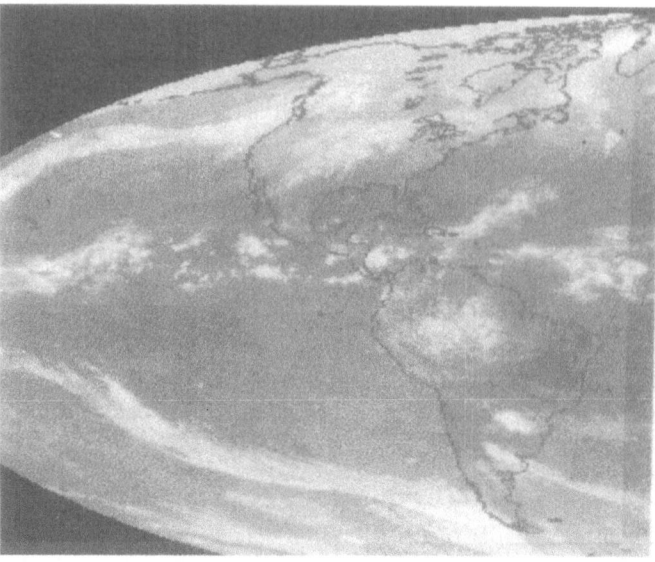

Fig 6.1. Infrared composite satellite image of November 9, 1994 (source National Oceanographic and Atmospheric Administration).

The leading edges of both of these cold-air outbreaks are marked by extensive cloudiness. When these cold-air outbreaks interact with mountains (e.g. in eastern Mexico, the Caribbean, and the western slope of the southern Andes), abundant precipitation can result.Much work remains to be done in understanding the details of where, when and how these tropical-mid latitude interactions take place in the Americas. It is, however, this type of interaction, that, when accumulated over time, creates climate anomaly patterns. Aceituno (1992) for instance, has shown evidence that patterns seen on a daily basis are often replicated on longer time scales and result in monthly and seasonal climate anomalies. It is this kind of linkage between time scales that provides a basis for extending knowledge of present-day weather to the analysis of longer time domains in climate and paleoclimate.

Present climate is composed of a "mix" of identifiable weather patterns. How this "mix" varies from month-to-month and year-to-year determines both the observed mean climate and climate anomalies. Key elements of these patterns include the strength and position of the ITCZ, subtropical high-pressure systems (and the associated trade wind belts), upper circulation features (trough and ridge patterns, polar and subtropical jet streams) and their surface manifestations (mid-latitude storm tracks, frequency of cold air outbreaks etc). Because of the steeper pole-equator temperature gradients in the Southern Hemisphere as well as the uneven distribution of land-masses between the hemispheres, the resulting circulation patterns of the hemispheres differ significantly. It is therefore not surprising that there are asymmetries in the meridional distribution of climatic elements. Figure 6.2 (Lawford 1993) shows average precipitation along the west coast of the Americas.

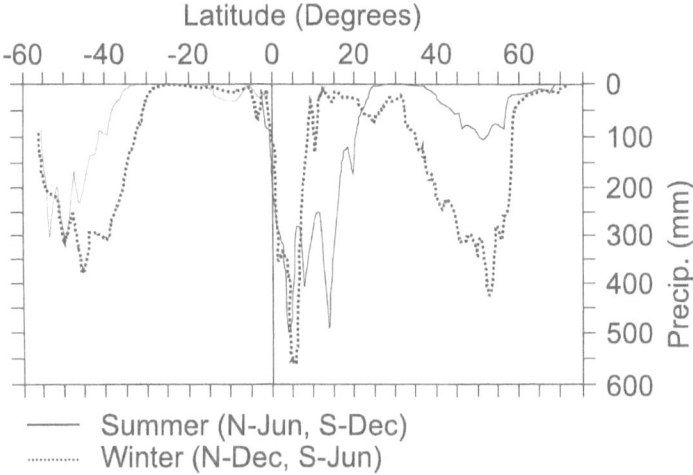

Fig 6.2. Variations with latitude of monthly precipitation for December and June for a 100 km wide transect along the west coast of the Americas from 71° N to 55° S (with permission from Lawford 1993).

Notice the ITCZ-related precipitation maximum located north of the equator and its strong seasonal variability (both in intensity and latitude), as well as the far stronger seasonal variability of the mid-latitude precipitation maxima in the northern compared to the southern hemispheres. At present, the most significant modulator of tropical climate and tropical-mid latitude interactions in the inter-decadal time scale is the El Niño-Southern Oscillation (ENSO) phenomenon. ENSO comprises a suite of anomalies in oceanic and atmospheric fields such as sea-surface temperature, salinity, upwelling intensity, surface pressure, winds and precipitation (Aceituno 1988; Aceituno et al. 1993). ENSO-related signals in these fields have been observed throughout the Americas (Fig. 6.3), based on climate analysis (Diaz and Kiladis 1992; Aceituno and Montecinos 1992; Compagnucci 1992, 1993; Cayan and Webb 1992) and historical records from Peru (Ortlieb and Macharé 1993) and Chile/Argentina (Prieto 1993). ENSO is the basic ocean-atmosphere rhythm, with oscillatory (aperiodic: 2 to 10 years) variations (Rasmusson et al. 1990). The main feature of ENSO is the remarkably coherent and out-of-phase fluctuations of atmospheric mass between the southeastern Pacific and western Pacific/Indian oceans. The different modes of the ENSO anomaly produce opposing climate signals: El Niño and La Niña.

El Niño is characterized by high SSTs in the central and eastern Pacific, negative SOI (Southern Oscillation Index), weak easterly trades, no upwelling along northwestern South America. La Niña is typified in the same locations by low SST, positive SOI, strong easterly trades, strong upwelling, etc.

ENSO-related tropical-extratropical climate teleconnection patterns have been the subject of a great deal of study in recent years. Several key areas have been identified as "centers of action" in these teleconnections (Wallace and Gutzler 1981; Diaz and Kiladis 1992).

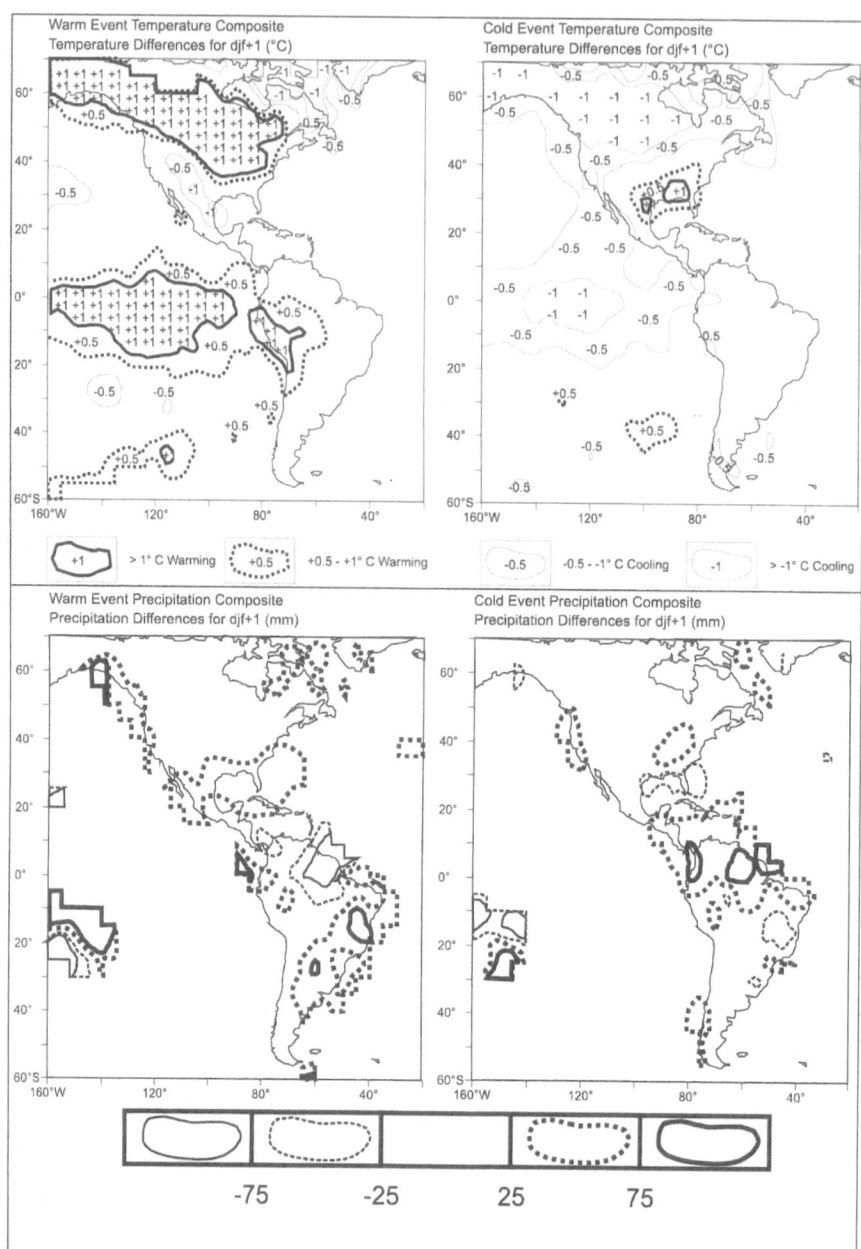

Fig 6.3. Precipitation and temperature anomaly patterns for Warm (El Niño) Events (December, January, February of year plus 1) and Cold (La Nina) Events (December, January, February of year plus 1). Positive anomalies are warmer (°C), respectively wetter (mm) and negative anomalies are colder (°C), respectively drier (mm). (Personal communication H. Diaz).

Of primary importance to the Americas in the Northern Hemisphere is the Pacific-North America (PNA) anomaly pattern (Hastenrath 1976; Murphree et al. 1992; Harr et al. 1992), and the Western Atlantic (WA) pattern. The PNA pattern is an alternating out of phase oscillation of pressure anomalies (and related weather) between the central tropical Pacific, the Aleutians, northwestern North ·America and the Southeastern US. The WA pattern is another out-of-phase oscillation of pressure between Labrador and the western tropical Atlantic. Other teleconnection patterns have been identified linking the tropics to the southern mid-latitudes (Ropelewski and Halpert 1986; Karoly 1989; Rutllant and Fuenzalida 1991; Aceituno and Montecinos 1992). An extensive analysis of global teleconnection patterns between different atmospheric and oceanic fields (Barnston 1993) points to the unique importance of tropical Pacific SST anomalies in influencing also northern hemisphere surface climate patterns especially for the winter and spring seasons. No other "predictor field" was found to correlate as strongly with mid-latitude anomalies. It is precisely this area in the tropical Pacific that is most directly influenced by ENSO fluctuations in SST. This accumulating evidence points to the importance of west–east variations in tropical Pacific SSTs to climate anomalies in the Americas generated through teleconnections (such as the PNA, WA and others). Indeed, Wolter and Timlin (1992) have incorporated west–east variations in a variety of oceanic and atmospheric indices along the tropical Pacific to improve the robustness of teleconnections to the climate of the Americas.

Fluctuations with quasi-periods of 10 to 30 years are characteristic for the 20th century climates in the Americas, and because of their temporal and spatial characteristics have been linked to changes in ENSO frequency (see e.g. Trenberth and Hurrell 1994). Mechanisms for these decadal and multi-decadal periodicities have been suggested, including internal oscillations of the atmosphere–ocean system (Schlesinger and Ramankutty 1994; Kushnir 1994), or changes in long-wave patterns alternating between both hemispheres (Allan and Haylock 1993). Relationships between ENSO and regional extratropical precipitation patterns, however, tend to be unstable when ENSO behaves aperiodic (Elliot and Angell 1988) or when the interannual ENSO-related oscillatory modes increase (Stahle et al. 1998). For example, the negative relation between SOI and rainfall in Bogota (Colombia) and Georgetown (Guiana) breaks down in the middle third of the century, when ENSO is less periodic (Aceituno and Montecinos 1992). On the other hand it is possible that a more robust ENSO index that combines several ENSO indicators will document a more robust teleconnection as well (Wolter and Timlin 1992; Dunbar and Cole 1993).

Along the pole-equator-pole transect the probably most complex region is Central America, because of the mixing of forcing, from the Atlantic (e.g. Garcia et al. 1978; Hastenrath 1984, 1991), as well as from the Pacific (e.g. Rogers 1987). Of especial interest is the analysis of intensity and direction of moisture flux across this region, because models have shown that they could influence thermohaline circulation in the North Atlantic, which in turn could influence global oceanic and atmospheric circulation (Stocker and Wright 1991).

6.3. Paleoclimate Records from the Americas

6.3.1. Low-Resolution Long-Term Late Quaternary Records

The majority of terrestrial paleoclimate proxy records from the Americas is based on changes in pollen assemblages, representative of changes in plant assemblages which in turn can be related to changes in specific climate conditions. In some cases, especially in the US Southwest, but more recently also in southern South America, plant macrofossil analyses help to further define past changes in plant assemblages (Betancourt et al. 1991; Markgraf et al. 1998). Far less common are records analyzed for multiple indicators, including pollen, diatoms, ostracodes, geochemistry, stable isotopes, etc., although it is self evident that the level of interpretation is greatly enhanced using a multidisciplinary, canonical approach.

Continuous long records on land that extend back through the entire Quaternary or at least for several glacial/interglacial cycles are rare. They are concentrated primarily in the unglaciated North American West (Tule Lake, Oregon; Searles Lake, Owens Lake, California; Willcox Playa, Arizona; San Augustin Plains, New Mexico; Lake Lahontan, Nevada; Lake Bonneville, Utah; etc.). There is only one such long record from South America (High Plains of Bogotá, Colombia); but several subsiding basins have the potential for containing the entire Quaternary record as well (Cuenca de Mexico; Lake Nicaragua, Lake Titicaca, saltlakes in western Argentina and Patagonia).

Two sites are mentioned here to exemplify the paleoclimate potential of such long records. These are from the High Plains of Bogotá (Hooghiemstra and Sarmiento 1991; Hooghiemstra et al. 1993; Hooghiemstra and Ran 1994) and Tule Lake (Adam et al. 1989; Bradbury 1991). Both records, covering the last 3 million years (my), were analyzed with a temporal resolution of 500 to 1000 years, and show comparable changes despite the vastly different setting and climate proxy analyzed (Fig. 6.4). Sabana de Bogotá, near the upper Andean treeline, was primarily analyzed for pollen, and interpreted to reflect temperature changes. Tule Lake was analyzed for diatoms, ostracodes, geochemistry, and pollen; the data were interpreted primarily in terms of past water level and water chemistry changes, that in turn can be interpreted as changes in precipitation and evaporation. Both records show three major shifts in climate modes during the last 3 my. The first shift occurred at about 2.4 my, suggesting a shift from warmer and moister to cooler and drier conditions. This point in time coincides with the initiation of large-scale glaciation in the Northern Hemisphere (Broeker and Denton 1989). The second major shift in climate mode occurred at 0.8 my after which the amplitude of climate change was greatly accentuated. The third shift occurred at 0.4 my, which marked the onset of the last two glacial-interglacial cycles that were substantially different from earlier cycles. Spectral analysis of climate changes documents comparable patterns during the entire time represented in both terrestrial records. Between 3 and 2.5 my, a 41 ky cycle dominates the spectrum. Between 2.5 and 1 my ago a 400 ky cycle, and after 1 my a 100 ky cycle dominates. All these spectral peaks are related to the Milankovitch insolation frequencies, suggesting that in fact the earth system does perceive insolation as a major low frequency forcing (Hays et al. 1976; Hooghiemstra et al. 1993).

Interestingly, the same types and frequencies of shifts are also seen in deep-sea records from the northern and central eastern Pacific (Sancetta and Silvestri 1986; Rea 1990). In the North Pacific at about 2.4 my subarctic conditions commenced and the subarctic gyre became increasingly decoupled from the subtropical gyre. Whereas the southern boundary of the subarctic water mass, the subarctic front, remained around 40° N throughout the Quaternary, the northern boundary of the subarctic front shifted stepwise towards the equator, first at 2.4 my, then at 1 my, 0.8 my, and finally at 0.35 my.

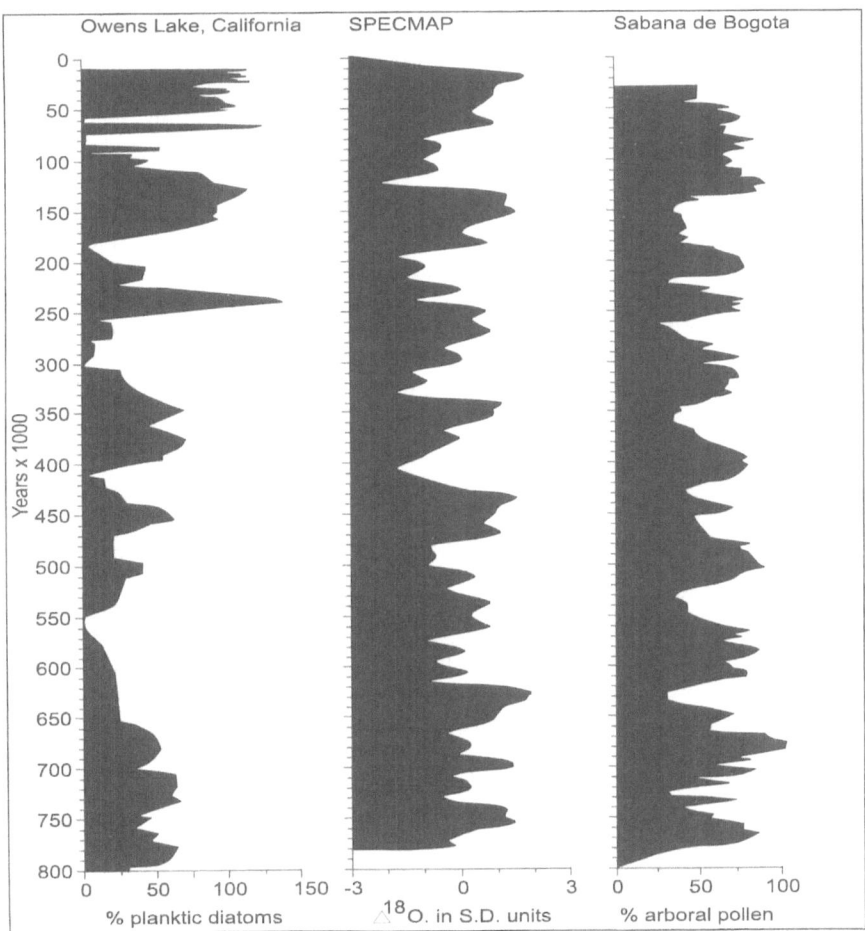

Fig. 6.4. Examples of paleoclimate records from the Americas for the last 800 000 years, from Owens Lake, California, based on diatoms (Bradbury 1997), and from Sabana de Bogotá, based on pollen records (Hooghiemstra, et al., 1993), in comparison to the SPECMAP marine isotope record (source National Oceanographic and Atmospheric Administration, National Geophysical Datacenter).

The latest shift is thought to relate to the onset of permanent ice cover in the Arctic Ocean (Sancetta and Silvestri 1986). In the central and eastern tropical Pacific, changes are recorded more or less synchronously to those in the northern Pacific. A shift to higher climate variability occurred at 0.8 my and is interpreted as an increase in intensity of the Southern Hemisphere trade winds, followed by an increase in frequency of glaciations (Rea 1990). The shift in the marine record at 0.3 my is interpreted as a latitudinal shift in the ITCZ.

6.3.2. Late Quaternary High-Resolution Records

Continuous records that extend to the previous interglacial are far more frequent and more evenly spaced throughout the Americas than those that contain the entire Quaternary, and records for the last late-glacial/postglacial are common, although the spatial coverage is still uneven. Correlations between records are reliable for the time interval back to about 50 ky, because several different dating techniques (radiocarbon, fission track, thermoluminescence, amino-acid dating, etc.) can be applied to establish the chronological frame.

Most of the records that contain upper Pleistocene sediments can be analyzed with a decadal to centennial resolution. Comparison of paleoclimatic records along the pole to pole transect shows marked differences between the timing, frequency, and amplitude of climate change. These differences suggest a number of interactive forcing parameters whose influences are spatially and temporally delimited. According to paleoclimate modeling experiments the asymmetrical accumulation of ice over the northeastern part of **North America** and its relatively slow disappearance was the strongest influence on past glacial and interglacial climates (Manabe and Broccoli 1985; Kutzbach and Guetter 1986; Kutzbach et al. 1993; Wright et al. 1993; Rind 1987; Broecker and Denton 1989). The models predicted that the first-order climate effect of the Laurentide ice dome was the development of a high-pressure center over the ice that blocked westerly zonal airflow over central and eastern North America. The high-pressure cell in turn leads to a split jet stream, with one branch positioned to the south of the ice (Kutzbach et al. 1993). As a consequence, the northwestern parts of the US came under the influence of dry easterly continental air masses (Anderson and Brubaker 1993; Ritchie and Harrison 1993; Webb et al. 1993). Under the influence of the southern branch of the split jet stream the westerly stormtracks were located farther equatorwards and the subtropical latitudes of the western United States and northern Mexico experienced conditions markedly wetter and cooler than today (R.S. Thompson et al. 1993; Bradbury 1998).

In North America the transitional interval between 12 and 9 ka is interesting in terms of climate controls because paleoclimate records and models reflect interaction and ultimate replacement of two climate forcing factors: the continuing, but diminishing influence of the Laurentide ice, and the growing influence of increased summer insolation and increased seasonality contrast. Air temperatures in North America increased and lead to the onset of summer monsoon circulation in the Southwest and atmospheric circulation, especially the location of the westerly stormtracks, continued to be displaced equatorwards (Barnosky et al. 1987; R.S. Thompson et al. 1993).

During full-glacial times the climatic influence of the Laurentide ice apparently extended as far equatorwards as **central and northern South America**. The locations of stormtracks across the North American continent were directly influenced by the Laurentide ice and indirectly through the North Atlantic Ocean, and its changes in SSTs, changes in polar oceanic heat transport, and in the thermohaline circulation. This is suggested by numerous terrestrial paleoclimatic proxy records from Mexico, central and northern South America, from high and low elevation environments (e.g. Markgraf 1989, 1993). Marine records from the eastern Pacific (Pederson et al. 1991; Byrne 1982; Horn 1985), the Gulf of Mexico and the Caribbean also show these changes (e.g. Overpeck et al. 1989; L. Peterson et al. 1994). Although the westernmost part of Mexico received some moisture even during glacial times, probably linked to the North Pacific forcing (Bradbury 1998), conditions along the eastern margin and throughout the circum-Caribbean region were apparently markedly drier than today, probably because of the dominant influence of the North Atlantic (Hubbs and Roden 1964; Bush and Colinvaux 1990; Grimm et al. 1993; Leyden et al. 1993; Hodell et al. 1991).

Interpretation of records and climate modeling experiments (Maasch and Oglesby 1990) suggests that the westerly stormtracks were located farther equatorwards and that polar air incursions were more frequent. This and the more equatorward location of the North Atlantic polar front, resulted in a more southwesterly position of the subtropical pressure cell (Bermuda/Azores high), weaker trades, and reduced precipitation in the eastern Caribbean. The ITCZ either must have shifted south, comparable to today when SSTs in the Gulf of Mexico are cold (Hastenrath 1991). More likely, however, the atmospheric system, including the ITCZ did not operate then as it does today (Bradbury 1998).

Judging from lowered snowlines in the mountains of central and northern South America (Schubert and Medina 1982; Heine 1989; Klein et al. in press) and lowered vegetation zones in the highlands and tropical lowlands (Lozano-Garcia et al. 1993; Bush and Colinvaux 1990; Liu and Colinvaux 1985; Bush et al. 1990; Hooghiemstra 1984), temperatures during glacial times must have been lower than today by about 5–10° C throughout the tropics. Recent estimates of tropical sea-surface temperatures of at least 4–6° C lower than previously proposed by CLIMAP (Lehman et al. 1992; Guilderson et al. 1994; Anderson and Webb 1993; Bush and Philander 1998) may help eliminate the long-standing debate over the cause of glacial climates in the tropics. Acceptance of cooler glacial tropical ocean temperatures would also alter our understanding of temperature and precipitation gradients in this region, which in turn would permit changes in the east-to-west freshwater flux across central America (Stocker and Wright 1991).

Higher temperatures and increased moisture after 10 ka suggest the influence of the North American ice sheet throughout Mexico and central America was replaced by insolation related forcing (Grimm et al. 1993; Hodell et al. 1991).

During glacial times the **southern-hemisphere sector of the Americas** along the Pacific as well as the Atlantic sector experienced shifts in environmental conditions that were comparable in magnitude to those of the northern hemisphere Hooghiemstra 1984; Villagran 1991; Heusser 1984; Markgraf 1993). Ice fields were greatly expanded, although perhaps greater prior to 25 ka than 18 ka (Schubert and Vivas 1993; Clapperton 1993), and aridity extended in the tropics and southern subtropics (Absy et al. 1991; Ledru 1993; Iriondo 1993). However, the circulation shifts were not symmetrical in both hemispheres, and perhaps not even symmetrical between eastern versus western South America. Apparently westerly stormtracks in the Southern Hemisphere moved equatorwards no more than 5–10° latitude (Markgraf et al. 1992; Villagran and Armesto 1993) compared to a 20° latitude shift in the Northern Hemisphere (R.S. Thompson et al. 1993). The asymetry in stormtracks suggests the temperature gradients in both hemispheres were asymmetrical. A combination of causes could explain this difference. Firstly, the Northern Hemisphere ice sheet had a greater equatorward influence on atmospheric circulation than Antarctica's augmented ice and sea-ice cover (Kutzbach et al. 1993). Secondly, the substantial changes in land area during lower eustatic sea-levels in the western tropical Pacific exerted a strong influence on the location of the southern hemisphere stormtracks (Markgraf et al. 1992). Thirdly, Southern Hemisphere insolation during full-glacial conditions was only slightly lower than today's, in contrast to the markedly lower insolation in the northern hemisphere.

During late-glacial times (14–12 ka) Southern Hemisphere atmospheric circulation ceased to be strictly zonal (Markgraf et al. 1992; Villagran, 1993) and as a consequence Antarctic cold air mass incursions east of the Andes increased in frequency, reaching into the Amazone and southern and central Brazil (Ledru 1993; Absy et al. 1991). The cold air masses also produced increased moisture in the subtropical latitudes of the Andes (Garleff et al. 1991; Grosjean 1994).

With the onset of Holocene times, the zonality of atmospheric circulation returned. However, this time the southern westerly stormtracks were focused at higher latitudes, most likely reflecting a poleward steepening of the temperature gradient. The reason for the poleward steepening of the temperature gradient is still not clear, but it possibly is related to the effect of insolation on Antarctic sea-ice. On the other hand, the apparent asynchronous climate changes between mid- and low latitudes suggest that the gradient was also variable through time and not as steep as previously thought. Records from Antarctica show air temperatures higher than today for the early Holocene (Ciais et al. 1992), implying overall reduced sea-ice in the circum-Antarctic region, and hence an air-temperature gradient located farther poleward than today. Insolation played only a minor role in early Holocene climate patterns in southern South America, because summer insolation was low and there was little seasonal contrast. Southern Hemisphere warm season insolation increased substantially during the late Holocene leading to the modern well-documented seasonal shift in atmospheric circulation and summer rains in the eastern Pampas (Zárate and Blasi 1991; Tonni et al. 1988; Gonzalez 1993). The late Holocene insolation maxima may also be part of the reason that ENSO only began to operate after 5 ka. Comparison of climate patterns in Holocene records from regions today highly related to ENSO climate forcing document no such correlation prior to about 5 ka (Martin et al. 1993; McGlone et al. 1992; Wells 1990; Villagran 1993; Sandweiss et al. 1996).

Recently evidence has accumulated from records in both hemispheres of the existence of repeated high-amplitude, short-term climate oscillations, especially during glacial (marine stages 3 and 4) and last full-glacial intervals. The evidence comes from ice cores in Greenland and Antarctica (Dansgaard-Oeschger oscillations, Dansgaard 1993), marine cores in the North Atlantic (Heinrich layers, Bond et al. 1993), and more recently also from terrestrial records in North and

South America (Florida: Grimm et al. 1993, Lake Estancia, New Mexico: Allen and Anderson 1993; Lowell et al. 1995). Shifts occurred from one climate mode to another in less than a decade, or perhaps even a season. A possible mechanism that could help explain the inter-hemispheric synchroneity of such past rapid changes in a linked ocean/atmosphere system is the salt-oscillator theory described by Bond et al. (1993) and MacAyeal (1993). This model requires internal instability of the northern-hemisphere ice sheets leading to repeated ice surges after ice build-up, which then produce marked changes in ocean salinity and deep water formation (Broeker and Denton 1989; Lehman et al. 1994).

During late-glacial times inter-hemispheric synchroneity is apparently more regionalized, suggesting more regionalized climate forcing. Crossdating of ice cores by gases trapped in the ice has shown that these late-glacial fluctuations were asynchronous between Greenland (GISP 2) and central Antarctica (Vostoc, Byrd) (Sowers and Bender 1995). In contrast, comparison of late-glacial fluctuations from a coastal Antarctic ice core (e.g. Taylor Dome, Steig et al. in press) with the Greenland ice core (GISP 2) and a peat record from Tierra del Fuego (Figge and White 1995; Markgraf and Kenny 1997) showed synchroneity. Thus, while the asynchroneity, the bi-polar temperature see-saw (Broeker 1998), apparently can be explained by paleoclimate models invoking changes in ocean heat transport (Crowley 1992), the existence of other records showing inter-hemispheric synchroneity calls for a different explanation.

6.3.3. Late Holocene High Resolution Record

The high-resolution focus of paleoclimate research for the Americas is to examine the changes in the signature of decadal modulation of interannual variability at all latitudes for specific time intervals. Temporal differences in amplitude and frequency of climate and regional contrasts should provide insights into the mechanisms and the teleconnections controlling decadal modulation of interannual variability. Studies of the last 500 to 2000 years of climate change with annual to decadal-resolution from the Americas include instrumental records and records from corals, tree-rings, varved-marine and lacustrine sediments, and ice cores and offer an example for the direction of an inter-hemispheric paleoclimate correlation-project. An early example for this approach is the correlation between below normal tree-ring growth and increase in albacore-tuna fishing records from the western United States (Clark et al. 1975). Although at that time a mechanisms was not known to explain the correlation, today we feel confident that changes in the El Niño/Southern Oscillation (ENSO) anomaly are the primary cause of climate variability in the Americas (e.g. Diaz and Markgraf 1992; Martin et al. 1993; Wells 1990).

Ocean–atmosphere interactions between the Pacific and North America have a strong expression in cumulative seasonal precipitation field over North America. Using the tree-ring network spatial resolution is excellent over the last 300 years for the conterminous United States, and excellent for the last 2000 years for western United States (e.g. Fritts 1991). The reconstructed time series of spatial anomalies of precipitation-related variables can be analyzed in terms of 1) the nature and stability of persistent modes of circulation as represented by indices, such as PNA and SO; 2) quasi-periodic phase-locking between regions sensitive to specific modes of atmospheric circulation to constrain forcing mechanisms for seasonal precipitation; and 3) moving correlations between time series on a subcontinental scale to document contraction and expansion of decadal-scale climate anomalies.

Analyzing a 17-year time series from 1968 to 1984 Ebbesmeyer et al. (1991) demonstrated a consistent step-like change at 1976 to persistently negative SOI conditions in 40 instrumental and environmental variables from a transect of sites in the eastern Pacific and the Americas. The time series included such variables as sea ice extent in the Bering Sea and Arctic, the Pacific North American Index of sea surface pressure, stream-flow patterns, and Amazon River discharge at Manacapuru. Certain other features of 20th century interdecadal climate variability have also been related to ENSO and its climate teleconnection (for another explanation see Schlesinger and Ramankutty 1994; Kushnir 1994); they include multi-year subcontinental-scale droughts such as the 1930s drought (Dust Bowl) in the Great Plains and the 1950s drought in the North American subtropics (Thomas 1962; Stahle and Cleaveland 1988; Karl and Koscielny 1982; Betancourt et al. 1993) and the general tripartite division of climate variability in the 20th century (roughly separated by the ENSO episodes of 1940/41 and 1957/58) (UNESCO 1993; Webb and Betancourt 1992; Elliott and Angell 1988; Zhang et al. 1997).

Analysis of individual time series spanning the last 100 years (Cole et al. 1993; Cayan and Webb 1992; Kahya and Dracup 1994) reveal not only some of these changes but additional shifts in the variance patterns. In the middle third of this century, e.g. variance at the annual cycle increased while the 3-year component weakened (Cole et al. 1993). A similar pattern of changes in frequency variance is found in a Patagonian ice field record (Aniya et al. 1992), in the Peruvian ice core record from Quelccaya (Thompson et al. 1992) and in tree ring records from New Mexico and northern Mexico (Michaelson and L. Thompson 1992) and Patagonia (Boninsegna and Villalba 1996). A 400-year long time series of stable isotope and coral growth data from the Galapagos Islands contains evidence for decadal modulation of interannual variability at the ENSO frequency bands as well as at periods ranging from 30 to 50 years long (Dunbar et al. 1994). Inter-hemispheric comparison of tree-ring records from Patagonia and Alaska show another feature of decadal modulation of climatic variability in the ENSO band. While for most of the 400-year long records the ENSO-related temperature correlation is excellent, the correlation disappears between AD 1860 and 1920 (Villalba et al. 1998). An increase in interannual ENSO-related oscillatory modes after AD 1850 could explain the disappearance of correlation (Stahle et al. 1998).

Based on the above results, we therefore can anticipate that the decadal modulation of interannual variability was not constant in the past, under different boundary conditions, and may not be constant in the future. Expanding time series of both the number and types of variables and linking their respective spatial and temporal signature is the most promising approach for a high-resolution inter-hemispheric climate study (see e.g. Baumgartner et al. 1989; Boninsegna and Hughes 1998; Kitzberger et al. 1998; Villalba et al. 1998).

References

Absy ML, van Cleef A., Fournier M., Martin L, Servant M., Siffedine A, Silva MFF, Soubies F, Suguio K, Turcq B, van der Hammen T (1991) Mise en évidence des quatre phases d'ouverture de la forêt dense dans le sud-est de l'Amazonie au cours des 60 000 dernières années. Comptes Rendus de l'Academie des Sciences, Serie 2, 312:673–678

Aceituno P (1988) On the functioning of the Southern Oscillation in the South American sector. Part 1: surface climate. Monthly Weather Review 116:505–524

Aceituno P (1992) Anomalías de precipitación en Chile Central relacionadas con la Oscillación del Sur: Mecanismos asociados. In: Paleo-ENSO Records, International Symposium, Lima Peru, 1992. pp. 1–6

Aceituno P, Fuenzalida H, Rosenblüth B (1993) Climate along the extratropical west coast of South America. In: H.A. Mooney, E.R. Fuentes, B.I. Kronberg (eds) Earth System Responses to Global Change: Contrasts between North and South America. Academic Press, pp. 61–69

Aceituno P, Montecinos A (1992) Analisis de la estabilidad de la relación entre la oscilación sur y la precipitación in America del Sur. In: Ortlieb L, Macharé J (eds) Paleo-ENSO Records ORSTOM–CONCYTEC, Lima, pp 7–13

Adam DP, Sarna–Wojcicki AM, Rieck HJ, Bradbury JP, Dean WE, Forester RM (1989) Tule Lake, California: the last 3 million years. Palaeogeography, Palaeoclimatology, Palaeoecology 72:89–103

Allen BD, Anderson RY (1993) Evidence from Western North America for rapid shifts in climate during the last glacial maximum. Science 260:1920–1923

Allan RJ, Haylock MR (1993) Circulation features associated with the winter rainfall decrease in southwestern Australia. Journal Climate 6:1356–1367

Anderson DM, Webb RS (1993) Ice-age tropics revisited. Nature 367:23–24

Anderson PM, Brubaker LB (1993) Holocene vegetation and climate histories of Alaska. In: Wright HE, Kutzbach JE, Webb III T, Ruddiman WF, Street–Perrott FA, Bartlein PJ (eds) Global Climates since the Last Glacial Maximum. University of Minnesota Press, pp 386–400

Aniya M., Naruse R, Shizukuishi M, Skvarca P.,Casassa G (1992) Monitoring recent glacier variations in the southern Patagonia icefield, utilizing remote sensing data. Internat. Archives Photogrammetry and Remote Sensing 29 B 7:87–94

Barnosky CW, Anderson PM, Bartlein PJ (1987) The northwestern U.S. during deglaciation; vegetational history and paleoclimatic implications. In: The Geology of North America, vol K-3: North America and adjacent oceans during the last deglaciation. The Geological Society of America, pp 289–321

Barnston A (1993) North American response to linear statistical short term climate predictor skills in the northern hemisphere. NOAA NWS, Technical Report, 49

Baumgartner TR, Michaelsen J, Thompson LG, Shen GT, Soutar A, Casey RE (1989) The recording of interannual climatic change by high-resolution natural systems: tree-rings, coral bands, glacial ice layers, and marine varves. In: Peterson DH (ed) Aspects of Climate Variability in the Pacific and the Western Americas. American Geophysical Union, Geophysical Monograph 55:1–14

Betancourt JL, Pierson EA, Aasen–Rylander K, Fairchild–Parks JA, Dean JS (1993) Influence of history and climate on New Mexico pinyon–juniper woodlands. In: Aldon EF, Shaw DW (eds) Managing Pinyon–Juniper Ecosystems for Sustainability and Social Needs. Proceedings Symposium, Santa Fe, New Mexico, April 1993. USDA Forest Service, Rocky Mountain Forest and Range Experimental Station, Gen Tech Rep. RM-236, Fort Collins, Colorado, pp 29–68

Betancourt JL, van Devender TR, Martin PS (1991) Pack Rat Middens: The Last 40,000 Years of Biotic Change. University of Arizona Press

Bond G, Heinrich H, Broeker W, Labeyrie L, McManus J, Andrews J, Huon S, Jantschik R, Clasen S, Simet C, Tedesco K, Klas M, Bonani G, Ivy S (1992) Primary production in the glacial North Atlantic and North Pacific oceans. Nature 360:245–249

Bond G, Broeker W, Johnsen S, McManus J, Labeyrie L, Jouzel J, Bonani G (1993) Correlations between climate records from north Atlantic sediments and Greenland ice. Nature 365:143–147

Boninsegna JA, Villalba R (1996) Dendroclimatology in the southern hemisphere: Review and prospects. Radiocarbon 1996 38:127–141

Bradbury JP (1991) The late Cenozoic diatom stratigraphy and paleolimnology of Tule Lake, Siskuyou Co. California. Journal of Paleolimnology 6:205–255

Bradbury JP (1997) A diatom-based paleohydrologic record of climate change for the past 800k.y. from Owens Lake, California. Geol. Soc. America, Special Paper, 317:99–112.

Bradbury JP (1998) Sources of glacial moisture in Mesoamerica. Quaternary International 43/44:97–110

Bradley RS Jones PD (1993) Climate since AD 1500. Routledge, London

Broeker WS (1998) Paleocean circulation during the last deglaciation: A bipolar seesaw? Paleoceanography 13:119–121

Broeker WS Denton GH (1989) The role of ocean–atmosphere reorganizations in glacial cycles. Geochimica Cosmochimica Acta 53:2465–2501

Bush ABG, Philander SGH (1998) The role of ocean–atmosphere interactions in tropical cooling during the last glacial maximum Science 279:1341–1344

Bush MB, Colinvaux PA (1990) A pollen record of a complete glacial cycle from lowland Panama. Journal Vegetation Science 1:105–118

Bush MB, Colinvaux PA, Wiemann MC, Piperno DR, Kam–Biu Liu (1990) Late Pleistocene temperature depression and vegetation change in Ecuadorian Amazonia. Quaternary Research 34:330–345

Byrne R (1982) Preliminary pollen analysis of Deep Sea Drilling Project Leg 64, Hole 480, cores 1–11. In: Curray, JR, Moore DG et al. (eds) DSDP Reports 64:1225 1237

Cayan DR, Webb RH (1992) El Niño/Southern Socillation and streamflow in the western United States. In: Diaz HF Markgraf V (eds) El Niño: Historical and Paleoclimatic Aspects of the Southern Oscillation. Cambridge University Press, Cambridge, pp 29–68

Charles CD, Fairbanks RG (1992) Evidence from southern ocean sediments for the effect of North Atlantic deep-water flux on climate. Nature 355:416–419

Ciais P, Petit JR, Jouzel J, Lorius C, Barkov NI, Lipenkov V, Nicolaiev V (1992) Evidence for an early Holocene climatic optimum in the Antarctic deep ice-core record. Climate Dynamics 6:169–177

Clapperton C (1993) Quaternary Geology and Geomorphology of South America. Elsevier, Amsterdam

Clark NE, Blasing TJ, Fritts HC (1975) Influence of interannual climatic fluctuations on biological systems. Nature 256:302–305

CLIMAP (1981) Climate: Long-Range Investigation, Mapping, and Prediction Project Members. Seasonal reconstruction of the earth's surface at the last glacial maximum. Map Chart Series, MC-36. Geological Society America

Cole JE, Fairbanks RG, Shen GT (1993) The spectrum of recent variability in the Southern Oscillation: results from a Tarawa Atoll coral. Science 260:1790–1793

Compagnucci R (1992) Are southern South America winters surface circulation normal during ENSO events? In: Ortlieb L, Macharé J (eds) PaleoENSO Records ORSTOM–CONCYTEC, Lima, pp 41–46

Compagnucci R, Vargas WM (1993) Snowfall in the Cordillera de los Andes and the ENSO events. In: 4th Internat. Conference on Southern Hemisphere Meteorology and Oceanography. American Meterological Society, preprints, pp .332–333

Cook ER, Bird T, Peterson M, Barbetti M, Buckley B, D'Arrigo R, Francey R (1992) Climatic change over the last millennium in Tasmania: Reconstruction from treerings. Holocene 2:205–217

Copley TB, Winograd IJ, Landwehr JM, Riggs AC (1994) 500,000-year stable carbon isotopic record from Devils Hole, Nevada. Science 160:361–365

Crowley T (1992) North Atlantic Deep Water cools the Southern Hemisphere. Paleoceanography 7:521–528

Dansgaard W, Johnsen SJ, Clausen HB, Dahl Jensen D, Gundestrup NS, Hammer CU, Hvidberg CS, Steffensen JP, Sveinbjornsdottir AE, Jouzel J, Bond G (1993) Evidence for general instability of past climate from a 250-kyr ice-core record. Nature 364:218–220

Hughes MK, Diaz HF (1994) Was there a 'Medieval Warm Period", and if so, where and when? Climatic Change 26:109–142

Diaz HF, Kiladis GN (1992) Atmospheric teleconnecions associated with the extreme phase of the Southern Oscillation. In: Diaz HF Markgraf V (eds) El Niño: Historical and Paleoclimatic Aspects of the Southern Oscillation. Cambridge University Press, Cambridge, pp 7–28

Diaz HF, Markgraf V (eds) (1992) El Niño: Historical and Paleoclimatic Aspects of the Southern Oscillation. Cambridge University Press, Cambridge

Dunbar RB, Cole JE (1993) Coral records of Ocean–Atmosphere Variability. Report from the Workshop on Coral Paleoclimate Reconstruction, La Parguera, Puerto Rico, 1992. NOAA Climate and Global Change Program, Special Report No. 10

Dunbar RB, Wellington GM, Colgan M, Glynn PW (1994) Eastern Pacific sea surface temperatures since AD 1600: the δO^{18} record of climate variability in Galapagos corals. Paleoceanography 9:291–315

Ebbesmeyer CC, Cayan DR, McLain DR, Nichols FH, Peterson DH, Redmond KT (1991) 1976 step in the Pacific climate: forty environmental changes beteen 1968–1975 and 1977–1984. In: Betancourt JL, Tharp VL (eds) Prodeedings of the Seventh Annual Pacific Climate (PACLIM) Workshop, April 1990. California Dep. Water Resources. Interagency Ecological Studies Program Technical Report 26:115–126

Elliot WP, Angell JK (1988) Evidence for changes in the Southern Oscillation relationships during the last 100 years. Journal of Climate 1:729–737

Figge RA, White JWC (1995) High-resolution Holocene and late glacial atmospheric CO2 record: Variability tied to changes in thermohaline circulation. Global Biogeochemical Sycle 9:391–403

Fritts HC (1991) Reconstructing large-scale climatic patterns from tree-ring data: a diagnostic analysis. University of Arizona Press, Tucson

Garcia O, Bosart L, DiMego G (1978). Monthly Weather Review 106:961–982

Garleff K, Schäbitz F, Stingl H, Veit H (1991) Jungquartäre Landschaftsentwicklung und Klimageschichte beiderseits der ariden Diagonale Südamerikas. Bamberger Geographische Schriften 11:359–394

Glantz MH, Katz RW, Nicholls N (1991) Teleconnections Linking Worldwide Climate Anomalies. Cambridge University Press

Gonzalez MA (1993) The exogen terrestrial system, the South American environmental system of mid-latitudes and the global changes. Report to PEP 1 Panama 1993

Grimm EC, Jacobson GL, Watts WA, Hansen BCS (1993) A 50,000-year record of climate oscillations from Florida and its temporal correlation with the Heinrich events. Science 261:198–200

Grosjean M (1994) Paleohydrology of the Laguna Lejia (Northchilean Altiplano) and climatic implications for lateglacial times. Palaeogeogrpahy, Palaeoclimatology, Palaeoecology 109:89–100

Guilderson TP, Fairbanks RG, Rubenstone JL (1994) Tropical temperature variations since 20,000 years ago: modulating interhemispheric climate change. Science 263:663–665

Harr PA, Chen J-M, Murphree T (1992) Relationship of western Pacific monsoon and tropical cyclone activity to North Pacific and North American climate anomalies. Proceedings 8th Annual Pacific Climate (PACLIM) Workshop, 1991. Technical Report No. 31, Interagency Ecological Studies Program, pp 99–106

Hastenrath S (1976) Variations in low-latitude circulation and extreme climatic events in the tropical Americas. Journal Atmospheric Science 33:202–215

Hastenrath S (1984) Interannual variability and the annual cycle: mechanisms of circulation and climate in the tropical Atlantic sector. Monthly Weather Review 112:1097–1107

Hastenrath, S. 1991. Climate Dynamics of the Tropics. Kluwer, Dordrecht

Hays JD, Imbrie J, Shackleton, NJ (1976) Variations in the earth's orbit: pacemaker of the ice ages. Science 194:1121–1132

Heine K (1989) Die letzteiszeitliche Vergletscherung mexikanischer Vulkane als Zeugnis hochglazialer Aridität in Mittelamerika. Acta Albertiana Ratisbonensia 46:93–106

Heusser CJ (1984) Late Quaternary climates of Chile. In: Vogel JC (ed) Late Cainozoic palaeoclimates of the Southern Hemisphere Balkema, Rotterdam pp 59–84

Hodell DA, Curtis JH, Jones GA, Higuera–Gundy A, Brenner M, Binford, MW, Dorsey KT (1991) Reconstruction of Caribbean climate change over the past 10,500 years. Nature 352:790–793

Hooghiemstra H (1984) Vegetational and climatic history of the high plan of Bogotá Colombia: A continuous record of the last 3.5 million years. Dissertationes Botanicae 79:1–368

Hooghiemstra H, Melice JL, Berger A, Shackleton NJ (1993) Frequency spectra and paleoclimatic variability of the high-resolution 30–1450 ka Funza I pollen record (Eastern Cordillera, Colombia). Quaternary Science Reviews 12:141–156

Hooghiemstra H, Ran ETM (1994) Late Pliocene–Pleistocene high resolution pollen sequence of Colombia: an overview of climatic change. Quaternary International 21:63–80

Hooghiemstra H, Sarmiento G (1991) Long continental pollen record from a tropical intermontane basin: Late Pliocene and Pleistocene history from a 540-meter core. Episodes 14:107–115

Horn SP (1985) Perliminary pollen analysis of Quaternary sediments from deep sea drilling project site 565, western Costa Rica. In: von Huene R, Auboin J et al. (eds) DSDP Reports 584:533–547

Hubbs CL, Roden GI (1964) Oceanography and marine life along the Pacific coast of Middle America. In: West RC. (ed) Handbook of Middle American Indians, v. 1:143–186. University of Texas Press, Austin

Iriondo M, Garcia NO (1993) Climatic variations in the Argentine plains during the last 18,000 years. Palaeogeography, Palaeoclimatology, Palaeoecology 101:209–220

Kahya E, Dracup JA (1994) The influences of type 1 El Niño and La Niña events on streamflows in the Pacific Southwest of the United States. Journal Climate, in press

Karl TR, Koscielny AJ (1982) Drought in the United States: 1895–1981. Journal Climatology 2:313–329

Karoly DJ (1989). Southern hemisphere circulation features associated with El Niño–Southern Oscillation events. Journal Climate 2:1239–1252

Kitzberger T, Swetnam TW, Veblen TT (1998) A comparison of fire histories and climatic change in the southwestern United States, and Northern Patagonia, Argentina. Abstracts: Pole–Equator–Pole Paleoclimate of the Americas. Meeting, Merida, Venezuela, 5 pp

Klein, A.G., Seltzer, G.O. and Isacks, B.L. (in press) Modern and last glacial maximum snowlines in the central Andes of Peru, Bolivia, and northern Chile. Quaternary Science Reviews.

Kushnir Y (1994) Interdecadal variations in North Atlantic sea surface temperatures and associated atmospheric conditions. Journal Climate 7:141–157

Kutzbach JE, Guetter PJ (1986) The influence of changing orbital parameters and surface boundary conditions on climate simulations for the past 18 000 years. Journal of Atmospheric Sciences 43:1726–1759

Kutzbach JE, Guetter PJ, Behling PJ, Selin R (1993) Simulated climatic changes: results of the COHMAP climate-model experiments. In: Wright HE, Kutzbach JE, Webb III T, Ruddiman WF, Street–Perrott FA, Bartlein PJ (eds) Global Climates Since the Last Glacial Maximum. University Minnesota Press, pp 24–93

Lawford RG (1993) Regional hydrologic responses to global change in Western North America. In: Mooney HA, Fuentes ER, Kronberg BI (eds) Earth System Responses to Global Change: Contrasts between North and South America. Academic Press, pp 73–99

Ledru M–P (1993) Late Quaternary environmental and climatic changes in central Brazil. Quaternary Research 39:90–98

Lehman SJ, Weicker N, Eglington T (1992) Surface-most temperature changes of the subtropical Atlantic during the last deglaciation. AGU 1992 Fall Meeting, Abstracts, EOS 73:259

Lehman SJ, Wright DG, Stocker TF (1994) Transport of fresh water into the deep ocean by the conveyor. NATO ASI Series 12:187–209

Leyden BW, Brenner M, Hodell DA, Curtis JH (1993) Late Pleistocene climate in the Central American Lowlands. In: Climate Change in Continental Isotopic Records, Geophysical Monograph 78:165–178

Liu K–B, Colinvaux PA (1985) Forest changes in the Amazon Basin during the last glacial maximum. Nature 318:556–557

Lowell TV, Heusser CJ, Andersen BG, Moreno PI, Hauser A, Denton GH, Heusser LE, Schlüchter C, Marchant D (1995) Interhemispheric correlation of late Pleistocene glacial events. Science 269:1541–1549

Lozano–Garcia MS, Ortega–Guerrero B, Caballero–Miranda M, Urrutia–Fucugauchi J (1993) Late Pleistocene and Holocene paleoenvironments of Chalco Lake, central Mexico. Quaternary Research 40:332–342

Maasch KA, Oglesby RJ (1990) Meltwater cooling of the Gulf of Mexico: a GCM simulation of climatic conditions at 12 ka. Paleoceanography 5:977–996

Manabe S, Broccoli AJ (1985) The influence of continental cie sheets on the climate of an ice age. Journal Geophysical Research 90:2167–2190

Markgraf V (1989) Palaeoclimates in Central and South America since 18,000 BP based on pollen and lake-level records. Quaternary Science Reviews, 8:1–24

Markgraf V (1993) Climatic history of Central and South America since 18,000 yr B.P.: Comparison of pollen recores and model simulations. In: Wright HE, Kutzbach JE, Webb III T, Ruddiman WF, Street–Perrott FA, Bartlein PJ (eds) Global Climates Since the Last Glacial Maximum. University Minnesota Press, pp 357–385

Markgraf V (1995) PEP 1: The Americas Transect. In: Paleoclimates of the Northern and Southern Hemispheres. The PANASH project. The Pole Equator Pole Transects. PAGES Series 95–1, pp.23–41

Markgraf V (1998) Present and past inter-hemispheric climate linkages in the Americas and their societal effects. EOS in press

Markgraf V, Betancourt J, Rylander KA (1997) Late-Holocene rodent middens from Rio Limay, Neuquen Province, Argentina. Holocene 7:325–329

Markgraf V, Dodson JR, Kershaw AP, McGlone MS, Nicholls N (1992) Evolution of late Pleistocene and Holocene climates in the circum-South Pacific land areas. Climate Dynamics 6:193–211

Markgraf V, Kenny R (1997) Character of rapid vegetation and climate change during the late-glacial in southernmost South America. In: B Huntley B, Cramer W, Morgan AV, Prentice HC, Allen JRM (eds) Past and Future Rapid Environmental Changes: The spatial and evolutionary Responses of Terrestrial Biota. NATO ASI Series 47:81–90

Martin L, Fournier M, Mourgiart P, Sifeddine A, Turcq B, Absy ML, Flexor J–M (1993) Southern Oscillation signal in South American Palaeoclimatic data of the last 7000 years. Quaternary Research 39:338–346

MacAyeal DR (1993) Binge/purge oscillations of the Laurentide ice sheet as a cause of the North Atlantic's Heinrich events. Paleoceanography 8:775–784

McGlone MS, Kershaw AP, Markgraf V (1992) El Niño/Southern Oscillation climatic variability in Australasian and South American paleoenvironmental records. In: Diaz HF, Markgraf V (eds) El Niño: Historical and Paleoclimatic Aspects of the Southern Oscillation. Cambridge University Press, Cambridge, pp 435–462

Meehl GA, Branstator GW (1992) Coupled climate model simulation of El Niño/Southern Oscillation: implications for paleoclimate. In: Diaz HF, Markgraf V (eds) El Niño: Historical and Paleoclimatic Aspects of the Southern Oscillation. Cambridge University Press, Cambridge, pp 69–92

Michaelson J, Thompson LG (1992) A compariosn of proxy records of El Niño/Southern Oscillation. In: Diaz HF, Markgraf V (eds) El Niño: Historical and Paleoclimatic Aspects of the Southern Oscillation. Cambridge University Press, Cambridge, pp 323–348

Murphree T, Chen J–M, Harr P (1992) The anomalies in North American climate: the south Asian-tropical west Pacific connection. Proceedings 8th Annual Pacific Climate (PACLIM) Workshop, 1991. Technical Report No. 31, Interagency Ecological Studies Program, pp 179–186

Ortlieb L, Macharé J (1993) Former El Niño events: records from western South America. Global and Planetary Change 7:181–202

Overpeck JT, Peterson LC, Kipp N, Imbrie J, Rind D (1989) Climate change in the circum North Atlantic during the last deglaciation. Nature 338:553–557

Pederson TF, Nielsen B, Pickering M (1991) Timing of late Quaternary productivity pulses in the Panama Basin and implications for atmospheric CO_2. Paleoceanography 6:657–677

Peterson CH, Barber RT, Skilleter GA (1993) Global warming and coastal ecosystem response: how northern and southern hemispheres may differ in the eastern Pacific ocean. In: Mooney HA, Fuentes ER, Kronberg BI (eds). Earth System Responses to Global Change: Contrasts between North and South America. Academic Press, pp 17–34

Peterson LC, Lin HL, Overpeck JT, Murray DW (1994) Synchronous climate change in the North Atlantic and Circum–Caribbean region: the conveyor belt connection. AGU Spring Meeting 1994. Abstracts, EOS

Prieto MR (1993) Reconstrucción del clima de America del Sur mediante fuentes historicas. Estado de la cuestión. Abstracts, Workshop Project 341 IGCP/IUGS/UNESCO: Southern Hemisphere Paleo- and Neoclimates. Mendoza, Argentina

Rasmusson EM, Wong X, Ropelewski CF (1990) The biennial component of ENSO variability. Journal of Marine Systems 1:71–96

Rea DK (1990) Aspects of atmospheric circulation: the late Pleistocene (0–950,000 yr) record of eolian deposition in the Pacific Ocean. Palaeogeography, Paleoclimatology, Palaeoecology 78:217–227

Rind D (1987) The components of the ice age circulation. Journal Geogphysical Research 92:4241–4281

Rind D (1998) Latitudinal temperature gradients and climate change. Journal Geophysical Research 103:5943–5971

Ritchie JC, Harrison SP (1993) Vegetation, lake levels, and climate in western Canada during the Holocene. In: Wrigth HE, Kutzbach JE, Webb III T, Ruddiman WF, Street–Perrott FA, Bartlein PJ (eds) Global Climates Since the Last Glacial Maximum. University Minnesota Press, pp 401–414

Rogers JV (1987) Variability in clouds and precipitation over South America and tropical Atlantic. Tropical Ocean Atmosphere Newsletter (University of Miami), 37:7–10

Ropelewski CF, Halpert MS (1986) North American precipitation and temperature patterns associated with El Niño/Southern Oscillation (ENSO). Monthly Weather Review 114:2352–2362

Rutllant J, Fuenzalida J (1991) Synoptic aspects of the Central Chile rainfall variability associated with the Southern Oscillation. Internatl. Journal Climatology 11:63–76

Sancetta C, Silvestri SM (1986) Pliocene–Pleistocene evolution of the North Pacific ocean–atmosphere system, interpreted from fossil diatoms. Paleoceangraphy 1:163–180

Sandweiss DH, Richardson III JB, Reitz EJ, Rollins HB, Maasch KA (1996) Geoarchaeological evidence from Peru for a 5000 years B.p. onset of El Niño. Science 273:1531–1533

Schlesinger ME, Ramankutty N (1994) An oscillation in the global climate system of periods of 65 to 70 years. Nature 367: 723–726

Schubert C, Medina E (1982) Evidence of Quaternary glaciation in the Dominican Republic: some implications for Caribbean paleoclimatology. Palaeogeography, Paleoclimatology, Palaeoecology 39:281–294

Schubert C, Vivas L (1993) El Cuaternario de la Cordillera de Mérida, Andes Venezolanos. Universidad de los Andes, Fundación Polar, Mérida, Venezuela.

Sowers T, Bender M (1995). Climate records during the last deglaciation. Science 269:210–214

Stahle DW, Cleaveland MK (1988) Texas drought history reconstructed and analyzed from 1698 to 1990. Journal Climate 1:59–74

Stahle DW, D'Arrigo RD, Krusik PJ, Cleaveland MK, Cook ER, Allan RJ, Cole JE, Dunbar RB, Therrell MD, Gay DA, Moore M, Stokes MA, Burns BT, Villaneuva–Diaz J, Thompson LG (1998) Experimental dendroclimatic reconstruction of the Southern Oscillation. Bulletin American Meteorological Society (in press)

Steig EJ, Brook EJ, White JWC, Sucher CM, Bender ML, Lehman SJ, Morse DL, Waddington ED, Clow, GD (in press). Synchronous climate changes in Antarctica and the North Atlantic. Science

Stocker TF, Wright DG (1991) Rapid transitions of the ocean's deep circulation induced by changes in surface water fluxes. Nature 351:729–732

Thomas HE (1962) The meteorological phenomenen of drought in the Southwest: 1942–1956. U.S. Geological Survey Professional Paper, No. 342A

Thompson LG, Mosley–Thompson E, Thompson PA (1992) Reconstructing interannual climate variability from tropical and subtropical ice-core records. In: Diaz HF Markgraf V (eds) El Niño: Historical and Paleoclimatic Aspects of the Southern Oscillation. Cambridge University Press, Cambridge, pp 295–322

Thompson RS, Whitlock C, Bartlein PJ, Harrison SP, Spaulding GW (1993) Climatic changes in the Western United States since 18,000 yr B.P. In: Wrigth HE, Kutzbach JE, Webb III T, Ruddiman WF, Street–Perrott FA, Bartlein PJ (eds) Global Climates Since the Last Glacial Maximum. University Minnesota Press, pp 468–513

Tonni EP, Bargo MS, Prado JL (1988) Los cambios ambientales en el Pleistoceno tardío y Holoceno del sudeste de la provincia de Buenos Aires a través de una secuencia de mamiferos. Ameghiniana, 25:99–110

Trenberth KW, Hurell JW (1994) Decadal atmosphere–ocean variations in the Pacific. Climate Dynamics 9:303–319

UNESCO (1993) Oceanic interdecadal climate variability. Intergovernmental Oceanographic Commission, Technical Series, 40

Villagran C (1991). Historia de los bosques templados del sur de Chile durante el Tardiglacial y Postglacial. Revista Chilena Historia Natural 64:447–460

Villagran C (1993) Una interpretación climática del registro palinológico del último ciclo glacial-postglacial en Sudamérica. Bulletin Institut Français Etudes Andines 22:243–258

Villagran C, Armesto JJ (1993) Full and Late glacial paleoenvironmental scenarios for the west coast of southern South America. In: Mooney HA, Fuentes ER, Kronberg BI (eds). Earth System Responses to Global Change: Contrasts between North and South America. Academic Press, pp 195–208

Villalba R, D'Arrigo RD, Cook ER, Wiles G, Jacoby GC (1998) Decadal-scale climatic variability along the extra-tropical western coast of the Americas over past centuries inferred from tree-ring records. Abstracts: Pole–Equator–Pole Paleoclimate of the Americas. Meeting, Merida, Venezuela 8 pp

Wallace JM, Gutzler DS (1981) Teleconnections in the geopotential height field during the northern hemisphere winter. Bulletin American Meteorlogical Society 109:784–812

Webb III T, Ruddiman WF, Street–Perrot FA, Markgraf V, Kutzbach JE, Bartlein PJ, Wright HE Jr., Prell WL (1993.) In: Wright HE, Kutzbach JE, Webb T III, Ruddiman WF, Street–Perrott FA, Bartlein PJ (eds), Global Climates Since the Last Glacial Maximum. University Minnesota Press pp 514–535

Webb RH, Betancourt JL (1992) Climatic variability and flood frequency of the Santa Cruz River, Pima County, Arizona. U.S. Geological Survey Water-Supply Paper, No. 2379

Wells LE (1990) Holocene history of the El Niño phenomenon as recorded in floodsediments of northern coastal Peru. Proceedings 6th Annual Pacific Climate (PACLIM) Workshop, 1989. Technical Report No. 23, Interagency Ecological Studies Program, pp 141–144

White JWC, Ciais P, Figge RA, Kenny R, Markgraf V (1994) A high-resolution record of atmospheric CO_2 content from carbon isotopes in peat. Nature 367:153–156

Wolter K, Timlin MS (1992) Monitoring ENSO in COADS with a seasonally adjusted principal component index. Proc. 17th Climate Diagnostics Workshop, Oklahoma 1992, pp 52–57

Wright HE, Kutzbach JE Jr, Webb III T, Ruddiman WF, Street–Perrott FA, Bartlein, PJ (eds) (1993) Global Climates since the Last Glacial Maximum. University of Minnesota Press

Zárate MA, Blasi A (1991) Late Pleistocene and Holocene loess deposits of the southeastern Buenos Aires province, Argentina. Geojournal, 24:211–22

Zhang Y, Wallace JM, Battisti DS (1997) ENSO-like interdecadal variability: 1900–93. Journal of Climate 10:1004–1020

Chapter 2: Historical and Modern Climates

This chapter focuses on case studies in the reconstruction of modern and historical climates. Emphasis is put here on both methodology and reconstruction.

Climate model data are of course useful for the assessment of climate change. For local questions however, the knowledge of young past analogs from either colder or warmer situations might in many cases show *how* an environment looks in case warmer or older situations occur again; for instance, *where* floods would be more likely, *where* droughts might occur, *which* vegetation is most likely to represent the ecological equilibrium (and thus grow better). Historical archives do not often deal with the physical parameters of climate change, but rather with the impact of climate change on agriculture and economy (famine, floods, lost harvests). Therefore, spatial and synoptic usage of historical sources, used not to address physical data, but rather to study past and potential ecosystems, is a treasure with a value that is in many cases underestimated. The study of Prieto et al. focuses on the last few centuries in South America. This is an example of the potential of historical science for the assessment of environmental situations.

Often the water supply of regions and cities relies on glaciers. A careful study of very young fluctuations of glacier lines in the Massive of Aconcagua (Llorens and Leiva) shows that even in regions of limited expanse, a considerable spatial variability of glacier-line fluctuations can be observed.

The question of glacier lines is one aspect of climate change. Glacier lines are equilibria with many factors, from precipitation in the catchment area to melting. Therefore the most important question for the ecosystem is not the extent of a glacier, but the availability of moisture. Wingenroth presents an example from a Quebrada in the Mendoza area of Argentina. This study covers many changes of moisture availability in the recent past, along with other aspects, and as the Quebrada B. Matienzo is studied with different methods by various authors, presents a key-site for high-resolution integrated studies of climate changes.

Previous works in this chapter use case studies with geological parameters or environmental conditions. Similar approaches are also possible to establish forecast systems for environmental hazards. Norte presents such a case study that can be applied to other parts of the world.

A topic of ongoing concern is ENSO, especially as precipitation changes are of great economical importance (floods, landslides) in many parts of the world. Compagnucci examines the impact of ENSO events on the hydrological system in the Andean region. Her study also includes the pre-industrial period. In this context, the attention of the reader is drawn to consider the works of Compagnucci, Villalba, Jacovkis, Prieto, Markgraf and Wingenroth synoptically as the same or similar time interval is addressed with different methods.

While many methods for reconstructing past climates and modeling future climates exist, one prime question for ongoing economical planning (for example the installation of powerstations, wind-turbines and other agricultural questions) is, "has the climate already changed?" If yes, "how?" Barros et al. present a study that shows synoptically ongoing climate change in South America. Areas of increased precipitation and areas of decreased precipitation can both be clearly seen. Although anthropogenic impact could be suspected, the presented tools could, independent of the cause of climate change, be applied to other areas of the world, especially in connection with both environmental analysis and investment planning as well. Of special importance in this study is the decrease of precipitation in areas adjacent to the Southern Amazonia.

Archival Evidence for some Aspects of Historical Climate Variability in Argentina and Bolivia During the 17th and 18th Centuries

M. R. Prieto, R. Herrera, P. Dussel
Unidad de Historia Ambiental, Departamento de Dendrocronologia e Historia Ambiental, IANIGLA–CRICYT, Casilla de Correo 330, 5500 Mendoza, Argentina
mprieto@lab.cricyt.edu.ar

Abstract: An overview of past climatic changes inferred from historical sources is presented for southern South America with particular reference to Argentina and Bolivia. The nature of historical sources in these countries is discussed and major climatic events in Argentina and Bolivia during the 17th and 18th centuries are presented. Historical records from Mendoza and Buenos Aires, Argentina, indicate that climate during the 17th was relatively stable with a lower proportion of climatic extreme events. In contrast to the 17th century, climatic conditions during the 18th century in the regions of Mendoza, Buenos Aires, northwestern Argentina and Bolivia were variable with marked wet and dry cycles. Between 1780 and 1810, simultaneous with great droughts in the Andean area, abundant rainfalls and extraordinary floods were frequently recorded in the Río de la Plata basin. By comparing the historical ENSO record developed by Quinn with our information about droughts and floods, we found some relationships between the ENSO occurrence and exceptional floods in northeastern Argentina, extraordinary droughts in northwestern Argentina and Bolivia, and heavy precipitation in Mendoza.

7.1. Introduction

The international Geosphere–Biosphere Program recently has begun to focus on comparisons of interhemispheric climate variability at key temporal scales from interannual fluctuations evident in instrumental weather records or tree-rings series to glacial interglacial cycles documented in a variety of proxy records. South American historians have underestimated the value of archival evidence, which can be used to extend the instrumental record(see also PAGES 1992).

Traditionally, the scientific community is interested in climatic studies only under socio–economic aspects, especially extreme events (droughts and floods). The exploitation of the archives as a tool to better understand the climate system is generally not a prime objective of history; however, some climatologists have made good use of historical sources. Promising examples from South America include Lamb (1972), Hastenrath (1981), Cobos and Boninsegna (1983), Politis (1983), Hamilton and Garcia (1986), and Quinn and Neal (1992).

We review some of the key South American literature, using examples of our own work in Argentina and Bolivia to illustrate the power of archival data when reconstructing historical climate variability. Our work is based on standard historical techniques, such as Content Analysis (Baron 1982) that captures only the highest quality of observations.

The objective consists in tracing climatic information (droughts, precipitation, winds, temperature) from historical sources to obtain the closest climate reconstruction since the middle of 16th century, when the Spaniards arrived in the territory. This information has been used to build time series of various meteorological and hydrologic variables and the phenomena derived from them. These series need to be not only extensive but also accurate, allowing for objective analysis and detection of fluctuations, climatic tendencies and extreme meteorological phenomena (extraordinary droughts, major flooding) that may have influenced the economy and society.

7.2. Nature of Historical Sources from South America

In South America and specifically in Argentina, there is a greater amount of climatic information during the colonial times (16th, 17th, 18th centuries) than the 19th century. In the latter period, which ends around 1880, political anarchy and the subsequent administrative disorganization hindered the development of an adequate historical record. It was only in the two last decades of the 19th century that there was continuity and greater interest, manifested by the record of meteorological events. Newspapers begin to systematically report on weather phenomena in relation to agriculture and cattle around 1850.

During Colonial times all hierarchical levels of the Spanish administration accumulated documentation in local, regional and national archives, mainly in the Archivo General de Indias (AGI), in Sevilla, Spain. Climatic information from the Colonial period did not appear simultaneously all over South America. Historical information is associated with the moment of the exploration, conquest and colonization of each particular region.

At the end of the 15th century the Spaniards arrived at Antilles and after half a century of constant and rapid advance they ruled most of the southern part of the New World. Curiously, southern South America, the most inaccessible part of the region, provides the earliest information, recorded by Magellan when passing the Strait bearing his name in 1520. As the conquests advanced, new lands were incorporated and new cities were founded. Valuable historical information started to emerge from cities such as Santa Marta (1527) and Cartagena de Indias (1553) in Colombia; Lima (1532) in Peru (Jimenez de la Espada 1965); Santiago (1541) in Chile; Asuncion (1541) in Paraguay; and Mendoza, Salta, Jujuy, Tucumán (1561–1580) and Buenos Aires (1580) in Argentina (Morales Padron 1973).

As a consequence of requirements imposed by the Spanish Government during the Colonial period, homogeneous information was produced throughout the empire. Historical climatic information can be retrieved from four large groups of sources:

a)   Specific reports on climatic events such as the "Semestral Relations of waters, harvests and other particulars" from the 18th Century.
b)   Chronicles and traveler's descriptions that show a panoramic view of a time or period from determined places. The analysis of these documents provides

general information about precipitation regimes, seasonal weather characteristics and streamflow.

c)   The reports of Spanish scientific expeditions that described the environment as accurately as possible, with the help of the first precision scientific instruments. Antonio de Cordoba, Malaspina and Felix de Azara were the first Spanish scientists who visited South America. Special attention shall be paid to the rich information available in logbooks from Spanish ships.

d)   Another set of documents focuses on aspects of the government procedures and population's daily life: records of the town council, account books or books on harvest expenditures, notes provided by the "haciendas" (ranch or farm), instructions about food supplies, compilations of agricultural products and cattle prices, lawsuits for land possession. The letters and reports that were sent to the Central Government of Spain by viceroys, governors, militaries and another colonial functionaries are significant sources. Extreme climatic events affecting the regional economy such as droughts, floods, abundant precipitation, were reported with major frequency.

Continuous and homogeneous sources from the same origin like annals or diaries from priests, religious orders, militaries or particulars, provide a more complete information including not only years with extreme events, that are usually mentioned, but years considered within the range of climatic normality. A large number of personal reports written from the members of religious orders are also available. These include the Jesuitical "Cartas Anuas", which are annual reports about the activity of the "Society of Jesus" during the 17th and 18th century. Among the essential sources for the history of climate in Latin America are the "Actas Capitulares" (Mayoralty), which weekly recorded the colonial cities activities. The climatic phenomena occurred during the week appear written down, especially those that had brought adverse economical consequences.

7.3. Major Climatic Events in Argentina and Bolivia during the 17th and 18th Centuries

To give an impression about the power of historical methods the following chapter describes historical climatic phenomena at selected places (Figs. 7.1–7.4). Finally the results are summarized in synoptic maps that give an impression about the climatic variability in southern South America during the last centuries.

7.3.1 Climate during the 17th Century

Climatic fluctuations in the western areas of Argentina inferred from historical records indicate that humid and cold climatic conditions were more frequent in the 17th than in the 20th century. Climate during the 17th century in the semiarid tempered Piedmont and extrandean Plains was relatively stable having a lower proportion of climatic extreme events. In Mendoza, only two extraordinary floods (1662 and 1680) were reported during this century. Severe droughts were also infrequent (Prieto 1983, Prieto et al. 1998b).

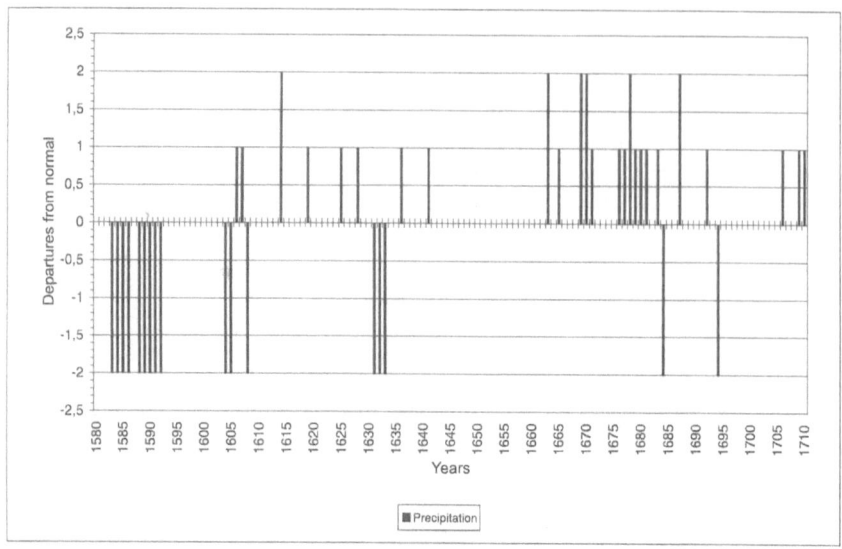

Fig. 7.1. Top: Summer temperature and precipitation variations in Mendoza, Argentina, during the 17th century. Dashed lines are used for temperature: 1 = warm, 0 = normal, –1 = cold, –2= very cold. Gaps indicate that no data is available. Bars are used for precipitation: 2 = very wet, 1 = wet, 0 = normal, –1 = dry, –2 = very dry.

Phenomenological data were used to infer intraseasonal variations in temperature and precipitation. Delayed or anticipated harvests (particularly grapevine and wheat) allowed the characterization of springs and summers climatic variations. Delays in vineyards harvest are related to cool and wet spring–summer conditions which do not permit a timely ripening of grapes. Wheat harvests in Mendoza are more dependent on temperature variations. However, fungus diseases in wheat were associated with rainy conditions during the growing season. Reports on plague of locust were also used as a spring humidity indicator. These data, together with direct information from the "Actas Capitulares" and other historical sources, were used to deduce records of temperature and precipitation fluctuations (Fig. 7.1).

Rainy conditions prevail in Mendoza during 17th century, from 1640 to 1691. Severe droughts were reported in 1626 and 1627, and at the end of the 17th century in 1694 and 1698. From a total of 41 years with temperature data, 17 were classified as cold years. Colder conditions during this century are consistent with a high frequency of early frosts. The climatic records appear to be consistent with descriptions from the same century referring to climate, such as this one from 1625:

"...what one admires most is that being the Cuyo lands...at the same height as in Chile, its /Cuyo/ climate is so different from the climate of Chile and Europe, that even though the winter is terribly cold and the frosts are so severe that the water freezes under cover and the animals usually freeze to death in the fields, nevertheless not a drop of rain falls during the winter..." (Ovalle 1889).

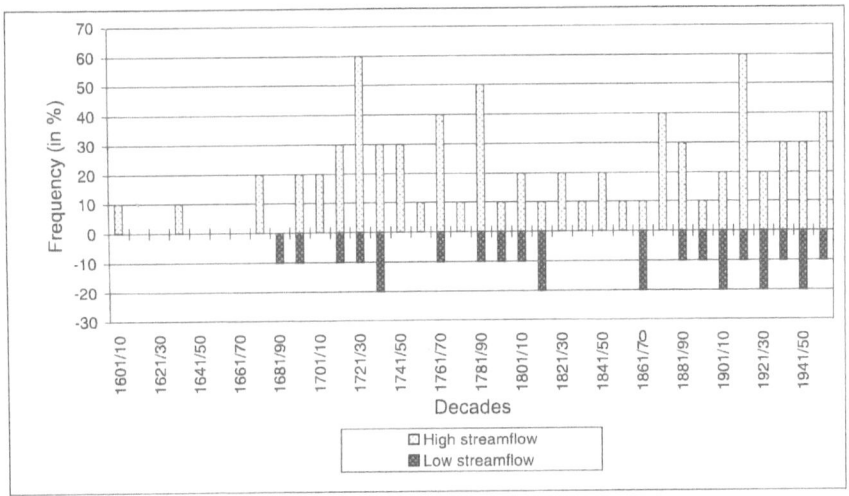

Fig. 7.2. Mendoza River high- and low-frequency streamflows by decades for the 17th, 18th, 19th and 20th centuries. Light dotted bars = high streamflow; Dark dotted bars = low streamflow.

The cold and frost that occurred in the western plains during 17th century could be correlated with the Andean events in the same interval.

We have analyzed Mendoza River runoff. This river springs in the Argentine–Chilean Central Andes (32° 27' to 33° 20' S). Glaciers and the snowmelt that accumulated during the winter feed the three main courses forming the river. The fluctuations in the streamflow respond to the variations in the winter snowfall. Historical series shows the existing of periods having higher or lesser streamflow during 350 years (Prieto et al. 1999).

The decadal-resolution series points out a long period without extreme events (1601–1670) and a gradual increase in the number of high streamflow since the middle of the 17th century (Fig. 7.2). The long-lasting interval up to 1670 with a high frequency of regular runoff would be concurrent with cold periods, probably related to the Little Ice Age that would have caused less snow melting due to low summer temperatures. Extraordinary and large streamflow events started in the 1671–1680 decade, peaked 1721–1730, and gradually decreased until 1761–1770.

Historical records from northwestern Argentina show a severe drought from 1580–1641. Since 1663 precipitation was abundant. Two droughts in 1684 and 1694 interrupted the wet period, which ends at the beginning of the 18th century (Figs. 7.3 and 7.5). Concurrent with this wet interval, flooding was common in the region. As a consequence of continuous flooding, some cities from the northwestern Argentina, such as San Miguel de Tucumán in 1678 (AGI, Charcas 121, 1679) had to be relocated.

In the semiarid Chaco, great droughts at the ends of 16th century occurred, followed by a long interval with a predominance of wet years during 17th century and part of the 18th century. Abundant precipitation forced the Salado River into a new bed in 1703 and 1709 (Herrera and Dussel, 1992).

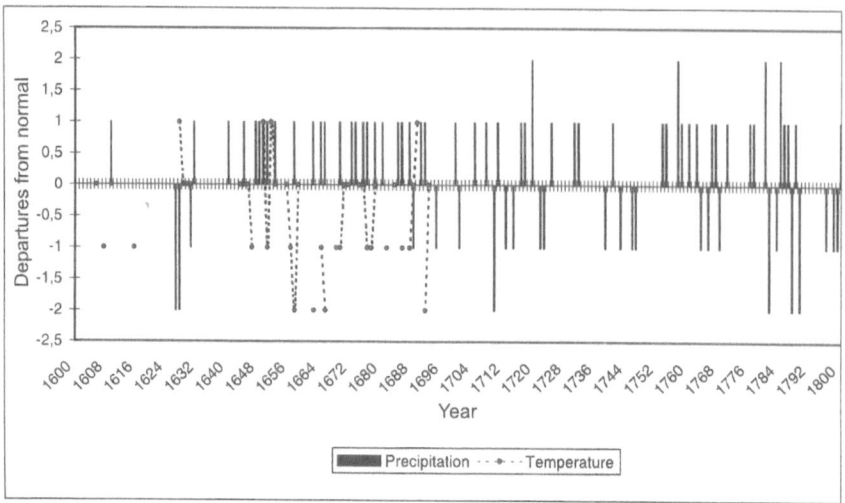

Fig. 7.3. Precipitation variations in northwestern Argentina (1580–1710) derived from different colonial historical sources. Classification: 2 = very wet, 1 = wet, 0 = normal, –1 = dry, –2 = very dry.

7.3.2. Climate During the 18*th* Century

In contrast to 17th century, climatic conditions during the 18th century in Mendoza were variable with marked wet and dry cycles (Prieto 1983). Periods of low precipitation were recorded 1709–1715, 1736–1751, and from the middle 1780's to 1800 (Fig. 7.4). Periods with abundant precipitation were also common, particularly 1750–1785.

Droughts prevail during 18th century, being particularly severe 1770–1803. In contrast, climatic conditions over the 16th and 17th centuries were more humid, similar to the current climate. The work of Politis (1983) about the Buenos Aires Pampas confirms this result. He highlights the long dry period during the period 1770–1803. According to this author, droughts during the 18th century would respond to large-scale climatic changes, concurrent with glacier advances (Little Ice Age) in the Patagonian Andes.

To determine if the climatic anomalies recorded in Cuyo and Buenos Aires were regional in nature or they responded to large-scale (continental) climatic changes, we searched for climatic information along the Central Andes from southern Peru to Central Argentina. Climatic variations from the adjacent plains (llanuras) to the Andes were also recorded. Tandeter and Wachtel (1984) showed a close correlation between food prices and climatic variations in Potosi and Charcas during the 18th century. Generally high food prices were simultaneous with several droughts. Twenty-four years of several droughts occurred in the first four decades of the 18th century, concurrent with anomalous climatic conditions in Mendoza and Cordoba, Argentina.

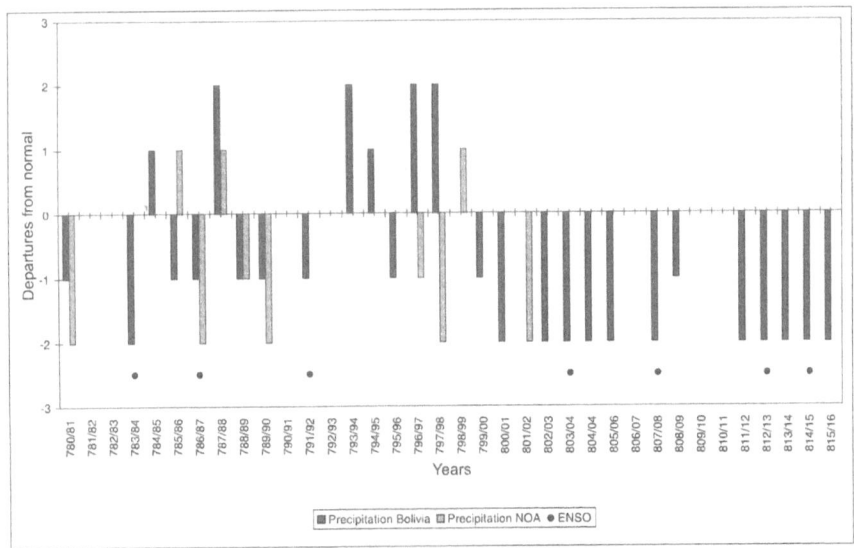

Fig. 7.4. Grey bars = precipitation variations in northwestern Argentina, 1780–1802. Black bars = precipitation variations in Bolivia, 1780–1816. Both curves are derived from the "Semestral Relations of waters, harvests and other particulars". Classification: 2 = very wet, 1 = wet, 0 = normal, –1 = dry, –2 = very dry.

After 1742 food price decreased. No extreme climatic events were recorded during the middle 18th century. Droughts were more frequent 1780–1805. In Santiago del Estero, Argentina, abundant precipitation from 1750 to around 1770 reverses to become drought conditions during the 1780's (Herrera and Dussel 1992).

The extreme climatic events, with drought predominant, show a wide spatial distribution 1780–1805. The "Semestral Relations of waters, harvests and other particulars" have permitted the study of these events both temporally and spatially (Prieto and Herrera 1992).

Based on the "Relations", two climatic series were developed, one for central southwestern Bolivia and another for the northwestern Argentina (Fig. 7.4). Records include information from Jujuy, Salta and Tucumán, whereas the Bolivian records include data from La Paz, Cochabamba, La Plata (Sucre), Potosi, Oruro and Tarija. The historical information is grouped into five categories: very dry (–2), dry (–1), normal (0), wet (1) and very wet (2). Normal years were considered those when no extreme events were recorded. Normality moves within a very wide range owing to the characteristics of the information used. This is because of the tendency of the population to register only catastrophic climatic events. The decade of 1780 was particularly dry. In 1781, Jeronimo Matorras wrote in his journal: "...this year the lack of rains has been felt more than others that not only the province of Tucumán, but also the provinces of Peru have suffered these plagues, as the Villa de Potosi, as I have news that the grinding mills had to be stopped at night" (Prieto and Herrera 1992).

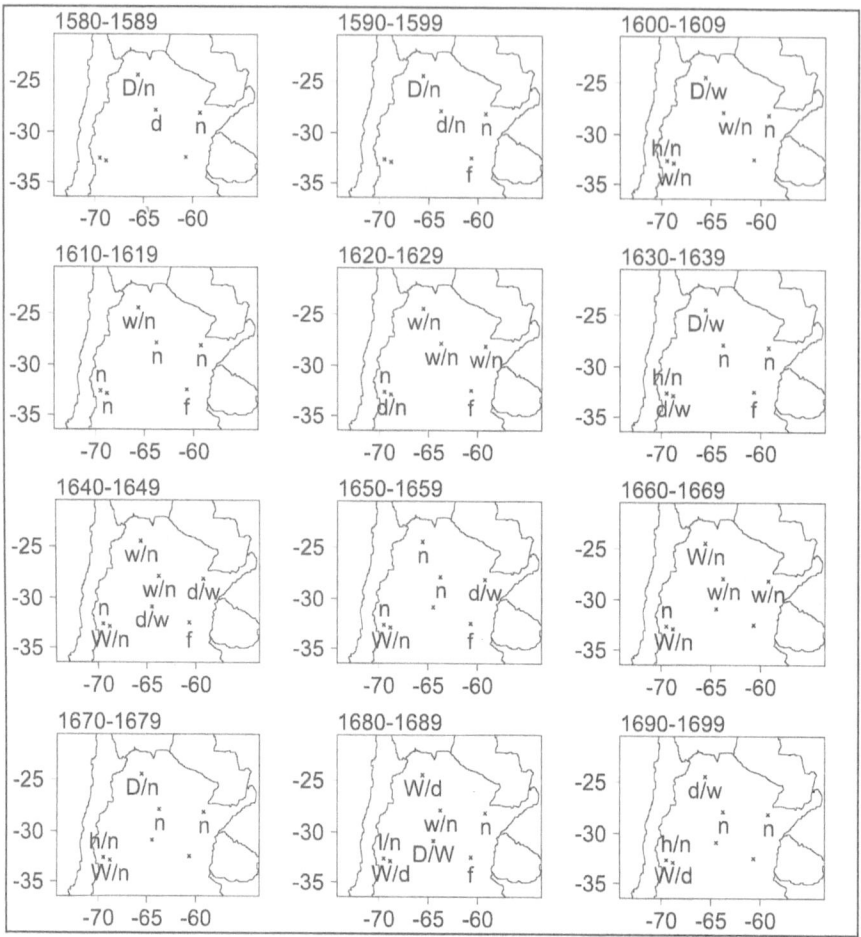

Fig. 7.5. Spatial variability of climatic conditions in northern Argentina during the 17th century. Classification: W = very wet, w = wet, n = normal, d = dry, D = very dry, f = some floods, F = massive floods. For the Mendoza river runoff: l = low, n = normal, h = high, H = very high. For comparison see the time interval 1970–1979 in Fig. 7.6. Note also the number of "wet" decades in the 17th century compared to that of the 18th century. See Fig. 7.7 for description of locations.

Although the agricultural year 1786–87 cannot be regarded as one of the most catastrophic seasons, it showed a marked decrease of precipitation both in Bolivia and northwestern Argentina. In the whole studied area, ranging from 300–700 mm, the drought conditions were for most Andean cultivations, with different intensities, more severe at the beginning of the agricultural season than at the end. The upper Andean Puna (Altiplano) was the one that suffered from the most severe lack of water (Prieto and Herrera 1992).

The years 1789–1790 were catastrophic for Tarija, Bolivia. Fray Francisco de Tamajuncosa told that in the Missions "...there followed...a horrible famine in all the cordillera, particularly in the years of 1789 and 1790, because of the great shortage of waters.../the cattle/...died in the fields for lack of pastures..." (Tamajuncosa 1969:112). Similar conditions were recorded in Catamarca, Tucumán, Salta and Santiago del Estero (AGI, Buenos Aires, 109).

Scarce precipitation was recorded in Cochabamba (Larson 1980) in 1792. The agricultural season of 1796–1797 also showed signs of droughts, but they were spatially more limited. In 1796 some months without water that "delayed the coca haciendas", were reported for Bolivia, particularly in La Paz. In contrast rains were reported in 1797 for La Paz and Cochabamba: "...this year has been abundant in waters and therefore the begetting of cattle has increased, especially cows and sheep" (AGI, Buenos Aires 383).

Contrary to Bolivia, northwestern Argentina and the Cuyo plains region suffered from the lack of water. In the Mendoza River, years with exceptionally low streamflow and other years with high streamflow occurred between the decades 1771/80 and 1821/30. Large streamflow events were rare between the decades 1831/40 and 1861/70 (one event per decade). From 1871/80 to 1901/10 large streamflow were more frequent (see Fig. 7.2).

The water deficit was very harmful in Tucumán. A drought starting in October 1796, affected not only the agricultural activities but also the raise of cattle. The wheat crop was already considered lost because of rot. The lack of water continues in December and January of the following year, reaching catastrophic levels: "There is such a constant drought experienced that the plentiful rivers have dried and several abundant water springs have retired, making the fields so dry and arid that many cattle have died" (AGI, Buenos Aires 21). On July 1797 a noticeable decrease in temperature made the situation worse: "The present winter is cruelly experienced because of the ice and drought...it has been noticed that most of the wells that maintain the houses have dried because of the lack of rains" (AGI, Buenos Aires 21).

In the semiarid Chaco, strong droughts occurred since 1770, a predominantly dry period, which lasted for more than 30 years. Dry years are predominant, but alternate with extremely rainy years periods due to extreme oscillations. The frequency analysis of great swellings per decades (since 1600) of the Sali–Dulce and Salado rivers, show at least one swelling every decade (up to four as a maximum) between the decades 1760/70 and 1780/90 (Prieto et al. 1998a).

A common pattern recorded during this dry period consisted of a year with a great water deficit, followed by a winter with extremely low temperatures. For the interval 1785–1800, frosts and very-cold weather were reported for eleven years along the Andes from Bolivia to Mendoza, Argentina. A list of these cold events is in Appendix I on the CD.

Droughts peaked again in the 1800–1801 agricultural cycle in Argentina and Bolivia (see Fig. 7.6). As a consequence of the intense lack of water the wheat crop was destroyed in Cochabamba (Larson 1980). The water deficit threatens the water supply of the mineral grinding in Potosi "...because of the lack of water in the last season and if copious rains do not happen next October it will suspend the course of the silver mills" (AGI, Buenos Aires 383). In 1802 the drought continued in Bolivia and northwestern Argentina (AGI, Buenos Aires 587). The culmination of this dry cycle occurred in the years 1803, 1804 and 1805 (see Figs. 7.4 and 7.6). In Cochabamba, two successive years of scarce water caused the loss of all the crops (Larson 1980). In Potosi the catastrophe was even worse.

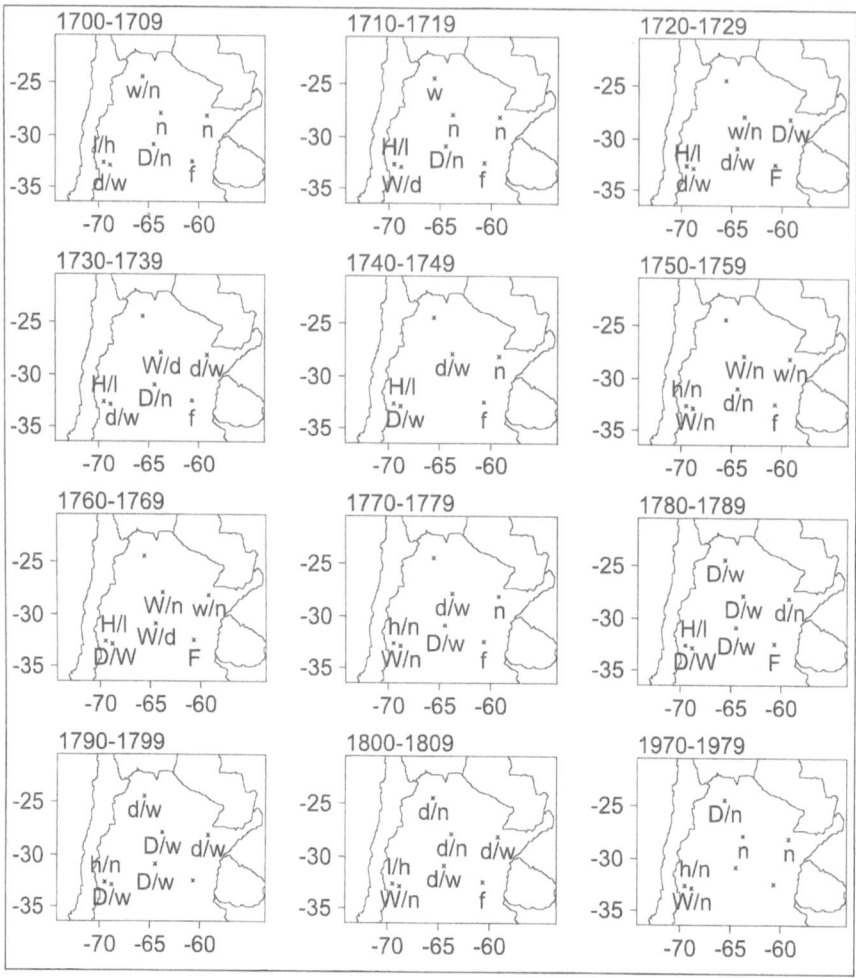

Fig. 7.6. Spatial variability of climatic conditions in northern Argentina during the 18th century. Classification: W = very wet, w = wet, n = normal, d = dry, D = very dry, f = some floods, F = massive floods. For the Mendoza river runoff: l = low, n = normal, h = high, H = very high. The time interval 1970–1979 is provided for comparison. See Fig. 7.7 for location descriptions.

As a consequence of the droughts in 1804 and 1805 the silver-grinding mills were stopped, the crops were lost, the prices increased and there was a general famine and a demographic crisis (Tandeter and Wachtel 1984).

The 18th century was also turbulent in the humid central zone of Argentina. Cordoba climate is representative of this region. The precipitation series shows an increase in the arid conditions in this area during the 18th century at the same time as similar behavior in the Andean zone. The Cordoba precipitation series shows: (1) A long drought at the beginning of the century. There were 7 extreme droughts, 1710–1731; (2) A period predominantly humid, 1732–1784; (3) Six droughts with different intensities, 1785–1797.

| 1. Norwestern Argentina |
| 2. Arid Chaco |
| 3. Humid Chaco |
| 4. Province of Córdoba (Humid plains) |
| 5. Province of Mendoza (Arid and semiarid plains) |
| 6. Mendoza River Streamflow |
| 7. Río de la Plata Basin floods. |

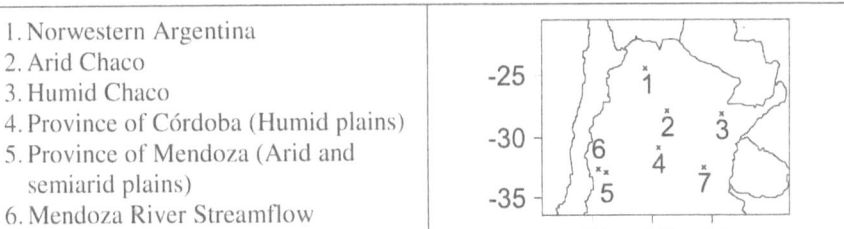

Fig. 7.7. Locations of discussed sites and regions in northern Argentina.

In the La Plata River Basin (mainly the Paraná, Paraguay and Uruguay Rivers) during the period 1780–1810, simultaneously with the great droughts in the Andean area and the central zone of Argentina, anomalous climatic situations of opposite signs occurred. Abundant rainfalls and extraordinary floods were frequently recorded (Prieto and Richard Jorba 1991). The exceptionally swelling of the Paraguay river and tributaries in 1785 was described by Felix de Azara (1941): "...The Paraguay river was so high that the most ancient men say they have never seen a flood of such size and duration...".

During the 17th and 18th centuries, periods with a high percentage of humid years or flooding occurred 1680–1710, 1740–1770 and 1790–1810 (see Table 7.1 for details). The location of the discussed sites and regions is displayed in Fig. 7.7.

Table 7.1. Spatial variation of droughts and floods in various Argentine regions, 1560–1806, in relation to the occurrence of El Niño and ENSO. See table bottom for explanations.

Year	1	2	3	4	5	6	7	El Niño	ENSO	Year	1	2	3	4	5	6	7	El Niño	ENSO
1560								m/S	S	1683	w	n	n			n			m+
1561								m/S	S	1684	D	n	n		w	n		m+	m+
1562						f				1685	n	n	n		w	n			
1563										1686	n	n	n			n	f		
1564										1687	W	n	n		w	n		S+	S
1565								m+	m+	1688	n	n	n		d	n			S
1566						f				1689	n	n	n		n	l			
1567						S+			S+	1690	n	n	n		w	h			
1568						S+			S+	1691	n	n	n		w	n			
1569										1692	w	n	n		n	n		S	m+
1570										1693	n	n	n		n	n			
1571										1694	D	n	n		d	n			VS
1572										1695	n	n	n		n	n		m	VS
1573										1696	n	n	n		n	n			
1574						S			S	1697	n	n	n		n	n		m+	m
1575										1698	n	n	n			n			
1576										1699	n	n	n		w	h			
1577										1700	n	n	n	n	d	l			
1578						VS			S	1701	n	n	n	n	n	h		S+	m
1579						f	VS		S	1702	n	n	n	n	n	n			
1580	n									1703	n	n	n	n	n	n	f		S
1581	n							m+	m+	1704	n	n	n	d	w	n		m	S

Year	1	2	3	4	5	6	7	El Niño	ENSO
1582	n							m+	m+
1583	D								
1584	D								
1585	D							m+	m
1586	D								
1587	n								
1588	D	d	n						
1589	D	d	n					m/S	S
1590	D	d	n					m/S	S
1591	D	d	n					m/S	S
1592	D	n	n						
1593	n	n	n						
1594	n	n	n						
1595	n	n	n						
1596	n	n	n			f		m+	m
1597	n	n	n						
1598	n	n	n		f				
1599	n	n	n						
1600	n	n	n	n	n			S	S
1601	n	n	n	n	n				S
1602	n	n	n	n	n				
1603	n	n	n	n	n				
1604	D	n	n	n	n			m+	S
1605	D	n	n	n	n				
1606	w	w	n	n	n				
1607	w	w	n	n	n			S	S
1608	D	n	n	n	n			S	S
1609	n	n	n	w	h				
1610	n	n	n		n	f			
1611	n	n	n	n	n				
1612	n	n	n		n				
1613	n	n	n		n				
1614	W	n	n		n	f		S	S
1615	n	n	n		n				
1616	n	n	n		n				
1617	n	n	n		n				
1618	n	n	n		n			S	m
1619	w	n	n		n			S	m
1620	n	n	n		n				
1621	n	n	W		n	f		m+	S
1622	n	n	n		n				
1623	n	n	n		n				
1624	n	n	n		n			S+	m+
1625	w	n	n		n				
1626	n	n	n	D	n				
1627	n	n	n	D	n				
1628	w	w	n	n	n				
1629	n	n	n	n	n				
1630	n	n	n	d	n			m	S+
1631	D	n	n	w	h				S+
1632	D	n	n		n				
1633	D	n	n		n				
1634	n	n	n		n				
1635	n	n	n		n			S	m

Year	1	2	3	4	5	6	7	El Niño	ENSO
1705	n	n	n	d	n	n			
1706	w	n	n	d	n	h			
1707	n	n	n	d	w	n		m/S	m
1708	n	n	n	d	n	n		m/S	m
1709	w	n	n	d	D	n		m/S	m
1710	w	n	n	d	w	h			
1711	n	n	D	n	h				
1712	n	n	d	d	n				
1713	n	n	d	n	n			m	m+
1714	n	n	D	n	n				m+
1715	n	n	D	n	h	f		S	S+
1716	n	n	n	w	n			S	S+
1717	n	n	n	w	n				
1718	n	n	n	n	n			m/S	m
1719	n	n	n	W	l	f			
1720	n	D	n	n	h			VS	m+
1721	n	D	D	d	h				
1722	n	W	n	d	h				
1723	n	n	n	n	h	f		M+	S
1724	n	n	w	w	n				
1725	n	n	n	n	n	f			m
1726	n	n	n	n	n				
1727	n	n	n	n	l				
1728	w	w	n	n	h			VS	m
1729	n	D	D	n	h	f			
1730	w	d	D	w	h				
1731	w	W	D	w	h				m+
1732	w	n	n	n	n				
1733	n	n	n	n	n				
1734	n	n	d	n	n			m	m
1735	n	n	n	n	n				
1736	n	n	n	n	n				
1737	n	n	n	n	n			S	S
1738	d	n	n	d	L	f			
1739	d	n	n	n	h				
1740	n	n	n	w	l				
1741	n	n	n	n	n				
1742	n	n	n	d	n				
1743	n	n	n	n	h				
1744	d	n	n	n	h			m+	m+
1745	w	n	n	d	n				
1746	n	n	d	d	n				
1747	d	n	d	n	H			S+	S
1748	d	n	W	n	n	f			S
1749	n	n	W	n	n				
1750	w	n	n	n	n				
1751	n	W	n	n	n			m+	m+
1752	w	n	n	n	n	f			
1753	n	n	n	w	n				
1754	n	n	n	w	n	f		m	S
1755	n	n	n	n	n			m	S
1756	n	n	n	n	n				
1757	n	n	n	W	H				
1758	w	n	n	W	n			m	m

Year	1	2	3	4	5	6	7	El Niño	ENSO
1636	w	n	n			n	f		
1637	n	n	n			n			
1638	n	n	n			n			
1639	n	n	n			n			
1640	n	n	D		w	n		m	S+
1641	w	n	n			n	f	m	S+
1642	n	n	n			n			
1643	n	n	w		n	n	f		
1644	n	n	n		w	n			
1645	n	n	n		n	n			
1646	n	n	n		n	n			
1647	n	n	n		w	n		m+	m
1648	n	w	n		w	n			
1649	n	n	n		w	n			
1650	n	n	n		w	n		M	S+
1651	n	n	n		n	n			
1652	n	n	W		w	n	f	S+	m
1653	n	n	n			n			
1654	n	n	n			n			
1655	n	n	n			n		m	m
1656	n	n	D		n	n			
1657	n	n	n		w	n			
1658	n	n	n		n	n			
1659	n	n	n		n	n			
1660	n	n	n			n			
1661	n	n	n			n		S	VS
1662	n	n	n		w	n			
1663	W	n	n		n	n			
1664	n	n	n		w	n			
1665	w	n	D		w	n			
1666	n	n	n			n			
1667	n	n	n		n	n			
1668	n	n	n		n	n			
1669	W	w	n		w	n			
1670	W	n	n		n	n			
1671	w	n	n		n	n		S	m+
1672	n	n	n		w	n			
1673	n	n	n		w	n			
1674	n	n	n		n	n			
1675	n	n	n		w	h			
1676	w	n	n		w	n			
1677	w	n	n		n	n			
1678	W	n	n		w	h			
1679	w	n	n		n	n			
1680	w	n	n		w	n			
1681	w	w	n			n		S	S
1682	n	n	n			n			
1759	w	n	d		N	n			
1760	n	n	n		w	n			
1761	w	n	D		n	h		S	S
1762	n	n	n		w	h			S
1763	n	n	n		d	n			
1764	w	n	n		n	h			
1765	n	n	W		d	n		m	m+
1766	w	n	W		w	n	f		m+
1767	w	W	W		w	L	f		
1768	n	W	w		d	n	f	m	m+
1769	n	n	n		n	h			m+
1770	n	n	n		w	n			
1771	n	n	n		n	n			
1772	w	n	D		n	n		m	m
1773	d	n	D		n	n			m
1774	n	n	D		n	n			
1775	n	n	D		n	n			
1776	n	n	D		w	h		S	m+
1777	n	n	n		w	n		S	m+
1778	n	n	n		n	n		S –E	m+–E
1779	n	n	W		n	n	f		
1780	D	n	n	n	W	n			
1781	n	n	n	n	D	n			
1782	n	n	n	n	n	n		S	VS
1783	n	n	n	w	d	H		S	VS
1784	n	n	n	d	W	H			VS
1785	w	d	n	d	w	h	f	m+	m+
1786	D	d	D	D	w	L	f	M+	m+
1787	w	d	n	w	D	n	f		
1788	d	w	n	w	w	H			
1789	D	w	n	w	D	h			
1790	n	w	n	d	n	n			VS
1791	n	n	n	n	n	n		VS	VS
1792	n	n	n	n	n	H			VS
1793	n	n	n	n	n	n			VS
1794	n	d	D	D	n	n			m+
1795	n	n	n	n	n	n			m+
1796	d	d	W	w	d	n			m+
1797	D	d	n	D	n	n			m+
1798	w	n	D	n	d	n			
1799	n	d	w	n	d	n			m
1800	n	n	W	n	w	L	f		
1801	D	n	D	n	n	L			
1802	n	d	D	w	n	n			S+
1803				n	w	h		S+	S+
1804				n	w	h	f	S+	S+
1805				w	n	n			
1806				D	w	n		m	m

Climate (col. 1-5, see Fig. 7.7): D = very dry, d = dry, n = normal, w = wet.
Mendoza river runoff (col. 6): L = very low, l = low, n = normal, h = high, H = very high.
Floods in the Paraná basin (col. 7): f = floods.
ENSO: m = medium, s = strong, S = very strong, VS = extremely strong. + and _ : emphasis

7.4. Discussion

At a first glance it appears that floods and extraordinary droughts may correlate with El Niño Southern Oscillation (ENSO) (see also the historical ENSO-reconstruction of Quinn and Neal 1992, diagrams above, Prieto and Boninsegna 1992, Villalba 1994 and Villalba in this volume). Although there might be some evidence for a possible relation between ENSO occurrence and exceptional floods in northeastern Argentina, extraordinary droughts in northwestern Argentina and Bolivia or heavy precipitation in Mendoza and Cordoba, we must conclude that from the historical perspective, it is difficult to establish direct relationships between regional climatic events and climatic forcing such as ENSO.

An analysis of anomaly events in relation to ENSO phenomena belonging to the studied period showed a high variability in response from a spatial perspective (Table 7.1). This fact confirms that local and regional factors may also affect changes of precipitation and temperature, and not only El Niño occurrence. El Niño events would explain only a percentage of the climatic anomalies that happened in the region.

The percentage varies also from a temporal point of view. For instance, from the end of the past century to the present, a strong relation has been detected between the occurrence of the "El Niño" and the Mendoza River streamflow magnitudes (Compagnucci and Vargas 1993). However, in the past, there are some intervals, like 1600–1670, where such a relation was not observed. On the contrary, there are periods where the coincidence between El Niño and high streamflow was very high, for instance reaching 82% during 1771–1830 (Prieto et al. 1999). In agreement with Ortlieb (1995, 1998) we think that the teleconnection system of the ENSO might have varied through time, in connection with longer scale climate variations like a Little Ice Age.

Additionally, the discussed historical data show a climatic situation differing from the present. They show the Little Ice Age and the subsequent adjustment to present-day situations. Therefore the preceding figures (Figs. 7.5 and 7.6) display synoptically the climatic evidence from southern South America. These maps show very clearly that (a) we are living today in a time with a remarkably low level of environmental obstacles (see present day map). Towards the cooler times (little ice ages) the number of unfavorable situations for culture increases. The maps show clearly that the colder interval at the end of the 16th and beginning of the 17th centuries is characterized by the following: a significant number of droughts in northwestern Argentina and the Arid Chaco, droughts in Central Argentina (Mendoza), normal streamflow in the Mendoza River, floods in the Río de la Plata Basin, and cold summers (late spring, early autumn) in Mendoza. This shows that the regional climatic gradients in South America had been much steeper in colder times than at present. If, based on the underlying physics, we utilize past scenarios as potential analogs for future equilibria, then we may expect that a shift to a colder climate (which at the moment seems to be unlikely) may result in climate patterns that are, depending on the region, either difficult or favorable for agriculture (see Figs. 7.5 and 7.6). For reasons of comparison and calibration to historical documents the time interval from 1970–1979 has been included as well. In this context also see the work of Barros et al. (this volume) who discuss recent precipitation changes in Argentina.

7.5. Future Perspectives

The above-mentioned results show the potential of historical data for climate analysis. In addition, such data may be used to calibrate climatic models, since when modeling past scenarios climate models should "predict" overall areas of drought correctly. Another aspect is that a wealth of historical data permits the deduction of the position of past glacier lines. As glaciers integrate the effects of longer time-intervals, the collaboration with other sciences may also provide past mass-balances and water budgets for the regions under study (see in this context Wingenroth, this volume).

The possibility that climate is changing towards unprecedented warming has, politically, resulted in the importance given to climate research. Longer climatic series records allow us to determine how unusual recent climatic events are.

If present climate is analyzed within a long-term perspective, both natural *and* anthropogenically induced climatic fluctuations can be handled better. In this context, the work of historians is of considerable importance. Their data not only allow the construction of time series, but also shows the real impact of climate change (droughts, high prices due to bad harvests, floods) on daily economic life. Floods are especially important, as they are the integration of precipitation, river runoff, vegetation, distribution of precipitation and topographic relief.

We can conclude that the high potential of historical data shows up when, in addition to fundamental physics data, assessments are made of the integrated effects of the whole ecosystem, including the impact of man.

References

Abbreviation: AGI Archivo General de Indias, Seville

Azara F de (1941) Viajes por la America Meridional. 1781 a 1801. Espasa–Calpe, Madrid
Baron W (1982) The reconstruction of eighteenth century temperature records through the use of content analysis. Climate Change 4:385–398
Cobos D, Boninsegna JA (1983) Fluctuations of some glaciers in the upper Atuel River Basin, Mendoza, Argentina. Quaternary of South America and Antarctic Peninsula, 1:61–82
Compagnucci R, Vargas W (1993) Snowfall in the Cordillera de Los Andes and the ENSO Events. Preprint 4th International Conference on Southern Hemisphere Meteorology and Oceanography. Hobart, Australia. eds. AM. Met. Soc, pp. 332–333
Hamilton K, Garcia R (1986) El Niño/Southern oscillations and their associated midlatitude teleconnections. 1531–1841. Bulletin American Meteorogical Society 67(11):1354–1361
Hastenrath S (1981) The glaciation of the Ecuadorian Andes. Balkema, Rotterdam
Herrera R, Dussel P (1992) Eventos climaticos extremos y ambiente en el Santiago del Estero de la segunda mitad del siglo XVIII. Cuadernos Proyecto NOA: El Noroeste Argentino como Región Histórica (Universidad de Sevilla–Junta de Andalucia). 3:7–31
Jimenez de la Espada M (1965) Relaciones Geograficas de Indias. Peru. Biblioteca de autores espanoles, T CLXXXIII, Madrid
Lamb HH (1977) Climate: Present, Past and Future. Vol. 2, Methuen, London
Larson B (1980) Ritmos rurales y conflictos de clases durante el siglo XVIII en Cochabamba. Desarrollo Economico, 20(78):183–214
Morales Padron F (1973) Historia del Descubrimiento y Conquista de America. Editora Nacional, Madrid
Past Global Changes Project (PAGES) (1992) Report No19: Propuesta de implementación de planes para actividades de investigación, IGBP, Stockholm

Ovalle A de (1889) Historica relacion del Reino de Chile (1625). Colección de Historiadores de Chile. T. XII. El Mercurio, Santiago de Chile

Ortlieb L (1995) Eventos El Niño y episodios lluviosos en el desierto de Atacama: el registro de los últimos dos siglos. Bull. Inst. fr. études andines, 24(3):519–539

Ortlieb L (1998) Historical reconstruction of ENSO events from documentary sources from Chile, Perú , Brasil and Mexico: evidences for variability of the teleconction regime in the last centuries. Extended Abstract. Pages Meeting, Quito–Eduador

Politis G (1983) Climatic variations during historical times in Eastern Buenos Aires Pampas, Argentina. Quaternary of South America and Antartic Peninsula 1:133–161

Prieto M del R (1983) El clima de Mendoza durante los siglos XVII y XVIII. Meteorologica (Universidad de Buenos Aires) XIV(1,2):165–174

Prieto M del R, Richard Jorba R (1991) Anomalias climaticas en la Cuenca del Plata y el NOA y sus consecuencias socioeconomicas durante los siglos XVI, XVII y XVIII. Leguas (Fac. Filosofía y Letras–Mendoza) 1:41–103

Prieto M del R, Herrera R (1992) Las perturbaciones climaticas de fines del siglo XVIII en el área andina. Cuadernos Proyecto NOA: El Noroeste Argentino como Region Histórica, (Universidad de Sevilla–Junta de Andalucia). 1:7–35

Prieto M del R, Boninsegna JA (1992) Evidencias dendrocronologicas e historicas de anomalias climaticas relacionadas con ENSO en Sudamerica Austral durante los siglos XVIII Y XIX. Abstracts of "Former ENSO Phenomena in Western South America" Symposium, Lima, Perú

Prieto M del R, Herrera R, Dussel P (1998a) Documentary evidences of Severe Droughts and Floods in Northern and Central Argentina During the 17th and 18th Centuries. Abstracts of Calibration of Historical Data for Reconstruction of Climate Variations Work shop. NOAA, PAGES and NSF. Barcelona, pp. 20–21

Prieto M del R, Herrera R, Dussel P (1998b) Clima y disponibilidad hídrica en el Sur de Bolivia y Noroeste Argentino entre 1550 y 1650. Los documentos españoles como fuente de datos ambientales. Bamberger Geographische Schriften, 15:36–57

Prieto M del R, Herrera R, Dussel P (1999) Historical evidences of the Mendoza River Streamflow fluctuations and their relationship with ENSO. In press.

Quinn WH, Neal VT (1992) The historical record of El Niño events. In: Bradley, R, Jones P (eds) Climate since A. D. 1500. Routledge, London, pp. 623–648

Tamajuncosa A (1969) Misiones a cargo del Colegio Nuestra Señora de los Angeles. In: Coleccion Pedro de Angelis. T VII, Ed. Plus Ultra , Buenos Aires

Tandeter E, Wachtel N (1984) Precios y producción agraria. Potosi y Charcas en el siglo XVIII. Estudios CEDES, Buenos Aires

Villalba R (1994) Tree-rings and glacial evidence for the Medieval Warm Epoch and the Little Ice Age in southern South America. Climatic Change 30:1–15

Recent Glacier Fluctuations in the Southern Andes

R. Llorens, J.C. Leiva
Instituto Argentino Nivologia, Glaciologia y Ciencias Ambientales (IANIGLA–
CRICYT), Casilla de Correo 330, 5500 Mendoza, Argentina
Jcleiva@lab.cricyt.edu.ar

Abstract: The water supply of several towns, in addition to the study area, depends on rivers that are fed by meltwater of glaciers. The fluctuation of various glaciers from the Anconcagua massive was studied. The period studied, 1909–1994, is a time of general shrinking; however, some advances, including surges, could be observed.

8.1. Introduction

The knowledge of the natural environment is vital for socioeconomic planning. This applies especially to key-factors, such as the water supply of cities located in desert conditions. The city of Mendoza, Argentina, is a key example. It is situated in a desert-like environment with an annual rainfall of 197 mm. The water supply of both the city (circa 900000 inhabitants) and its surrounding agriculture depends highly on the waters of the Rio Mendoza and its tributaries. These rivers are mainly fed by glaciers of Aconcagua (6959 m) and surrounding massifs such as Mt. Tupungato, Mt. Tupungatito and Mt. San Juan (see Fig. 8.1). The extension of these glaciers has been measured systematically for the period 1982–1994 using satellite images, supplemented by photos for 1963–1974. Complementary cartographic and historical information (see also Prieto et al., this volume) have been used for periods back to 1909.

All images have been geometrically corrected (see for example Jensen 1986). The Landsat TM data have been processed using standard techniques on a SUN SPARC 10 workstation with ERDAS-IMAGINE 7.5 software at the Instituto de Investigaciones Aplicadas de Ciencias Espaciales at the CRICYT research center. A detailed description of methodology can be found in Llorens and Leiva (1994a).

8.2. Glacier Fluctuations

The location of the glaciers studied can be found in Fig. 8.1. Figures. 8.2 to 8.5 show the relative position of the discussed glacier fronts with reference to their position in the year 1909 (Fig 8.2) and 1963 (Figs. 8.3 to 8.5). While the glacier fronts of many glaciers retreat, some glaciers advance significantly. This shows that simple relations between suspected "global warming" and glacier shrinkage do not exist. Around 1982 several glaciers of the study area start to advance. These advances are either gradual or rapid.

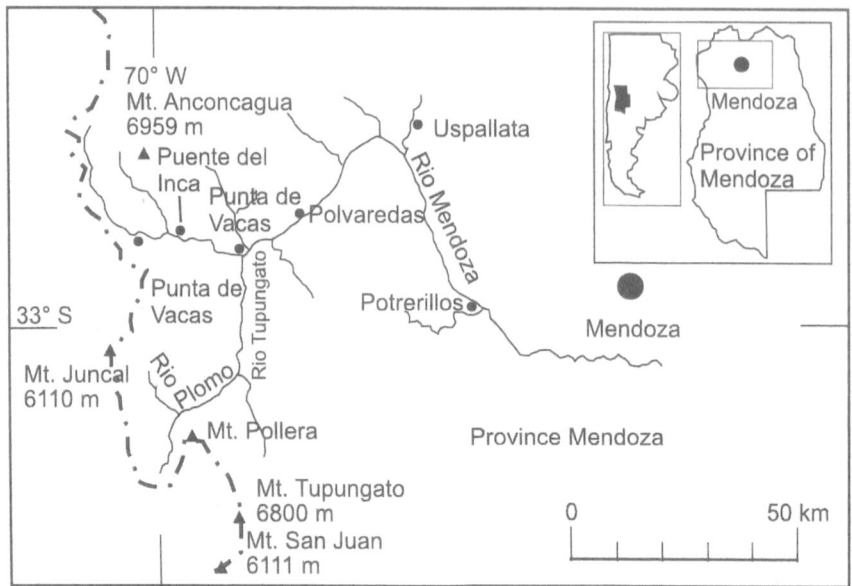

Fig. 8.1. Location map showing the glaciated massifs discussed in the text.

8.2.1. The Lower Horcones Glacier

The lower Horcones (Fig. 8.2) glacier, which, together with the Upper Horcones glaciers feeds the Rio Horcones and other tributaries to the Rio Mendoza, is characterized by considerable recent dynamics. The major quantity of snow originates from the high altitudes of Mt. Aconcagua (6963 m), the highest peak of the Americas. Air-photos from the year 1963 show that the southern foot of the glaciers is characterized by advanced ablation, widespread thermokarst and numerous holes distributed over the whole surface of the glacier.

A photo from 1974 shows that the frontal part of the glacier thickened and advanced about 150–170 m. Although the surface of the glacier is characterized by ablation holes and other indications of thermokarst, the relative area of ablation did not extend compared to the data of 1963. The details of this process are the subject of ongoing and future studies. On a satellite image from 1982 the upper part of the glacier (above 3900 m) appears darkened. Closer investigations show that this is caused by a chaotic surface in this part of the glacier. Thus, together with other indications, the advance is interpreted as a surge. In this context it must be noted that the authors follow the definition of surges as given by Paterson 1994. According to this definition surges cover a time interval ranging between several days and several years. Whether these phenomena are responses of past times with heavy precipitation (middle ages ENSO, end of "little ice age, see Prieto et al. this volume) is a question for future studies. Satellite data from 1986 show that these indications do not only spread over the whole area but also exceed the former terminal part of the glacier by about 200 m. According to Lenzano (pers. comm.), photos of this area show a surge in December 1984.

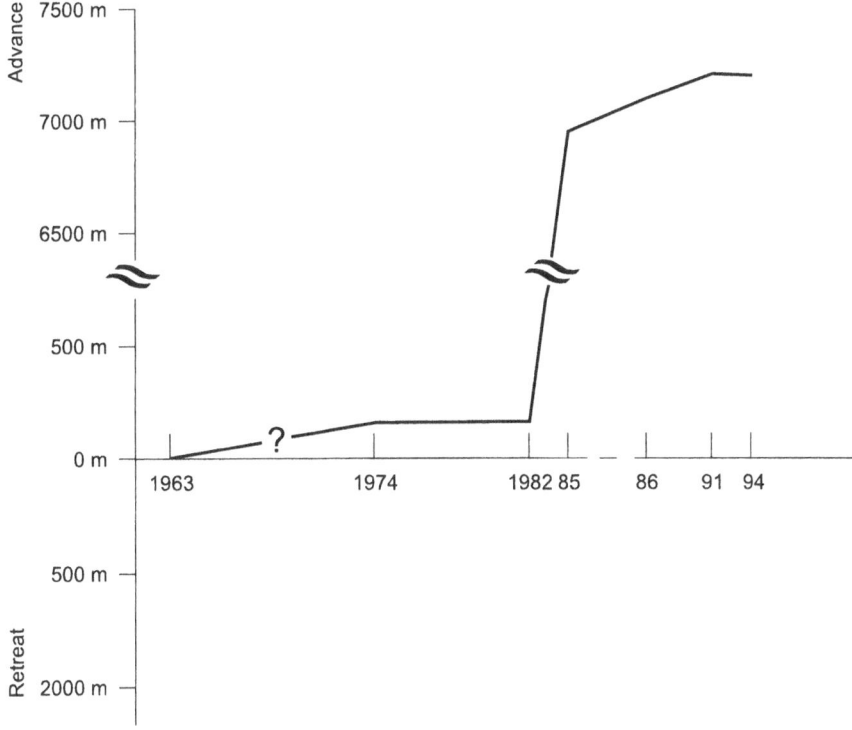

Fig. 8.2. Advance of the Lower Horcones Glacier, 1963–1994. Note the interrupted vertical axis.

According to 1991 photos, the advance continues by about 300–400 m while the lower part of the glacier itself remains in a state of ablation. The photos from 1993–1994 do not show changes of the terminal part of the glacier. In addition, the ablation amount does not change with respect to 1991.

8.2.2. The Glaciers of Mounts Tupungato, Tupungatito and San Juan

The mountains of Tupungato, Tupungatito and San Juan are covered by glaciers that feed the river valley of the Rio Tunujan, resulting in extensive agriculture. Historical information about these glaciers, including maps or photos, does not exist. The ice masses east of Mt. Tupungato however are described by Helbling (1919, 1935). The studies of Helbling are generally accepted as precise and reliable, therefore his data are included as background information. Using the data of Helbling (1919, 1935) and considering available data from the glaciers feeding Rio Plomo, which are still one single ice-mass, the authors infer indirectly that the retreat of the glacier amounts to about 3500 m compared to its position in 1909. Today only four glaciers remain (Fig. 8.3, parts A–D).

Glacier A shrinks about 150 m during 1963–1982. The thickness however increases in the lower 2.5 km of this glacier.

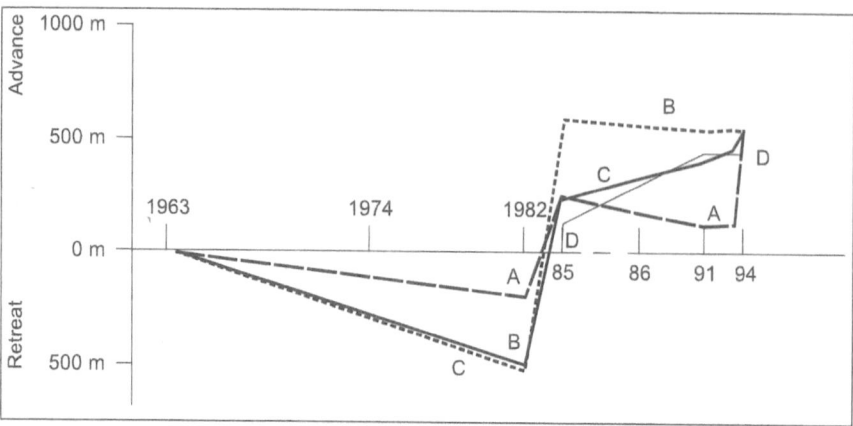

Fig. 8.3. Fluctuations of the glaciers of the mountains Tupungato, Tupungatito and San Juan.

In 1985 an advance by 550 m can be observed, accompanied by formation of a fracture system perpendicular to the flow-line. From 1986 on, a minor retreat of about 50–100 m can be observed. After 1993 no major changes can be found. The above-mentioned fracture-system, which characterized the movement of an internal glacial wave, has advanced by 400–500 m.The terminal line of the glacier however remains constant. Glaciers B–D show generally the same behavior. A retreat during 1963–1985, is followed by a more or less rapid advance during 1985–1986; in the case of glaciers C and D this continues to 1994 (see Fig. 8.3).

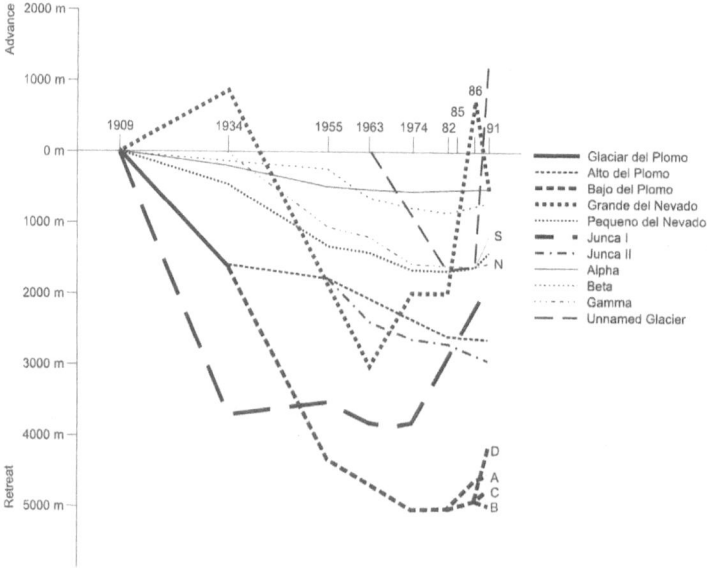

Fig. 8.4. Glacier fluctuations in the massive of Anconcagua, 1909–1991. Note that from 1982 on many glaciers started to advance.

8.2.2 The Glaciers feeding Rio Plomo

Figures 8.4 and 8.5 sketch the relative advances of the glaciers feeding Rio Plomo (see also Razza 1935). Figure 8.5 shows a detailed view of selected glaciers for the time-interval 1963–1994.

The figures show that all glaciers shrank considerably between 1909 and about 1982 (Fig. 8.4). It is however evident that from about 1982 on, several glaciers advanced. Some of them, such as the glacier Grande del Nevado, advanced quite rapidly (Fig. 8.5). The mechanisms are overall comparable to those of the Lower Horcones Glacier and the glaciers of the mountains at Tupungato, Tupungatito and San Juan (for a detailed analysis see Leiva 1986, 1989; Leiva and Cabrera 1995; Llorens and Leiva 1994b).

A closer look at the younger time interval, 1963–1991, shows that the advance started around 1982 and ended around 1990. As during this period also some glaciers thickened, these advances cannot be interpreted straightforward as phenomena of glacier decay.

Considering internal glacier physics, meteorological, historical and isotope data have to be sought that permit linking the young glacier advances to historical and prehistorical meteorological events. Finite element models running various temperature and viscosity scenarios as well as various shear regimes at the base of the glaciers might help to clarify these phenomena.

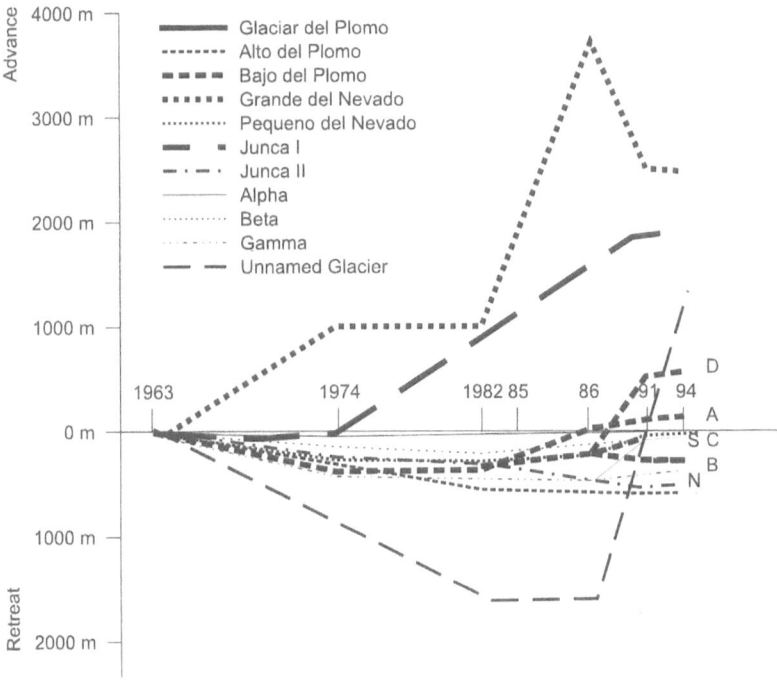

Fig. 8.5. Glacier fluctuations with respect to the terminal lines of 1963. Note the considerable advance of some glaciers.

8.3. Conclusions

The objective of this study is a contribution to a safety analysis of the water supply of Mendoza and its adjacent agricultural areas that relies on the waters of the glaciers discussed. While the causes of retreats of glaciers are known (negative mass-balances), the reasons of glacier advances are of a more complex nature. The terminal line of a glacier is (simplified) the expression of an equilibrium consisting of precipitation, ice-formation, subsequent push, relief, ice-temperature, internal friction, friction at the base of the glacier (temperature dependent) and finally ablation and melting processes.

Two opposite processes may cause a glacier advance:

1) Accumulation in the area above the snow line and subsequent ice formation. If the accumulation is rapid enough and if the internal glacier temperatures are sufficiently high, then this may cause an advance of km per year, which, strictly speaking cannot be called a surge. If internal threshold-values are exceeded, the advance may be rapid. If the overall atmospheric temperatures remain constant, which means a constant meltwater-production, then this scenario is *favorable* for the water supply of a city. If temperatures decrease and if thus meltwater-production reduces, *unfavorable* conditions for the water supply occur, although the glacier advances.
2) Rise of the temperatures at the glacier base, reduction of internal and base-frictions followed by a rapid advance until a new rheologic equilibrium is reached. Advances of this type may be a bad indication as this indicates a *shrinking of the reservoir* without sufficient replenishment.

Therefore, ongoing and future studies have to focus on:

(a) The general velocities of the glaciers located in the study area.
(b) Time series showing precipitation variations in the Andean region on a N–S transect.
(c) Time series showing temperature variations in the Andean region on a N–S transect.
(d) Calculated and observed time series of the flow of ice-particles in the glaciers. For older times this requires isotope dating of samples along the flow lines and subsequent velocity and mass-balance estimates.

Barros et al., Compagnucci, Markgraf, Prieto et al. and Villalba (all this volume) provide for different time scales and different age background data on the differential nature of climate change. This applies not only to young and ongoing climate change, but also to historical climate changes.

Due to safety considerations, it is necessary to be alert about glacier shrinking caused by global warming. The key site of the Massive of Aconcagua and adjacent massive; however, shows that analyses of glacier fluctuations – if executed under the aspect of water-supply – including glacier shrinking must consider all aspects involved, not only the most recent ablation data. In addition, this study shows that even in times of global warming, glaciers may shrink or expand for different reasons. This shows that only the *methods* can be transferred to other areas of the world, not the data or conclusions as glacier shrinking or advance in each area has its own genuine history which requires individual study.

References

Helbling R (1919) Beitrage zur Topographischen Erschliessung der Cordilleras de los Andes zwischen Aconcagua und Tupungato: Sonderabdruck aus dem XXIII Jahresbericht des Akademischen Alpenclub. Zürich 1918

Helbling R (1935) The origin of the Río Plomo ice-dam. The Geographical Journal 8(1):41–49

Jensen J (1986) Introductory digital image processing. Prentice Hall, Englewood Cliffs, New Jersey

Leiva JC (1986) El surge del glaciar Grande del Nevado del Plomo. Informe elevado al MOSP de la Provincia de Mendoza

Leiva JC (1989) Variations of Rio Plomo Glaciers, Andes Centrales Argentinos. In: Oerlemans J (Ed) Glaciers Fluctuations and Climatic Change, pp 143–151

Leiva JC, Cabrera G (1996) Glacier mass balance analysis and reconstruction in the Cajón del Rubio, Mendoza, Argentina. Zeitschrift für Gletscherkunde und Glazialgeologie Vol.31(1996) p. 1 - 7.

Llorens R, Leiva JC (1994a) Glaciological Studies in the High Central Andes through Digital Processing of Satellite Images. In press.

Llorens R, Leiva JC (1994b) Fluctuaciones de los frentes glaciarios de las Nacientes del Río Plomo a través de sensores remotos. In: Actas del 3rd. International Symposium on High-Mountain Remote Sensing Cartography

Paterson WSB (1994) The Physics of Glaciers 3rd edition. Elsevier Science, Oxford

Razza L (1935) El glaciar de Nevado del Plomo. Revista Geográfica Americana 4(25):221–238. Buenos Aires

Palynological Data Indicating Glacier Growth and Environmental Conditions

Monica Cristina Wingenroth
Instituto Argentino de Nivología, Glaciología y Ciencias Ambientales, Palinología, Casilla de Correo 131, 5500 Mendoza, Argentina
Wingenro@lab.cricyt.edu.ar

Abstract: At the Quebrada Benjamín Matienzo, a high valley in the Cordillera Principal of the Central Andes, an organic mineral deposit has been examined. Fossil pollen assemblages from the past 6400 years have been compared with modern the pollen assemblage characteristic for the region. Environmental conditions of the past 6400 years were inferred, including paleowind directions, glacier growth and water availability. In order to trace potential environmental gradients, the inferred environmental conditions have been compared with the environmental conditions, especially glacier advances and retreats, from other key sites inside and outside South America.

9.1. Introduction

The Quebrada Benjamín Matienzo is located in the Cordillera Principal of the Andes, in Las Cuevas, west of the city of Mendoza, Argentina. The Cordillera is a geographic, morphostructural and stratigraphic unit (Yrigoyen 1979), most recently shaped by Quaternary events. Today, glaciers are present in the upper mountain peaks, but geological and geomorphological studies in the area suggest larger extensions of Pleistocene glaciers in the region (e.g. Corte 1954–1957; Espizúa and Corte 1981; Espizúa 1989, 1993; Suarez 1983). Due to its geographic setting and availability of various data types, including palynology (Wingenroth 1990, 1992, 1998), geomorphology (Suarez 1983; Videla 1996), glacial geology (Espizúa 1989, 1993), dendrochronology (Villalba et al. 1990), glaciology (Leiva et al. 1986; Leiva and Cabrera 1995; Leiva in press), and climatology (De Fina 1964; Miller 1976), the area has the potential to be a key site for studying the impact of glacier-line changes on water availability.

We have studied 572 cm of sediment. The main questions addressed are (1) Are there relations between past pollen assemblage modifications, physical and chemical sediment features and glacier volume? (2) If yes, what is this association like? Other research and palynological studies in South America (Caldenius 1932; Auer 1965; Mercer 1965, 1972, 1973, 1983, 1984) contributed to Southern Hemisphere glacier advance and deglaciation knowledge. This work contributes research on the present and past pollen assemblages, physical and chemical sediment features of an organic mineral body, and inferred environmental parameters in relation to glacier growth at the Quebrada Benjamín Matienzo.

9.2. Geography and Geology

The Las Cuevas River flows through the Quebrada Benjamín Matienzo. This valley is about 35 km long at an altitude of 3000–3900 m asl (above sea level). It is located at 32° 35' S to 32° 50' S, 70° 06' W in the Cordillera Principal. Glaciers are found at the summit of the mountains. The glaciers are the source of the Las Cuevas River, through ice and snow melting. This mountain chain has a great glacial influence. It diminishes in altitude to the south of the study area (Regairaz et al. 1991). Marine and continental sediments are linked to its evolution. Limestones and gypsum are present, as well as lutite, andesite, porphyrite, trachyte and basaltic rocks (Yrigoyen 1969).

9.3. Glaciological Background

Glacier events (advances and recessions) have been observed in the region (Corte 1954–1957; Espizúa and Corte 1981; Espizúa 1989, 1993; Suarez 1983). For the Quebrada Benjamín Matienzo, Suarez (1983) proposed four glacier advances. One of the glaciers that contributes through melting ice to the total flow of Las Cuevas River is the Piloto Glacier, at the Cajón del Rubio. This glacier was studied from 1979 to present. Despite heavy snowfall at the Cordillera de Los Andes associated with known ENSO events, the accumulative mass balance is negative between 1979 and present (Leiva et al. 1986; Leiva and Cabrera 1995; Leiva, in press).

9.4. Climatological Background

The nearest climatological station is at Cristo Redentor, 3832 m asl, 32° 50' S, 70° 05' W. According to the Servicio Meteorológico Nacional (1958, 1975, 1981, 1986) of Argentina, the mean temperatures in January and July were, respectively, +4.1° C and –6.9° C (1941–1950), +3.9° C and –7.3° C (1951–1960), +3.8° C and –6.9° C (1961–1970), and +4.1° C and –6.6° C (1971–1980). In the years 1971–1980, the mean annual temperature was –1.8°C. At Las Cuevas, the mean temperature for January and July (1980–1981) was +4.67° C and –6.55° C. Between 1941–1950, the Servicio Meteorológico Nacional (1958) observed an average precipitation of 368 mm. Miller (1976) observed 8 mm precipitation for January, 56 mm for July and an annual average of 357 mm. Rain in this area seldom occurs. It can be observed only in the lowest section of the Quebrada. Precipitation falls mostly as snow or snow pellets.

The Servicio Meteorológico Nacional (1986) reported that for 1971–1980, prevalent winds are from the S/SW, were less frequent from the N/NE, and came the least frequently from the W/NW and E/SE.

9.5. Vegetation

The Quebrada Benjamín Matienzo (Fig. 9.1) is located in the Cordillera de Los Andes and included in the Altoandean phytogeographical region discussed by Cabrera (1971). The Altoandean vegetation is characterized by high species diversity (Hauman 1918; Wingenroth and Suarez 1984; Ambrosetti et al. 1986), but very low plant coverage (Hauman 1918; Wingenroth 1990, 1992).

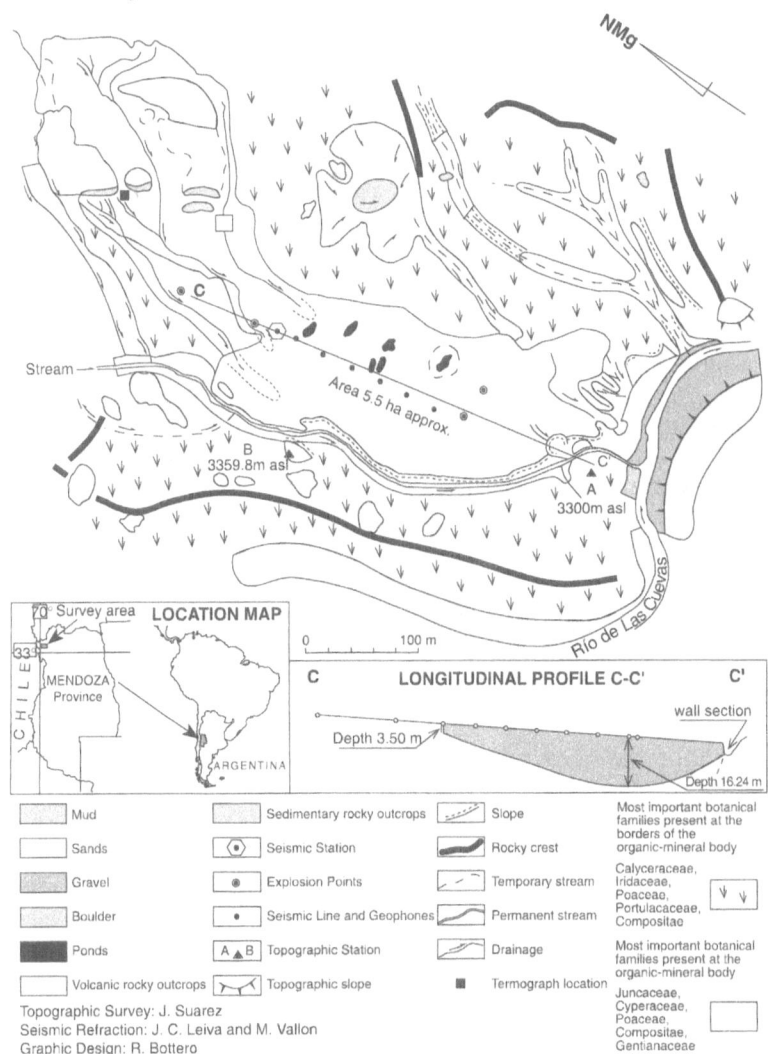

Fig. 9.1. Length and depth of the organic mineral deposit at the Quebrada Benjamin Matienzo.

Presently (1979–1996) various plant communities can be distinguished at the mountain slopes. The *Poa holciformis-Adesmia subterranea* community thrives on stable terrain up to 3700 m asl, with low species diversity and the highest plant frequency and coverage for the slope communities.

The *Poa holciformis-Perezia carthamoides* community thrives upon unstable terrain up to 3700 m asl. It has a very low plant frequency and coverage but the highest species diversity of the three slope communities. Species like *Adesmia, Leuceria salina, Nassauvia uniflora, Phacelia secunda, Senecio crithmoides* and *Senecio volkmannii* are present on the slopes. The *Poa holciformis-Nassauvia lagascae* community flourishes on unstable terrain between 3600–3900 m asl, with low species diversity and the lowest observed plant abundance. At the base of the slopes, between 3300–3400 m asl, the community of *Oxychloe mendocina* and *Carex incurva* thrives upon swampy terrain and shows the highest plant frequency and coverage of all the communities in the Quebrada. It is characterized by high species diversity, with taxa belonging principally to the Cyperaceae, Juncaceae, Compositae, Gentianaceae and Poaceae families. The community of *Poa holciformis-Nastanthus agglomeratus* thrives at the banks of the rivers up to 3600 m altitude, characterized by very low plant abundance and low diversity of taxa. In this community species like *Arenaria andicola, Chaetanthera pulvinata, Chaetanthera spathulifolia* are present. The community of *Poa holciformis-Calceolaria luxurians* flourishes at the margins of streams, with very low plant abundance and scant species diversity. In the two communities mentioned last, the taxa belong principally to the Poaceae, Calyceraceae, Iridaceae, Portulacaceae and Compositae botanical families (Fig. 9.1).

9.6. The Organic Mineral Deposit

At the Quebrada Benjamín Matienzo, ground water flowing down the slopes fills a depression 16.24 m deep. It was most likely generated during a phase of glacier retreat (Suarez, pers. comm.). During several thousands of years it filled with water, peat and mineral sediment. Generally parallel to the Las Cuevas River (Fig. 9.1) a visible section of the organic mineral body, 5.62 m thick and 6400 yrs BP old, shows 174 layers of sediment.

9.7. Material and Methods

This study consists of two parts. In the first part of this study, present-day floral communities have been analyzed, both at the site of the pollen profile and throughout the Quebrada. This allows for the identification of "local" taxa (pollen from taxa thriving at the Quebrada) and "non-local" taxa (taxa living elsewhere but occurring in the pollen assemblages). The terms "local" and "non-local" will be used frequently in the following paragraphs. In the second part of this study, the sequence of fossil pollen throughout the last 6400 years was studied with an emphasis on environmental change.

After identification by botanical specialists, plant species thriving presently at the Quebrada were assigned to xerophytic, hygrophytic and hydrophytic groups. In addition, they were grouped according to altitude (Wingenroth and Suarez 1984). Recent plant communities are identified. Species presence or absence in each community were surveyed using a measuring tape and a wind compass. The recent "local" pollen (from plants at the Quebrada Benjamín Matienzo) have been described (Wingenroth and Heusser 1984). Recent pollen assemblages from the plant communities have been identified (for more detail see Wingenroth, 1992, 1998). In these assemblages, non-local pollen (from plants not in the area of the Quebrada Benjamín Matienzo) have also been observed. As mentioned above, the present vegetation dwelling on the organic mineral body, that has approximately an area of 5.5 ha (Fig. 9.1), is a community characterized largely by the presence of *Oxychloe mendocina* and *Carex incurva* (together with other Cyperaceae). The present surface pollen of this community has been compared with the pollen assemblages of each of the 174 studied layers. In addition, correlations with physical and chemical sedimentological features were performed. Samples of pollen from the surface had been analyzed quantitatively and from the organic mineral body analyzed both qualitatively and quantitatively.

After the present pollen assemblage from the *Oxychloe mendocina* and *Carex incurva* community was recognized, a fossil layer with a similar pollen assemblage was identified. This is regarded as a time with environmental conditions similar to those of today. Other environmental conditions characterized by different pollen assemblages have been identified as well.

The applied methodology permits the identification of two variables: The first is defined by the sum of the pollen averages consisting of local species that presently thrive on the slopes and river banks. The second consists of averages of the Cyperaceae pollen (plants that thrive on spongy soils). Both variables have been related to chemical and physical characteristics of the sediments (for instance, pH), the presence of carbonates, color and organic material. They were also related to the pollen diversity and the habitat of those plants whose pollen appeared in the analyzed sediments.

In the studied regional context the variables are interpreted to represent water availability and temperature modifications of the study area. They allowed, at least for the last 6400 years, the identification of cold (dry, mesic and wet), temperate (dry, mesic and wet) and warm (dry and wet) times (Wingenroth 1992). In addition, seven environmental conditions in relation to the present (Wingenroth 1998) could be identified: paleovegetation, paleotemperature, paleowater availability and predominant paleowinds.

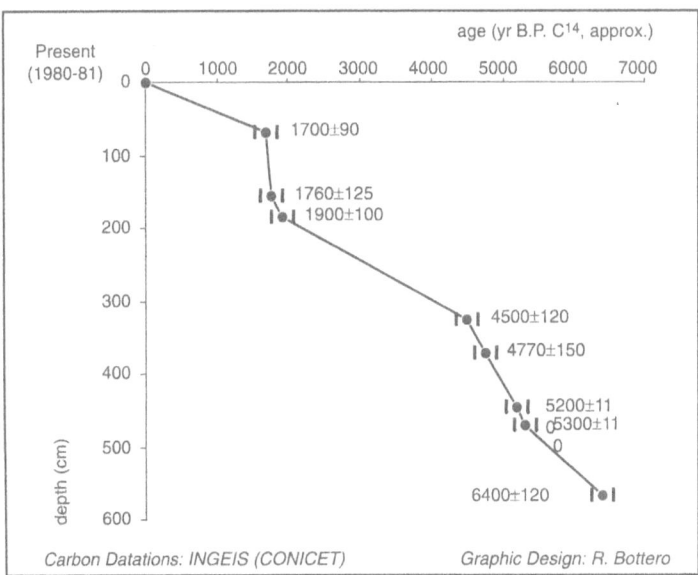

Fig. 9.2. Sedimentation rate of the analyzed section at the Quebrada Benjamin Matienzo.

In order to compare paleoclimatic discussions, a time-scale was introduced assuming linear sedimentation rates between the eight non-corrected ^{14}C dated samples. This approach is common practice within the context of the Ocean Drilling Program (Smolka, pers. comm.; see also e.g. Moullade 1987:502). Both the observed sedimentation rate (Fig. 9.2.) and field evidence (Heusser, pers. comm.) support this approach. The relative sequence of the discussed environmental changes will not be affected by potential future reassessments and recalibrations of the time-scale.

Temperature was recorded at the organic mineral body during six consecutive months from December 1981 to May 1982, using a Grant thermograph (Fig. 9.3). Five sensors were distributed, one above ground, two in the mud at 10 and 30 cm depth, and two in the ground water at 5 and 20 cm depth. Recent temperature measurements (see Fig. 9.3) show that when the air temperatures (1981–1982) were relatively cooler or warmer, groundwater temperatures were respectively higher or lower, and mud temperatures at 10 cm depth, were respectively lower or higher.

9.8. Results

9.8.1. Palynological and Sedimentary Features

Three main associations (environments) have been identified. These are described by both pollen assemblages and sedimentary features. While one association (environment) type (A) characterizes the conditions of today, other association (environment) types (B and C) can only be found in the fossil strata.

9.8.1.1. Local Pollen Assemblage and Sedimentary Features Type A:

Association (environment) Type A is characterized by the following conditions:
The pollen diversity from the local species that presently thrive upon the slopes and river banks is higher than four, the sum of their pollen percentages varies between 57.0 and 58.5% with an average of 48.5. This includes exceptionally high percentages and averages for the Poaceae (41.5–49.5%, avg. 41). The pollen percentages from the Cyperaceae presently thriving upon the morasses vary between 28.4 and 36.8% with an average of 23.5. Details can be found in Table 9.1a. Pollen species always present are *Adesmia, Nastanthus agglomeratus, Perezia carthamoides, Poa holciformis, Senecio crithmoides* and *Senecio volkmannii*. The sedimentary features of assemblage Type A (Table 9.2) are dark brown or light brownish gray colors, pH of 6, weak reaction of the carbonates and a content of organic matter around 10.6%. Association Type A describes overall the present-day conditions (see Table 9.4).

Table 9.1. Sum of pollen percentages and averages coming from local species that presently thrive upon the slopes and river banks (Poaceae included) and pollen percentages and averages from the Cyperaceae, that presently thrive upon the morasses.

	Assemblage A	Assemblage B	Assemblage C
Angiospermous species :	48.5 (avg.) 57.0–58.5%	54.0–279.2 (avg.) 59.4–93.4%	68.0–301.0 (avg.) 22.2–64.4%
Poaceae	41.0 (avg.) 41.5–49.5%	49.75–276.0 (avg.) 57.3–93.0%	65.0–292.0 (avg.) 21.8–62.2%
Cyperaceae	23.5 (avg.) 28.4–36.8%	0.75–21.0 (avg.) 0.0–19.3%	65.5–459.0 (avg.) 25.4–70.5%

Table 9.2: Physical and chemical features of the sediments for assemblages A, B and C.

Ass.	pH	Color as defined in Munsell Soil Color Charts (Chart 10YR, Charts 5YR, 2.5YR and 7.5YR)	Reaction of carbonates	Organic matter (%)
A	6.0	Light brownish gray, dark brown	Weak	10.6
B	2.0–2.5 6.5*	Dear brown, dark gray, gray-dark-yellowish brown grayish brown, pinkish gray, yellowish brown, grayish brown, pale brown, very dark grayish brown, dark grayish brown, pinkish brown	Weak, strong*	4.12–60.60
C	5.5–6.0 3.0*	Dark reddish brown, yellowish brown, dark yellowish brown, dark grayish brown, grayish brown, dark brown	Feeble, strong	15.80–34.20

*In exceptional cases

9.8.1.2. Local Pollen Assemblage and Sedimentary Features Type B:

Compared to assemblage (environment) Type A, pollen diversity is similar or higher, if one considers only those pollen of the local species that today thrive upon the slopes and river banks. Overall, the sum of their pollen percentages and averages, that include the Poaceae, is frequently higher, reaching values of 59.4–93.4% for percentages and 54–279.2 for averages. Pollen values from the Cyperaceae are lower (0.0–19.3% and 0.75–21.0, see Table 9.1). Pollen species always present are *Adesmia, Arenaria andicola, Leuceria salina, Nassauvia uniflora, Nastanthus agglomeratus, Phacelia secunda, Poa holciformis* and *Senecio crithmoides*. Sediment features of this pollen assemblage are characterized by lower pH, generally darker colors, higher organic matter contents, and coincident weak reactions of the carbonates in relation to the sediments in the association Type A (Table 9.2).

9.8.1.3. Local Pollen Assemblage and Sedimentary Features Type C:

Compared to assemblage (environment) Type A, pollen diversity is generally similar or higher owing to pollen coming from the local species that presently thrive upon the slopes and river banks. In the fossil strata, the sum of their pollen percentages (22.2–64.0%) is lower, similar or higher than that of the assemblage (environment) Type A. The sum of their averages (68–301) is always higher than that of assemblage (environment) Type A. The pollen percentages and averages of the Cyperaceae (25.4–70.5% and 65.5–459 for averages) are mostly higher (Table 9.1). Pollen species present are *Adesmia, Arenaria andicola, Chaetanthera pulvinata, Leuceria salina, Nastanthus agglomeratus, Phacelia secunda, Poa holciformis, Senecio crithmoides* and *Senecio volkmannii*. Compared to the association Type A (Table 9.2), sediment features of this association type have a similar pH, similar carbonate reactions, generally darker colors and higher content of organic matter.

9.8.1.4. Non-Local Pollen in Assemblages Type A, B, C

In assemblage (environment) Type A, the non-local pollen is characterized by the absence of *Fitzroya*, scanty averages and percentages for *Austrocedrus, Nothofagus, Podocarpus*, Umbelliferae, Chenopodiaceae, Pteridophyta and *Typha*, and slightly higher percentages and averages of *Ephedra* (Table 9.3).

Table 9.3. Averages and percentages of non-local pollen. The upper number(s) describe the average of the species (individuals), the lower number their percentages in the corresponding assemblage.

Assemblages: Subassemblages:	A	B b1	B b2	B,C b3, c3	B,C b4, c4	B,C b5,c5	B b6
Taxa							
Fitzroya	0 0	0–0.25 0–0.3	0 0	0 0	0 0	0 0	0 0
Nothofagus	0.25 0.3–0.7	0 0	0 0	0–1.5 0–0.9	0 0	0 0	0 0
Podocarpus	0.25 0–0.3	0 0	0 0	0 0	0–1 0–0.1	0–0.5 0–0.6	0 0
Austrocedrus	1.75 0.1–2.1	3–4 3.8–4.4	0.75–1 1.1–1.5	0.25–19 0.4–16.5	6–25 0.8–10	2–12 1.3–4.9	5 18
Ephedra	3 2.8–3.6	1.5–2 1.9–2.2	1–2.75 1.5–4.1	3–8.7 0.7–6.7	12–27 4.1–12.2	9–15 2.8–11.3	39 14
Umbelliferae	0.75 0.7–0.9	0.25–4 0.3–4.4	0–0.75 0–1.5	0–2 0–2.6	0–2 0–0.8	0–5 0–2.4	0 0
Chenopodiaceae	0.75 0.3–0.9	1.25–2 1.6–2.2	0.25–1.25 0.4–1.9	0–3.5 0–2.7	0–2 0–0.5	3.3–7 1.4–5.3	16 57
Pteridophyta	1.5 0–1.8	0–0.75 0–0.9	0 0	0–4.7 0–3.9	0–8 0–3.2	0–4 0–1.8	1 0.3
Typha	2 0.4–2.4	0 0	1–2.75 0.4–0.7	0–1 0–0.8	0–2 0–2	0–4 0–1.2	2 0.7

Differences within the non-local pollen of assemblage (environment) Type B and C lead to the definition of subassemblages (subenvironments). The non-local pollen of these subassemblages are compared with the non-local pollen of assemblage Type A, including the recent non-local pollen. Consequently, subassemblage Type b1 is characterized by the occasional presence of *Fitzroya*, lower averages and percentages for *Ephedra* and *Typha* and values that are higher for *Austrocedrus* and Chenopodiaceae. In addition they are sometimes lower and sometimes slightly higher for Pteridophyta and Umbelliferae. In relation to recent non-local pollen, the subassemblage Type b2 is generally characterized by lower averages and percentages for *Ephedra, Austrocedrus,* Pteridophyta and *Typha* and lower or slightly higher percentages and averages for the Umbelliferae and Chenopodiaceae. In relation to recent non-local pollen the fossil non-local pollen from subassemblages Type b3 and c3 display lower, similar or higher percentages and averages for *Nothofagus, Austrocedrus,* Umbelliferae, Chenopodiaceae and Pteridophyta, always similar or higher averages and lower or higher percentages for *Ephedra,* and generally lower averages and percentages for *Typha.* The presence of *Podocarpus* is sometimes observed in the non-local pollen of subassemblages Type b4 and c4, with, in relation to recent non-local pollen, generally very high averages and percentages for *Austrocedrus* and *Ephedra,* similar percentages for Chenopodiaceae, Umbelliferae, Pteridophyta and *Typha,* and lower or generally higher averages for Chenopodiaceae, Umbelliferae, Pteridophyta and *Typha.* In subassemblages b5 and c5 *Podocarpus* is seldom observed. The non-local pollen is generally characterized by much higher averages and percentages for *Austrocedrus, Ephedra* and Chenopodiaceae, and similar, higher or lower averages and percentages for Pteridophyta, *Typha* and Umbelliferae. At least one sample, assigned to subassemblage b6, is characterized by generally much higher averages and percentages of *Austrocedrus, Ephedra* and Chenopodiaceae and by the absence of Umbelliferae. In addition, again in relation to recent non-local pollen (see Table 9.3) lower or similar averages, and lower or higher percentages were observed for Pteridophyta and *Typha.*

9.9. Interpretation and Discussion

9.9.1. Water Availability

Hauman (1918) and Wingenroth (1980–1996, field surveys) found that plant diversity and abundance at the slopes and river banks depends, in the high Andes, on water availability. Birkeland (1984) mentioned feeble carbonate reaction, dark colors and high organic-matter contents as sedimentary features pointing to wetter environmental conditions. In the studied region, water availability depends on snow and/or ice melting (see "Climatological Background"). Therefore pollen of assemblage B and C, and related sedimentary features suggest, in relation to recent conditions, similar plant distribution at the slopes and river banks with higher species abundance and similar or higher species diversity. Thus, for these environments a higher water availability is inferred.

9.9.2. Temperature

For morass grasses, Ruthsatz (1995) observed that bud, shoot and leafs delayed their appearance. In addition they sometimes disappeared with lower temperature. Air temperatures at the high altitudes of the study area are low in both the summer and winter (Servicio Meteorológico Nacional 1986). In addition, at 10 cm depth in the organic mineral body of the Quebrada Benjamín Matienzo, where the root system of the Cyperaceae species develops, these temperatures correlate with air-temperatures. This correlation is even more pronounced for meteorological situations with low temperatures: When the air temperature is for a prolonged period low, the soil temperature at 10 cm depth is generally much lower and vice versa (see Fig. 9.3). As temperatures at this altitude level are low (see section on climatological background), the Cyperaceae thriving at the morass are directly affected. Therefore the amount of *Carex incurva*, *Carex goodenoughii* and *Eleocharis* (Cyperaceae) pollen can be used as an indicator for air temperature.

When compared with the recent assemblage, higher amounts of the Cyperaceae pollen observed in assemblage Type C possibly suggest the flourishing of the Cyperaceae at the morasses, probably as response to higher temperatures. Lower amounts of the Cyperaceae pollen as documented in assemblage Type B thus indicate lower temperatures. It should be noted that this assignment is only of local and regional significance. While the approach can be applied to other areas of the world, the calibration has to be done within a regional context.

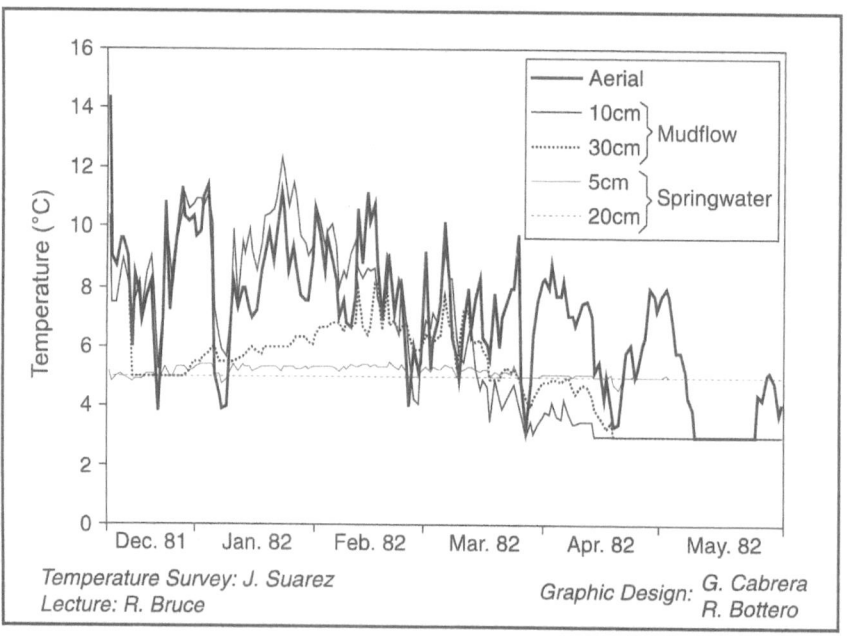

Fig. 9.3. Observed temperatures (°C) at the organic mineral deposit at the Quebrada Benjamín Matienzo.

9.9.3. Deduction of Predominant Wind Directions

Presently (1980–1981, assemblage Type A) wind blows predominantly from the south and southwest (Servicio Meteorológico Nacional 1970–1980). This shows up in the present pollen assemblage, that consists of pollen possibly coming from (1) the latifoliate forest and the Andean steppe (Quintanilla 1985), and (2) the Litre-Cipres forest (Oberdorfer 1960).

Compared with recent pollen, non-local pollen from subassemblage b2 points to environmental conditions with predominant winds probably blowing through the above mentioned steppe and forests, but from the western lower latitudes. During the deposition of subassemblage b1, the predominant winds blew generally from higher southwestern latitudes as (compared with recent pollen) the winds possibly blew (1) through the Alerce and Litre-Cipres forest (Oberdorfer 1960) and (2) through the Andean steppe (Quintanilla 1985).

Past environmental conditions documented by subassemblages b3 and c3, suggest, through the comparison with recent pollen, predominant southeasterly winds. They possibly blew through the latifoliated temperate forest (Quintanilla 1985), the Litre-Cipres forest (Oberdorfer 1960), the Altoandean pulvinate vegetation, and through the high Pampean and piedmont communities (Roig et al. 1998).

In relation to recent pollen assemblages, subassemblages b4 and c4 suggest environmental conditions with predominant winds coming from the south. Winds probably blew through the *Austrocedrus chilensis* forest, occasionally through the pulvinated communities with Umbelliferae (observed by Roig 1998), through the high Pampean and piedmont communities with *Ephedra*, and seldom through the shrub steppes with the Chenopodiaceae described by Roig et al. (1998).

In relation to recent pollen, non-local pollen from subassemblages b5 and c5 suggest environmental conditions with predominant winds from the east and northeast. They probably blew through the Litre-Cipres forest (Oberdorfer 1960), occasionally through the Umbelliferae pulvinate communities, through the high Pampean and piedmont regions, through the psamophyll communities of Travesía de Guanacache and after turning back to the east, through the halophyll communities of the Huayquerías (reported by Roig et al. 1998).

Subassemblage b6, when related with recent non-local pollen, points to environmental conditions with predominant winds from the north and northeast, probably blowing through the Litre-Cipres forest (Oberdorfer 1960), through the Andean steppe (Quintanilla 1985), through the Altoandean vegetation, through the psamophyll communities of the Travesía de Guanacache and the Huayquerías and Barreales (Roig et al. 1998), and perhaps also through the Salinas Grandes halophyll communities (described by Ragonese 1951).

9.9.4. Glacier Growth

Water availability depends on the melting of snow and ice in this region. Consequently, at this altitude and related to the present environmental conditions, higher plant abundance and diversity on the slopes indicate higher water availability and lower Cyperaceae abundance at the morasses indicates lower temperatures. Both features together denote higher snowfall and suggest glacier growth in the valley at this altitude, with water availability possibly the principal factor effecting glacier growth component in the area. In the Cordillera de Mérida, Malavé (1989) stated that glacier growth depends principally on lower temperatures.

The environment with the present pollen assemblage (1980–1981) is characterized by July mean temperatures of –6.55° C, January mean temperatures of +4.67° C and predominant winds blowing from the south and southwest. The average water flow in the Las Cuevas River in July is 4.22 m³/s and in January is 14.38 m³/s (Agua y Energía Eléctrica, 1961). A similar pollen assemblage was observed 4635 yrs BP. Consequently, similar environmental conditions are deduced for that time.

In relation to today's pollen assemblages, we find higher amounts of Cyperaceae pollen and local taxa pollen for the periods 3520–3420, 3323–3284, 2955–2936, 2570–2551, 2522–2493, 2445–2416, 1490–1413, and 950–884 yrs BP. Cyperaceae are taxa that presently are thriving at the morasses. Local taxa are thriving today at the slopes and river banks. The higher quantities of these taxa indicate that during these periods in the past, this area had higher temperatures and water availability compared to today. A much higher presence of *Austrocedrus* and *Ephedra* during 3520–3420, 2445–2416, and 950–884 yrs BP indicates the predominance of southern winds:. Additionally, the very high presence of Chenopodiaceae indicates the predominance of easterly and northeasterly winds. Predominant winds came from the southeast during the period 1490–1413 yrs BP, characterized by slightly higher amounts of *Ephedra* pollen.

For the periods 4885–4852, 3284–3072, 2676–2637, 2281–2020, 1900–1704, 1038–950 and 884–807 yrs BP (with occasional short interruptions), higher water availability and lower temperatures than the present ones prevailed. This is indicated by higher amounts in the past than today of local taxa pollen that nowadays flourish at the slopes and river banks of the Quebrada, and by a lower quantity of Cyperaceae pollen, that presently thrive at the morass.

In addition, very high quantities of *Austrocedrus, Ephedra* and Chenopodiaceae pollen indicate predominant winds from the east and northeast during 3284–3072, 2039–2020, 1811–1799 and 884–807 yrs BP. Very high abundances of *Austrocedrus and Ephedra,* the highest abundance of Chenopodiaceae, and the absence of Umbelliferae, indicate predominant winds from the north and northeast, 2676–2637 yrs BP. Higher abundances of *Ephedra* and *Austrocedrus* indicate prevailing winds from the south, 1038–950 yrs BP. Slightly higher abundances of *Ephedra* and slightly lower or higher presence of *Austrocedrus* indicate prevailing winds from the southeast during 4870–4852, 2281–2242, 2136–2078, 1859–1811, 1799–1729 (with short interruptions) and 1709–1704 yrs BP. Generally, lower amounts of *Austrocedrus* and *Ephedra* indicate prevailing winds from the west, 4885–4879 yrs BP. Finally, higher amounts of *Austrocedrus* and slightly higher or lower amounts of *Ephedra* indicate prevailing southwesterly winds, 1900–1888 and 1717–1711 yrs BP. It should be noted that although this method can be applied worldwide, these specific findings are based on regional calibrations and regional interpretations of pollen associations in areas around the study area.

The vertical sequence of the paleoenvironmental conditions discussed above is shown in Table 9.4.

Table 9.4. Paleoenvironment sequence through time. Ages have been linearly interpolated between the non-corrected ^{14}C ages (Fig. 9.2). Sedimentary evidence supports this approach.

Depth (cm)	Age BP	Ass. Subass.	Las Cuevas River water quantity (1980/1981)		Temperature		Wind
0–0	Present	A	July 4.22 m³/s	January 14.38 m³/s	July −6.55°C	January 4.67°C	S SW
36.5–40.0	807–884	B5	H		L		E NE
40.0–43.0	884–950	C4	H		H		S
43.0–47.0	950–1038	B4	H		L		S
64.0–67.0	1413–1490	C3	H		H		SE
81.0–83.5	1704–1706	B3	H		L		SE
83.5–88.5	1706–1709	B3	H		L		SE
92.0–100.5	1711–1717	B1	H		L		SW
117.5–123.0	1729–1733	B3	H		L		SE
131.0–135.5	1738–1741	B3	H		L		SE
152.5–154.0	1752–1753	B3	H		L		SE
155.5–157.0	1754–1755	B3	H		L		SE
157.0–165.0	1755–1760	B3	H		L		SE
165.0–175.0	1760–1799	B3	H		L		SE
173.0–175.5	1799–1811	B5	H		L		E NE
175.5–185.5	1811–1859	B3	H		L		SE
191.5–194.0	1888–1900	B1	H		L		SW
200.5–201.5	2020–2039	B5	H		L		E NE
203.5–206.5	2078–2136	B3	H		L		SE
212.0–214.0	2242–2281	B3	H		L		SE
221.0–222.5	2416–2445	C4	H		H		S
225.0–226.5	2493–2522	C5	H		H		E NE
228.0–229.0	2551–2570	C5	H		H		E NE
232.5–234.5	2637–2676	B6	H		L		N NE
248.0–249.0	2936–2955	C5	H		H		E NE
255.0–266.0	3072–3284	B5	H		L		E NE
266.0–268.0	3284–3323	C5	H		H		E NE
273.0–275.5	3420–3468	C4	H		H		S
275.5–278.5	3468–3520	C4	H		H		S
351.5–355.5	4635–4659	A	SIM.		SIM.		SIM.
388.0–391.0	4852–4870	B3	H		L		SE
392.5–393.5	4879–4885	B2	H		L		W

Explanation: Sim.: Similar; Ass. and Subass.: Assemblages and subassemblages as discussed in the text; Water availability: H: Higher; Temperature: H: Higher; L: Lower; Predominant Wind Direction: S: South; SE: Southeast; SW: Southwest; W: West; E NE: East Northeast.

In this context, the comparison of the environmental conditions observed in the Quebrada Benjamín Matienzo with known glacier advances and retreats from other parts of the world is interesting. This shows which climatic signals are of global importance and which document local and regional conditions. The details can be found in Table 9.5. This table shows in the left column the depth of the respective sediments from the Quebrada Benjamín Matienzo. The second column shows the interpreted age. The third column shows environmental phenomena of comparable ages. Environmental phenomena that are not explicitly described are glacier advances.

Table 9.5. Correlation of paleoenvironments observed at the Quebrada Benjamín Matienzo with glacier advances in other areas.

Depth (cm)	¹⁴C and approximate calculated ages (yrs BP)	Glacier Advances in Other Areas Author (Year); Location; [age in yrs BP]
11.5–13.0	260–284	Röthlisberger (1986); Río Manso Glacier; [300 ± 85]
		Seltzer and Wright (1989); 11° 51' S / 75° 06' W; [250 ± 60], (Minimum date for glacier recession)
		Vianna et al. (1989); 4° 54' – 5° 05' S / 35° 30' – 35° 15' W; [250]
36.5–40.0	807–884	Seltzer and Wright (1989); 11° 51' S / 75° 06' W; [650 ± 60], peat between two tills, (Maximum age for at least one Glacial advance)
43.0–47.0	950–1038	Seltzer and Wright (1989); 11° 51' S / 75° 06' W; [920 ± 200], Minimum date for glacier retreat
77.0–100.5	1700–1709 1709–1710 1711–1717	Stingl and Garleff (1985); Cerro Domuyo 36° 38' S / 70° 26' W; [2300–1500]
117.5–123.0 129.0–135.5	1729–1733 1737–1741	Stingl and Garleff (1985); Cerro Domuyo 36° 38' S / 70° 26' W; [2300–1500]
152.5–194.0 200.5–201.5	1752–1900 2020–2039	Stingl and Garleff(1985); Cerro Domuyo 36° 38' S / 70° 26' W; [2300–1500]
		Mercer and Ager (1983); Uppsala Glacier, 48° – 51° S; [2000]
203.5–206.5	2078–2136	Stingl and Garleff (1985); Cerro Domuyo 36° 38' S / 70° 26' W; [2300–1500]
		Mercer and Ager (1983); Hammick, 48° 57' S / 74° 13' W; [2070 ± 95]
		Ten Brink and Weidick (1974); 66° 30' – 67° 00' N; [2000–2500]
212.0–214.0	2242–2281	Stingl and Garleff (1985); Cerro Domuyo, 36° 38' S / 70° 26' W; [2300–1700]
		Stingl and Garleff (1978); Río Atuel, 34° S; [2290 ± 60]
		Heusser (1974); Puerto Octay, 41° S / 73° W; [2700–2200]
		Mercer and Ager (1983); Hammick, 48° 57' S / 74° 13' W; [2300 ± 110]
		Ten Brink and Weidick (1974); 66° 30' – 67° 00' N; [2000–2500]
232.5–234.5	2637–2676	Heusser (1974); Puerto Octay, 41° S / 73° W; [2700–2200]
		Mercer (1982); Hammick, 48° 57' S / 74° 13' W; [2800 ± 100]
255.0–266.0	3072–3284	Heusser (1983); Alerce, 41° 25' S / 72° 54' W; Taiquemo, 42° 10' S; [Before 3200]
		Mercer (1982); Hammick, 48° 57' S / 74° 13' W; [2800]
		Garleff et al. (1991); 27° 30' S; [2990 ± 70] (Wet)
351.5–355.5 388.0–391.0 392.5–393.5	4635–4659 4852–4870 4879–4885	Heusser (1960); San Rafael lake, 46° 40' S / 74° 00' W; [Around 5000]
		Heusser (1983); Alerce, 41° 25' S; Taiquemo, 42° 10' S; [Around 5000]; Washington [4700 ± 300]; Swiss Alps [4600 ± 80]
		Mercer and Ager (1983); East of the Cordillera, 48° – 51° S; [5300], (Glacial maximum)
		Campbell (1989); North of La Paz; [Approx. 5000]

Deduced higher water availability in the Las Cuevas River, lower temperatures and inferred predominant wind directions during the same time intervals (Table 9.4) suggest that during the deposition of environment Type B, wetter winds blew very frequently from the southeast and occasionally from easterly-northeasterly directions. This is in agreement with Sayago and Collantes (1991), Fox and Strecker (1991) and Servant et al. (1989), who emphasize the importance of the easterlies as a moisture source for glaciation in the Andes. Occasionally, wet winds also arrived from the west and southwest, and rarely from the south and the north-northeast. Servant et al. (1989) observed also low lake levels and minor glacier extension during times of predominance of westerlies. Thus, times characterized by environment Type B are also times of glacier growth.

Higher water availability at Las Cuevas River, higher temperatures in the studied area and predominant winds from the south, east-northeast and once, the southeast, are intervals that are surprisingly congruent with times of outwash or deglaciations mentioned in other regions (see Table 9.6). With the present knowledge it is not possible to interpret growth or wastage of glacier volume in the area for the episodes that are characterized by environment Type C.

Table 9.6. Correlation of the paleoenvironments observed at the Quebrada Benjamín Matienzo with deglaciations in other areas.

Depth (cm)	Times of higher water availability and temperatures in the study area (yrs BP)	Outwash, deglaciation, alluvial deposits, ridges, sea and lake-level rise in other places Author (year); description/location of paleoenvironment; [age in yrs BP]
0.0–11.5	Present–260	Malagnino and Strelin (1992); Herminita Peninsula, Glacier retreat; [260]
40.0–43.0	884–950	Ortlieb et al. (1989); Northwestern coast of Peru; [960 ± 230] (Ridges)
		Stine and Stine (1990); Lake Cardiel, 48° S / 71° W; [865]
64.0–77.0	1413–1700	Seltzer and Wright (1989); 11° 51' S / 75° 06' W; [1290 ± 85 yrs BP] (Minimum dates for glacier recession)
		Hope and Peterson (1975); 4° N / 136–140° W; [1550–1350], deglaciation
221.0–222.5 225.0–226.5 228.0–229.0	2416–2445 2493–2522 2551–2570	Ortlieb et al. (1989); Northwestern coast of Peru; [2170 ± 300, 2080 ± 560, 2040 ± 400, 2550 ± 500, 2510 ± 250] (Ridges)
248.0–249.0	2936–2955	Knoppers et al. (1989); Fluminense coastline, between Río de Janeiro and Cabo Frío; [2700]
		Wijmstra and van der Hammen (1966); Laguna de Aguas Sucias, Colombia; [3000], high lake levels
266.0–268.0	3284–3323	Ortlieb et al. (1989) Northwestern coast of Peru 3170 ± 300, 3020 ± 250, 2890 ± 250 yrs BP
273.0–275.5 275.5–278.5	3420–3468 3468–3526	Dominguez et al (1989); Doce River beach, east coast of Brazil; [3800–3500] (sea level rise)
		Suguio et al. (1989); Southeastern Brazil; [4000–3000], alluvial deposits

9.10. Conclusions

Palynological and sedimentary observations have been integrated to set up a time-series of paleoenvironmental events. Interpreted phenomena include water availability, temperature, paleowind direction, and times of inferred glacier growth. Furthermore, the observed paleoenvironments have been correlated with known glacier advances and retreats from other areas of the world.

Compared to present day conditions, environments with higher water availability and temperatures were observed at the Quebrada Benjamín Matienzo. During such times, winds blew predominantly from the south (3520–3420, 2445–2416 and 950–884 yrs BP), east-northeast (3323–3284, 2955–2936, 2570–2551 and 2522–2493 yrs BP) and seldom from the southeast (1490–1413 yrs BP). Other authors interpret that at different areas for similar time intervals, there exists outwash. For these time-intervals ice growth cannot be inferred.

In relation with recent environmental conditions, times of higher water availability and lower temperature were also observed at the Quebrada Benjamín Matienzo. During these times, winds blew during one period from the west (4885–4879 yrs BP); seldom from the southwest (1900–1888, 1717–1711 yrs BP); very often from the southeast (4870–4852, 2281–2242, 2136–2078, 1859–1811, 1799–1729, with some interruptions, and 1709–1704 yrs BP); once from the south (1038–950 yrs BP); more frequently from the east-northeast (3284–3072, 2039–2020, 1811–1799, 884–807 yrs BP); and once from the north-northeast (2676–2637 yrs BP). Ice growth was inferred for all of these environments. This coincides generally with glacier advances detected in other areas.

Acknowledgements: I thank Arturo Corte, Wolfgang Volkheimer, Peter Smolka and Fidel Antonio Roig for review and critical comments. I am indebted to Liliana Andrada, Maria Elena Soler and Jorge Suarez. I also thank very much Richard Branham, Juan Carlos Leiva, Gabriel Cabrera, Robert Bruce, Luis Baigorria, Rafael Bottero, Dario Soria and Jorge D'Angello. I gratefully acknowledge the Consejo Nacional de Investigaciones Científicas y Técnicas, Centro Regional de Investigaciones Científicas y Tecnológicas, and Instituto Argentino de Nivología, Glaciología y Ciencias Ambientales and all their members.

References

Agua y Energía Eléctrica (Empresa del Estado) (1961) Anuario Hidrológico (1953–1958). T.II. Buenos Aires

Ambrosetti J, Del Vitto L, Roig F (1986) La Vegetación del Paso de Uspallata, Provincia de Mendoza, Argentina. Veröff Geob Inst 91:141–180

Auer V (1965) The Pleistocene of Fuego–Patagonia, Part IV: Bog Profiles. Ann Acad Scien Fenn Series A, III. Geol-Geogr 80

Birkeland P (1984) Soils and geomorphology. Oxford University Press, Oxford, pp. 372

Cabrera A (1971) Fitogeografía de la República Argentina. Bol Soc Arg Bot 14 (1–2)

Caldenius C (1932) Las Glaciaciones Cuaternarias en la Patagonia y Tierra del Fuego. Geogr Ann Bd. XIV

Campbell K (Jr.) (1989) The Late Pleistocene of South America: A New Approach. International Symposium on Global Changes in South America during the Quaternary: Past–Present–Future. Special Publication no.1. ABEQUA–INQUA (ed), São Paulo, Brazil

Corte A (1954–1957) Sobre Geología Glacial Pleistocénica de Mendoza. Anal Depart Invest Cient Sec Geofis T.II Fasc. 2. Univ. Nac. de Cuyo, Mendoza

De Fina A, Giannetto F, Richard A, Sabella L (1964) Difusión Geográfica de Cultivos. Indices en la Provincia de Mendoza y sus causas. Instituto Nacional de Tecnología Agropecuaria, Pub (83)

Dominguez J (1989) Ontogeny of a strandplain: Evolving concepts on the evolution of the Doce river beach-ridge plain (East coast of Brazil). International Symposium on Global Changes in South America during the Quaternary: Past–Present–Future. Special Publication no 1. ABEQUA–INQUA (ed) São Paulo, Brazil

Espizúa L, Corte A (1981) Inventario de Glaciares del Río Mendoza. IANIGLA (ed) CONICET, Mendoza

Espizúa L (1989) Climas Cuaternarios de América del Sur. Publicación Especial no. 1, Project 281 IGCP UNESCO

Espizúa L (1993) Quaternary Glaciations in the Río Mendoza Valley, Argentina Andes. Quat Res 40(2):150–162

Fox A, Strecker M (1991) Pleistocene and modern snowlines in the Central Andes (24–28° S). Bamb Geograph Schr Bd. 11:169–182

Garleff K, Schäbitz H, Stingl H, Veit H (1991) Jungquartäre Landschaftsentwicklung und Klimageschichte beiderseits der Ariden Diagonale Südamerikas. Bamb Geograph Schr Bd. 11:359–394

Hauman L (1918) La Vegetation des Hautes Cordillères de Mendoza. Imprimerie et Edition "Coni", Buenos Aires

Heusser C (1960) Late Pleistocene environments of the Laguna de San Rafael, Chile. Geogr Rev 50:555–577

Heusser C (1974) Vegetation and Climate of the Southern Chilean Lake District During and Since the Last Interglaciation. Quat Res 4:290–310

Heusser C (1983) Late Quaternary Climates of Chile. In: Vogel JC (ed) Late Cenozoic Paleoclimates of the Southern Hemisphere. A. Balkema, Rotterdam Boston

Hope G, Peterson J (1975) Glaciation and Vegetation in the High New Guinea Mountains. Quat Studies. Suggate RP and Cresswell MM (eds), Wellington, pp 155–162

Knoppers B, Machado E, Moreira P, Turcq B (1989) A physical and biogeochemical description of Lagoa de Guarapina, a subtropical Brazilian Lagoon. International Symposium on Global Changes in South America during the Quaternary: Past–Present–Future. Special Publication no.1. ABEQUA–INQUA (ed) São Paulo, Brazil

Leiva J, Cabrera G, Lenzano L (1986) Glacier mass balances in the Cajón del Rubio, Andes Centrales Argentinos. Cold Reg Sc Techn 13:83–90

Leiva J, Cabrera G (1995) Glacier mass analysis and reconstruction in the Cajón del Rubio, Mendoza, Argentina. Zeitschr Gletsch Glazialgeol Bd. 31:1–7

Leiva J (1995, in press) Present situation of the Andean Glaciers. In: IANIGLA, Rev 25 años del IANIGLA, Mendoza

Malavé S (1989) Topoclimates and Quaternary glaciers in Cordillera de Mérida, Venezuela. International Symposium on Global Changes in South America during the Quaternary: Past–Present–Future. Special Publication no.1. ABEQUA–INQUA (ed) São Paulo, Brazil

Malagnino E, Strelin J (1992) Variations of Upsala Glacier in southern Patagonia since the late Holocene to the present. In: Naruse R, Aniya M (eds) Glaciological Researches in Patagonia, 1990. Ministry of Education, Science and Culture, Japan

Mercer J (1965) Glacier variations in the Andes. Geogr Rev 55:390–413

Mercer J (1972) Chilean Glacial Chronology 20, 000 to 11, 000 Carbon-14 Years Ago: Some Global Comparisons. Science 176:1118–1120

Mercer J (1973) Glacier in Chile Ended a Major Readvance about 36, 000 Years Ago: Some Global Comparisons. Science 182:1017–1019

Mercer J (1982) Holocene glacier variations in southern South America. Striae 18:35–40

Mercer J (1983) Cenozoic Glaciation in the Southern Hemisphere. Ann Rev Earth Plant 11:99–132

Mercer J (1984) Simultaneous Climatic Change in both Hemispheres and similar Bipolar Interglacial Warming: Evidence and Implications. In: Amer Geophys Union (ed.) Climate processes and Climate Sensitivity

Mercer J, Ager T (1983) Glacial and Floral Changes in Southern Argentina since 14, 000 Years Ago. Nat Geog Soc Res Rep 15:457–477

Miller A (1976) The Climate of Chile. In: Schwerdtfeger (ed) World Survey of Climatology: Climate of Central and South America. Elsevier, Amsterdam

Moullade M (1987) Distribution of Neogene and Quaternary planktonic foraminifers from the upper continental rise of the New Jersey Margin (western North Atlantic), Deep Sea Drilling Project Leg 93, Site 604: Sedimentary and paleoceanographic implications of the biostratigraphy. In: vam Hinte JE, Wise SW Jr. et al. (Eds) Init. Repts. DSDP 931, Washington (US Govt. Pronting Office)

Oberdorfer E (1960) Flora et Vegetatio Mundi. Pflanzensoziologische Studien in Chile. Reinhold Tüxen (ed), Verlag von J. Cramer, Weinheim

Ortlieb L, Machare J, Fournier M, Woodman R (1989) Late Holocene Beach Ridge sequences in Northern Peru: Did they register the Strongest Paleo-El Niños?. International Symposium on Global Changes in South America during the Quaternary: Past–Present–Future. Special Publication no.1. ABEQUA–INQUA (ed) São Paulo, Brazil

Quintanilla Perez V (1985) Carta Fitogeográfica de Chile Mediterráneo. Contrib Cient Tec Area Geocs IV, Aäo V (70)

Ragonese A (1951) La Vegetación de la República Argentina II. Estudio fitosociológico de las Salinas Grandes. Rev Investig Agric V (1–2):1–233

Regaraiz A, Zambrano J (1991) Unidades morfoestructurales y fenómenos neotectónicos en el norte de la provincia de Mendoza (Andes Centrales argentinos entre 32° y 34° latitud sur). Bamb Geogr Schr Bd. 11:1–21

Roig F (1998) Vegetación de la Patagonia. In: Maevia Correa (ed) Flora Patagónica. vol. 1

Roig F, Carretero E, Mendez E (1998) Mapa de vegetación de la Provincia de Mendoza. Programa Fitocartográfico Mendocino. IADIZA–CRICYT (ed), Mendoza

Röthlisberger F (1986) 10, 000 Jahre Gletschergeschichte der Erde. Verlag Sauerländer, Salzburg

Ruthsatz B (1995) Vegetation und Ökologie tropischer Hochgebirgsmoore in den Anden Nord-Chiles. Phytocoenologia 25(2):185–234

Sayago J, Collantes M (1991) Evolución paleogeomorfológica del valle de Tafí (Tucumán, Argentina) durante el Cuaternario superior. Bamb Geograph Schr Bd. 11:109–124

Seltzer G, Wright H (Jr) (1989) Radiocarbon-dated glacial deposits in the Central Peruvian Andes. International Symposium on Global Changes in South America during the Quaternary: Past–Present–Future. Special Publication no 1. ABEQUA–INQUA (ed) São Paulo, Brazil

Servant M, Argollo J, Oliveira-Almeida F, Servan-Vildary S, Wirrmann D (1989) Paleohydrology in the Bolivian Andes during the last 15, 000 Years: Paleoclimatic Scenarios. International Symposium on Global Changes in South America during the Quaternary: Past–Present–Future. Special Publication no.1. ABEQUA–INQUA (ed) São Paulo, Brazil

Servicio Meteorológico Nacional (Ministerio de Aeronáutica) (1958) Estadísticas Climatológicas (1941–1950). B1(3), Buenos Aires

Servicio Meteorológico Nacional (Fuerza Aérea Argentina, Comando de Regiones Aéreas) (1975) Estadísticas Climatológicas (1951–1960). B(6), Buenos Aires

Servicio Meteorológico Nacional (Fuerza Aérea Argentina, Comando de Regiones Aéreas) (1981) Estadística Climatológica (1961–1970). B(35), Buenos Aires

Servicio Meteorológico Nacional (Fuerza Aérea Argentina, Comando de Regiones Aéreas) (1986) Estadísticas Meteorológicas (1971–1980). Estad (36), Buenos Aires

Stine S, Stine M (1990) A record from Lake Cardiel of climate change in southern South America. Nature 345 (6277)

Stingl H, Garleff K (1978) Gletscherschwankungen in den subtropisch semiariden Hochanden Argentiniens. Z Geomor N F Suppl (30):115–131

Stingl H, Garleff K (1985) Spätglaziale und Holozäne Gletscher und Klimaschwankungen in den argentinischen Andes. Zbl Geol Paläont I(11 / 12):1667–1677

Suarez J (1983) Rasgos del modelado glaciario en la Quebrada Benjamín Matienzo, Andes Centrales, Cordillera Principal. IANIGLA (ed), Mendoza

Suguio K, Turcq B, Servant M, Soubiäs F, Fournier M (1989) Holocene fluvial deposits in southeastern Brazil: Chronology and Paleohydrological Implications. International Symposium on Global Changes in South America during the Quaternary: Past–Present–Future. Special Publication no.1. ABEQUA–INQUA (ed) São Paulo, Brazil

Ten Brink N, Weidick A (1974) Greenland Ice Sheet History since the Last Glaciation (1). Quat. Res. (4):429–440

Vianna M, Solewicz A, Cabral A (1989) Early Holocene sea-level stillstands in the Brazilian northeast mapped by satellite. International Symposium on Global Changes in South America during the Quaternary: Past–Present–Future. Special Publication no.1. ABEQUA–INQUA (ed) São Paulo, Brazil

Videla M (1996) Fluctuaciones del Glaciar Horcones Superior, Región del Cerro Aconcagua, Mendoza, Argentina y la variabilidad climática actual. Dissertation, Universidad Nacional de Cuyo

Villalba R, Leiva J, Rubulis S, Lenzano L (1990) Climate, Tree Ring and Glacial Fluctuations in the Río Frias Valley, Río Negro, Argentina. Arctic and Alpine Res 22(3):215–232

Wijmstra T, van der Hammen T (1966) Palynological data on the history of tropical savanna in Northern South America. Leid Geol Meded (38):71–90

Wingenroth M, Suarez J (1984) Flores de Los Andes, Alta Montaäa de Mendoza, Quebrada Benjamín Matienzo, Mendoza. IANIGLA (ed), Mendoza, pp 144

Wingenroth M, Heusser C (1984) Polen en la Alta Cordillera, Quebrada Benjamín Matienzo, Mendoza. IANIGLA (ed), Mendoza, pp 195

Wingenroth M (1990) Historia de la vegetación y del clima en la Quebrada Benjamín Matienzo, Cordillera de Los Andes, Mendoza. Dissertation, Universidad Nacional de Buenos Aires

Wingenroth M (1992) La Quebrada Benjamín Matienzo, su naturaleza presente y pasada. Ediciones Culturales de Mendoza (ed), Mendoza

Wingenroth M (1998) Palynological data and environmental modifications at the Quebrada Benjamín Matienzo, High Andean Cordillera, Mendoza, Argentina. (Reviewed)

Yrigoyen M (1979) Cordillera Principal. Seg Simp Geol Reg Arg, Acad Nac Cs, Córdoba. I:651–694

A Hybrid Expert System for the Prediction of Extreme Meteorological Situations such as the Zonda Wind in the Cuyo Region, Western Argentina

F. Norte, IANIGLA-CRICYT, Casilla de Correo 330, 5500 Mendoza, Argentina
Fnorte@lab.cricyt.edu.ar

Abstract: The principal objectives of the Regional Program of Meteorology (RPM) in Argentina, are to study the climate, its variations, the meteorological phenomena of the Cuyo region (in the west of Argentina) and their effect on the socio-economical activities of the area. The objectives of the RPM include regional climatology and the diagnosis and forecasting of special phenomena such as the Zonda wind, frost, hail, snow, severe convective storms, urban environmental pollution. The Mendoza Province/Cuyo Region is a natural laboratory for carrying out research on significant meteorological phenomena. The Stepwise Discriminant Analysis (SDA) predictive model for the Zonda wind is presented. The results permit forecasting Zonda phenomenon with a high degree of certainty (75% to 90%). Furthermore, analysis of the Zonda wind leads to the establishment of an early alert system for these sudden and sometimes harmful meteorological phenomena.

10.1. Introduction

This work focuses on the prediction of important meteorological phenomena that are harmful for the socio-economic activity of the region studied. Although this work is specific to the Cuyo region in Argentina, the methods and results obtained can be applied worldwide. The study of meteorological phenomena in the west of Argentina is important to the agricultural activities of the region, industry, the prevention of avalanches and floods, tourism, andinism (alpinism), transportation, health and climate, and the management of water resources. Data used are from regional, national and international databases. Several regional phenomena are considered.

10.2. Background

The province of Mendoza constitutes a natural laboratory for the study of important meteorological phenomena. At the present time, and because of the alerts caused by a possible global climatic change, the need to develop meteorological programs, like the one existing in the province, is recognized. One important, often overlooked aspect is, that *regional* programs are needed because many meteorological phenomena that are important for day-to-day decision-making occur below the spatial resolution of large-scale atmospheric circulation models.

The analysis we provide allows for the characterization of regional circulation and its climatology, vital for the Mendoza Province. In addition, our analysis is a good reference-parameter for the study of paleo-winds. Similar approaches can be applied to other parts of the world. Therefore primary data, intermediate data, software and results, are also available in the NORTE-Directory on the compact disk.

10.3. Methods

The methodology we use is based on the theoretical framework established in Norte (1988a) and. Additional information can also be found in Aceituno (1987), Compagnucci (1988), Norte (1988b), Norte and Silva (1990), and Selucchi (1993) The following summarizes selected aspects of our methodology: In mountainous zones, a warm, dry and occasionally strong wind can be observed descending to the valleys and plains. It has regionally different names: Chinook in Canada and the United States, Föhn in the European Alps and ZONDA in Argentina. Predictive models for the Zonda wind for the two most populated cities of West Argentine (San Juan and Mendoza cities) are presented. Sources of information used are the zonal and meridional pressure indices, surface pressure fields at 12 UTC, variables from the rawinsonde of Mendoza Airport and the "geopotential height fields" at 500 and 1000 hPa (the altitude where a pressure of 500 and 1000 hPa occurs, expressed in meters) and the thickness of cold and warm air.

The results of our methodology permit the forecasting of phenomenon with a high degree of certainty (75% to 90%). This success is the result of a tragedy that happened in 1992, where the Argentine National Weather Service forecast a Zonda wind on September 16, 1992, but was unable to say whether it would be a severe event. The wind killed nine persons and caused a lot of damage. Some days later, when facing the probability of another Zonda event, the local government decided to suspend all activities. The phenomenon occurred but it was moderate, did not last long and did not cause damages. In both cases, the forecasting was correct, but there were no tools available to establish the degree of severity. Therefore the RPM decided to elaborate a new predictive model of the probable wind intensity using the SDA (Stepwise Discriminant Analysis).

A Zonda event is considered severe when the wind gusts are equal or faster than 15 meters/second. Fortunately, the frequency of severe cases is only 38% in San Juan and 20% in Mendoza. Analyzing the mean pressure fields on the surface for days with severe and non-severe Zonda events, the following phenomena are observed: (1) In severe cases, a low cut-off in the surface above the center of Argentina. (2) In moderate cases, a trough axis above this territory or a low cut-off located in lower latitudes, in the northwest of the country. (3) Comparing satellite images of days with severe Zonda events vs days with moderate Zonda events, it can be seen that during severe Zonda events that there is a formation of cirrus clouds aligned over the studied region, associated with the jet stream.

A statistical analysis (see compact disk) shows that 16 meteorological stations from 84 analyzed, are sufficient for prediction using SDA. These stations are located in the center, east and north of Argentina. The absence of discriminatory variables from the south (Patagonia) is noticeable, especially because in other models even the pressure from the Antarctica influences the forecasting of Zonda winds. This would indicate the possibility that the cause of severity is the intensity and location of the pressure field on the surface combined with the position of the jet-stream in the altitude. The SDA method has up to a 95.7% degree of success. It was efficiently tested in later opportunities with satisfactory results. It permits a forecasting of the wind intensity up to 48 hours in advance.

The regional Zonda process involves not only the wind itself, but also other associated phenomena such as the thermal inversion, which is more acute during the days before Zonda wind. In addition, such weather conditions have several other effects: (1) The Zonda results in higher levels of environmental pollution in urban zones of the area. (2) The presence of strong winds in the middle levels of the atmosphere affecting the high mountains causes rainfall and snow precipitations in the Chilean Central and Southern regions. (3) The Zonda results in increased snow and storms in the mountains. (4) The Zonda also results in the passage of frontal systems, followed by frost in the agricultural zones of the region.

Readers who would like more details about SDA and would like to apply our method to other areas of interest are referred to the files on the compact disk in the NORTE directory.

Acknowledgements: To Silvia Simonelli, Martin Silva, Julio Cristaldo and Nicolas Heredia, people who belong to the RPM and that helped to prepare this work and Marcela Lingiardi who helped us in the typewriting.

References

Aceituno P (1987) On the interannual variability of South American climate and the Southern Oscillation. Ph.D. Thesis, University of Chile. Santiago de Chile

Compagnucci R.H. (1988) Climatologia sinoptica de las precipitaciones en Cuyo. Ph.D. Thesis, University of Buenos Aires, Argentina

Norte F (1988a) Caracteristicas del viento Zonda en la regionde Cuyo. Ph.D. Thesis, University of Buenos Aires, Argentina

Norte F (1988b) Metodos de pronostico del viento Zonda en Mendoza (Argentina).-- Proceeding of III Congreso Interamericano de Meteorologia y III Congreso Nacional Mexicano de Meteorologia. Mexico, D.F. noviembre 1988. p. 114-118

Norte F, Silva M (1990) Comparacion de la efectividad del pronostico del Viento Zonda y el de las nevadas en Mendoza (Argentina) usando Analisis Discriminante Escalonado.-- Proceedings of IV Congreso Interamericano y II Colombiano de Meteorologia. Bogota, 17 al 21 de setiembre de 1990. p. 71-75

Seluchi ME (1993) Estudio del comportamiento de los sistemas migratorios en la Argentina.-- Phd., University of Buenos Aires. Argentina

Impact of ENSO Events on the Hydrological System of the Cordillera de los Andes during the last 450 Years

Rosa Hilda Compagnucci
CONICET – University of Buenos Aires Ciudad Universitaria, Pabellón 2, Departamento de Ciencias de la Atmosfera, 1427 Capital Federal, Argentina
Rhc@at1.fcen.uba.ar

Abstract: The correlation between river runoff and ENSO-events in the Cuyo region, western Argentina was studied. It could be shown that most "wet" years are associated with an ENSO-event (preceding or same year) while only about 67 percent of ENSO events cause high river runoff in the study-area. Thus ENSO is one, but not a definite criterion for high precipitation / high river runoff.

11.1. Introduction

In South America, for latitudes higher than 30° S, some peaks of the Cordillera de los Andes reach 7000 m. The orientation of the Cordillera is perpendicular to the atmospheric circulation. For latitudes lower than 35° S, the continent is under the influence of the semipermanent anticyclones of the Pacific and Atlantic Oceans. South of this latitude, the westerlies are predominant; additionally the mean height of the Cordillera decreases.

The melting of the accumulated snow during the Southern Hemisphere winter is the main cause of the river runoff during the summer; the runoff reaches its minimum in winter. Therefore, the interannual variability of the river runoff is connected with variations of the general atmospheric circulation.

The interannual variability of the accumulation of snow in relation to the El Niño years was studied by Pittock (1980) and Quinn and Neal (1982). They noted a tendency for exceptionally abundant subtropical rainfall in Chile during El Niño years. More recently, Aceituno (1988), Aceituno and Vidal (1990), Waylen and Caviedes (1990), Rutllant and Fuenzalida (1991) amongst others, widely examined the response of the precipitation and river runoff for the western area of the Cordillera de los Andes at middle latitudes to the El Niño/Southern Oscillation events.

11.2. River Runoff and Precipitation in the Cordillera de los Andes

The snow accumulation rates on the highest peaks of the Cordillera are closely related to the precipitation on its western slope as well as to the water content of the snow in the Cuyo river basin on the eastern slope. This was documented in the "Primeras Jornadas de Nivo-Glaciologia, Mendoza, Argentina, 1969". In order to specify the area with similar interannual variability in the winter precipitation, Vargas and Compagnucci (1985) analyzed the behavior of the total winter precipitation (April to September) in the Cordillera de los Andes.

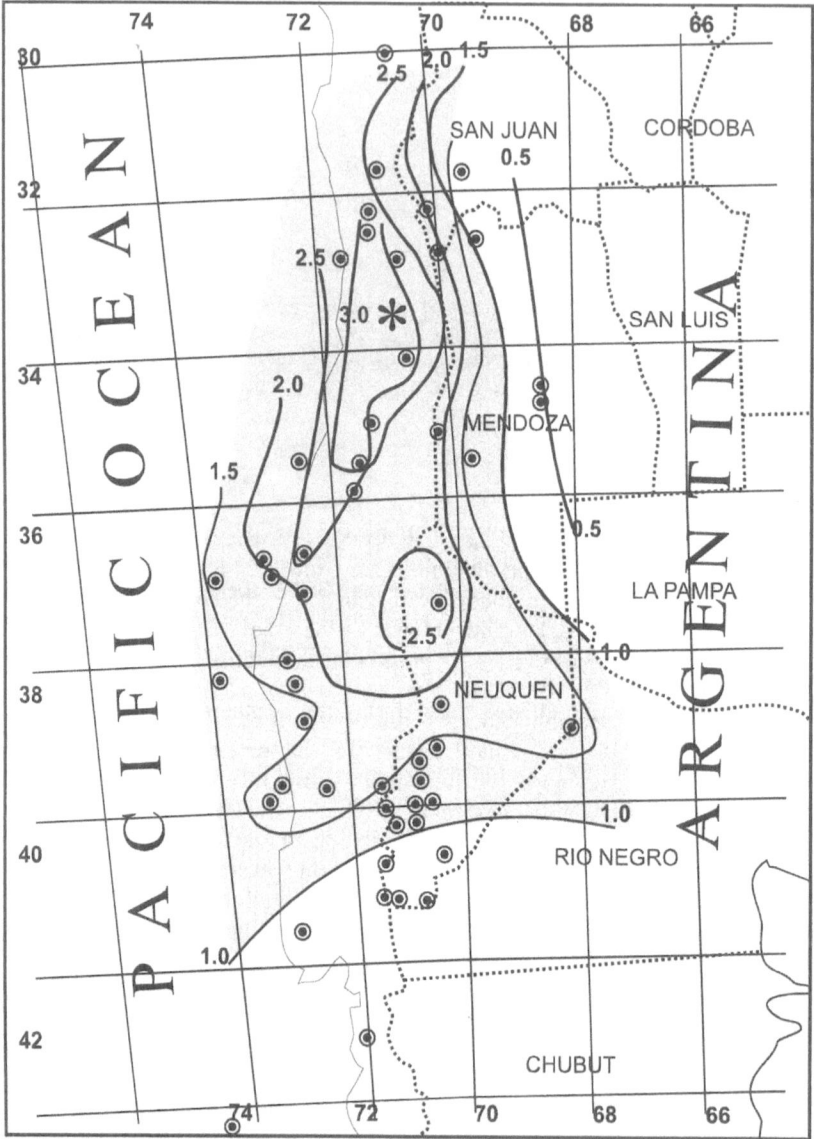

Fig. 11.1. Ratio *r/rc*. The variable *r* is the correlation coefficient between the precipitation at Santiago de Chile and 52 other stations between 30–44° S. The variable *rc* is the critical value of *r* above the significance level of 0.05 using the Student t-test (Vargas and Compagnucci 1985).

Information from 53 stations located on both sides of the Cordillera was studied (see Fig. 11.1). The linear correlation coefficients, r, between the precipitation of Santiago de Chile (33° S, 71° W) and the corresponding series in each of the other stations were calculated. Figure 11.1 shows the ratio between r and its critical value rc, defined as the critical values of the sample correlation coefficient in a non-correlated universe. The coefficient rc was obtained using the Student t-distribution to judge the significance of a coefficient computed from n independent pairs of observations using 0.05 as significance-level (Technical Note No 71, WMO 1966).

The shaded area, r/rc greater than one, indicates the winter-precipitation area, which is significantly related to the precipitation of Santiago de Chile. The region with homogeneous interannual variability is located between 30–40° S. The relation with the Santiago de Chile precipitation-signal disappears south of 40° S and east of 69° W.

The snowfall accumulation area in the basins of San Juan, Mendoza, Tunuyan, Diamante, Atuel, Colorado and Neuquen rivers corresponds to the homogeneous interannual precipitation area showed in Fig. 11.1. These results agree with those obtained by Menegazzo de Garcia and Radicella (1982). They found correlation coefficients higher than 0.82 between Santiago de Chile winter precipitation and the runoff of San Juan, Mendoza, Atuel and Neuquen rivers. The coefficient for the Limay river however amounts only to 0.60. This result is understandable since snowfall accumulation in the Limay river basin is located south of 40° S.

11.3. Atmospheric Circulation in Southern South America and ENSO Events

The interaction between the El Niño (EN) event and the Walker Circulation anomaly named Southern Oscillation (SO) is well known. The coupled ocean and atmospheric phenomenon is known as an ENSO event but, as indicated by Trenbreth and Shea (1987), the SO and EN are not necessarily linked on a one-to-one basis. The warm episode may occur without an SO swing, such as during 1979.

The anomalies in the Sea Surface Temperature (SST) for the Equatorial Pacific Ocean determine the El Niño intensity and the SO behavior would be evaluated by the Southern Oscillation Index (SOI) as proposed by Trenberth (1984). It is the difference between standardized anomalies of the sea level pressure in Tahiti and Darwin.

Recently, Philander (1990) summarized what is currently known about El Niño, La Niña and the Southern Oscillation, and the interactions between the oceans and the atmosphere that causes this phenomenon. Furthermore a comprehensive compilation and new results for understanding the physical mechanisms of the ENSO were presented in the book of Diaz and Markgraf (1992).

The influence of ENSO events in the atmospheric circulation over Southern South America was studied by Compagnucci (1989, 1992) for winters between 1972 and 1983. Daily surface pressure fields (12:00 TMG) were analyzed using un-rotated principal component analysis with a T-mode correlation input matrix. The methodology and its mathematical formulation is explained in Green (1978).

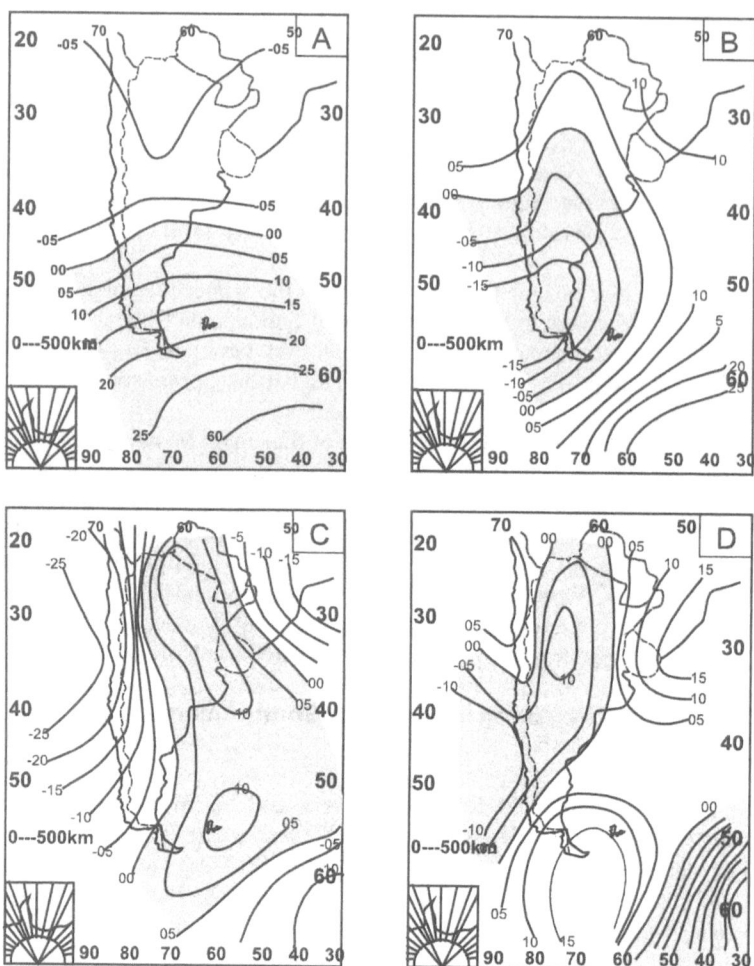

Fig. 11.2. First four principal component patterns from the analysis of daily surface-pressure fields in winter (Compagnucci 1992).

More than 80% of the variance was explained by the first four principal component patterns shown in Fig. 11.2.

The first pattern A exists only in the shaded area under low-pressure conditions. That is, it describes zonal westerlies south of 40° S and subtropical anticyclones over the coast north of 40° S. This basic flow accounted for more than 50% of the variance.

The remaining patterns are associated with the perturbations of the basic flow. Patterns B and C with high pressures in the shaded areas characterize post-frontal anticyclones and advection of cold air masses. Furthermore, situations with cold fronts and low-pressure systems are sketched by patterns B, C and D; the latter for low-pressure conditions in the shaded areas.

The results, for each winter of the period 1972 to 1983, showed that negative SOIs and positive SSTs (El Niño) are generally associated with disturbed surface circulations. In other words, they are associated with a higher variance explained by patterns B, C and D and a lower variance explained by pattern A.

These perturbations are associated with frequent cold fronts which cross the continent north of 40° S and, therefore, with precipitation higher than normal in Santiago de Chile. Additionally, positive SOI are related to a strong persistence of the basic flow, since a higher variance is explained by pattern A. In addition positive SOI are related to precipitation below the average values in Santiago de Chile.

11.4. Runoff in the Cuyo Region and ENSO Events

The Cuyo-region is located at middle latitudes comprising the slope and the plain on the east side of the Cordillera de los Andes. The relation between the surface circulation, the ENSO indicators (SOI and SST) and the runoff of the Cuyo rivers was studied by Compagnucci and Vargas (1993).

Considering the results obtained by Menegazzo de Garcia and Radicella (1982), time series for the Mendoza river gauged in Cacheuta (33° 01' S, 69° 07' W, 1238 m asl) have been taken into account as representative information about the characteristics of the Cuyo river runoff

During the analyzed period, 1910–1997, twenty-one ENSO events were considered according to the compilation and classification made by Quinn et.al. (1978), Rasmusson and Carpenter (1982), and recent information from the TAO/TOGA project available over INTERNET.

Four different groups were found according to the occurrence of ENSO events:

(1) In ten years, abundant runoff (larger than the mean) occurred at the same time as the austral summer warm event (WEY(0), warm event year zero). Such an event means abundant snowfall during the preceding winter.

(2) In four cases, abundant runoff was observed in the summer following WEY(0) (i.e., abundant snowfall occurred during winter following WEY(0)).

(3) Conversely, during three austral summers (1922/23, 1934/35 and 1978/79) the runoffs were higher than one standard deviation without ENSO events occurrence during the previous or the following year.

(4) Finally, there were seven ENSO events in which less than the mean runoff occurred during the WEY(0) and in the following summer. That is, precipitation less than the mean was registered in the preceding and in the following winter to the WEY(0).

From the above groups, fourteen ENSO events (67%) are linked to summer runoff higher than normal (i.e. group 1 and 2). Seven of them (33%) are characterized by river runoff below the average (i.e. group 4).

11.5. Historical Record of El Niño and Subtropical Rainfall in Chile for the Last 450 Years

Compilations of El Niño occurrences and SO proxy data have been produced by Quinn et. al. (1978) and Quinn (1992). Table 11.1, according to Quinn (1992), shows the El Niño classification for the last 498 years. Furthermore, reconstruction of the subtropical Chile rainfall from 1535 to 1930 using historical documents is given by Taluis (1934) (see Fig. 11.3 as well as Prieto, this volume, and Villalba, this volume). This time series is one of the most homogeneous long historical records in the world (Mitchell et al. 1966). The long-term character of these data permits their use for testing the above-mentioned relationship between ENSO events and the response of the hydrological system between 30–40° S in the Cordillera de los Andes. According to the previous results the precipitation classes for the years 1535–1930 given by Taulis (1934) shown in Fig. 11.3. could be regarded as classes of summer river runoff for the Cuyo's rivers.

Twenty-five of the thirty-one (80.6%) years that are classified as very wet (Fig. 11.3), i.e. with runoff above normal values, correspond to years with moderate, strong or very-strong ENSO events (using Quinn's compilation and the classification shown in Table 11.1.). For the remaining six very wet years (19.4%) we can see relations with ENSO events. Three of them (1609, 1829 and 1851) are preceded by strong or moderate ENSO events and the other three (1746, 1764 and 1843) are followed by strong, moderate and very strong events respectively.

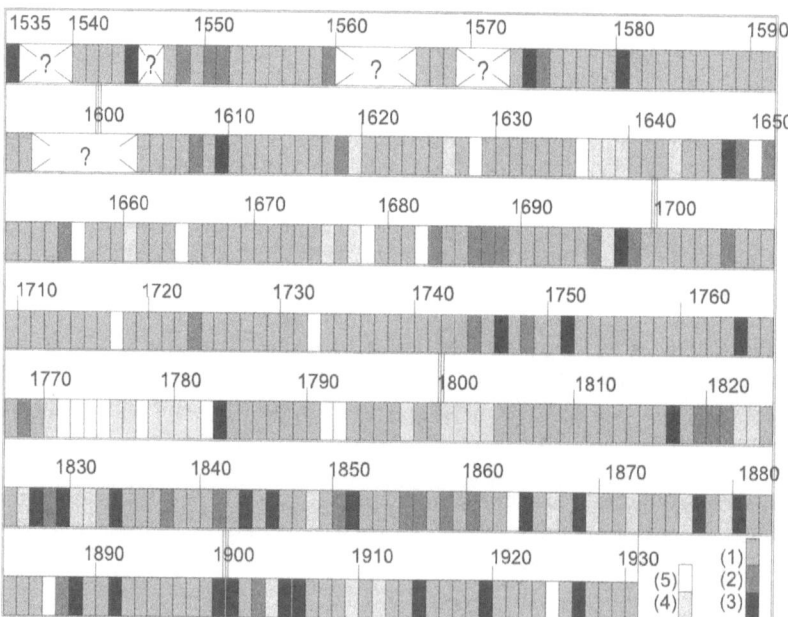

Fig. 11.3. Classification of the precipitation in Central Chile between 1535 and 1931: (1) Normal, (2) Wet, (3) Very Wet, (4) Dry and (5) Very Dry (according to Taulis 1934).

Table 11.1. ENSO events of the last 450 years. Years in which large-scale El Niño/Southern Oscillation (ENSO) events occurred. For some years, additional information is given using the terms E (early), M (mid), or L (late). Strengths (Str) are moderate (M), strong (S), or very strong (VS) with a + or - added for intermediate values. Information from Quinn (1992). Recent ENSOs are added to the time-series and can be found at the end of the table.

Year	Str.	Year	Str.	Year	Str.	Year	Str.
1535	M+	1683-84	M+	1799	M	1891	M
1539-41	S	1687-88	S	1802-04	S+	1896-97	M+
1544	M+	1692	M+	1806-07	M	1899-M1900	VS
1546-47	S	1694-95	VS	1810	M	L1901-02	S+
1552-53	S	1697	M	1812	M+	1904-05	S
1558-E61	S	1701	M	1814	S	1907	M+
1565	M+	1703-04	S	1817	M+	1911-12	M+
1567-68	S+	1707-09	M	1819	M+	M1913-M15	S+
1574	S	1713-14	M+	1821	M	1918-E20	S+
1578-E79	S	1715-16	S+	1824-25	S	1923	M
1581-82	M+	1718	M	1827-28	S+	1925-26	S
1585	M	1720	M+	1830	M	L1929-E31	M+
1589-91	S	1723	S	1832-33	S+	1932	M+
1596	M	1725	M	1835-36	M	1939	M
1600-01	S	1728	M	1837-39	S	1940-41	VS
1604	S	1731	M+	1844-E46	VS	1943-44	M
1607-08	S	1734	M	1850	S	1951-E52	M+
1614	S	1737	S	1852-53	M	1953	M
1618-19	M	1744	M+	1854-55	S	1957-58	S
1621	S	1747-48	S	1857-59	M+	1965-66	S
1624	M+	1751	M+	1860	M	M1968-69	M-
1630-31	S+	1754-55	S	1862	M-	1972-73	S+
1635	M	1761-62	S	1864	S+	1976-77	M
1641	S+	1765-66	M+	1865-66	M+	1983-83	VS
1647	M	1768-69	M+	L1867-E69	S+	M1986-87	M
1650	S+	1772-73	M	1871	M		
1652	M	1776-e78	M+	1873-74	M+	More recent	ENSOs
1655	M	1782-84	VS	L1876-78	VS	1991-94	M
1661	VS	1785-86	M+	1880-81	M+	1997	VS
1671	M+	1790-93	VS	1884-85	M+		
1681	S	1794-97	M+	L1887-E89	S		

When the wet years are also taken into account, only two of them (1550 and 1841) are *not* related to the ENSO events.

On the other hand, considering the ENSO events during the period 1535–1931, 105 ENSO events happened according to Quinn (1992). Only 25 of them (23.8%) are not related to normal or above-normal precipitation (or runoff) values. This means that such ENSO events occur but do not show up as wet or very-wet precipitation classes during the present or following year. Extreme situations correspond to the ENSO events for the years 1772–1773, 1790–1793, 1802–1804 and 1911–1912. These years show up in the "dry to very-dry" precipitation classes. A total of 76.2% of the ENSO events were linked to years classified regarding precipitation as with "normal" or "very wet".

It is interesting to compare the changes of the ENSO event frequency during the last four centuries: During the 17th century 22 ENSO events have been observed, in the 18th century 27 events, the 19th century 33 events, and finally during the 20th century (up to 1998) 24 events. The last century contains the largest frequency of ENSO events. In addition, the 19th century was the wettest period in which "wet" and "very wet" years were more frequent. Only two years are classified as "very dry". The 19th century is also regarded as part of the Little Ice Age (Crowley and North 1991).

These results confirm the relation between the summer runoff of the Mendoza river and the ENSO events obtained by Compagnucci and Vargas (1993) using the instrumental data for 1910–1991 years above. Furthermore, the lower frequency of ENSO events in the first half of the current century is in agreement with the negative trend detected in the runoff of the Cuyo river since the beginning of the 20th century up to 1970 (Minetti and Sierra 1989)

Overall, years that are classified as "very wet" are very likely associated with an ENSO event. An ENSO event however may – at least in the study area – not necessarily cause a wet year. Overall however, the data indicate that the changes observed for the hydrological system of the Cordillera de los Andes are significantly explained by changes in the strength and frequency of the ENSO events. Nevertheless, ENSO events occurred that did not increase precipitation (or river runoff) to values above the normal level.

11.6. Conclusions and Ideas for Future Work

It is very likely that when the rivers of the Cuyo region have discharge records above normal during summers, then ENSO events are present. This is due to the fact that during the winters of years that *precede* a "warm event" in the South Hemisphere summer, WEY(0), snow falls in larger quantities than normal in latitudes between 30–40° S. It is also possible, although less frequent, that during the winter *following* WEY(0), snowfall above normal is recorded, and therefore discharges are larger than their mean values in the following summer (the summer following the *ENSO event*).

Nevertheless, it is important to note that records for the Mendoza river show that for some years *before* and others *after* an ENSO event, the runoff is *less* than normal. In short, river runoffs above normal are related to ENSO events, but the existence of ENSO events does not guarantee the occurrence of discharges above normal.

Taking into account the proximate nature of the data, the 19th century has the largest frequency of ENSO events. Furthermore it is the wettest period. This however could be expected as the 19th century is regarded as the end of the Little Ice Age. The sea-surface temperature patterns for 1879 in the subpolar North Atlantic reconstructed by Lamb (1979), the winter circulation pattern in the upper atmosphere over North America and Europe depicted by Crowley (1984) and annually dated ice-core data from Antarctica and Greenland studied by Kreutz et al (1997), indicate that the Little Ice Age was characterized by a considerable variability of the meridional circulation strength. It appears that the meridional atmospheric circulation intensity increased in both polar hemispheres with similar magnitude and timing during the 19th century. The results of the surface atmospheric circulation analysis, using daily pressure fields for the period from

1972 to 1983, confirm that in the warm phase (EL Niño) of the ENSO events the circulation is more perturbated by *meridional* components.

Taking this evidence into account, some potentially fruitful strategies for future paleoclimatic studies focusing on ENSO and the Cuyo hydrological system can be set up. Furthermore two different hypotheses of the causes of ENSO could be tested:

The first hypothesis is the possible enhancement of ENSO events due to volcanic eruptions in the equatorial region. Here, the study of varved sediments of the Los Orcones lagoon (Mendoza) or other lagoons in the high Central Andes following the technical process of del Valle et. al. (1993) combined with the theoretical method given by Compagnucci et. al. (1993) could be a tool. This has the following background:

Since three of the more recent large volcanic eruptions, that of Mount Agung in 1963, El Chichon in 1982 and El Pinatubo in 1991, coincided with major ENSO events in the present century, our objective in the future is to test the "Volcanic Hypothesis" given by Handler et al. (1990). This hypothesis states that the warm (El Niño) and the cold (La Niña) events are caused *only* by volcanic eruptions. This hypothesis is a point of discussion particularly after the year 1997, which was a year with the strongest ENSO during the current century. Before the beginning and during the first months of the development of this event the volcanic eruptions registered were insignificant to produce such a big and strong ENSO. Nevertheless, Kirchner and Graf (1995) studied both the volcanic forcing and El Niño forcing. They concluded that the combined signal is different from a simple linear combination of the separate signals. It leads to a climate perturbation stronger than forcing of El Niño or stratospheric aerosol alone and to a somewhat modified pattern.

The second hypothesis to be tested is non-linear coupling in the climate system. If longer paleoclimatic records were obtained, it would be possible to test the hypothesis of Vallis (1986). This hypothesis says that El Niño could be the response to a non-linear dynamic system. The temperature difference between the west and the east of the Equatorial Pacific Ocean would generate a chaotic signal with characteristics similar to those of the Southern Oscillation. Other results that support this theory are Vallis (1988), Munnich et al. (1991), Jin et al. (1994), Tziperman et al. (1994), Chang et al. (1994) and Wang and Fang (1996). Testing both hypotheses is a question of future work.

References

Aceituno P (1988) On the functioning of the Southern Oscillation in the South American Sector. Part I: Surface Climate. Mon. Wea. Rev. 116:505–524

Aceituno P, Vidal F (1990) Variabilidad interanual en el caudal de los rios andinos en Chile Central en relación con la temperatura de la superficie del mar en el Pacífico Central. Rev. Chilena de Ing. Hidráulica, 5(1):7–19

Chang PB, Wang B, Li T, Ji L (1994) Interactions between the seasonal cycle and ENSO-frequency entrainment and chaos in a coupled atmosphere–ocean model. Geophys. Res. Lett. 21:2817–2820

Compagnucci RH (1991) Influencia del ENSO en el desarrollo socio-económico de Cuyo. Anales del CONGREMET VI. 23–27 Sep. 1991, Buenos Aires, Argentina. Eds. Centro Argentino de Meteorólogos, 95–96

Compagnucci RH (1992) Are Southern South American winters surface circulations normal during ENSO events? Extended Abstracts "Paleo ENSO Records" Intern. Symp. Lima, Perú. Eds. L. Ortlieb and Macharé, 41–46

Compagnucci RH, Vargas WM (1993) Snowfall in the Cordillera de los Andes and the ENSO events. Preprints "4th International Conference on Southern Hemisphere Meteorology and Oceanography" 28 March – 2 April 1993, Hobart, Australia. Eds. Am. Met. Soc. 332–333

Compagnucci RH, Giraldes A, del Valle, RA (1993) Climatic interpretation from Mascardi Lake cyclic deposits of the Late Pleistocene. Proceedings "Southern Hemisphere Paleo- and Neoclimates: Review of the State of the Art" 11–14 October, 1993, Mendoza, Argentina. Eds. Project 341 IGCP/IUGS/UNESCO

Crowley T J (1984) Atmospheric circulation patterns during glacial inceotion: A possible candidate. Quat. Res. 21:105–110.

Crowley T J, North GR (1991) Paleoclimatology. Oxford University Press, New York

del Valle RA., Amos A, Bianchi, MM, Cusminsky G, Lirio JM, Martinez JC, Macchiavello J. Masaferro HL, Nuñez CA, Rinaldi C A, Vallverdú, R (1993): Late Pleistocene–Holocene Sedimentary Core from Mascardi Lake Nahuel Huapi National Park, Argentina. Proceedings "Southern Hemisphere Paleo- and Neoclimates: Review of the State of the Art" 11–14 October, 1993, Mendoza, Argentina. Eds. Project 341 IGCP/IUGS/UNESCO

Diaz FH, Markgraf, V (1992) El Niño: Historical and Paleoclimatic Aspects of the Southern Oscillation. Cambridge University Press, Cambridge

Handler P, Andsager K (1990): Volcanic Aerosols, El Niño and the Southern Oscillation. Int. Jour. of Clim. 10:413–424

Jin FF, Ghil Neelin JD, Ghil M (1994) El Niño on the Devil's Staircase: Annual subharmonic steps to chaos. Science, 264:70–72

Kirchner I, Graf HF (1995) Volcanos and El Niño: signal separation in Northern Hemisphere winter. Clim. Dynamics. 11:341–358

Kreutz KJ, Mayewski PA, Meeker LD, Twickler MS, Whitlow SI Pittawala II (1997) Bipolar Changes in Atmospheric Circulation During the Little Ice Age. Science, 227:1294–1296.

Lamb HH (1979) Climatic variation and changes in the wind and ocean circulation: the Little Aice Age in the North Atlantic. Quat. Res. 11:1–20

Menegazzo de Garcia MI, Radicella S (1982) Variación climática–hidrológica en la región cordillerana andina. Meteorológica, XIII(1):49–62

Minetti JL, Sierra EM (1989) The influence of general circulation patterns on humid and dry years in the Cuyo Andean region of Argentina, Int. Journ. Of Clim, 9:55–68.

Mitchell JM, Dzerdzeevskii B, Flohn H, Hofmeyr WL, Lamb HH, Rao KN, Wallen CC (1966) Climatic Change. T.N. 79, W.M.O., N° 195, TP 100, 79

Munnich M, Cane MA, Zebiac SE (1991) A study of self excited oscillations of the tropical ocean–atmosphere system. Part II: Nonlinear cases. J. Atmos. Sci. 48:1238–1248

Philander SG (1990) El Niño, La Niña and the Southern Oscillation, Academic Press Inc. New York, USA

Pittock AB (1980) Patterns of Climate variation in Argentina and Chile I: Precipitation 1931–1960. Mon. Wea. Rev., 108:1347–1361

Primeras Jornadas de Nivo- Glaciologia, 1969, Mendoza, Argentina, available in SECECOM, CRICYT–ME, Mendoza, Argentina

Quinn WH, Zopf DO, Short KS, Kou Yang RTW (1978) Historical trends and statistics of the Southern Oscillation El Niño and Indonesian droughts. Fishery Bull., 76:663–678

Quinn WH. Neal VT (1982): Long-term variations in the Southern Oscillation, El Niño and the Chilean subtropical rainfall. Fish. Bull., 81:363–374

Quinn WH (1992): A study of Southern Oscillation related climatic activity for A.D. 622–1990 incorporating Nile River flood data. In: Diaz HF, Markgraf V (eds) El Niño: Historical and Paleoclimatic Aspects of the Southern Oscillation. Cambridge University Press, pp 119–149

Rasmusson EM, Carpenter TH (1983) The relationship between Eastern Equatorial Pacific Sea Surface Temperatures and Rainfall over Indian and Sri Lanka. Mon. Wea. Rev., 111(3):516–528

Rutllant J, Fuenzalida H (1991) Synoptic aspects of the Central Chile rainfall variability associated with the Southern Oscillation. Int. Jour. of Clim., 11:63–76

Taulis EM (1934) De la distribution des pluies au Chili. La periodicité des pluis depuis 400 ans. In: Matiriaux pour l'etude des Calamités. Genevé Société de Géographis, N° 33, 1, 1934

Trenberth KE (1984) Signal versus noise in the Southern Oscillation. Mon. Wea. Rev., 112:326–332

Trenberth KE, Shea DE (1987) On the evolution of the Southern Oscillation. Mon. Wea. Rev.,115:3078–3096

Tziperman EL, Stone M, Cane M, Jarosh H (1994) El Niño Chaos: Overlapping of resonances between the seasonal cycle and the Pacific Ocean–atmosphere oscillator. Science, 264:72–74.

Vargas WM, Compagnucci RH (1985) Relaciones del régimen de precipitación entre Santiago de Chile y las series de la región cordillerana. Geoacta, 13(1):81–93

Vallis GK (1986) El Niño: A Chaotic dynamical system? Science, 232:243–245

Vallis GK (1988) Conceptual models of El Niño and the Southern Oscillation. J. Geophys. Res. 93C:13979–13991

Wang B, Fang Z (1996) Chaotic Oscillations of Tropical Climate: A Dynamic System Theory foe ENSO. J. Atmos. Sci. 53(19):2786–2802

Waylen PR, Caviedes CN (1990): Annual and seasonal fluctuations of precipitation and streamflow in the Aconcagua river basin, Chile. Jour. of Hydrol., 120:79–102

W.M.O. (1966): Statistical analysis and prognosis in meteorology. T.N. N° 71

Recent Precipitation Trends in Southern South America East of the Andes: An Indication of Climatic Variability

Vicente Barros, Maria Elizabeth Castañeda, Moira Doyle
Department of Atmospheric Sciences, University of Buenos Aires Pabellon II, 2nd Piso, Ciudad Universitaria, 1427 BuenosAires, Argentina
Barros@at1.fcen.uba.ar, eliza@at1.fcen.uba.ar, doyle@at1.fcen.uba.ar

Abstract: Positive trends in precipitation were observed during 1916–1991, especially since the fifties, over most of the Argentine territory. The seasonal variation of the climatic parameters, including precipitation between 1956–1991, can be summarized as a displacement of the positive nucleus of precipitation to the northeast from summer to winter and a less systematic return to the southwest from winter to summer. Correlation studies between annual precipitation and hemispheric indices show that the correlation with the mean meridional gradient of temperature (MMTG) is of the same importance and in some areas greater than the correlation with the Southern Oscillation Index (SOI). Furthermore, the correlation field strongly suggests that the precipitation trends observed in the last 35 years are due to the decrease of the MMTG. In fact, the MMTG decreases around 1.5° C during that period. According to the theory of baroclinic instability this implies a displacement to higher latitudes of the planetary circulation systems. A displacement of 3° in latitude to the south has been reported by Gibson (1992) for the mean position of the maximum wind at 200 hPa in the Southern Hemisphere. Since 1976 a similar displacement of the Atlantic Subtropical High also can be inferred from data of the Atlantic coast. The shift to the south of the general circulation features produces trends in precipitation because of the close connection between precipitation and the latitude of the circulation systems. The data show relationship between the precipitation field and the latitude of the maximum wind speed at the altitude of 200 hPa. Consequently, the observed trends in the precipitation fields could be explained largely by a 5° displacement to the south of this latitude of maximum wind during the last 35 years. This means that in the study area an important component of the global circulation system changed its position in a statistically provable manner.

12.1. Introduction

In the last 20 years, the damages caused by the floods in Argentina have caused social and economic problems in several regions. In the western part of the province of Buenos Aires and in the depressed basin of the Salado River the floods are becoming more frequent and intense. There are frequent reports of catastrophes of the same kind in the low lands along the Paraná river and its tributaries. On the other hand, the more favorable social and economical manifestation of what seems to have the same cause can be seen silently but continuously: The *increase* in the rainfall in the semi-arid zone of the country has permitted the expansion of the agricultural frontier considerably westward. The economic benefits have been very important for the landowners of that zone and for the national economy although they have not yet been duly measured. In other words, natural and anthropogenic climate change can cause both, loosers and winners. The magnitude of precipitation changes impacting a great part of the Argentine territory, with adverse *and* favorable consequences, has not been unnoticed by the inhabitants of the most affected zones.

Despite these observations there are few studies focusing on the recent trends of rainfall in Argentina. Relevant in this sense are the papers of Hoffmann et al. (1987) and Hoffmann (1991), showning a displacement of isohyets westward by approximately 200 km during the last century. The trends and fluctuations before the decade of the seventies have been discussed by several authors. Weber (1951) studied the dry decades for the years 1931–1950 in the subtropical continental region. Schwerdtfeger and Vasino (1954) showed the existence of low-frequency fluctuations in the regional circulation and suggested a link between the intensity of the "west flow" (mid-latitude circulation) and low-frequency variations in the rainfall over subtropical Argentina. They suggested that the negative trend in the subtropical region for the period 1931–1950 could be caused by a decrease of the westerly flow over Patagonia. Díaz (1959) found significantly negative correlations between the west flow and the rainfall in the agricultural zone of Argentina, supporting the hypothesis of Schwerdtfeger and Vasino (1954).

Barros and Mattio (1977) have documented trends and low-frequency fluctuations of rain in Patagonia. During the 1940's, in the arid zone of Chubut, there was an anomalously high precipitation. Due to a simultaneous decrease of the zonal circulation over Patagonia, the authors estimated that this could be caused by a greater frequency of cases with high-pressure centers south of Chubut, which favored the inflow of wet air masses from the Atlantic. The humid zone of Chubut next to the Andes showed a trend toward greater precipitation during the period 1920–1965 that was more pronounced in winter. Minetti and Vargas (1983a) associated the anomalously low latitude of the maximum pressure on the coast of the Pacific during the decade of the 1940's with the abundant rainfalls windward of the Andes in the latitudes 35–42° S and droughts on their leeward side. Minetti and Vargas (1983b) related important changes of the rainfall between 1941–1950 and 1951–1960 with generalized cooling in continental Argentina.

Krepper et al. (1987, 1989) found a general increase of the rainfall in the southwest of the Pampa region during a large part of this century. Vargas (1987) has documented an evident increase of the annual rainfall northwest of Buenos Aires. In the northeast of Argentina, Moyano et al. (1991) studied the response of the Paraná river to occurrence of periods with soil water excesses and deficits. For the same region, using data from Formosa, Posadas and Corrientes, Vargas (1987) did not find evidence that supports large alterations of the rainfall regime.

The above mentioned evidence shows that during the last 60 years significant precipitation changes occurred at least in *some* regions of the country. This applies especially to the humid Pampa, the heart of the Argentine grain production area. Figure 12.1, taken from Castañeda and Barros (1994a) shows an increase from a mean of 850 mm during the 1920's to 1150 mm in the 1980's. Most of this increase took place after 1960.

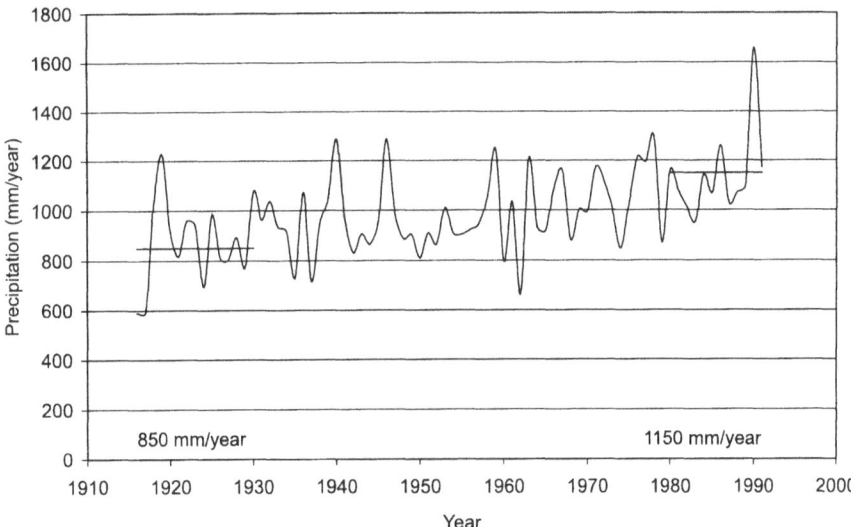

Fig. 12.1. Mean annual precipitation in the humid pampa. Average of stations Paraná, Rosario, Concordia, Buenos Aires, Las Flores, Azul and Mar del Plata. Note the two horizontal bars that show the increase of the average precipitation between the 2nd decade of the century and the decade 1980–1990.

Finally, Castañeda and Barros (1994b) analyzed the precipitation trends over the southern part of South America east of the Andes for 1916–1991. Most of the Argentine territory presented positive trends during this period as a whole. But, when analyzing the different decades, they observed areas of negative trends in central-western Argentina during the first four decades of the twentieth century. In the late fifties the positive trends become more important.

In the northeast of Argentina and Paraguay an inverse behavior was observed with a marked increase in the rainfalls until approximately 1960 and a subsequent decrease until the present. Figure 12.2, taken from Castañeda and Barros (1994b), shows the impressive positive trends over most of subtropical Argentina, Uruguay and southern Brazil.

The variation of the trends for the period 1956–1991 during the year can be summarized (Castañeda and Barros 1994b) as a displacement of the positive nucleus toward the northeast from the summer until the winter, and a subsequent return less continuous to the southwest from the winter to the summer. The agreement of these seasonal displacements with those of the monthly isohyets is an indicator that some features of the circulation that influence the seasonal variation of the rainfall have been changed during the last 35 years. In this paper, it will be shown that positive trends in precipitation over subtropical Argentina were mainly due to the different trends in the hemispheric temperatures at different latitudes. Regarding the connection between the patterns of rainfall variation and hemispheric indexes, Pittock (1980) found significant correlations between hemispheric indexes and the interannual variation of principal precipitation patterns in Argentina and Chile.

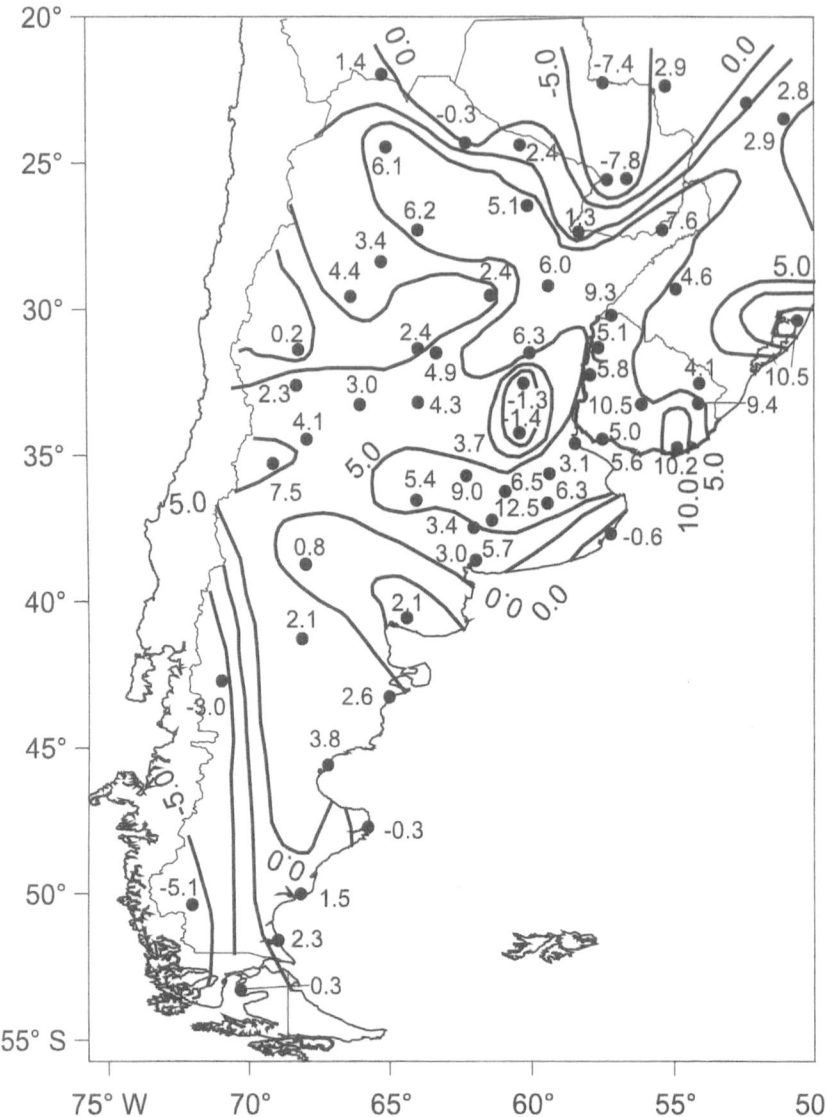

Fig. 12.2. Linear precipitation trends, 1956–1991, in mm/year.

It should however be noted that no trends can be seen for 1931–1960, the time-interval Pittock (1980) studied. In the following pages, it will be shown that there are significant correlations between the principal precipitation patterns and the indexes, both circulation and thermal, between 1956–1991. In addition, these links seem very important in the low-frequency variation of the precipitation fields over Argentina.

12.2. Principal Component Analysis of Annual Precipitation

Two approaches will be used to discuss the dominant patterns of the interannual variability of precipitation and to relate them to variations of the atmospheric circulation. One approach is to correlate precipitation series at each individual station with hemispheric indexes. This approach will be addressed in the next section. The other approach, the principal component analysis, has been applied to climatological data by many authors. The method used here is based on a so-called T-Mode correlation-matrix (see Green and Carol 1978). Simply speaking, this is a method that correlates between maps of annual parameters. Previously, the anomaly values at each station were calculated with respect to their mean value.

Table 12.1: Location of the stations used in this study.

Place	Lon (W)	Lat (S)	Place	Lon (W)	Lat (S)
La Quiaca	65° 36'	22° 06'	Rivadavia	62° 54'	24° 10'
Salta	65° 29'	24° 51'	P. Roque Saenz Peña	60° 27'	26° 49'
Corrientes	58° 49'	27° 28'	Santiago del Estero	64° 18'	27° 16'
La Rioja	66° 49'	29° 49'	Ceres	61° 57'	29° 53'
Córdoba	64° 13'	31° 19'	San Juan	68° 33'	31° 36'
Paraná	60° 29'	31° 47'	Mendoza	68° 51'	32° 53'
Río Cuarto	64° 14'	33° 07'	San Luis	66° 21'	33° 16'
Nueve de Julio	60° 53'	35° 27'	Malargüe	69° 35'	35° 30'
Las Flores	59° 06'	36° 04'	San Carlos de Bolivar	61° 06'	36° 15'
Azul	59° 50'	36° 45'	Coronel Suarez	61° 53'	37° 26'
Mar del Plata	57° 35'	37° 56'	Bahía Blanca	62° 10'	38° 44'
Maquinchao	68° 44'	41° 15'	Trelew	65° 16'	43° 12'
Puerto Deseado	65° 55'	47° 44'	Santa Cruz	68° 34'	50° 01'
Río Gallegos	69° 17'	51° 37'	Esquel	71° 09'	42° 56'
Bariloche	71° 10'	41° 09'	San Rafael	68° 24'	34° 35'
Junín	60° 55'	34° 33'	Pehuajó	61° 56'	35° 52'
Ponta Pora	55° 42'	22° 30'	Londrina	51° 12'	23° 24'
Brusque	48° 54'	27° 06'	Alegrete	55° 48'	29° 48'
Santa Victoria do Palmar	53° 24'	33° 30'	Puerto Casado	57° 54'	22° 18'
San Juan Bautista	56° 24'	31° 12'	Salto	58° 00'	32° 12'
Paso de los Toros	56° 30'	33° 18'	Mercedes	58° 06'	33° 12'
Colonia	57° 54'	34° 30'	Rocha	54° 18'	34° 30'
Punta Arenas	70° 54'	53° 00'	Las Lomitas	60° 35'	24° 42'
Catamarca	65° 46'	28° 36'	Comodoro Rivadavia	67° 30'	45° 47'
Monte Caseros	57° 39'	30° 16'	Lago Argentino	72° 18'	50° 20'
Concordia	58° 02'	31° 23'	Reconquista	59° 42'	29° 11'
Pilar	63° 53'	31° 40'	Posadas	55° 58'	27° 22'
Rosario	60° 47'	32° 55'	Neuquén	68° 08'	38° 57'
Observatorio Central	58° 29'	34° 35'	Curitiba	49° 12'	25° 24'
Trenque Lauquen	62° 44'	35° 58'	Porto Alegre	51° 12'	30° 06'
Santa Rosa	64° 16'	36° 34'	Asunción	57° 36'	25° 48'
Pigüe	62° 23'	37° 36'	Melo	54° 18'	32° 48'
San Antonio Oeste	64° 57'	40° 44'	Treinta y Tres Orientales	54° 36'	34° 30'
Punta del Este	55° 00'	35° 00'			

Though various criteria for statistical significance have been suggested (Anderson 1963; Preisendorfer and Barnett 1977), the discussion in this paper is confined to the first two eigenvectors, which, as stated by Pittock (1980), are clearly significant by any of the usual criteria. Table 12.1 shows the locations of the stations used in this study.

12.2.1. The Period 1916–1991

Due to the lack of available data from Uruguay and Brazil before 1956, the analysis of this period is restricted to Argentina and some stations of Paraguay. Figures 12.3a and 12.3b show the patterns of the first two eigenvectors of annual precipitation. They account for 19.0 and 12.4% of the total variance respectively. The first eigenvector has a remarkable good agreement with the spatial pattern of the mean annual precipitation over subtropical Argentina. This suggests that in this period a significant part of the interannual variance is due to a general increase or decrease of the precipitation. This is largely proportional to the mean annual precipitation at each location. This could be an indication that the observed pattern is largely governed by processes of a broader scale, that is to say, processes of regional or even hemispheric scale.

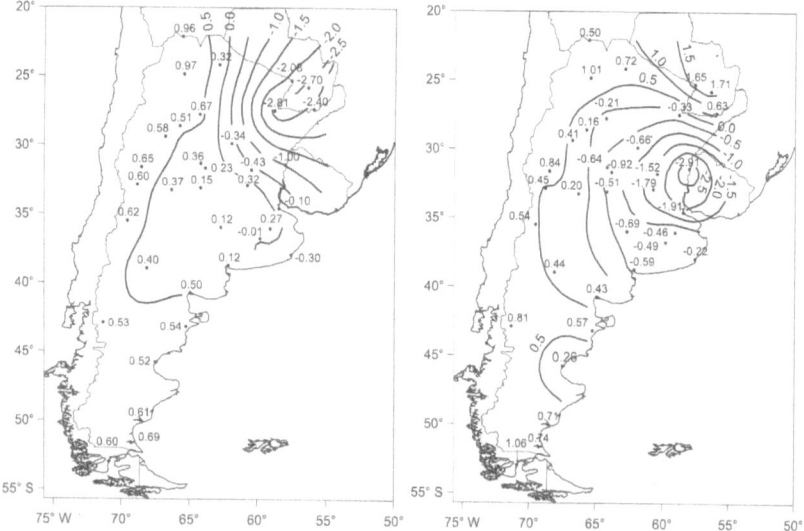

Fig. 12.3. Factor scores of precipitation for 1916–1991. Shown is the spatial distribution of (a) the first eigenvector (above left) and (b) the second eigenvector (above right).

Though no data are available north of Paraguay, the pattern of the second eigenvector (Figure 12.3b) suggests a bipolar structure, i.e., a positive nucleus in Paraguay and a negative one in Argentina. Therefore, when precipitation is registered in Paraguay, dry conditions should be expected in eastern Argentina, and vice versa.

Figures 12.4a and b show the time series corresponding to the patterns of Figs. 12.3a and b. Both series have a negative trend that according to Figs. 12.3a and b means positive trends of the precipitation over most of eastern Argentina. These trends are more evident since about 1960, especially if the second eigenvector is considered. The correlation coefficients of each time series of the eigenvectors with hemispheric subtropical surface temperature and surface temperature at Orcadas are shown in Table 12.2.

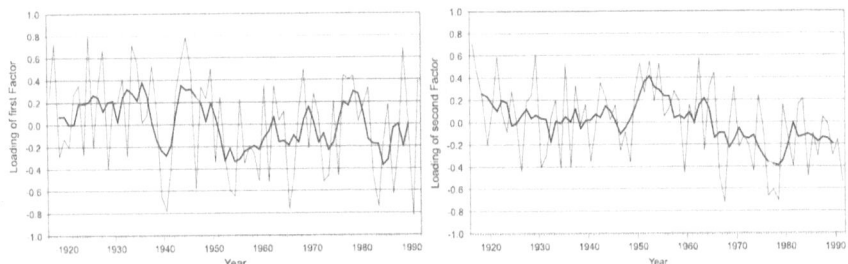

Fig. 12.4. Factor loadings of precipitation, 1916–1991. Shown is factor loading of (a) the first eigenvector (above left) and (b) the second eigenvector (above right). In both cases, moving averages over 5 years are also shown.

The numbers in parenthesis are the significance level. The subtropical temperatures were taken from Jones et al. (1991). These data are globally averaged. Instrumental records do not permit the calculation of an index of surface temperature for polar regions before 1960.

Only for Orcadas Islands (starting around 1900) does a complete record exists (Jones 1990). The results of Orcadas show a negative correlation (–0.3) with the surface temperature of the polar region (60–90° S) elaborated by Angell (Boden et al. 1994). According to this, Orcadas surface temperature should not be considered as a hemispheric index. However, some evidence not shown here, suggests a positive correlation in low frequency between both series. Therefore, the results of Orcadas correlating the eigenvectors of the precipitation field with the temperature field are also included in Table 12.2.

Table 12.2. Correlation coefficients with the factor loadings of the annual precipitation field between 1916 and 1988.

Mean temperature at	Loadings 1st factor	Loadings 2nd factor
Subtropics (10–30° S)	–0.15 (0.15)	–0.20 (0.15)
Orcadas Islands	0.00	–0.02

As shown in Table 12.2, both the first and the second eigenvector correlate negatively with the subtropical temperature. The data suggest that a warmer hemispheric temperature *increases* the precipitation over most of eastern Argentina, probably due to the increase of water vapor.

12.2.2. The Period 1956–1991

As demonstrated above, after 1960 both eigenvectors show a pronounced trend. Also, the precipitation of the humid pampa (see Fig. 12.1) shows a pronounced positive trend starting around 1960. Since 1956 more data are available in Argentina and Paraguay. Furthermore, new data from Uruguay and south of Brazil could be added to the analysis. Both reasons justify running a principal component analysis that focuses on the period 1956–1991. Figures 12.5a and b show the fields of the first and second eigenvectors of the annual precipitation. They account for 15.7 and 11.5 % of the total variance respectively. Thus they account cumulatively for 27.2% of the total variance. In contrast to Fig. 12.3a, the first eigenvector (Fig. 12.5a) has a clear bipolar structure with opposite centers in Paraguay and at the border of Argentina with Uruguay. Considering the area common to both analyses, the pattern agreement with the pattern of the second eigenvector for the 1916–1991 period is remarkable. Data from Uruguay, Brazil and northern Paraguay allow one to see both centers much clearer, as in Figs. 12.3a and b. Similar observations can be made comparing the first eigenvector for the period 1916–1991 (Fig. 12.3a) with the second for the period 1956–1991 (Fig. 12.5b).

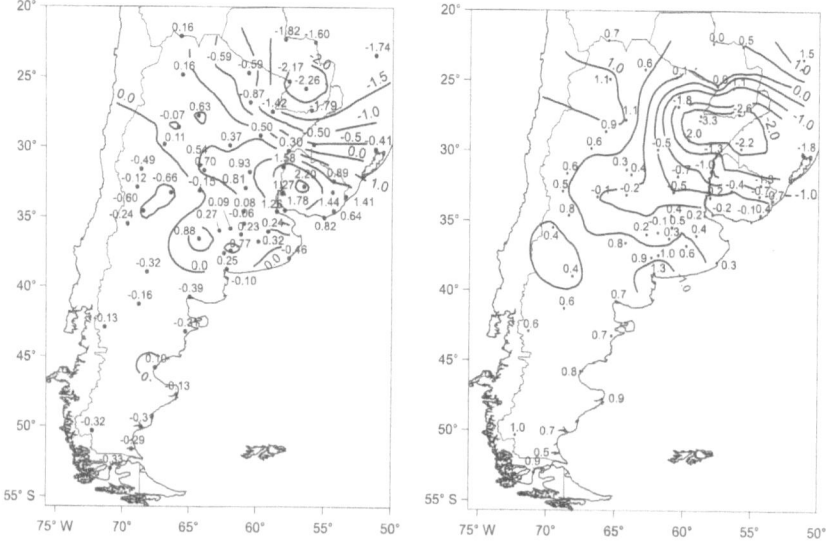

Fig.12.5. Factor scores of precipitation, 1956–1991. Shown is the spatial distribution of (a) the first eigenvector (above left) and (b) the second eigenvector (above right).

In this case the similarity of both fields is not so striking. In both cases there is, however, a prevalence of gradients in the zonal direction over subtropical Argentina with a more detailed structure in the 1956–1991 period.

As can be expected from (a) the similarity of the fields of the first two eigenvectors during both periods and (b) the trends observed since 1960, both eigenvectors show definite trends in the period 1956–1991 (see Fig. 12.6). A closer look shows that the positive trend of the first eigenvector can be seen before 1970, keeping during the time that follows the mean value established around 1970. According to the spatial pattern of this eigenvector (Fig. 12.5a), a contribution to an *increase* in precipitation in eastern Argentina, Uruguay and Southern Brazil and a *decrease* in northeastern Argentina and Paraguay should be expected. This is in remarkable agreement with the trend field observed in the period 1956–1991 (Fig. 12.2). This demonstrates that at least parts of those trends have been produced by the trend in the first eigenvector. The second eigenvector shows not only a trend but also a shift of the baseline during the first years of the 1980's.

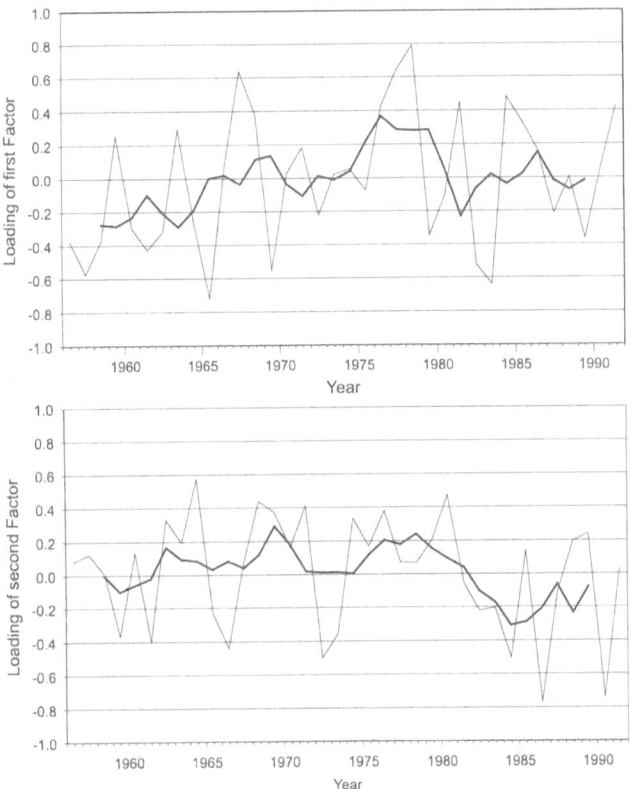

Fig. 12.6. Factor loadings of precipitation, 1956–1991. Shown is the factor loading of (a) the first eigenvector (top) and (b) the second eigenvector (bottom). In both cases moving averages over 5 years are also shown.

According to Figure 12.5b, a similar baseline shift can be expected for the time series of the precipitation in northeastern Argentina and the south of Paraguay. Instrumental data (see Castañeda and Barros 1994b) show that the above theoretical expectation is correct.

Tables 12.3 and 12.4 show the correlation coefficients between the time series of each eigenvector and hemispheric indices (such as the Southern Oscillation Index).These indices were taken from data elaborated by Angell and published by Boden et al. (1994). The numbers in parenthesis mean the same as in Table 12.2. In contrast to the period 1916–1991, the first eigenvector for this period correlates better with temperature at high latitudes. It correlates best with the temperature-difference between the subtropical and the polar region.

The negative correlation indicates that weak meridional gradients will lead to increased precipitation in eastern Argentina and Uruguay. In Paraguay rainfall will decrease under this conditions. The opposite behavior should be expected when circulation gradient is strong; i.e., dry conditions in Argentina and Uruguay opposite to wet conditions in Paraguay.

The second eigenvector is for all latitudes negatively correlated with hemispheric temperature. This suggests that this phenomenon is related to the amount of water vapor in the atmosphere, which in turn depends on temperature. The highest correlation of the eigenvectors can be seen with subtropical surface temperature. This is consistent with the observation that in subtropical Argentina most of the water vapor is transported from the north at low levels (it is advected not at high altitudes but closely to the surface).

Table 12.3. Correlation coefficients between the mean temperature with the first factor loading of the annual precipitation field for the period 1956–1991.

SOI 0.13	Surface level	850–300 hPa
Subtropical (10–30° S)	−0.13	0.02
Middle Latitudes (30–60° S)	−0.09	−0.08
High Latitudes (60–90° S)	0.21 (0.15)	0.15
Subtropical minus High Lat.	−0.30 (0.10)	−0.16

Table 12.4. Correlation coefficients between the mean temperature with the second factor loading of the annual precipitation field for the period 1956–1991.

SOI 0.39 (0.05)	Surface level	850–300 hPa
Subtropical (10–30° S)	−0.35 (0.05)	−0.23 (0.15)
Middle Latitudes (30–60° S)	−0.12	−0.27 (0.10)
High Latitudes (60–90° S)	−0.18	−0.28 (0.10)
Subtropical minus High Lat.	−0.02	0.09

As discussed before, the spatial pattern of this eigenvector resembles the general pattern of the mean annual precipitation with a prevailing zonal gradient. It seems natural that a higher (lower) availability of water produces an increase (decrease) of precipitation more or less proportional to the annual rain over subtropical Argentina. On the other hand, the correlation with the Southern Oscillation Index (SOI) is almost 0.4. This is in part due to the high negative correlation of the SOI with subtropical temperature, which is caused by the well-known fact that El Niño events coincide with warmer temperatures. It should be noted that, as will be discussed later, this is not the only physical mechanism that is responsible for the correlation of the precipitation field and El Niño.

Concluding this section, it can be said that the two principal components expressed as eigenvectors of the interannual variation of precipitation over the east of the Andes have suffered a substantial change around the year 1960. In fact, after that date, the dominant component (eigenvector) starts to be of a bipolar structure over eastern Argentina, well correlated with the meridional gradient at a hemispheric scale. The second, subdominant mode shows a structure which is similar to the main annual precipitation field. It is correlated both with hemispheric temperature and the SOI. After 1960 it explains a smaller amount of variance than before. Finally, both eigenvectors show a trend after the sixties, especially the first one having above-mentioned bipolar structure.

12.3. Correlation of the Precipitation Field with Hemispheric Indices

As shown in section 12.1, some components of the principal component analysis do not contribute much to the explanation of the interannual precipitation field. However, the correlation of these "minor" components with some hemispheric indices is significantly different from zero. Since these PCA components only explain a small fraction of the total variance the question arises whether these hemispheric indexes are correlated with the precipitation field.

Figure 12.7a shows the correlation field of annual precipitation and the SOI. There is a quite high correlation in northeastern Argentina, Paraguay and southern Brazil. A high correlation can also be seen in a corridor extending from southeast of Buenos Aires to about Mendoza in western Argentina. There is also some correlation on the Atlantic Patagonian coast (SE Argentina), with a remarkable maximum of 0.56 in Trelew (35° W, 38° S). The more striking feature of this field is the lack of correlation in a band oriented from SE to NW that covers Uruguay, the area north of Buenos Aires, Córdoba and the northwest of Argentina. According to Fig. 12.7a , El Niño events related to the negative phase of the SOI are associated to positive anomalies of precipitation over great part of the region. During the warm events of the eastern Pacific, surface temperatures also show an increase in South America and the tropical Atlantic (Aceituno 1988). This warming is observed in the troposphere north of 40° S (Barros et al. 1994) over Argentina and Chile. This regional warming could be a cause for the increased precipitation over the region, especially in the subtropical part in high altitudes. In addition, during the warm phase of the ENSO, a negative anomaly center in the altitude of the geopotential field was observed in the high troposphere over Patagonia with a trough axis extending over subtropical South America (Karoly et al. 1989).

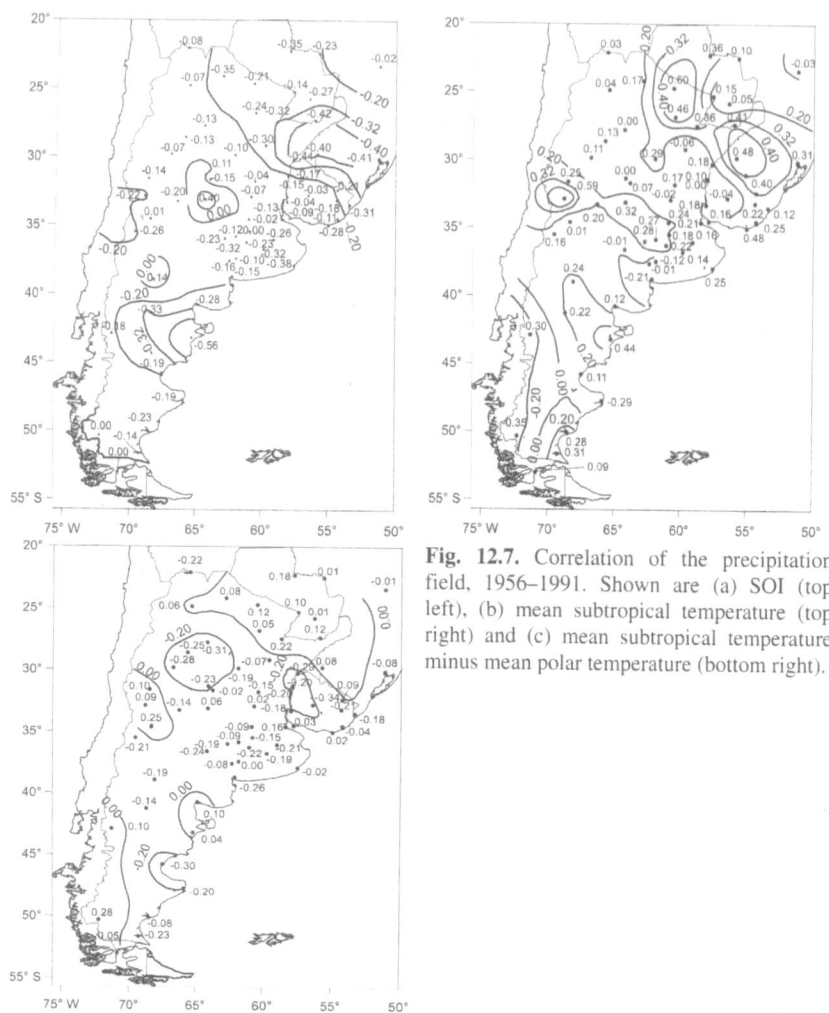

Fig. 12.7. Correlation of the precipitation field, 1956–1991. Shown are (a) SOI (top left), (b) mean subtropical temperature (top right) and (c) mean subtropical temperature minus mean polar temperature (bottom right).

This anomaly extends barotropically through the troposphere during all seasons except summer (Barros and Scasso 1994; Barros et al. 1994). Both the circulation features and the warming of the subtropical and tropical troposphere favor the increase of precipitation over the region. Thus the question arises why this does not happen in above-mentioned corridor that extends from Uruguay to northwestern Argentina. In order to understand this, the correlation of the precipitation field with the SOI and tropospheric temperatures (Fig. 12.7a–c) must be considered. Figure 12.7b shows the correlation of the annual precipitation with the subtropical temperature. Since this index is highly (negatively) correlated with the SOI (0.7) it is not a surprise that Fig. 12.7b resembles largely Fig. 12.7a . Of course, there are some differences, mainly at high latitudes.

Figure 12.7c shows the correlation with the difference of the mean temperature between subtropical and polar temperature. This difference can be regarded as the hemispheric meridional gradient in temperature. The correlation pattern in the subtropical region is highly complementary to of those of Figs. 12.7a and 12.7b. It should be noted that the region with significant correlations is almost exactly that region that does not show correlation with the SOI or the subtropical mean temperature. This is true not only for the corridor that extends from Uruguay to the northwestern Argentina but even for a detail of a smaller scale in the western part of the province of Buenos Aires as well. According to Figs. 12.7a and b this area is a kind of gap between two areas of high correlations east and west of it. We will call this area the area of complementary pattern (ACP): a combination of (a) low but non-significant correlation with the SOI and (b) significant correlation with the meridional gradient of temperature.

The complementary pattern of Fig. 12.7c compared with Figs. 12.7a and b can be explained by an interference of the influence of the El Niño and the meridional gradient on the annual rainfall. Let us assume that the pattern showed in Fig. 12.7c is caused by processes related to the meridional gradient. Since this meridional gradient is calculated as the difference of the subtropical and polar temperatures, an increase of subtropical temperature contributes to a higher meridional gradient. According to Fig. 12.7c this means that because of the increase of the meridional gradient, in statistical terms, less precipitation should be expected over the ACP. On the other hand, due to the high correlation between the subtropical temperature and the SOI, a positive anomalous precipitation over the whole region as it is observed outside the ACP should be expected. So, during El Niño years there is in statistical terms an opposite effect of the meridional gradient over the ACP, which could explain the lack of correlation over the wide area shown in the Fig. 12.7a.

12.4. Precipitation over Subtropical Argentina and the Meridional Gradient of Temperature

At this point of the discussion some additional climatic phenomena of the region must be considered. North of 40° S, the circulation is frequently under the influence of the South Atlantic quasi-stationary high, with prevailing light northeastern winds. Most of the time, there is a low-pressure center in the western and central part of the country generated by the topography of the Andes and by surface heating. This low appears also in climatic maps. The western low shows up in every monthly climatic mean map all over the year although it is less intense in winter (Hoffmann 1991; Taljaard et al. 1969). This pattern is often modified by polar fronts approaching from the south. These polar fronts are less frequent in summer when many fronts do not progress north of the 40–35° S latitude band. South of 40° S, the circulation is very intense and persistent from the west. There it is only occasionally altered by deep perturbations of the west flow.

Because of the prevailing circulation, most of the humidity comes from the north. Therefore it could be expected that in subtropical Argentina precipitation is lower than normal in those periods that are characterized by a westerly flow *north* of its mean position. On the contrary, when the westerly flow remains south of its mean position, the number of days with warm and humid advection from the north or northeast is higher than normal. This means that prefrontal and frontal rains are more abundant.

To check the climatic consequences of the described processes, we study how the rain responds to the latitude of the circulation systems. In the region under study many synoptic patterns occur that are difficult to explain if only data from surface analysis are considered. Therefore to understand the circulation system better, we look at the latitude of the maximum wind at the 200 hPa level.

Above South American at the latitudes covered by Argentina, the mean wind flow in 200 hPa is almost zonal. In addition small meridional shear can be observed. The maximum of this small meridional shear occurs in a zone that is about 1000 km wide. This zone changes its mean position during the year. Many authors have documented these characteristics (Berbery 1993; Karoly et al. 1986). Doyle (1994) studied the geographical positions of the regional maxima of the wind in 200 hPa over South America. She found that the mean latitude of the maximum winds in 200 hPa oscillates from 25° S in winter to 35° S in summer. In the monthly mean fields there is only one zone of maximum winds in South America and in the greater part of the Southern Hemisphere. The Australian sector is an exception with two differentiated branches. In spite of these mean fields, synoptic observations show that in the South American sector quite frequently more than one jet occurs simultaneously.

In order to assess *subtropical* rainfall in those cases that are characterized by two maxima of wind, the one which is located at a lower latitude was included in the analysis. This criterion does not assure that these maxima correspond all to the subtropical jet. Anyhow, it serves as an indicator of the northern bound of the westerly flow, and as such, of the latitude of the main features of the circulation over southern South America. The latitude of the regional maximum winds was calculated on a day-to-day basis from 1980 until 1988 using analyses of the European Center for Medium Range Weather Forecasts (ECMWF).

The mean annual rainfall field in the subtropical Argentina has a great variability ranging from the arid zone in the west, to values exceeding 1000 mm per year in the east. These characteristics, with some modifications, are present in most of the monthly means (Hoffmann 1975). In a field with so much spatial variation, the association between the rainfall and a given feature of the circulation must be based on the analysis of the individual data (stations, anomalies). So, for each locality and for each day, the rainfall anomaly was calculated with respect to the average monthly rainfall in that locality.

For each day stations were grouped into regions as a function of the distance to the latitude of the maximum wind considering six zonal bands of 5° of latitude. One band was centered at the latitude of the maximum wind; northward of this band we consider three bands of 5° width. Southward of the latitude of maximum wind two bands of the same width are considered. The daily rainfall anomaly of each region was calculated averaging the anomalies of the localities of that particular region (band) which for the respective day presented meteorological information either reporting rainfall or the lack of it. The annual anomaly of each band was calculated adding up daily anomalies. This calculation was done for every year between 1980 and 1988 yielding mean annual anomalies for each band.

12.4.1. The Rainfall Fields in Relationship to the Latitude of the Maximum Wind in 200 hPa

Figure 12.8 shows that the annual precipitation *anomaly* reaches a maximum annual mean value of 160 mm in the band located between approximately 800 to 1400 km north of the latitude of the maximum wind in 200 hPa. Anomalies are smaller in the bands south and north of this latitude but still positive. In the band centered at the jet, the average anomaly amounts to –30 mm. In the two bands south of the central band the anomalies are negative as well. The precipitation anomaly reaches a minimum in the band located some 300 to 800 km south of the latitude of the maximum wind in 200 hPa. These annual anomalies are the result of similar structures in the summer and the autumn, with a smaller contribution during spring and a very small contribution during winter.

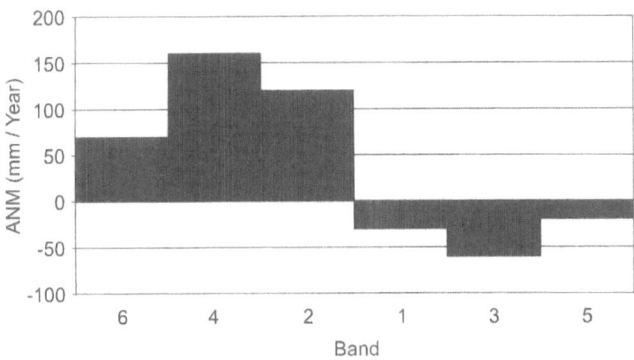

Fig. 12.8. Mean annual anomaly of the precipitation as a function of the distance from the latitude of the daily maximum wind at the 200 hPa level.

According to Fig. 12.8 a northward (southward) displacement of the circulation will produce a negative (positive) precipitation anomaly over a region of at least 1000 km width with a maximum north of the mean position of the subtropical maximum wind at 200 hPa. In addition it will produce a decrease (increase) of precipitation further north. This behavior is consistent with the evidence discussed in section 12.2.2, the analysis of the eigenvectors of the precipitation field.

The first eigenvector of the precipitation for the period 1956–1991 and the second eigenvector of the precipitation for the period 1916–1991 show a similar pattern. This suggests that the pattern of the above-mentioned precipitation-field at the surface is caused by the interannual variation of the circulation well above this area. That is, changes of the jet-stream. This will be discussed with more detail further below.

It is interesting to check if the latitude of the systems has some dependence of the meridional gradient of temperature as the pattern of the eigenvectors (see for example Fig. 12.5) suggests. We know that this happens at hemispheric scale. Smagorinski (1963) suggested a dependence for the latitude (ϕ) of the subtropical highs belt of this type: $tg\phi \propto 1/\partial\theta/\partial y$, where $\partial\theta/\partial y$ is the meridional gradient.

Flohn (1980) found a good agreement of this expression with mean monthly climatic values of both hemispheres. In South America, the wind monthly mean maximum is not located at single latitude, but rather between a band of latitudes. Therefore the area of wind maxima is wide. Because of this, the particular latitude of the mean maximum could not represent the latitudinal displacement of the westerly flow well (see Figs. 12.9a and b). A better parameter of the latitudinal position of the systems seems the ratio between the wind value at 15–30° S and at 30–45° S. This ratio is higher in winter than in summer indicating the more frequent position of the subtropical jet at lower latitudes. Therefore, we use the seasonal anomaly of this ratio to check its correlations with the anomaly of the meridional gradient. This correlation is 0.24 with a level of significance of 15%. This means that stronger meridional gradients are associated in statistical terms to lower latitudes of the maximum winds.

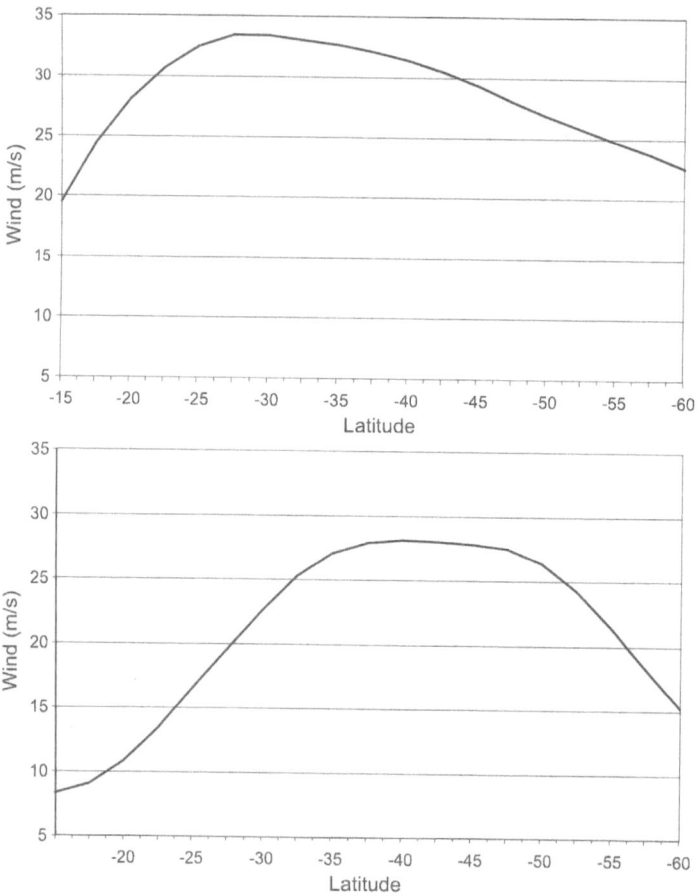

Fig. 12.9. Mean zonal flow over the South American sector for (a, top) Southern Hemisphere winter and (b, bottom) Southern Hemisphere summer.

The meridional gradient of temperature influences the mean latitude of the circulation systems over South America. Therefore, because of the dependence of the anomalous precipitation on the latitude of the systems (see Fig. 12.8), the interannual meridional gradient of temperature is at least partially responsible for the bipolar mode of the interannual variation of precipitation.

12.5. Trends of Hemispheric Indices

Figures 12.10a–d show the temporal evolution of that hemispheric indices which correlate with the annual precipitation. Figure 12.10a displays the SOI showing a negative trend. A closer inspection of Fig. 12.10a shows that more than just a gradual change of the SOI occurred and shows a baseline-shift of the Southern Oscillation Index around the year 1975. This is a phenomenon that, due to its worldwide importance, needs further discussion. Figure 12.10b shows that the mean subtropical temperature had a steep trend during the period 1977–1983 keeping a more or less constant value since that time. As was shown above, both SOI and mean subtropical temperature are well correlated. The better correlation however exists in the high-frequency domain with some phase-shift characterizing the low-frequency domain. If a comparison is made with the second eigenvector (see Fig. 12.4b), which is the one with better correlation with these indices, the best agreement with the subtropical mean temperature occurs in the low frequency-domain. The reason for this behavior might be the trend this eigenvector shows for the period 1979–1983. The mean temperature of high latitudes (see Fig. 12.10c) shows an impressive trend of about 1° C for the whole period, 1958–1991.

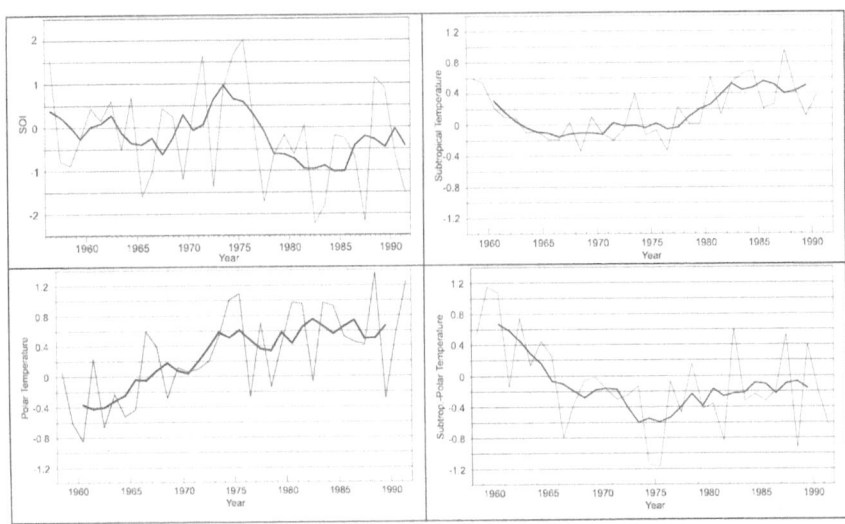

Fig. 12.10. Evolution of hemispheric indexes, 1958–1991. Shown are (a) SOI (top left), (b) subtropical temperature (top right), (c) polar temperature (bottom left), and (d) difference between subtropical and polar temperature (bottom right).

The difference between the mean subtropical and polar temperature (see Fig. 12.10d) has diminished to almost 1.5° C between 1960 and 1967 keeping this value since then due to a comparable warming in both latitudinal bands. In this context the behavior of the first eigenvector, which correlates with the SOI (see Fig. 12.4a) is very interesting. In the graph of this eigenvector the trend of the period appears exactly in the same time-interval as the trend of the meridional gradient. This is regarded as additional evidence for the dependence of the meridional gradient on large-scale circulation patterns. The interpretation of the interannual variation of the hemispheric indexes (Figs. 12.10a–d) combined with the first two eigenvectors of the precipitation pattern over Southern South America east of the Andes leads to two conclusions:

1) Both, the subtropical mean temperature and the meridional gradient of temperature have had a good correlation in the low frequency-domain with the principal modes of the annual variation of the precipitation pattern.
2) Due to above correlation, the trends of both hemispheric indexes (the SOI subtropical temperature, polar temperature, difference between subtropical and polar temperature) seem responsible for the major proportion of the precipitation trends over the region.

12.6 The Poleward Shift of the Circulation Systems

According to section 12.4, a decrease of the meridional gradient of temperature should be accompanied of a poleward displacement of the hemispheric circulation. Gibson (1992), using the Australian data set, showed a poleward shift amounting to 3° of the maximum wind in 500 hPa in 1976–1991. For almost the same period (1973–1989), Van Loon et al. (1993) calculated the latitude of the zonally averaged subtropical ridge. Although the authors only approximated the position of this ridge, it has a latitudinal trend of nearly 2° for that period.

There are no studies about trends of the latitude of the atmospheric circulation systems over South America. There are however indications for a stronger shift than in the hemispheric scale. Camilloni (1995) found that the maximum of the subtropical height over South America shifted poleward more than 5° of latitude during the 1970's. Barros and Scasso (1994) found a positive trend in the pressure of about 45° S, which is consistent with a poleward shift of the westerly flow over Patagonia. Finally, Duarte (1993) has commented that the troposphere over Buenos Aires has become more tropical in its thermodynamical features.

It is interesting to note that according to Fig. 12.9, a southward shift of the latitude of the maximum wind at 200 hPa during 1956–1991 could produce a zonally averaged meridional profile of the anomalous precipitation which is very similar to the one that can be calculated from the observed trends (see Fig. 12.2). In addition, the correlation of the first mode of the variability of the annual precipitation with the meridional gradient of temperature in the low frequency-domain (see sections 12.3 and 12.5) *and* the trend of the meridional temperature-gradient (section 12.4), indicate, that the *recent* poleward shift of the atmospheric circulation was the main cause for the *observed* changes of precipitation over subtropical Argentina.

12.7. Conclusions

Interannual variability of the precipitation pattern in southern South America has suffered a change since 1960. Not only are there strong positive trends since then over the subtropical region to the east of the Andes, but also the mode with a bipolar structure with its main action over Paraguay and the east of Argentina became more important. This mode is correlated with the meridional gradient of temperature and produces, over a band that goes from Uruguay to the northwest of Argentina, an interference with the El Niño signal on the annual precipitation.

The meridional gradient of temperature influences the mean latitude of the circulation systems over South America and because of this, is in part responsible for the bipolar mode of the interannual precipitation variability. The decrease of the meridional gradient over the last 35 years seems the first cause for the precipitation trends over subtropical South America to the east of the Andes.

Acknowledgements: This paper was possible due to a research grant given to the first author by the University of Buenos Aires.

References

Aceituno P (1988) On the functioning of the Southern Oscillation sector. Part. I: Surface climate. Mon. Wea. Rev. 116:505–524

Anderson TW (1963) Asymptotic theory for principal component analysis. Ann. Math. Stat. 34:122–148

Barros VR, Mattio HF (1977) Tendencias y fluctuaciones en la precipitación de la región patagónica. Meteorológica 8/9:237–246

Barros VR, Scasso LM (1994) Surface pressure and temperature anomalies in Argentina in connection with the Southern Oscillation. Atmósfera. 94(7):159–171

Barrros VR, Duarte ML, Scasso L (1994) Southern Oscillation response in Southern South America. Submitted to J. of Climatol

Berbery EH (1993) Estadísticas para el Hemisferio Sur en 200 hPa en base a 10 años de análisis del ECMWF. Meteorológica 18:13–22

Boden TA, Kaiser DP, Sepansky R, Stoss FW (1994) Trends '93: A compendium of data on global Change. Carbon Dioxide Information Analysis Center. Oak Ridge National Laboratory

Camilloni Y (1995) La influencia de la isla urbana de calor en las tendencias seculares de la temperatura media anual en la Argentina Subtropical. Dissertation, Univ. of Buenos Aires

Castañeda ME, Barros, V (1994a) Las tendencias de la precipitación en en el Cono Sur de América al Este de los Andes II Congreso Latinoamericano e Ibérico de Meteorologia. Belo Horizonte. Brasil. 1:207–211

Castañeda ME, Barros V (1994b) Las tendencias de la precipitación en en el Cono Sur de América al Este de los Andes. Meteorológica 19(1):23–32

Díaz EL (1959) Fluctuaciones de la continentalidad y en las lluvias. Anal. Soc. Cient. Tom. CLXVII:73–97

Doyle ME (1994) Los máximos de viento en 200 hPa y 300 hPa en Sudamérica y Atlántico adyacente. Dissertation (Tesis a Licenciatura), University of Buenos Aires

Duarte ML (1993) Temporal variations of the vertical profile of temperatures of Argentina. Int. J. of Climat. 13:437–445

Flohn H (1980) World Climate Conference. WMO. 537

Gibson TT (1992) An observed poleward shift of the Southern Hemisphere Subtropical wind maximun - a greenhouse symptom? Intl. J. of Climatology. 12:637–640

Green PE, Carol JD (1978) Analysing multivariate data. The Dryden Press, Hinsdale, Illinois

Hoffmann JAJ (1975) Atlas Climático de América del Sur. OMM-UNESCO. Impreso en Cartagraphia, Budapest, Hungaria

Hoffmann JAJ (1991) Las variaciones de la temperatura del aire en la Argentina y estaciones de la zona sub-antártica adyacente desde 1903 hasta 1989 inclusive. Seminario sobre el Cambio Climático Global. IPCC. Buenos Aires. Argentina

Hoffmann JAJ; Nuñez, S, Gómez, A (1987) Fluctuaciones de la precipitación en la Argentina, en lo que va del siglo. II Congreso Interamericano de Meteorología.V Congreso Argentino de Meteorología. Buenos Aires, Argentina

Jones PD (1990) Antartic temperatures over the present century: A study of the early expedition record. J. of Climate 3:1193–1202

Jones PD, Raper SCB, Cherry BSG, Goodess CM, Wigley TML, Santer B, Kelly PM, Bradley RS, Díaz HF (1991) An updated global grid point surface air temperature anomaly data set: 1851–1990. Environmental Sciences Division, Publication N° 3520

Karoly DJ, Kelly GAM, Le Marshall JF, Pike DJ (1986) An atmospheric climatology of the Southern Hemisphere based on 10 years of daily numerical analysis (1972–1982). WMO TD N° 92, Report Series N° 7

Karoly D., Plumb BA, Ting M (1989) Examples of the horizontal propagation of quasi-stationary waves. J. Atmos. Sci. 46:2802–2811

Krepper CM, Scian BV, Pierini JO (1987) Variabilidad de la precipitación en la región Sudoccidental Pampeana. II Congreso Interamericano de Meteorología- V Congreso Argentino de Meteorología, Buenos Aires, Argentina

Krepper CM, Scian BV, Pierini JO (1989) Time and Space Variability of Rainfall in Central-East Argentina. Journal of Climate 2

Minetti JL, Vargas WM (1983a) Comportamiento del borde anticiclónico subtropical en Sudamérica. Meteorológica XIV(1–2)

Minetti JL, Vargas WM (1983b) El enfriamiento de la década de 1950 en la República Argentina. Meteorológica XIV(1–2)

Moyano MC; García N; Almeira G, Dente M (1991) Respuesta del Río Paraná ante fluctuaciones climáticas. Congremet VI:97–98

Pittock AB (1980) Patterns of climatic variation in Argentina and Chile- I. Precipitation, 1931–60. Mon. Weath. Rev. 108:1347–1360

Preisendorfer RW, Barnett TP (1977) Significance tests for empirical orthogonal functions. Fifth Conf. Probability and Statistics, Las Vegas, Amer. Meteor. Soc. pp 169–172

Smagorinsky J (1963) General circulation experiments with the primitive equations: 1. The basic experiment. Mon. Wea. Rev. 91:99–165

Schwerdtfeger W, Vasino CJ (1954) La variación secular de las precipitaciones en el este y centro de la República Argentina. Meteoros. 4(3):174–193

Taljaard JJ, Van Loon H, Crutcher HL, Jenne RL (1969) Climate of the upper air. Southern Hemisphere. NCAR and Weather Bureau, Republic of South Africa, NCAR and NWRC

Van Loon H, Kidson JW, Mullan AB (1993) Decadal variation of the annual cycle in the Australian Data Sets. J. of Climate 6:1227–1231

Vargas WM (1987) El clima y sus impactos. Implicancias en las inundaciones del noroeste de Buenos Aires. Boletín Informativo Techint. N° 250. Noviembre-Diciembre

Weber TFA (1951) Tendencias de las lluvias en la Argentina en lo que va del siglo. IDIA 48. INTA

Chapter 3: Quaternary Climates

The Quaternary is a time with widespread glaciations in the Northern Hemisphere and is often regarded as a model for understanding colder and more humid environmental conditions. A closer look at Quaternary paleoenvironmental conditions shows that Quaternary studies may contribute to questions brought about by the growing public concern for the deforestation of tropical rainforests. In this context, it is suspected that continuing deforestation might affect the global climate significantly. This is also important because the work of Barros et al. (chapter 2) shows that areas located immediately south of Amazonia are already characterized by reduced precipitation.

The contribution of Latrubesse (Amazonia), the first study in this chapter on the Quaternary climates, shows that during the last glacial maximum there was a considerable retreat of the Amazon rainforest and a partial replacement by dunes and desertic conditions. Thus a study of Quaternary conditions may, if adjusted for present day temperatures, serve as analog for the impact of potential expected future deforestations.

Works of Anhuf and Runge, focus on central Africa, and independently show a reduction of the extension of tropical rainforests. Anhuf examines vegetational history with respect to climate change for 10° N to 10° S in Africa, for the time period 18 to 15 thousand years ago. Runge examines enviornmental data with respect to corresponding climate history for an area of Zaire, from 40 thousand years ago to the present.

Corbella et al. present paleoclimate results from new drillholes in volcanic maars in southern South America. They use lacustrine sediments from these cores and palynological studies to interpret paleoenvironmental changes from the region.

Del Valle et al. focus on climatic conditions during the last deglaciation. For this study, they use lacustrine sedimentary cores from the Nahuel Huapi National Park in Argentina.

Finally, in the last contribution for this chapter, Espizua presents data on Quaternary glacier lines in both Argentina and Chile.

The Late Pleistocene in Amazonia: A Paleoclimatic Approach

Edgardo M. Latrubesse
Universidade Federal de Goiás, IESA, Laboratory of Geology, Campus Samambaia, 74001–970, Goiânia, GO, Brazil
latrubes@virtualhouse.com.br, latrubes@iesa.ufg.br

Abstract: The Amazon rainforest at the time of the last glaciation, during the late Pleistocene, suffered strong changes in its biogeography and landscape. The review of multidisciplinary data (geology, palynology, vertebrate paleontology) permits one to obtain some palaeoenvironmental pictures. The best information that we have about Pleistocene chronology, paleoecology and geology is related to the time before 24 000 yr BP, possibly corresponding to the middle Pleniglacial (isotopes stage 3 and early stage 2). During that period the climate was apparently influenced by high precipitation at the Andes and a continuing change towards dry conditions in the lowlands. The Amazonian rivers were characterized by alluvial sedimentation, associated with the changing climate. The aridity reached its climax during the upper Pleniglacial. The resulting eolian sedimentation extended over parts of the central and north Amazonia. At the same time, the savanna vegetation reached its maximum extension. The trade winds were more intense and dry than today. Beginning at about 14 000 yr BP, the sedimentation in the alluvial belts was influenced by the deglaciation. Probably this sedimentation phase culminated with the marine transgression of the Middle Holocene.

13.1. Introduction

South America, with an area of 17.8 million km^2, the large Andes mountain chain, deserts, semiarid regions, savannas and tropical and nontropical forests, is a fascinating continent for Quaternary study. In South America, Quaternary research has focused on the Andes, Pacific and Atlantic coasts, as well as on the large plains of middle latitudes. Nevertheless, almost three fourths of South America is situated between the tropics, making it basically a tropical continent. The responses of the tropical regions to Quaternary climate changes are not well understood. A little more is known about the climatic changes that occurred during the last glaciation, and more specifically during the last 50 ky (thousand years). In this chapter, both already existing and new data collected in the Amazon region will be considered, with the objective of studying climatic changes that occurred during the late Pleistocene.

13.2. The Amazon Rainforest

The Amazon Rainforest occupies more than 5 million km^2 and is the largest tropical rainforest in the world. Although the rainforest extends into various countries, 70–80% of the total area is located in Brazil (Oldeman 1981). The Amazon vegetation is dominated by closed rainforest canopy with trees that reach 50 m height. In reality, the Amazon region is a complex mosaic of different vegetation types, including flooded forests, campina forests, savannas, bamboo forests, mangroves, montane and sub-montane forests, and liana forests (Murcia–Pires 1984; Oldeman 1981; Witmore and Prance 1987).

13.3. Climate

A humid tropical climate prevails. Rainfall, averaging 2000 mm/yr increases in the northwest to more than 5000 mm/yr (Fig. 13.1). The Intertropical Convergence Zone (ITCZ) shifts within the Amazon region. During the Southern Hemisphere summer, the ITCZ is located between 10–15° S. In this period, Southern Amazonia receives the most rainfall. The ITCZ shifts northwards, reaching its extreme position from July to August and located over Venezuela and Colombia. Trade winds blow from east to west on both sides of the ITCZ. During winter, the area situated to the south of the Solimoes–Amazon River has a marked "dry season". Between June and August, the total average rainfall for the three months decreases to 100–140 mm. At this time, cold air masses enter Southwestern Amazonia, producing strong drops in temperature, southerly winds and an increase in air pressure. This phenomenon is known as "friagem" in Brazil and "Surazo" in Peru.

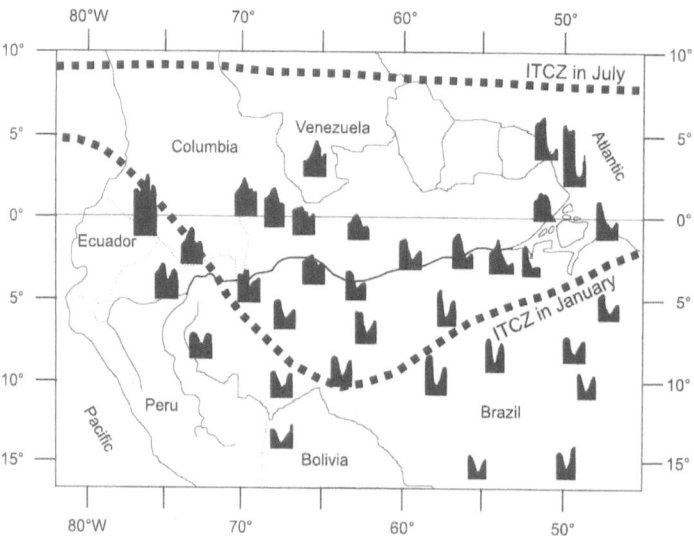

Fig. 13.1. Present annual rainfall distribution in Amazonia. The dotted lines indicate the locations of the intertropical convergence zone (modified according to Salati et al. 1978).

Therborg (1983) described a sharp temperature decrease in July 1975 when at Iquitos (latitude 3° S) temperatures of 8° C were recorded for three consecutive nights. According to the author, the normal average temperature during the "friagem" would be 14–16° C. In the Bolivian plains, strong drops in temperature occur during the "friagem". The minimum temperature reaches 0–5° C having an important biological effect. In addition, this causes economic losses in agriculture (Ronchail 1992).

13.4. Materials and Methods

Different materials and methods are used to reconstruct Quaternary climatic changes in Amazonia. Classical geological studies are integrated with satellite image interpretations, radar mosaics (SLAR), and, amongst others, aerial photographs at different scales. Geological and vertebrate paleontology studies are carried out along river banks and road cuts. The most complete vertebrate collection of Neogene Amazon vertebrates belongs to the Paleontology Laboratory of the Universidade Federal do Acre in the city of Rio Branco in the state of Acre, Brazil. For limnology and palynological studies, samples are obtained by drilling in lakes. Absolute chronologies are commonly obtained through radiocarbon dating. Thermoluminiscence (TL) dating procedures were used by some researchers.

13.4.1. Palynology

Palynological records came principally from lacustrine sediments situated in the states of Para and Rondonia (Brazil), in the south of the Brazilian Shield (Fig. 13.2).

The Rondonia record indicates that savannas replaced rainforest during different periods in the late Pleistocene (Absy and van der Hammen 1976). Dry periods were registered between 41.0–18.5 ky BP. (van der Hammen and Absy 1994). In the state of Para, in a little lake on the Carajás Plateau, dry periods were recorded with replacement of rainforest by savanna at 60 ky, 40 ky and 23–11 ky BP (Absy et al. 1991). However, a hiatus in the lacustrine sedimentation is recorded between 22–13 ky BP and, in consequence, no pollen was recorded during this period.

In the Ecuadorian rainforest, earlier works claimed that a cool climate existed during the late Glacial period before 26 ky BP and a lowering of montane Podocarpus vegetation of nearly 700 m (Liu and Colinvaux 1985). These data were strongly questioned in light of geomorphologic and pedological evidence (Heine 1994).

Palynological data obtained in fluvial sediments of the Middle Caqueta River by van der Hammen et al. (1992a), show an interval, during the Middle Pleniglacial (about 65–25 ky BP), when drier and more open vegetation must have covered a more extensive area than today. However, Amazon rainforest vegetation types could have been the dominant vegetation. According to van der Hammen et al. (1992a) and van der Hammen and Absy (1994), a reasonable estimate is that in Amazonia, during the Pleniglacial, temperatures were 2–6° C lower than today.

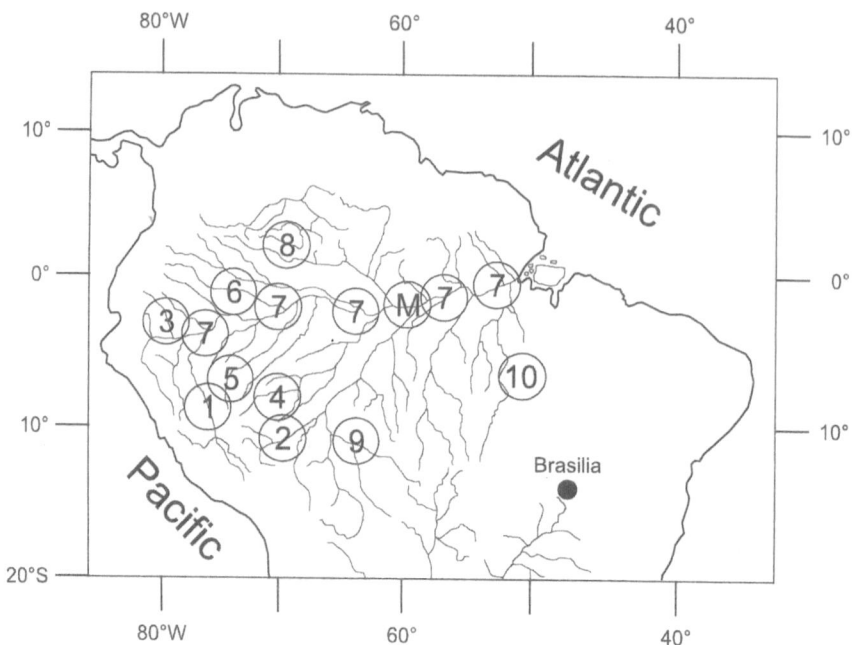

Fig. 13.2. Location map. Geological and palynological localities discussed in this contribution. 1: Ucayali river; 2: Madre de Dios river; 3: Pastaza river (and Pastaza fan); 4: Upper Purus fluvial basin; 5: Upper Juruá fluvial basin; 6: Middle Caquetá (Japurá) river; 7: Solimoes–Amazon river; 8: Upper Rio Negro fluvial basin; 9: Palynological sites (Capoeira and Catira) from the State of Rondonia (Brazil); 10: Carajas palynological site (State of Para, Brazil); M: Manaus.

13.4.2. Vertebrate Paleontology

Abundant large mammals of the Lujanean Mammal Age are found in Southwestern Amazonia (Simpson and Paula Couto 1981; Latrubesse and Rancy, 1998; Rancy 1991). Rancy (1991) reviewed vertebrate data from western Amazonia, studied the fossils found in the upper Jurua river and described findings in the Purus and Madeira rivers. Previous studies have noted the presence of large Late Pleistocene mammals in the Napo and Ucayali rivers. More detailed studies came from the upper Jurua basin. The fossils are principally found in the conglomerate facies of the Late Pleistocene sediments (Latrubesse and Rancy, 1998) which was defined as the "Bone-bearing conglomerate" (see Fig. 13.4) by Simpson and Paula Couto (1981). A rich Late Pleistocene vertebrate fauna was collected in the Madeira river (Sant´Anna et al. 1996). The majority of the fossil mammals are indicative of a savanna environment during the last. It was assumed that this fauna (see Table 13.1. and Fig. 13.3) inhabited Southwestern Amazonia in times close to the end of the Middle Pleniglacial and early stages of the Upper Pleniglacial (Latrubesse and Rancy 1998).

Table 13.1. Occurrence of Pleistocene mammals from western Amazonia and rivers (data source, Rancy 1992 and Sant´Anna et al. 1996).

TAXA/RIVERS	NAPO	JURUA	UCAYALI	MADEIRA
EDENTATA/PILOSA	X	X	X	X
Megatheriidae	X	X	X	X
Ocnopus	–	X	–	–
Eremotherium	X	X	X	X
Mylodontidae	X	X	–	–
Glossotherium	–	X	–	–
Lestodon	–	X	–	–
Scelidotherium	–	–	–	–
Mylodon	X	–	–	–
Megalonychidae	–	X	–	–
Megalonyx	–	X	–	–
EDENTATA/CINGULATA	X	X	X	X
Pampatheriidae	–	X	–	X
Pampatherium	–	X	–	X
Dasypodidae	–	X	X	–
Euphractus	–	X	–	–
Propraopus	–	X	–	–
Dasypus	–	X	–	–
Glyptodontidae	X	X	X	X
Hoplophorus	–	X	–	X
Neuryurus	–	X	–	–
Panoctus	–	X	–	–
Glyptodon	X	X	X	–
PROBOSCIDEA	X	X	X	X
Haplomastodon	X	X	X	X
Cuvieronius	X	X	–	–
ARTIODACTYLA	X	X	X	–
Camelidae	–	X	–	–
Vicugna	–	X	–	–
Palaeolama	–	X	–	–
Tayassuidae	X	X	X	–
Tayassu	X	X	X	–
NOTOUNGULATA	X	X	X	X
Toxodon	–	X	X	X
Mixotoxodon	–	X	–	–
PERISSODACTYLA	–	X	X	X
Tapirus	X	X	–	X
CARNIVORA	–	X	–	–
Eira	–	X	–	–
SIRENIA	–	–	–	X
Trichechidae	–	–	–	X
Trichechus	–	–	–	X
CETACEA	–	–	–	X
Odontoceti	–	–	–	X
Platanistidae	–	–	–	X
Iniinae	–	–	–	X
Inia	–	–	–	X

Explanation: Species is present (X); Species is absent (–).

Fig. 13.3. Paleontological record (black circles) of late Pleistocene mammals (Lujanean Mammal age) in Southwestern Amazonia (adapted from Rancy 1991). Solid dashed lines and extended areas indicate detailed surveys along the rivers of Southwestern Brazilian Amazonia.

The fauna recorded in the rivers of Southwestern Amazonia indicates open environments. *Glyptodon*, *Eremotherium*, *Mylodon*, *Toxodon*, *Haplomastodon*, and *Cuvieronius* were well-adapted groups for an open country savanna or savanna-like habitat (Webb and Rancy 1996). The record of *Vicugna* and *Paleaeolama* in the Jurua River, provides the stronger paleontological evidence for aridity in Amazonia at this time (Latrubesse and Rancy 1998).

13.4.3. Geology

Geological research on the Quaternary in Amazonia is scarce. Traditionally, geological surveys on the Quaternary were carried out in alluvial sediments of the fluvial belts and by pedological studies related to chemical weathering. However, extensive eolian fields in the northeastern and middle Amazon were discovered during the recent years. We will analyze the geological record differentiating alluvial from eolian sedimentation.

13.5. Sediments

13.5.1. Fluvial Sediments

Quaternary alluvial sediments are found in fluvial belts, frequently forming fluvial terraces. For the best understanding of the fluvial processes, we can classify the Amazon rivers into three groups:

1) Fluvial systems with headwaters in the Andes chain (Ucayali, Marañón, Madre de Dios, Caquetá Japurá, Pastaza).
2) Fluvial systems with headwaters in sedimentary lowlands (Purus Juruá, Javarí).
3) Fluvial systems with headwaters in the cratonic areas (Xingú, Negro, Tapajos)

In the rivers of western Amazonia a fluvial terrace with a height of 8–14 m above the water level of the rivers in the dry season is found. Sandy and conglomeratic deposits were deposited in Middle Pleniglacial times during the Late Pleistocene in the Ucayali, Madre de Dios, Caquetá, Purus and Jurua fluvial systems (Fig. 13.2). In the Ucayali river, Dumont et al. (1992) found that alluvial gravels (up to 10 times coarser than the present sandy bedload) were deposited between 32 ky and more than 40 ky BP. Along the Madre de Dios, in Peruvian Amazonia, wood material collected from the lowest terrace suggest an age around 36–38 ky BP (Rasanen and Linna 1993; Rasanen et al. 1990).

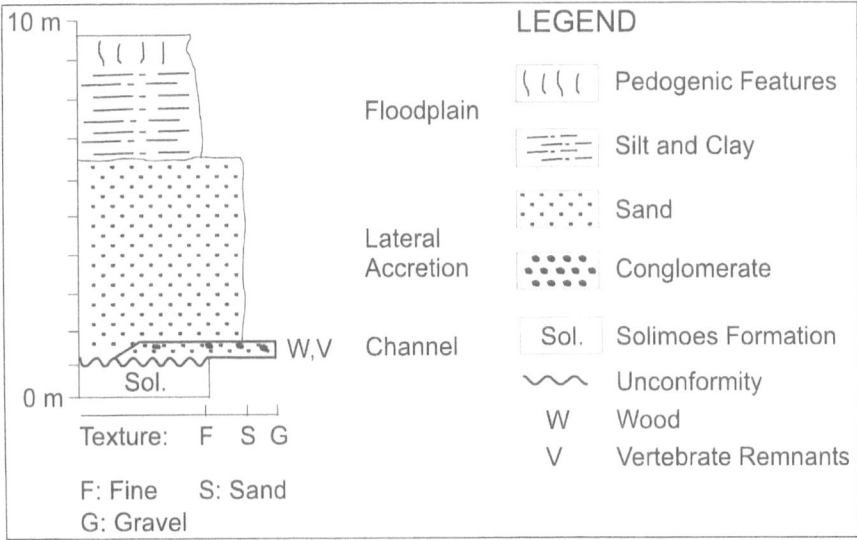

Fig. 13.4. Profile 6 in the upper Jurua river (left bank, 8° 57.43' S and 72° 45.34' W). A radiocarbon dating in wood, 32 300 ± 1600 yr BP (LP-652), was obtained from the conglomerate fossiliferous facies (Latrubesse and Rancy 1998).

The Caquetá river, in Colombian Amazonia, shows a low terrace formed by sandy and gravel deposits with some lenses of clay and peaty material. Radiocarbon analyses ranged between 30 ky BP and infinite age (van der Hammen et al. 1992a).

In the Ecuadorian and Peruvian Amazon, on the northern part of the Pastaza Marañón basin, widespread quaternary alluvial sediments cover an area of about 50 000 km^2. The alluvium was deposited by a megafan system (Pastaza megafan) probably during the late Pleistocene (Rasanen et al. 1992; Iriondo 1994).

The rivers with headwaters in the lowlands of Southwestern Amazonia have been studied by the author since 1989. Fieldwork was carried out in the upper and middle Acre, lower Iaco, Purus (near Boca do Acre), Moa and upper Jurua rivers (Fig. 13.3). Quaternary conglomerates were found in some rivers (Acre, Jurua, Purus and Madeira rivers).

The more precise Quaternary data came from the upper Jurua basin. In this region, Latrubesse and Rancy (1998) describe the stratigraphy and sedimentology of the fluvial sediments. The more important fossiliferous level is the "bone-bearing conglomerate" of Simpson and Paula Couto (1981) or conglomerate facies of Latrubesse and Rancy (1998). Radiocarbon dates ranging between 32–29 ky BP were obtained in this level.

The conglomerate is vertically and laterally discontinuous. It is facially related to sandy deposits. The prevailing colors range from black to red brown, due to precipitation of iron oxides; most pebbles reach 10–15 cm in diameter. The matrix is sandy to clayey sand. The "bone-bearing conglomerate" is found in the lower terrace, facially representing a lag facies or a short period of strong morphogenetical channel activity with the formation of gravel channel bars. The terrace sequence is typically fining upward. Occasionally strata of leaves and stems, some of which are decimeters in thickness, are interstratified with sandy and silty sediments (Fig. 13.4).

A rich vertebrate fauna of the Lujanean Mammal Age is found in the deposits, principally in conglomerates (Simpson and Paula Couto 1981; Latrubesse and Rancy 1998). Latrubesse and Rancy (1998) assign a late Pleistocene (Middle Pleniglacial-early stages of the Upper Pleniglacial) age to the fauna of the Juruá river. In the Moa river, an affluent of the Juruá river, an absolute dating for the lower terrace with an infinite age (more than 37 ky BP) was obtained (Fig. 13.5).

For cratonic areas, the quaternary alluvial sedimentary record is scarce. We have only some data (see Fig. 13.2) from the upper Negro river basin (Latrubesse and Franzinelli 1998). In the Tiquié, Vaupés and Curicuiarí rivers, late Pleistocene alluvial sediments frequently forming a terrace level of about 14 m thickness, are found. The sediments are sandy with planar and trough cross bedding. Coarse materials are found less frequently compared to sandy facies. The gravel reaches sometimes up to 3 cm in diameter. Plant fragments (stems and leaves) organic matter and impregnations of iron oxide are common. Radiocarbon dating of trunks and organic matter found in this sediment, range between 27 ky to more than 40 ky BP (infinite age). This episode of sedimentation was correlated with caution with the Middle Pleniglacial (Latrubesse and Franzinelli 1998). At that time, the paleohydrological regime of the basin was similar to that of the present, but it was morphogenetically more active.

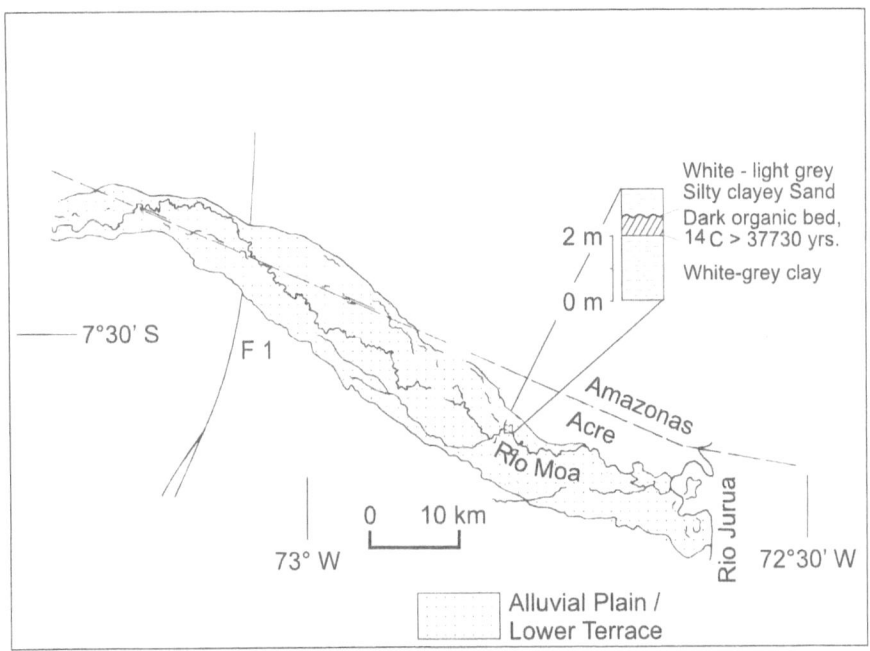

Fig. 13.5. Lower terrace in the Moa river, Juruá fluvial basin, Southwestern Brazilian Amazonia and vertical section (left bank) with radiocarbon dating of organic sediment (*). (Sample: Beta Analytic 52460; Age: more than 37 730 yr BP; Location: 7° 36.134' S and 72° 48.989' W).

In the Upper Negro Basin and on the watershed between Caquetá and Negro rivers, drier and more open vegetation covered a larger area than today, but probably that Amazonian forest was the dominant type of vegetation.

A younger late Pleistocene sedimentation phase was described in some rivers of Amazonia. In Southwestern Amazonia this sedimentation phase was recognized for the end of the Pleistocene. Analyzing the morphology of ancient channels of the Ucayali, Dumont et al. (1992) concluded that at 13 ky BP, the river discharge was seven to ten times smaller than at present. In the middle Caquetá (Japurá) river, late Pleistocene sediments (after 14 ky BP) were recorded (van der Hammen et al. 1992b). In the cratonic area the lateglacial sedimentation phase was recognized as well. In the Upper Rio Negro Basin sedimentation occurred at this time (Latrubesse and Franzinelli 1998).

13.5.3. Aeolian Sediments

Almost all of the research about eolian sediments in the Amazonia is preliminary. Eolian sediments can be differentiated into sand fields and silty loam deposits. Sand fields are covered by a "campina" vegetational association, composed of low, sparse trees and abundant tall grass. The largest sand field in Amazonia is the "Pantanal Setentrional" covering an area of nearly 100 000 km^2 between the middle Negro and Branco rivers. Parabolic dunes, probably of Holocene and recent age, were described by Carneiro Filho (1992), Nelson (1994), Santos (1992) and Santos et al. (1993). No geomorphologic and stratigraphic relationships were offered by these authors. The dune orientation is ENE–WSW and NE–SW. Smaller sand fields are numerous between Manaus and the Atlantic coast. Some of them were described by Iriondo and Latrubesse (1994).

To the north, in the state of Roraima (Brazil) and in western Guiana, the Boa Vista Formation is found. The upper section of this formation consists of silty and sandy deposits nearly two meters thick. These deposits are spread over more than 10 000 km^2. These sediments were interpreted as deposits of an eolian system during the Late Pleistocene (Latrubesse and Nelson, submitted). Dissipate longitudinal and parabolic dunes with NE–SW orientation are distinguishable to the north of the Tucano range area and in the Cauame area. Silty deposits were deposited peripherally to the west of the eolian system and in intramontane valleys.

In the area between Manaus and Parintins, in middle Amazonia, the Parintins Formation is found (Iriondo and Latrubesse 1994). This formation covers the landscape as an irregular mantle. It overlies the older Alter do Châo (Cretaceous) and Trombetas (Silurian) formations unconformably. In general, the sediments of the Parintins Formation have the granulometric composition of loam, varying locally to silty loam, clayey loam and clayey sand. The upper sections formed by massive loam were interpreted to be primarily of eolian origin (Iriondo and Latrubesse 1994).

13.6. Discussion

The best information that we have about Pleistocene chronology, palaeoecology and geology, is related to the period before 24 ky BP, possibly corresponding to the middle Pleniglacial (ca. 65–24 ky BP). Some authors suggest that sedimentation in the fluvial belts occurred during this period in rivers with headwaters in the Andes and in the lowlands of Southwestern Amazonia (Dumont et al. 1992; Latrubesse and Franzinelli 1995, 1998; Latrubesse and Ramonell 1994; Rasanen et al. 1990, 1992; van der Hammen et al. 1992a). For the majority of the authors (Dumont et al. 1992; Rasanen and Linna 1993; van der Hammen et al. 1992a), the alluvial sedimentation could have been directly associated with glacial advances in the central and northern Andes. Terraces formed by sandy and gravel deposits were associated with strong rains that fell in the Andes (Dumont et al. 1992; Rasanen 1993; Rasanen and Linna 1993; van der Hammen et al. 1992a). This explanation could be valid for rivers with headwaters in the Andes. However, this explanation cannot be accepted for rivers with headwaters in the lowlands of Southwestern Amazonia (Purus and Jurua basins) and those with headwaters in cratonic areas like Negro river. In the lowlands, the occurrence of pebbles (10–15 cm in diameter) and megafauna found in the conglomerate deposits clearly indicate the magnitude of hydrological changes. Thus, according to the available record, the Middle Pleniglacial would have been characterized by high precipitation in the Andes and a continuing change toward dry conditions in the lowlands. The sedimentological record of the Upper Rio Negro Basin, in the cratonic area, is indicative of stronger seasonal rainfall contrasts during the Middle Pleniglacial than at the present.

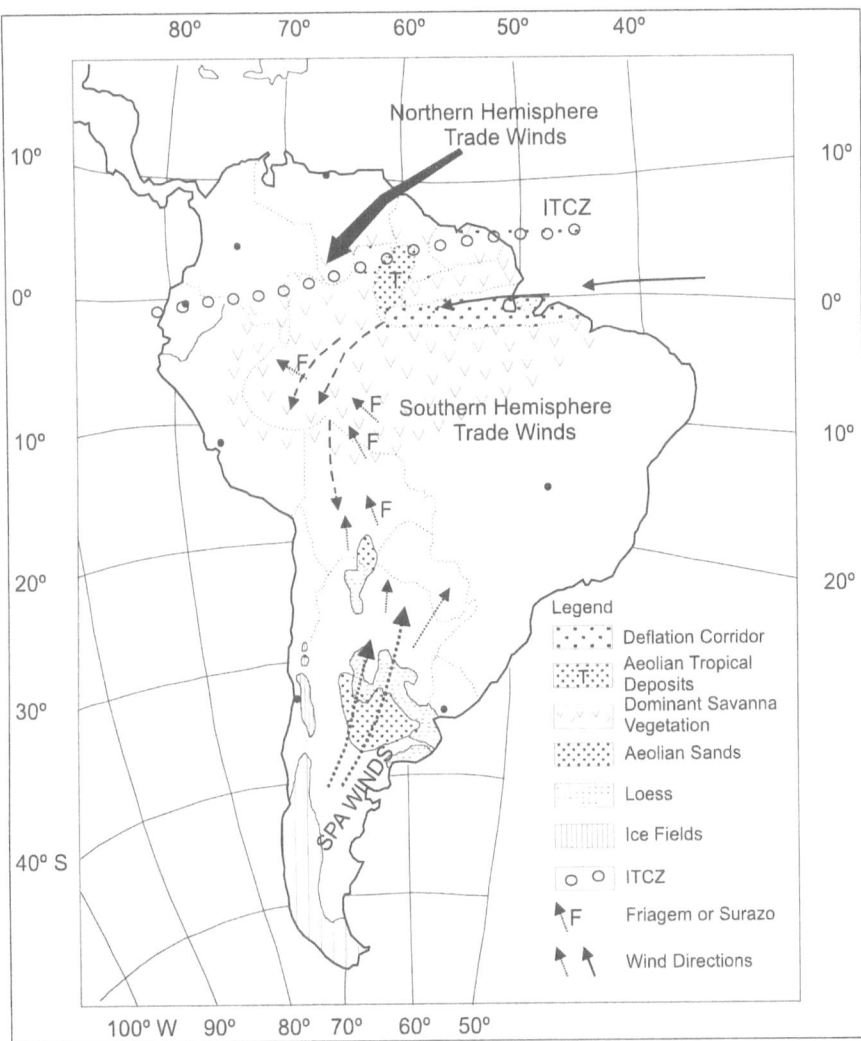

Fig. 13.6. Wind circulation model for Amazonia (data sources: Iriondo 1990; Iriondo and Latrubesse 1994; Latrubesse and Ramonell 1994; Ramonell and Latrubesse 1991).

The aridity in Amazonia reaches its climax during the Upper Pleniglacial. Probably at that point the eolian sedimentation extended over parts of central and northern Amazonia. At the same time, savanna vegetation reached its maximum extension.

Wind-circulation models were proposed for Amazonia for times close to the late Pleistocene (ca. 40–14 ky BP, see Iriondo and Latrubesse 1994; Latrubesse and Ramonell 1994). It was demonstrated that in the middle Amazonia, second-order changes in regional climate dynamics would have been sufficient for the occurrence of a dry climate phase (Iriondo and Latrubesse 1994). Trade winds would have been stronger and drier than at the present, producing extensive deflation corridors and eolian sedimentation. Northeastern trade winds were dominant approximately north of 2° N, removing sand and silt in "Pantanal Setentrional" and in Roraima. Southeastern trade winds, originated in the anticyclonic circulation of the Southern Hemisphere would have dominated the south of 0–2° N. Considering the above mentioned data, the palynological record in Carajas and Rondonia and the palaeomammals record of Southwestern Amazonia show that Amazonia was characterized by a dominant savanna environment during the Upper Pleniglacial with a dry season that was more pronounced and prolonged than that of the present.

For Southwestern Amazonia, it is inferred that cold fronts, originating in winter times from the cold air masses of the South Pacific Anticyclonic circulation (SPA) moved into Amazonia causing strong drops in temperature (Latrubesse and Ramonell 1994). These air masses were dominant in the Argentinean Pampas where the climate was dry and cold (Iriondo and García 1993; Ramonell and Latrubesse 1991). These SPA winds should have been strengthened by katabatic winds coming from the ice field of the Patagonian Andes (Iriondo and García 1993). In addition, the "friagem" or "surazo" phenomenon should have been more frequent and intensive than today, having great importance as a bioregulating factor (see Fig. 13.6) because of the accompanying temperature decreases (Latrubesse and Ramonell 1994).

After 14 ky BP, sedimentation in the fluvial belts occurred in response to climatic changes associated with the last deglaciation. At this time, the paulatine recuperation of the rainforest occurred. Probably this sedimentation phase culminated with the marine transgression of the Middle Holocene.

13.7. Concluding Remarks

In the last two decades, the more common explanation for the biodiversity and Amazon rainforest distribution during glacial times was that of the forest refuge (Haffer 1969, 1982; Prance 1982). Refuges were defined in terms of endemic centers of butterflies, lizards, woody angiosperms, birds, etc. (Brown 1982; Haffer 1969, 1982; Prance 1982; Vanzolini and Willams 1970). Haffer (1981:411) claimed that "forest and nonforest refugia should be not be identified only on the basis of biological data such as center of endemism, but rather on the basis of geoscientific data".

In the context of more recent botanical, geological and paleoecological advances, the refuge models (including the question of the validity of the models themselves and the distribution of the individual refuge areas) were strongly criticized (Rancy 1991, 1992; Latrubesse 1992; Rasanen et al. 1992; Nelson et al. 1993; Schubert et al 1994).

Amazonia suffered deep biogeographic and landscape modifications in times close to the last Glacial. Paradoxically, more detailed paleogeographic and palaeoclimate reconstructions exist for the Upper Pleniglacial in spite of the fact that this period is characterized by scarce paleontologic and geological record. On the other hand, a reasonable record for times before the Upper Pleniglacial exists. However, paleogeographic reconstructions for this period are almost non-existent.

The knowledge of Quaternary Amazonia advanced significantly in recent years. The discovery of extensive and widespread aeolian fields in Amazonia opened a new horizon in Quaternary research. Aeolian sediments and landforms can be the future key for the reconstruction of wind patterns and for the determination of different times of maximum aridity in Amazonia. In order to reach this objective, absolute dating of eolian sediments to determine the periods of eolian activity during the late Quaternary is essential.

Radiocarbon dating should be used with caution particularly for data obtained from a period close to the chronological boundary of the method. Apparently, contamination of organic layers by seepage water in an open system could be common (Heine 1994).

Palynological estimates of temperature variability should be taken with caution as well. It was demonstrated here that different areas could have experienced very different and variable temperatures. Thus, for example, Southwestern Amazonia would have suffered, at least seasonally, from a colder climate than other regions. In recent years "Quaternarists" and biologists preoccupied with explaining the wonderful Amazon biodiversity asked themselves: What happened to the forest during Glacial times? Was it remaining in the refuges? Did significant modifications in rainforest distribution occur or not?

In light of the new data, I think that another question is open to the scientist for the upcoming years. It is, "Where was the Amazon rainforest during Glacial times?" Quaternary research will have to offer a response in the future.

13.8. Future Perspectives

In the light of the work done already, three main points can be identified:

1) Amazonia suffered a high variability of climatic change during the Last Glacial. This includes the replacements of forest by savannas, differentially in time and space. Based on these results it can be deduced that deforestation and/or overgrazing in specific areas such as the "Pantanal Setentrional" or the Savannas of Roraima, could cause the reactivation of aeolian activity and the "detonation" of an induced desertification process.

2) Today, the necessity to conserve the rainforest's genetic heritage is accepted. Geological studies show that, once induced, environmental change such as dune field formation, can affect large areas. Conservation areas cannot continue to be delimited in function of actualistic models. Forest refuge or models based on present-day records of biodiversity have been demonstrated to be insufficient criteria for determining future conservation areas. Therefore, ecological paleo-scenarios determining high response-variability of different Amazonia sectors to natural or man-made induced climatic changes need to be established as an aid to finding the best strategy for conservation policies.

3) About 20% of the world's free water carried by rivers to the oceans comes from the Amazon fluvial system. This system drains the largest tropical forest of the world. Therefore, paleohydrological studies and the development of an integrated synthetic approach based on paleohydrological and present-day fluvial data are fundamental essentials for the assessment of the consequences of deforestation and other human induced impacts on the hydrological cycle.

Acknowledgement: I would like to thank Dr. Louis Richie for the English review of this manuscript.

References

Absy ML, van der Hammen T (1976) Some palaeocologica data from Rondonia, southern part of the Amazon basin. Acta Amazónica 6(3):293–299

Absy ML, Cleef ALM, Fournier RM, Martin L, Servant M, Sifeddine A, Da Silva MF, Soubies F, Suguio K, Turq B, van der Hammen T (1991) Mise en evidence de quatre fases d'ouserture de la fôret dense dans le sud-est de l' Amazonie au cours des 60000 dernieres annes. Priemiére comparásion osec d'autre regions tropicales. C.R.Acad. Sci. Paris, II, 312:673–678

Brown KSJ (1982) Palaeocology and regional patherns of evolution in neotropical butterflies. In: Prance GT (ed): Biological Diversification in the Tropics. Columbia University Press, New York

Carneiro Filho A (1992) Observaçiones preliminares das dunas do Río Negro – Resumos e Contribuiç·es Científicas. Simp. Int. Quat. Amazonia y 4 Reunión PICG 281. In: Franzinelli E, Latrubesse E (eds), UFAM, Manaus, p 166

Dumont JF, Deza E, Garcia F (1991) Morphostructural provinces and neotectonics in the Amazonian lowlands of Peru. Journal of South American Earth Sciences 4(4), 373-381

Dumont JF, Garcia F, Fournier M (1992) Registros de cambios climáticos para los depósitos y morfologías fluviales en la Amazonia Occidental – Paleo ENSO records. Intern. Symp., Extended Abstracts In: Ortlieb, L Macharé J (eds) ORSTOM: pp 87–92

Dumont JF, Lamotte S, Fournier M (1988) Neotectónica del Arco de Iquitos (Jenaro Herrera, Perú). B.S.G. Perú 77:7–17

Haffer J (1969) Speciation in the Amazon Forest birds. Science 165: 131–137

Haffer J (1981) Aspects of Neotropical bird speciation during the Cenozoic In: Nelson G, Rodson DE (eds) Vicariance biogeography. A critique. Columbia University Press, New York, p 371–412

Haffer J (1982) General aspects of the refuge Theory. Biological diversification in the tropics. In: Parnce GT (ed) Biological diversification in the tropics. Columbia University Press, New York, p 6–24

Heine K (1994) The Mera site revisited: Ice Age Amazon in light of new evidence Quaternary International 21:113–120

Iriondo M (1994) The Quaternary of Ecuador. Quaternary International 21:101–112

Iriondo M, Garcia NO (1993) Climatic variations in the Argentine plains during the last 18,000 years. Paleogeography, Palaeoclimatology, Palaeoecology 101:209–220

Iriondo M, Latrubesse E (1994) A probable Scenario for a dry Climate in Central Amazonia during the late Quaternary. Quaternary International 21:121–128

Latrubesse E (1992) El Cuaternario fluvial de la cuenca del Purus en el estado de Acre, Brasil. Dissertation, Universidad Nacional de San Luis, Argentina, pp219.

Latrubesse E, Franzinelli E (1995).Cambios climáticos en Amazonia durante el Plesitoceno tardio-Holoceno. In: Argollo,J and Mourguiart, Ph (eds) Climas Cuaternarios en américa del Sur. ORSTOM, p77-93.

Latrubesse E, Franzinelli E (1998) Late Quaternary Alluvial Sedimentation in the Upper Rio Negro Basin, Amazonia, Brazil: Palaeohydrological Implications. In: Benito G, Baker V and Gregory K (eds.) Paleohydrology and Environmental Change, John Wiley & Sons Ltd., p 259–271.

Latrubesse E, Ramonell C (1994) A Climatic Model for Southwestern Amazonia at Last Glacial times. Quaternary International 21:163–169

Latrubesse E, Rancy A (1998) The Late Quaternary of the Upper Jurua River, Southwestern Amazonia, Brazil: Geology and Vertebrate Paleontology. Quaternary of South America and Antartic Peninsula, 11:27–46.

Liu K, Colinvaux P (1985) Forest Changes in the Amazon basin during the last Glacial Maximum. Nature 311: 556–557

Murca–Pires J (1984) The Amazonian forest. In: Sioli H (ed) The Amazon. Dr. W. Junk Publishers, Dordtrecht, p. 581–601

Nelson B (1994) Natural Forest disturbance and change in the Brazilian Amazon. Remote Sensing Review, 10:105–125

Nelson BW, Ferreira CAC, Silva MF, Kawasaki ML (1993)Endemism centres and botanical collection density in Brazilian Amazonia. Nature 345 (6277): 714-716

Oldeman RAA (1981) Tropical America In: Jacob M (edited by Remke Krub et al) p 100–124

Prance GT (1982) Forest refuges: Evidences from woody angiosperms. In: Prance GT (ed) Biological Diversifiaction in the:, Columbia University Press, New York, p 137–158

Ramonell C, Latrubesse E (1991) El loess de la Formación Barranquita: comportamiento del Sistema Eólico Pampeano en la provincia de San Luis, Argentina. Third Meeting, IGCP 281, Lima, "Quaternary Climates of South América", Spec. Publ., N$^{\circ}$3:69–81

Rancy A (1991) Pleistocene mammals and palaeocology of the western Amazon. Dissertation, University of Florida, pp 151

Rancy A (1992) Western Amazon Paleomammals and the forest refugia model. In: Franzinelli E, Latrubesse E (eds): International Symposium on the Quaternary of Amazonia, UFAM, Manaus, Abstracts and Sci. Contr. p 45–48

Rasanen M (1993) Geologia: la Geohistoria de la Amazonia Peruana. In: Kalliola R, Puhakka M, Danjoy W (eds) ONERN Universidad de Turku, Turku (Finland), p 43–68

Rasanen M, Linna A (1993) Late Pleistocene alluvial terrace deposits and their dating in Peruvian Southwestern Amazonia. In: Franzinelli E, Latrubesse E (eds) International Symposium on the Quaternary of Amazonia, Manaus, Abs. and Sci. Contrib, pp 122

Rasanen M, Salo JS, Jungnert H, Romero Pitman L (1990) Evolution of western Amazon lowland Relief: impact of Andean Foreland dynamics. Terra Nova 2:320–332

Rasanen M, Neller R, Salo J, Jungens H (1992) Recent and ancient fluvial deposition system in the Amazonian foreland basin Peru. Geol. Mag. 129(3):293–306

Ronchail J (1992) Funcionamiento de los surazos en América del Sur y sus efectos climáticos en Bolivia: algunos resultados. Simposio de PHICAB, Actas, pp.11

Salati E, Marques J, Molion J (1978) Origem e distribuição das chuvas na Amazonia–Interciencia 3:200–205

Sant´Anna MJ, Trinidade A, Marques M (1996) Mamíferos fósseis do Quaternário de Rondônia. In: Latrubesse, E (ed) Southwestern Amazonia Paleo and Necolimates, Field Conference IGCP 341,UFAC, Rio Branco, 26–36

Santos JOS (1992) O Pantanal setentrional e os campos de dunas da Amazonia.In: Franzinelli E, Latrubesse E (eds) Intern. Symp. on the Quat. of Amazonia, UFAM, Manaus, Abstracts and Sci. Contr., p 110

Santos O, Nelson B, Giovannini CA (1993) Campos de dunas: Corpos de areia sob leitos abandonados de grandes ríos. Ciencia Hoje 16(93):22–25

Schubert C, Fritz P, Aravena R (1994) Late Quaternary paleoenvironmental studies in the Gran Sabana (Venezuelan Guyana Shield). Quaternary of South America, 21:81-90

Simpson G, Paula Couto C (1981) Fossil mammals from the Cenozoic of Acre, Brasil III; Pleistocene Edentata, Pilosa, Proboscidea, Sirenia, Perissodactyla and Artiodactyla. Iheringhia, Serie Geológica 6:11–73

Terborg J (1983) Five new world primates. A study in comparative Ecology. Princeton University Press, Princeton, N.J,260pp

van der HammenT, Absy, ML (1994) Amazonia during the Last Glacial. Palaeogeography, Palaeoclimatology,Palaeoecology 109:247–261

van der Hammen T, Duivenvoorden JF, Lips JM, Espejo N, Urrego L (1992a) The late Quaternary of the middle Caquetá River area (Colombian Amazonia). Journal of Quaternary Science, 7:45–55

van der Hammen T, Urrego L, Espejo N, Duivenvoorden J, Lips J (1992b) Late Glacial and Holocene Sedimentation and fluctuation of river water level in the Caquetá river area (Colombia, Amazonia). Journal of Quaternary Science, 7(1):57–67

Vanzolini PE, Williams E (1970) South American anoles: Geographic differentiation and evolution of the Anolis Crysolepis species group (Sauria, Iguanidae). Arquivos de Zoología 19:1–219

Webb SD, Rancy A (1996)Late Cenozoic Evolution of Neotropical Mammal Fauna. In: Jackson J B C, Budd, A B and Coates A G (eds). Evolution and Environment in Tropical America, The University of Chicago Press, pp 335–358

Witmore TC, Prance GT (1987) Biogeography and Quaternary history in tropical America. Clarendon Press, Oxford

Vegetation History and Climate Changes in Africa North and South of the Equator (10° S to 10° N) During the Last Glacial Maximum

Dieter Anhuf
Geographical Institute, University of Mannheim, L 9, 1–2, D-68131 Mannheim, Germany
anhuf@rumms.uni-mannheim.de

Abstract: This work presents vegetation maps and paleoenvironmental data for central Africa, covering both the last glacial maximum (LGM) and present times. The most striking feature is that the overall environmental conditions during the LGM have been much drier in central Africa than present-day conditions. Investigations focused on reconstructing former vegetation patterns for the Sudanian, Zambezian, and Guinean savannas, as well as tropical semi-evergreen and evergreen rainforests. A number of research projects concerning changes of vegetation cover have shown that even tropical regions have been affected by considerable climatic oscillations during the last 20,000 years. The increasing effect of human influence on appearance of the African vegetation cover imposes a significant problem for the study of past vegetations. Thus, all investigations studying temporal change in tropical ecosystems are confronted with the fact that areas that have not been influenced by man can rarely be found. The primary question is which vegetation forms allow us to draw conclusions applicable to a nearly natural vegetation. A first step is the estimation of potential quasi-natural vegetation formations under present climatic conditions. Derivation of West African paleovegetation is based on published paleoclimate and paleovegetation information, including palynology, deep-sea cores and isotope analysis. Using numerical relations between natural vegetation and climate under present conditions, models linking analogous vegetation and climate items permit subsequent assessment of paleoclimatic conditions. The present distribution of quasi-natural vegetation is linked closely to the climatic water-budget of the African continent. The amount of precipitation, as well as duration and intra-annual distribution of rain, also depends on the condition of important water-vapor sources for Africa, the SE Atlantic on Africa's west coast and the Indian Ocean on the continent's east side. The availability of well-resolved and dated profiles from the West African coast enables a reconstruction of sea-surface temperatures (SSTs) for the LGM. As there are significant connections between SSTs of oceans and the precipitation system today, such connections must have also existed during the LGM. The relations between the two allow establishment of transfer functions. These permit reconstructing the continent precipitation system for the LGM on the basis of SSTs. At first, the precipitation amount with its annual distribution is reconstructed. On the basis of reconstructed climatic water budget and in relation to present-day climate and vegetation conditions for the LGM, the paleovegetation is reconstructed and compiled in a map. Finally, the reconstructed vegetation is critically compared with respect to paleoenvironmental conditions for the LGM.

14.1. Introduction

On January 1, 1995, the BMBF (German Federal Ministry for Education and Research) began a special study into "climate variability and signal-analysis." An attempt was made to present cartographically the vegetation cover for Africa between 10° N and 10° S during the LGM (last glacial maximum) (15,000–13,000 ^{14}C-yr BP/18,500–16,000 cal-yr BP). The resulting map showed that it is possible to quantitatively reconstruct vegetation from defined and well-dated time series in past regional patterns, enabling reliable controlling methods for climate modeling. However, this does not mean that future climatic developments, and developments in vegetation, will change based on patterns already presented. Comparing glacial periods to themselves shows clear regional and chronological differences within their patterns of vegetation development. However, man's impact today on the vegetation and climate system has never had such a lasting effect.

A vegetation map for 18,000 yr BP (LGM) needed to be drawn. Before beginning, we had to investigate potential quasi-natural West African vegetation formations under present climatic conditions. A problem facing nearly all research work involving change in tropical ecological systems is that the registration of contemporary vegetation rarely encompasses the natural vegetation. The overall question is, which types of vegetation indicate a more or less natural vegetation?

14.2. Natural Vegetation

Geo-botanical analyses were done in the Ivory Coast, a country which can be divided into two floristic areas: the Guinea Region in the south and the Sudan Region in the north. The tropical ombrophile forests which dominate the floristic Guinea Region can be further subdivided into evergreen ombrophile forests, in which the annual dry season does not exceed one to two months, and into semi-deciduous ombrophile forests with not more than three to four months of dry season (Anhuf and Frankenberg 1991, see Fig. 14.1).

Dry forests dominate the Sudan Region. Most trees shed their leaves during the arid season, which lasts for more than four months. The northward decrease in precipitation combined with the long arid season allows yet another subdivision of these forests into dense dry forests (forêts denses sèches) and open dry forests (forêts sèches); the latter grow in regions in which the arid season lasts longer than half a year (Anhuf and Frankenberg 1991).

Estimating the potential forest types while considering contemporary climatic conditions was the first step of our analysis. The tree as such represents an essential element for the stability of the landscape within certain forest types, as well as being an important indicator for the West African Savanna. An important aspect of the vegetation change with respect to landscape degradation is described by the changing patterns of tree species.

But why is this question so important? Is it not enough to simply equalize vegetation and precipitation zones in order to define the borders of landscape areas? The idea that the character as well as the borders of a landscape can be drawn by only one of numerous natural factors, automatically leads to the idea of a shrinking or perhaps expanding vegetation for the present existing vegetation formations.

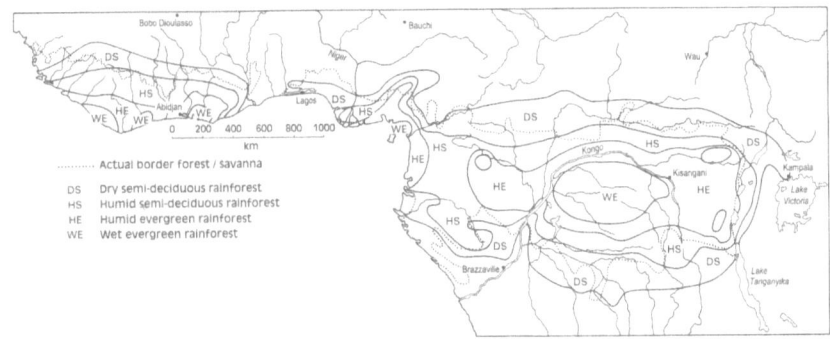

Fig. 14.1. Distribution of tropical ombrophile forests in West Africa.

Investigations in Senegal (Anhuf and Frankenberg 1993) and the Ivory Coast, as well as other works (Schulz and Pomel 1992; Neumann 1988) show that in many climate and vegetation reconstructions, man has seldom been considered an important factor for paleoenvironmental change. Actually, man has influenced the paleoenvironmental development of West Africa in three main phases. As early as 7000 yr BP (but more after 4000 yr BP), includes sustainable impacts of man influencing the savanna and forest regions of sub-saharan West Africa. Man's impact during this time changed and effected vegetation. At first, hunting and grazing fires were of limited effect. Later, increases in livestock caused essential changes in vegetation. This led to a more open landscape with a reduction of trees and shrubs, and a different spectrum of plant species. The distribution of trees and shrubs like *Acacia albida* and *Acacia laeta* was reduced due to zoochoric animals.

Not only grazing and burning are responsible for less dense vegetation. The main cause lies in specific selection and support of different trees and shrubs that serve human nutrition, also true for grasses and herbs. Out-of-the-tree savannas and open dry forests, so-called "park-savannas", have emerged because trees and shrubs have been (and are) important for human diet. Park-savannas are integrated land-use systems with fruit trees and crops. Typical park-savanna trees include *Balanites aegyptiaca, Acacia albida, Sclerocarya birrea, Butyrospermum parkii, Adansonia digitata, Parkia biglobosa, Elaeis guineensis, Borassus aethiopum, Daniellia oliveri* and *Lophira alata*. A further massive impact on Sudanian and Guinean Zone vegetation occurred 3000–2500 yr BP, when iron-ore extraction and iron processing began. Iron production was based on charcoal: the smelting process required 100 kg of wood to obtain 5 kg of metal (Kadomura 1989).

In order to create a map with potential quasi-natural vegetation formations in West Africa, testing areas with sizes of 1 ha were chosen in the Sudan and Guinea Domains in the Ivory Coast. All tree species were identified and counted. The inventories in anthropogenic- or zoogenic-influenced testing areas were compared with those of quasi-unaffected areas using PCA (Principal Component Analysis) (Anhuf 1994). Comparison of 229 savanna and forest test-plots allowed for numerical quantification of the anthropogenic-influenced vegetation and in addition, a reconstruction of the vegetation cover without human impact.

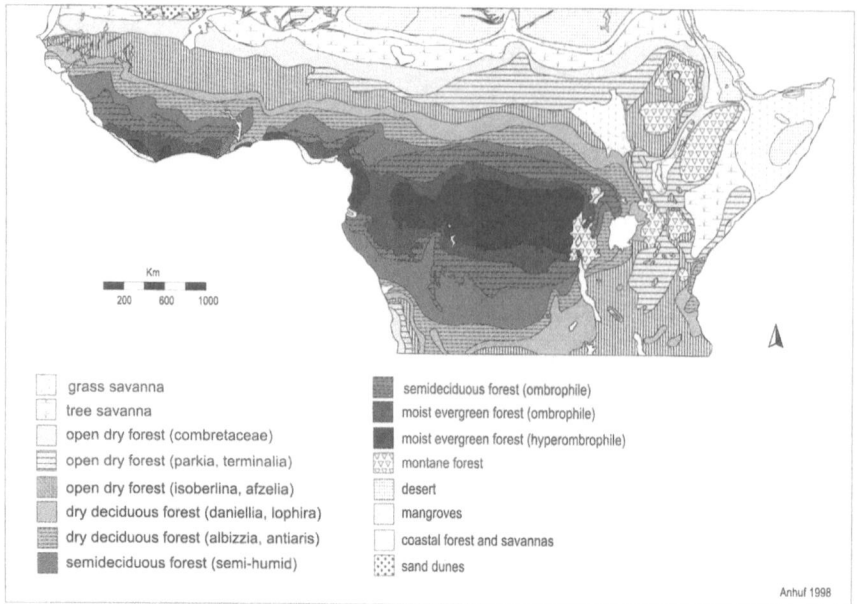

Fig. 14.2. African natural vegetation cover under present climatic conditions, north and south of the equator.

Based on research results in the Ivory Coast (Anhuf 1994) as well as literature, a map of potential quasi-natural forest types with modern-day climate parameters was constructed for Africa between 10° N and 10° S (Fig. 14.2). The hyper-ombrophile forest cannot exist in regions with less than 1800 mm/yr precipitation and/or a dry period not exceeding one month. This type of forest can only be found in the farthest southwest and southeast of the Ivory Coast, in the west of Cameroon, in Equatorial Guinea, and in the surroundings of Libreville in Gabon. Further areas with hyper-ombrophile evergreen forests grow in the center of Zaire between Mbandaka and Kisangani, as well as on the eastern side of the Congo Basin, and at the beginning of the East African Highlands between Bukavu at Lake Kivu and Lake Edward, 250 km further in the north.

Most regions with ombrophile evergreen forests have about 1500–1800 mm/yr of rainfall. Here too, dry periods do not last longer than one or two months. Generally, northern boundaries of ombrophile forests coincide approximately with the 1500 mm (minimum) line of annual precipitation. In West Africa, this forest type follows forest formations mentioned above. In Central and Equatorial Africa, these forests form a continuous forest belt, which reaches about 5° N and 4° S, at the latitude of Kinshasa. In the east of the continent, the ombrophile evergreen forests have their natural border along 30° E.

The continuation of the ombrophile forests is the transitional zone of dense semi-deciduous ombrophile forests that are characterized by a humid season lasting up to eight to nine months. Its northern expansions reach areas with a maximum dry season of four months. Furthermore, this zone is recognized as the frontier between the floristic Guinea and floristic Sudanian Zones in the north and between the floristic Guinea and Zambezian Zones in the south.

Within the Sudanian and Zambezian Zones, dry forests dominate the landscape. Forest types following the semi-deciduous ones are dense dry forests (forêts denses sèches) and represent a climax vegetation within the transitional zone of the above described floristics. The dry season is likely to last up to six months. Areas with an even longer dry season (> 6 months), which at the same time have a precipitation of 1000–1100 mm per year, are a transition zone leading to open dry forests (forêts sèches). In the Ivory Coast, this forest type might also be found (e.g. Parc Nationale de la Comoé) in the farthest northeastern parts near the border of Burkina Faso, as open dry forests can survive with only 4–5 humid months a year.

Today, many of these forests have disappeared almost entirely as they are very sensitive to fire. With the exception of the East African Highlands, their main distribution area used to be in a zone called the "Forest-Savanna-Mosaic" (Keay 1959). Further to the south, Miombo Forests follow a dryer variation of dense dry forests. According to international agreements (Menaut 1983), these types are not classified as belonging to forests anymore; they are now considered savanna formations (woodland/forêt claire). North of the equator, open dry forests (woodlands) are dominated by *Isoberlinia doka, Uapaca togoensis, Terminalia macroptera*, and *Monotes kerstingii*. In the Miombo Forest *Julbernardia globiflora, Brachystegia spiciformis, B. boehmii*, and other genera (e.g. *Monotes, Terminalia, Combretum*, and *Acacia*) are the leading genera. Big areas in East Africa were open dry forests with different species. Today, there are only some places where the trees, mostly *Acacia* and *Commiphora* species, form open dry forests (woodland). In the farthest east of the "Horn of Africa" there exist tree and/or shrub savannas. Species of *Acacia* and *Commiphora* occur here, with *Salvadora* and *Leptadenia pyrotechnica* also typical.

A special characteristic of East Africa includes extensive areas of afro-mountainous levels, due to strong tectonic and volcanic activity since the Tertiary. The afro-mountainous levels here are different from West Africa (Guinean Highlands), starting at 1500-1800 m (Knapp 1973). In West Africa the cool winter NE trade winds push the mountainous vegetation down to an altitude of 800 m. Mountainous altitudes in East Africa are covered by dry forests, of which *Podocarpus falcatus, P. latifolius, and Juniperus procera* are the main species. Average forest rainfall is 1000–1500 mm/yr, with the humid season lasting up to 6 months. At dryer locations in Tanzania, Kenya and Somalia, there are often patches of sub-mountainous Leguminosae such as *Albizia gummifera* and *Acacia abyssinica*. With further decreases in precipitation (down to 500 mm/yr) the *Olea-Juniperus (procera)* forests starts to occur. The semi-deciduous humid mountainous forests with *Entandophragma excelsum* (Mahogany), *Podocarpus latifolius, Aningeria adolfi-friedericii* and *Chrysophyllum gorungosanum*, are located at Mount Kenya, southern Ethiopia, and in the Usambara Mountains, Tanzania. There exists a humid season here of seven to eight months.

Even more humid, and therefore comparable in climatic needs with lowland rain forests, are evergreen mountain rain forests. They are much smaller in stature; their growth is restricted by a combination of low temperature and unpredictable rainfall. They only exist in southwest Ethiopia, at the east side of Mount Kenya as well as the west side of the East African Highlands leading down to the Congo basin between the Ruwenzori area in the north, west of Lake Edward, west of Lake Kivu and further to the south up to the latitude of Bujumbura. They reach altitudes of only 1500 and 2800 m on the west side of Lake Tanganyika. Species of the genera *Ekebergia, Chrysophyllum, Macaranga, Ocotea*, and *Schefflera* (Knapp 1973) dominate these forests. At altitudes between 2700–3200 m, and partly up to 3500 m, in the upper cloud forest belt, Kosso trees (*Hagenia abyssinica*) linked to *Hypericum revolutum* and to the African mountain-bamboo (*Arundinaria alpina*) characterize the landscape. Above the forest growth limit (3200 m) an Ericaceous belt follows, reaching as high as 3600 m (at Mount Kenya 4200 m) before the afro-alpine level starts. Average temperatures are about 7° C at the lower borderline and 2° C at the upper borderline, though there is frost almost every night. The afro-alpine vegetation is dominated by very open formations, from tussock grass formations to small shrubs. Probably the most representative species at this level are the Lobelia- and Senecio-group: *Lobelia bambuseti* in Kenya, *L. mildbraedii* in Rwanda, *Senecio johnstonii* in Kenya and Rwanda.

14.3. Paleobotanical Samples in Africa Representing the LGM

Maps reflecting vegetation changes were constructed using standardized methods. An ecological vegetation analysis of already mentioned sites, combined with published material on paleovegetation and paleoclimates, allowed for comparing past and present conditions. Information used included palynology, paleobotany, deep-sea cores and isotope analysis. Shown in Fig. 14.3 are localities which have dated materials, giving direct information about vegetation cover during the LGM. The data density varies considerably over the whole study area. The numerous pollen profiles from the Atlantic west of Western and Central Africa were very helpful. In East Africa however, there is no usable data from the Indian Ocean.

14.4. Materials and Methods for Constructing Paleovegetation Maps

One of the results of this work is a database describing both vegetation and climate for the LGM. In order to assess the paleovegetation, several steps have been performed: (1) Numerical relations between climatic parameters and vegetation have been established. (2) These relations are applied using the present-day *relations* in connection with the LGM *data*. The result is a vegetation map for the LGM. (3) The derived map is compared with independent evidence from pollen spectra. Background data can be found in Adams and Tetzlaff 1985, Agwu and Beug 1982, Anhuf 1997, Brenac 1988, Butzer 1976, Fischer and Hinkel 1992, Fredoux 1994, Gasse and Street 1978, Gasse et al. 1980, Hooghiemstra 1988, Hooghiemstra et al. 1986, Hooghiemstra 1987, Hurni 1982, Jahns et al. 1998, Lezine 1991, Lioubimtseva et al. 1996, Livingstone 1971, Lutze 1988, Messerli et al. 1980, Petit-Maire et al. 1987, Richardson and Richardson 1972, Roche and Bikwemu 1989, Rossignol-Strick and Duzer 1979, Runge 1996, Runge 1998, Schulz 1987, Schulz et al. 1990, Servant and Servant-Vildary 1980, Sowumi 1981, Stager 1988, Tiercelin et al. 1981, van Campo et al. 1982, Verschuren 1996, Vincent 1972, Webb et al. 1997, Wefer 1988 and White 1983. The inclusion of further pollen spectra is a task for future work.

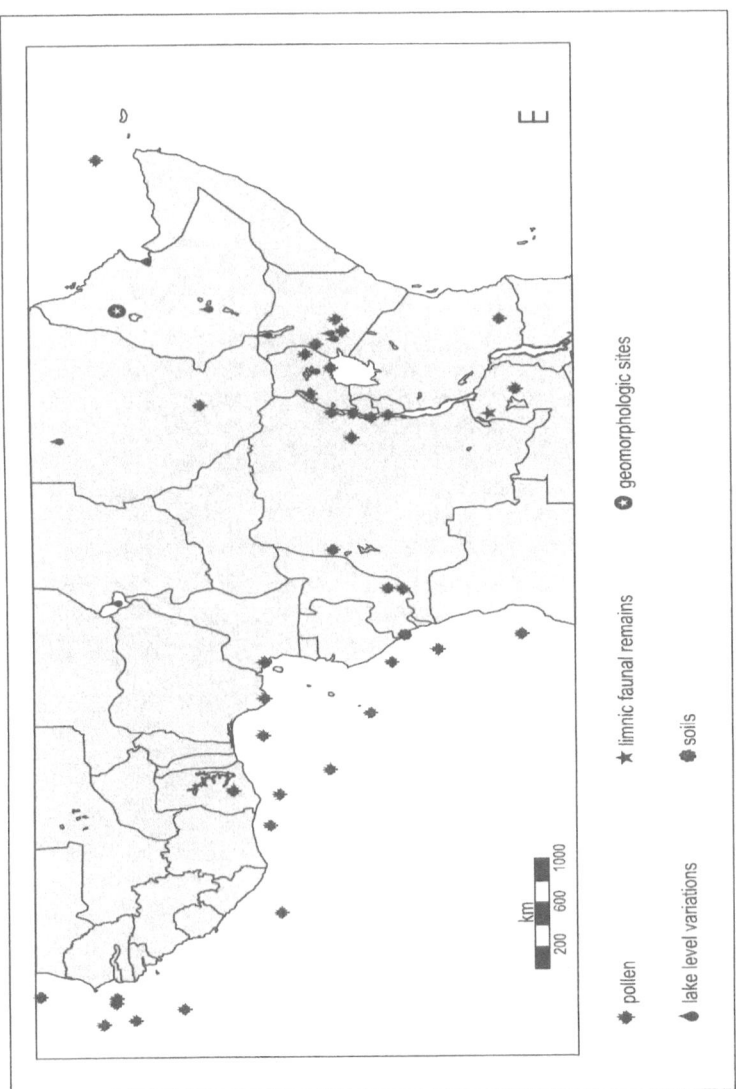

Fig. 14.3. Map of the locations of pollen-profiles for the LGM published in literature.

14.4.1. Sea-Surface-Temperatures as a fundamental Element for the Rainfall Distribution on the African Continent

Vegetation depends vitally on both temperature and precipitation. To assess past precipitations an integrated approach linking marine and terrestrial data is applied: The present distribution of the quasi-natural vegetation is linked very closely to the climatic water-budget of the African continent.

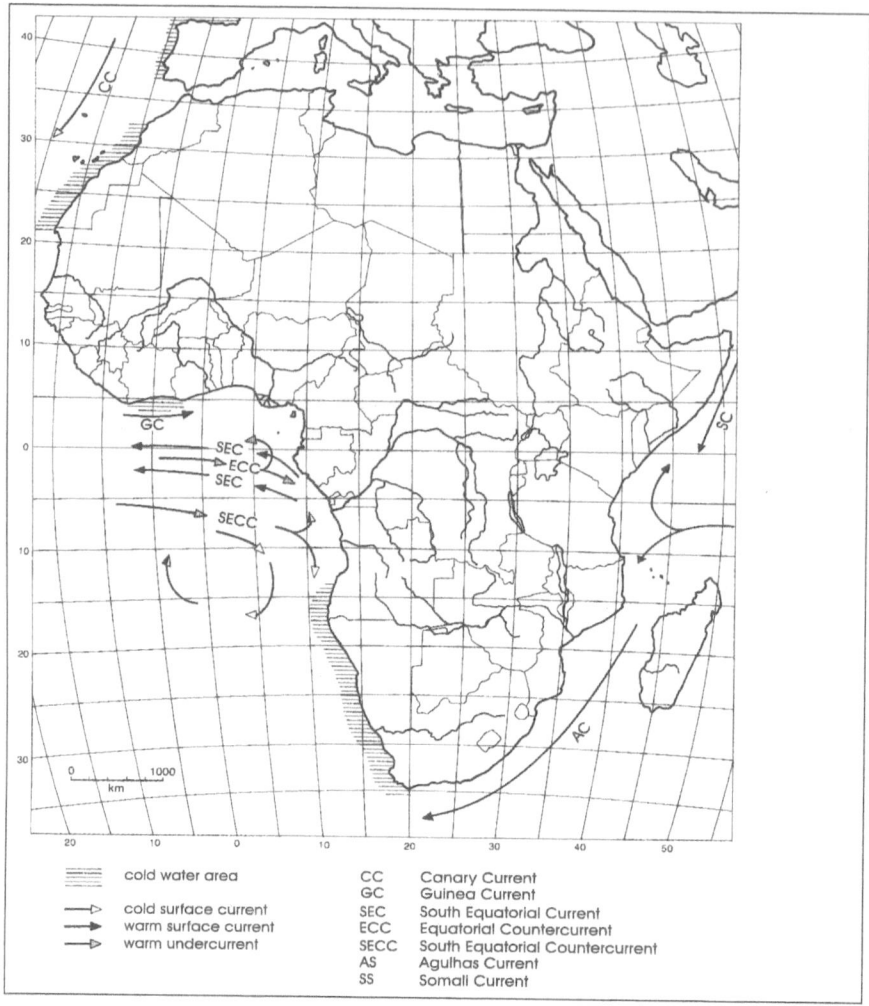

Fig. 14.4. Ocean currents west and east of Africa.

The climatic water-budget is based on the atmospheric circulation above Africa. The amount of precipitation as well as the duration and the annual distribution of rainfall depends on the very important water vapor sources for Africa, the SE Atlantic on the west coast of Africa, and the Indian Ocean on the continent's east side (Fig. 14.4). A major influence regarding the water vapor over the oceans comes from the different warm and cold ocean currents. On the surface of the South-Atlantic we have a southeaster current driven by the St. Helena anticyclone. The east side of this current in front of the southwest African coast is formed by the Benguela Current (BC = cold water area), which runs parallel to the coast. On its way to the north, and before reaching the equator, the current turns towards the west, hence becoming the Southern Equatorial Current (SEC) (Fig. 14.4).

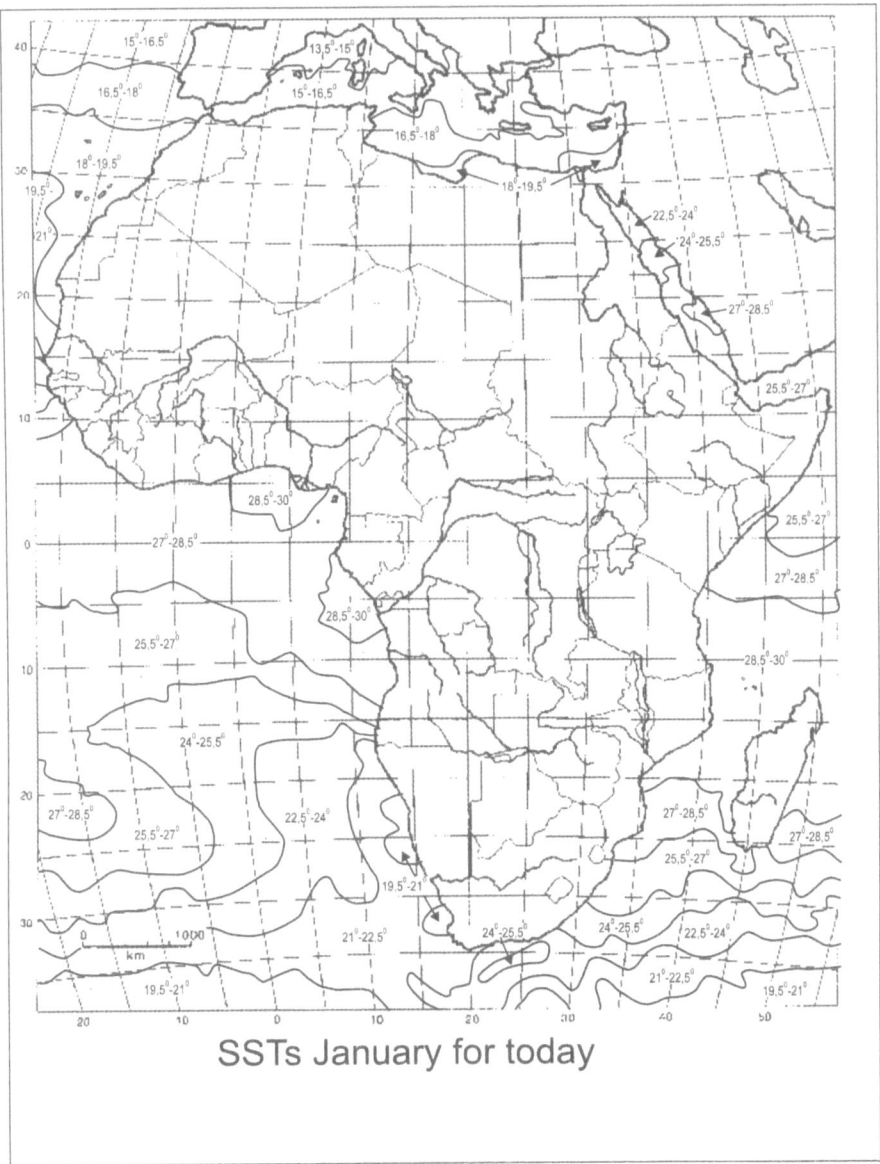

SSTs January for today

Fig. 14.5. SSTs of the Atlantic Ocean during winter (NH) for today.

The Equatorial Countercurrent (ECC), which has its origin approximately at 50° W, flows in an eastward direction. Out of the ECC arises the Guinea Current (GC), which reaches along the Guinea coast until the bay of Biafra (southeast Nigeria). The highest sea surface temperatures (SSTs) of the Atlantic near the equator are reached during winter in the Northern Hemisphere (NH) with 26–28° C. In contrast, the temperatures reach only 20° C during summer (NH).

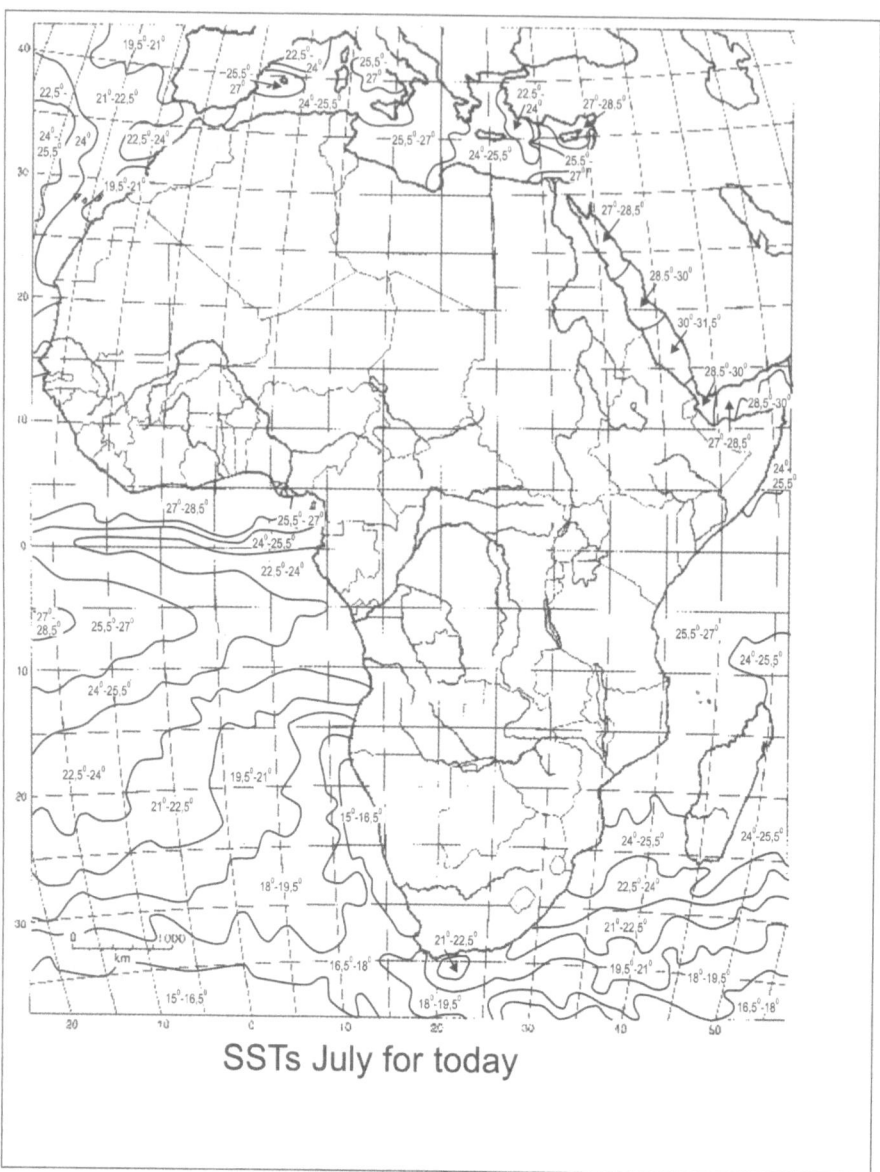

SSTs July for today

Fig. 14.6. SSTs of the Atlantic Ocean during summer (NH) for today.

East of the continent the conditions change. While the Benguela Current transports cool waters northwards to the equatorial area, the Agulhas Current (AC) flowing southwards off Africa's east coast leads warm waters to the sub-tropical regions. Although the Agulhas Current is a warm ocean current, eastern Africa is known as a region with considerably less rainfall compared to the western side of the continent, despite the fact that it lies in the midst of the inner tropics.

The reason for this is the divergent character of both the northeast and the southwest monsoons, the shallow depth of the southwest monsoon, and the strong meridional flow in all but the transition seasons (Griffiths 1972). Due to the shifting of the entire circulation system throughout the year, SSTs are also subject to variations. The highest temperatures of the western Indian Ocean, measuring 26–28° C, occur during winter (NH). Accordingly, only 22–24° C is reached during summer. The use of SSTs is well regarded because of well-resolved and dated profiles off the West African coast (Figs. 14.5 – 14.7.), permitting reconstruction of SSTs for the LGM.

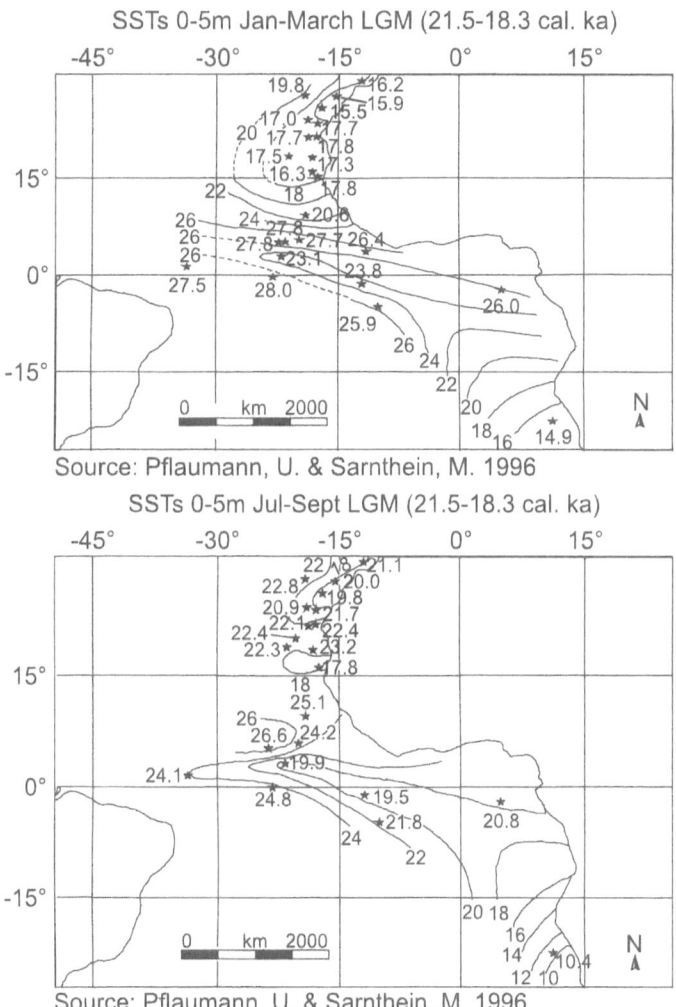

Source: Pflaumann, U. & Sarnthein, M. 1996

Source: Pflaumann, U. & Sarnthein, M. 1996

Fig. 14.7. SSTs of the Atlantic Ocean during winter and summer (NH) for the LGM (Pflaumann and Sarnthein 1996, see acknowledgements).

Since significant connections between ocean SSTs and the precipitation system exist today, such connections must have also existed during the LGM. Relations between the two allow the establishment of transfer-functions, permitting reconstruction of the continent precipitation system for the LGM on the basis of SSTs. The amount of precipitation with its annual distribution is reconstructed first. On the basis of reconstructed climatic water-budget, and in relation to present climate-vegetation conditions, a map showing the vegetation distribution for the LGM is derived (Fig. 14.9). We then critically examine whether the vegetation results for the LGM match reconstructed vegetation conditions.

14.5. Derivation of Precipitation Conditions for the LGM

Using Schneider (1991), Anhuf (1994) demonstrates that SST reconstructions following dated foraminifers and radiolarians after 18,000 yr BP show that the band of strongest buoyancy and that of the ECC must have been positioned similar to today, except that phenomena were more intense. In Figs. 14.5 – 14.7. we present data from Pflaumann and Sarnthein (unpublished, Dec. 1986) who developed a draft for an SST map for winter/summer in the Atlantic for the LGM.

Permission to use this data is greatly appreciated by the author. During the winter (NH) temperatures reached 24–26° C, about 2° C lower than today. In summer (NH) the temperatures were approximately 5–6° C lower than today, average 17–18° C. In particular, for stations near the coast between 5° S and 5° N, this means a clear reduction of the water vapor content during summer. Areas between 5–10° S are characterized by a smaller reduction of water vapor as the higher SSTs of the SH-summer coincide with the rainy season of the southern hemisphere, and therefore the precipitation losses must have been approximately 20–30% lower (Lautenschlager 1991).

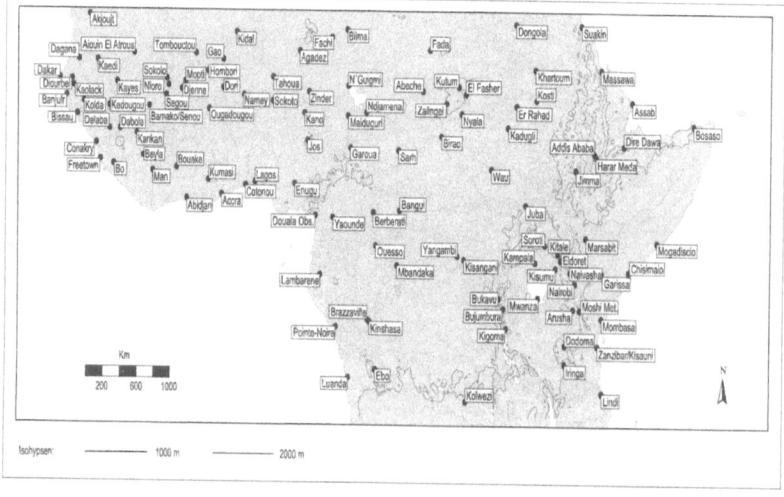

Fig. 14.8. Topography and meteorological stations in Africa north and south of the equator.

Table 14.1. Meteorological stations in central and equatorial Africa. Shown are the present rainfall distribution (first line) and the reconstructed rainfall amounts for the LGM (second line).

	Jan	Feb	Mar	Apr	May	Jun	Jul	Aug	Sept.	Oct	Nov	Dec	Σ	% of recent annual rainfall
Kinshasa	129	139	180	211	136	5	1	3	31	136	224	164	1374	
	97	104	144	169	95	3	1	1	9	57	157	123	960	70
Ouesso	55	100	171	112	159	135	64	137	203	239	166	81	1622	
	41	75	137	90	111	81	32	41	61	96	116	61	942	58
Libreville	204	290	282	386	229	60	3	11	109	389	422	263	2648	
	153	218	226	309	160	36	2	3	33	156	295	197	1788	68
Isiro	18	61	132	216	226	231	208	201	160	213	213	79	1958	
	14	46	106	173	158	139	104	60	48	85	149	59	1141	58
Bangassou	30	37	107	149	228	195	202	208	189	266	98	38	1747	
	23	28	86	119	160	117	101	62	57	106	69	29	957	55
Ilebo	152	163	211	180	74	13	20	48	170	178	249	218	1676	
	114	122	169	144	52	8	10	14	51	70	174	164	1092	65
Luanda	30	37	81	129	16	0	0	1	2	6	30	23	355	
	40	28	65	103	11	0	0	0		2	21	17	288	81
Reduction factor	0.75	0.75	0.80	0.80	0.70	0.60	0.50	0.30	0.30	0.40	0.70	0.75	--	--

The data determines the boundary conditions for derivation of precipitation conditions around 18,000 yr BP. At first, the stations of the Congo Basin area must be considered as they are characterized by a high amount of precipitation during summer (NH). The stations Kinshasa, Ouesso, Libreville, Isiro, Bangassou, Ilebo and Luanda in Angola were chosen and examined closer (Fig. 14.7). Table 14.1 shows the present distribution of precipitation, the entire yearly sum for each station and the reconstructed precipitation distribution for the LGM. The initial situation for reconstruction of precipitation was the SSTs of the summer and winter half-year as shown in Fig. 14.5. In order to calculate the distribution of precipitation for the whole year of the LGM, the present SSTs with their inter-annual variations were also considered (Figs. 14.5 and 14.6). In accordance with the guidelines given by Flohn (1985), evaporation rates over the Atlantic can be approximated based on the above mentioned glacial SSTs. Accordingly, the daily evaporation rates are about 4 mm over a ≥ 27° C warm water body and 1 mm over a water body with 17–18° C (Baumgartner and Reichel 1975, see Table 14.2). What is valid for the western side of the continent, namely the dependence of the amount of water vapor depending on different SSTs during the LGM, must consequently also apply to the eastern side of the continent for the section between the Horn of Africa and 10° S. Although, when comparing the values of the SSTs over the year between the Horn of Africa and the northern tip of Madagascar, the seasonal variations within the spectrum of surface temperatures become obvious. The idea of constant temperatures during the LGM (Coetzee 1987) must be dismissed since the system of the summer monsoon and phases of upwelling north of the Horn of Africa must have also existed at that time, a fact shown by Sirocko et al. (1991) and Sirocko (1996) using biogenic and lithogenic sediments.

Table 14.2. Evaporation values obtained from Baumgartner and Reichel (1975). Mean sea-surface temperature fields derived from Beckel (1996) and Nicholson and Nyenzi (1990).

SSTs (°C)	Evaporation Depth (E, mm/day)
> 27°	≈ 4.5 mm
25°	≈ 4.0 mm
23°	≈ 3.6 mm
21°	≈ 3.3 mm
19°	≈ 2.4 mm
17°	≈ 1.4 mm

Hutson (1980) and Winter and Martin (1990) come to similar results when examining the late Quaternary history of the Agulhas Current. They found that temperature decreases of SSTs in front of Africa's east coast amounted to only 2° C (winter) and 5° C (summer) (NH). Although maps of the Indian Ocean for SSTs during the LGM are not available, SSTs and the appropriate evaporation for the time of 18,000 yr BP can nevertheless be derived using the method discussed above for West and Central Africa.

Table 14.3. Meteorological stations in East Africa. Shown are the present rainfall distribution (first line) and the reconstructed rainfall amounts for the LGM (second line).

	Jan	Feb	Mar	Apr	May	Jun	Jul	Aug	Sept	Oct	Nov	Dec	Σ	% of recent annual rainfall
Lodwar	6	6	21	41	24	8	13	8	3	6	8	8	152	
	5	5	17	33	17	5	7	4	2	4	6	6	111	73
Marsabit	25	13	57	192	121	11	13	26	13	83	116	72	742	
	19	10	46	91	85	64	93	95	8	58	87	54	547	74
Eldoret	18	48	58	114	122	107	185	190	79	30	48	28	1027	
	14	36	46	91	85	64	93	95	47	21	36	21	649	63
Nanyuki	19	24	49	99	107	67	78	72	50	68	89	41	763	
	14	18	39	79	75	40	39	36	30	48	67·	31	516	68
Kisumu	38	79	142	182	146	81	64	81	62	53	85	98	1111	
	29	59	114	146	102	49	32	41	37	37	64	74	784	70
Nairobi	41	57	117	207	144	60	17	25	29	54	119	74	944	
	31	43	94	166	101	36	9	13	17	38	89	56	693	73
Makindu	43	30	86	122	28	3	2	2	2	28	173	119	638	
	32	23	69	98	20	2	1	1	1	20	130	89	486	76
Voi	33	30	76	97	33	8	3	8	13	23	97	130	551	
	25	23	61	78	20	5	2	4	8	16	73	98	413	75
Malindi	10	8	30	150	340	163	95	51	43	64	48	23	1025	
	8	6	24	120	259	98	48	26	26	45	36	17	713	70
Reduction factor	0.75	0.75	0.80	0.80	0.70	0.60	0.50	0.50	0.60	0.70	0.75	0.75	--	--

When applying reduction factors from this method to the monthly precipitation for the East Africa region, compared to today for the East African Highlands, a reduction of annual precipitation by about 25–27% can be deduced. The only exception is the area north of Lake Victoria, near the stations of Eldoret and Nanyuki, because the precipitation there is mainly controlled by water vapor that is advected from the Congo Basin. At these two stations, the precipitation reduction factor amounted to 50% during July and August (see Table 14.3). The statements made here are valid for areas reaching as far as the northern border of Kenya. For the highlands of Ethiopia a reduction factor of approximately 33% was estimated using the works of Wickens (1982), particularly in the upper Nile-river area (today Sudan).

14.6. Vegetation Map of Africa for the LGM

On the basis of the precipitation calculations, the following aspects of the water-budget and therefore the ecological conditions of the vegetation can be deduced (Fig. 14.9). The Niger outflow was blocked by the dune range near Mopti after 18,000 yr BP (Alayne–Street 1981). At 18,000 yr BP, the transition from the open dry forests, today found in the central and southern Sahelian areas, to the open tree savannas, was shifted to 13° N at the western coast, almost to 10° N in the northeastern edge of Ivory Coast, and to 12° N in the region of Kaduna (Nigeria). This means that it lies further south than today (3–4°). The transition from the grass savannas to the diffuse vegetation of the southern Saharan border lies at 15° N in Senegal and nearly 14° N in the central part of today's Sudanian Zone (north of Kano, Nigeria). This northern frontier of the grass savanna in West Africa is comparable to the southern border of moving sand dunes during the Pleistocene (Talbot 1984). The northern coast area reaching from 5° N until just south of the equator has quite a lot of rainfall during the winter months (NH), which, with reference to the situation at 18,000 yr BP, can be interpreted such that the present 2000 mm isohyet could describe the 1400 mm precipitation limit during the LGM. The 1800 mm precipitation limit of the LGM most probably coincides with the present locations of the 2600 mm isohyetal line. Thus, there was only a very narrow strip of land east of Douala (Cameroon) reaching southward until Libreville, in which ombrophile and hyper-ombrophile evergreen rainforests could survive. The borderline of the semi-deciduous rainforests may have extended to a line approximately halfway between Douala and Yaounde at about 4° N. Southwards, these forests were able to expand until Port Gentil (Gabon) along the coast, whereas towards the interior of the country they reached as far as Brazzaville (Congo). Parts of these rain forests extended over the western edge of the Western Rift of the Congo southwards until the geographical latitude of Point Noire. In the section south of Libreville the easterly borderline of the semi-deciduous rainforests lay along the border between Gabon and Congo and expanded towards the south into the Republic of Congo northwest of Brazzaville. Also in the eastern central area of the Congo Basin, the conditions must have been similar with enough humidity for the existence of semi-deciduous rainforests. The same can be said for the southeastern edge of the Congo Basin on the west side of Lake Kivu as well as for the area of Bukavu (Zaire).

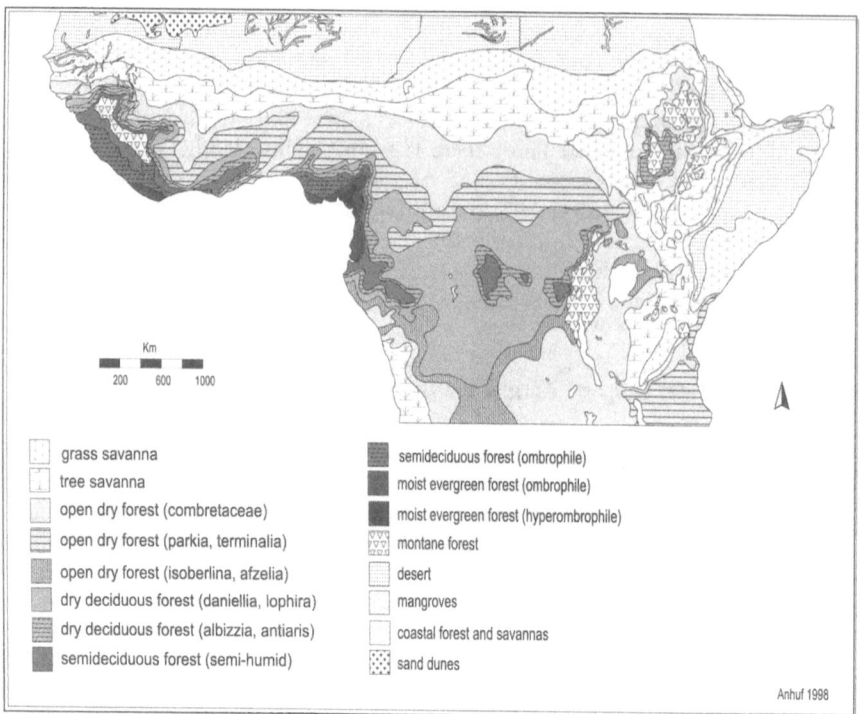

Fig. 14.9. Vegetation Map of Africa for the LGM.

The major part of the Congo Basin, in accordance with the reconstruction of the precipitation, was dominated by open and partly by more dense dry forests. Open grass expands into the area of river systems that might be interpreted as flooded tributaries of the Congo and Ubangi. Open dry forests, in their physiognomy comparable with the Miombo forests of today, dominated the area of Bangui in an approximately 400–600 km wide strip along the coast of Angola as well as in the province of Katanga in the southeast of Zaire.

It should be noted that, due to the reduced SSTs during summer (NH) which is a general characteristic for the LGM, the amount of precipitation changed compared to today. In the area of the escarpment reaching from the eastern of Gabon to Congo (Brazzaville), which used to be dominated by afro-montane *Podocarpus*-forests during the LGM (Dupont 1996), the precipitation reduction slowly increased from west to east to 40% in comparison to the annual sum of today. In the northeastern area of the Congo Basin, in the triangle formed by Bangassou, Kisangani and Isiro, the precipitation reduction even reached 45%. The increasing continentality of the coast towards the East is also recorded in today's distribution of precipitation. Particularly, the continental stations of the Congo Basin have contemporaneously an extreme high amount of precipitation during the summer months of the northern hemisphere. During the LGM, temperatures have been extremely low over the Atlantic. Therefore, only a fraction of today's water vapor originating from the Atlantic could reach the Congo Basin.

Open dry forests dominated the entire East African Highlands. Along the edge of the Western Rift and northeast of Lake Victoria, remains of semi-deciduous rainforests could be found during the LGM. The generally dry highland with tree and grass savannas is interrupted only by islands of dry montane forests. The aridity increases towards the east. The Danakil-Plain, the Ogaden, large parts of Somalia, as well as the northeastern part of Kenya, were dominated by a desert-like climate. The sub-alpine and alpine zone of the mountain region was clearly shifted down to 1000 m asl, covering considerably larger areas than today (Fig. 14.2). According to Sirocko (1996), the coastal strip north of the Horn of Africa may even have been more humid than today, although there is no direct proof for the LGM supporting this interpretation.

14.7. Comparison of the Reconstructed Vegetation with Palaeobotanic Samples

We now need to test the presented climatic reconstructions using macro-remains and pollen-spectra (see Fig. 14.3. and Tab. 14.4). In the eastern part of Ivory Coast, near the border with Ghana, only small relics of the semi-deciduous rainforest have survived the last glacial maximum. In contrast, this forest type remained in the west in the highlands of Guinea, as well as in the area of Cape Palmas at the south peak of the West African continent. Even reduced precipitation between Greenville (5°01' N, 9°03' W) and Cape Palmas permitted ecological climatic conditions that allowed the survival of a band of evergreen rainforests along the coast. These conditions can also be found along the Niger Delta and even further to the east, around Douala, and its hinterland that again allowed the survival of evergreen rainforests. In the central part of Ivory Coast, as well as in the area of the "Dahomey Gap", dry forests with a high percentage of graminees almost reached the Guinean coast. The area around Accra showed only open tree savannas comparable to the Sahelian type of today around 18,000 yr BP.

The orographic situation of the highlands of Guinea supported the survival of rather humid forest formations even in its northern parts (Fouta Djalon). Because of the general reduction of temperature by four to six degrees, the low mountain range of the Guinean Highlands was affected by maximum rainfall (Lauer 1989). Therefore, the evergreen and the dense dry forest types in the presented map should be described as cloud forests and mountain forests (Fig. 14.2).

The Palynology of Lake Bosumtwi in Ghana, now west of the "Dahomey Gap", showed the existence of a grass-rich savanna with mountainous elements (*Olea hochstetteri*) at low altitudes (Maley and Livingstone 1983; Maley 1987; Talbot et al. 1984). Tree pollen was limited to 4–5%. These tree pollen however, do not descend from the Sudanian savannas, they mainly came from the area of semi-deciduous rainforests (Dupont and Agwu 1992; Fredoux and Tastet 1988). Further west, sediments of a marine core situated off the Ghanaian margin recorded the permanence of rain forest on the adjacent continent during the LGM (Lezine and Vergnaud–Grazzini 1993; Lezine and Le Thomas 1995). For the West African region, the vegetation reconstructions are in accordance with the paleobotanical samples.

The pollen diagram from Lake Barombi-Mbo (NE of Mount Cameroon at 300 m altitude) also shows continual presence of rain forests in this area during the last 25 ky (thousand years). Between 20 and 14 ky BP a drier climate was interrupted from high percentages of grass pollen which are evidence of a more open forest. Furthermore, four other pollen profiles show that the lowland rain forest was still present at the LGM but with a different species composition - namely with more dry elements than today (Anhuf 1994). These pollen profiles are M 16856 south of Lagos; M 16867 southeast of São Tomé (Dupont 1995); KW 23 northwest of Pointe Noire (Bengo and Maley 1991); and GeoB 1008 in front of the estuary of the Zaire (Jahns 1996).

Rain forests also persisted during the last 24 ky around the "Mare de Ngamakala" and the southern "Batéké Plateau" NE of Brazzaville (Elenga 1992; Elenga et al. 1994; Schwartz et al. 1995). The profile of the Mare de Ngamakala (Gama 2) shows that from 22,170 until 6500 yr BP there was an undiminished high amount of *Sapotaceae* and *Syzygium*. The amount of Gramineae and Cyperaceae clearly starts to increase not before 1,610 ± 200 yr BP, which is a definite indication for anthropogenic-zoogenous influences (Fig. 14.9). These results clearly indicate that dense dry forests existed in the Zaire savannas during the LGM. These conditions are consistent with the presented map (Fig. 14.9). In addition to that, the $\delta^{13}C$ analyses of further Congo savannas, today covering more than 40% of the Congo area, prove, that these savannas developed themselves over the last 3000 years from former forests. In the "Bois de Bilanko" (Batéké Plateau) at about 700 m altitude, mountainous elements, such as *Podocarpus, Olea hochstetteri* and *Ilex mitis*, record the presence of afro-montane forests during the LGM, indicating that temperatures were about 5° C lower than today (Elenga et al. 1991).

Preuss (1990) examined the inner basin of Zaire, east of Mbandaka, more closely. The pollen analyses of the profiles Hv 11776 and Hv 13724 for the time sections of 17,735 ± 135 yr BP and 16,675 ± 140 yr BP indicate dry forests. But the great amount of *Syzygium* also shows that dense gallery-forests existed along the rivers in the Congo Basin, where also evergreen and semi-deciduous species survived. These conditions are consistent with the present map as well. Runge and Runge (1995) worked in the east of Zaire in Walikale (1°25' S, 28°05' E), in the province of Kivu. Today this region is dominated by evergreen rainforests with an annual precipitation of 2400 mm. New pollen analyses suggest a relatively open vegetation type like an "open woodland (or forest) with gallery forests" during the LGM (Runge and Runge 1995:118). According to our own reconstructions, the precipitation may have been just above 1000 mm. The dry season lasted 5 months, 3 months during winter (NH) and 2 during summer. Given that the rainy seasons were interrupted, only open dry forests were able to thrive at that time. Our own precipitation and vegetation reconstructions indicate that only a few kilometers eastward in the direction of Goma at the edge of the Western Rift, semi-deciduous mountain forests are likely to have survived, a fact, which Runge and Runge (1995:120) indirectly confirm: "To explain the obviously very fast re-colonization of the Walikale area with rain forest vegetation it can be stated that the refuge areas of the rain forest must have been local to Walikale."

For the reconstruction of the palaeobotanic conditions of East Africa, the presented results must first be seperated into two groups, pollen from the East

African high mountains and pollen from the highland lakes (e.g. Lake Victoria, Lake Mobutu, Lake Tanganyika).

The pollen profiles of the East African high mountains (Simen Mountains, Bale Mountains, Mt. Kenya, Mt. Elgon, Ruwenzori, Kilimanjaro) document a clear lowering of the snow line of 700–1000 m. Therefore, the vegetation limit is lowered also about 700–1000 m. The result of this lowering is the obvious increase of areas with afro-alpine vegetation in comparison to today. The Sacred Lake at Mt. Kenya for instance, lies today within a semi-deciduous mountain forest at an altitude of 2400 m. The upper limit of forest growth at the transition Ericaceous-belt today lies at 3100 m altitude. The pollen profiles of the LGM mainly reflect species of the Ericaceous belt. In southwest Rwanda there is the pollen profile of the Kamiranzovu Swamp (1950 m altitude) belonging to the Rugege Forest (Hamilton 1982). Its surrounding area is today covered by evergreen mountain forest. At the time of the LGM there were increased values of Gramineae, Ericaceae, *Podocarpus* and *Sphagnum*. This corresponds to the lowering of the vegetation belts by 600–800 m. In the Burundi highlands, the Kashiru pollen sequence (3°28' S, 29°34' E) provides paleoclimatic information for the LGM. Today, the site lies in the humid montane forest vegetation belt. This is consistent with the map of quasi-natural vegetation of Africa (Fig. 14.2). The most characteristic trees of these forests are *Entandophragma, Prunus, Chrysophyllum, Neoboutonia*. During the LGM, grassland with afro-alpine indicators expanded down to 2500–2000 m altitude (now above 2600 m), "the montane forest persisted in refuges, although much reduced" (Bonnefille and Riollet 1988:S.19).

A selection of further pollen profiles regarding the present-day vegetation as well as the vegetation for the LGM is illustrated in Table 14.4. Overall, the lowering of vegetation-belts amounted to 700 m in the dry high mountains (for instance the Simen Mountains, the Bale Mountains, Mt. Elgon and Kilimanjaro), whereas in the humid high mountains it amounted to 1000 m. With an average vertical temperature gradient of 0.85° C per 100 m in the first case and one of 0.6° C per 100 m in the second case, the lowering of temperatures during the LGM amounts to 5–6° on average, which again matches the reduction of the SSTs above the western Indian Ocean. Next to the high mountain areas of East Africa, the lakes also provide different pollen sequences, allowing derivation of the vegetation cover during the LGM. Kendall (1969) assumes open savanna-like vegetation at the outlet of the Nile near Jinja from pollen of Lake Victoria. *Artemisia afra* is found in the dry forest savanna mosaics.

Vincens (1991) examined a pollen profile from the South-Tanganyika-Basin. The main vegetation communities were open dry Zambezian woodlands prior to 15,000 yr BP with locally included patches of afro-montane elements such as *Podocarpus* and Ericaceae. This proves that the Miombo forest basically has not changed much, it was merely more open and dominated by Gramineae. For Lake Mobutu (Lake Edward), no usable pollen data are available for the LGM. From the pollen data prior to 25,350 yr BP. and after 14,700 yr BP, Sowunmi (1991) concludes that climatic conditions during the LGM were very arid (see for Amazonia also Latrubesse, this volume).

Table 14.4. Quaternary pollen sites from East African mountains.

location	highland sites	Altitude (m)	altitudinal belt recent	altitudinal belt 18 000 yr BP	Reference
Mt. Kenia	Sacred Lake	2400	Humid Montane Forest	Ericaceous-Belt	Coetzee (1967)
Mt. Kenia	Lake Rutundu	3140	Ericaceous-Belt	Afroalpine-Grassland	Coetzee (1967)
Ruwenzori	Mahoma Lake	2960	Forest Boundary *Rapanea-Hagenia-* Zone	Afroalpine-Grassland	Livingstone (1967)
Ruwenzori	Kitandara Lake	3990	Afroalpine-Grassland	Area of the Omurubaho Glaciation	Livingstone, (1967)
Rukiga Highlands	Muchoya Swamp	2256	*Arundinaria alpina-* Forest	Forest Boundary *Hagenia abyssinica* -Ericaceous-Belt-.	Morrison (1968)
Cherangani Hills	Kaisungor Swamp	2900	Dry Montane Forest	Afroalpine-Grassland *Artemisia, Chenopodiaceae*	Coetzee (1967)
Mt. Elgon	Laboot Swamp	2880	Dry Montane Forest	Ericaceous-Belt	Hamilton (1987)

Other results from Lake Mobutu (1°31' N, 30°34' E) show a very open vegetation of the type of a savanna or woodland for 15,000 yr BP (Ssemmanda and Vincens 1991). *Olea capensis* is mainly presented in the arboral pollen. Pollen analysis of a sediment core from Lake Naivasha (0°45' S, 36°20' E), central Rift Valley of Kenya, reveals that the vegetation during the LGM was dominated by tree savannas or woodlands (open dry forests). Maitima (1991) however emphasizes that already shortly after 17,000 yr BP, the pollen profile shows a relatively slight increase of lowland forest vegetation (species of the dry forests like *Celtis*).

14.8. Discussion

Overall the available pollen sequences are consistent with the presented vegetation map for the LGM. It is obvious that there are significant connections between the SSTs of the Atlantic, the Indian Ocean and the water budget of present-day Africa. The relations between the three allow a derivation of transfer-functions, from which the precipitation input into the continent for the LGM can be reconstructed. On this basis and in relation to present-day ecological conditions a map showing the distribution of vegetation for the LGM is presented. The validation of the map using selected pollen spectra showed consistent results between the map and the data. Therefore, good reasons exist to regard the methodology, at least in general, as a promising and overall reliable tool.

Acknowledgements: The author is indebted to Michael Sarnthein und Uwe Pflaumann of the Geological-Palaeontological Institute of the University of Kiel for their kind cooperation to publish their maps of SSTs for the LGM. I thank Rainer Malmberg, Barbara Müller, and Birgit Schröder for their collaboration both in the institute and in the field. I am indebted to Anja Kamolz for translating the manuscript. This work was supported by the German Federal Ministry for Education and Research (BMBF); Project Nr.: 07 KFT 57/9.

References

Adams LJ, Tetzlaff G (1985) The extension of Lake Chad at about 18000 yr BP. Zeitschrift für Gletscherkunde und Glazialgeologie 21:115–123

Agwu CO, Beug, H-J (1982) Palynological studies of marine sediments of the West African coast. Meteor-Forschungsergebnisse C 36:1–30

Alayne–Street F (1981) Tropical palaeoenvironments Progress. Geography 5/23:157–185

Anhuf D (1994) Zeitlicher Vegetations- und Klimawandel in der Côte d'Ivoire. Erdwissenschaftliche Forschung 30:7–299

Anhuf D (1997) Palaeovegetation in West Africa for 18000 BP. and 8500 BP. Eiszeitalter und Gegenwart 47:112–119

Anhuf D, Frankenberg P (1991) Die naturnahen Vegetationszonen Westafrikas. Die Erde 122:243–265

Anhuf D, Frankenberg P (1993) Etude du changement végétal saisonnier au Sénégal. Cahiers d'Outre-Mer 46/183:297–324

Baumgartner A, Reichel E (1975) Die Weltwasserbilanz. München, Wien

Beckel L (ed) (1996) Global Change. München, Stuttgart

Bengo MD, Maley J (1991) Analyses des flux polliniques sur la marge sud du Golfe de Guinée depuis 135000 ans. C.R. Acad. Sci. Paris 313, 2:843–849

Bonnefille R, Riollet G (1988) The Kashiru pollen sequence (Burundi) palaeoclimatic implications for the last 40000 yr B.P. in Tropical Africa. Quaternary Research 30:19–35

Brenac P (1988) Evolution de la végétation et du climat dans l'Ouest-Cameroun entre 25000 et 11000 ans BP. Inst. fr. Pondichéry, trav. sci., techn. 25:91–103

Butzer KW (1976) The Mursi, Nkalabong, and Kibish Formations, lower Omo basin, Ethiopia. In: CoppensY, Howell FC, Isaac GL, Leakey REF (eds) Earliest Man and Environments in the Lake Rudolf Basin. Chicago, London, pp 12–23

Coetzee JA (1967) Pollen analytical studies in East and Southern Africa. Palaeoecology of Africa 3:1–146

Coetzee JA (1987) Palynological intimations on the East African mountains. Palaeoecology of Africa 18:231–244

Dupont LM (1995) Lowland rain forest and afromontane forest in West Equatorial Africa during the Middle and Late Pleistocene. In: 2nd Symposium on African Palynology, Publ.Occas. CIFEG 1995/31: 87–98 Tervuren (Belgium)

Dupont LM (1996) Vegetation and climate in West-Equatorial Africa (0–700000 yr BP). Unpublished Habilitation, University of Göttingen, Germany

Dupont LM, Agwu COC (1992) Latitudinal shifts of forest and savanna in NW Africa during the Brunches chron: further marine palynological results from site M 16415 (9° N 19°W). Vegetation History and Archaeobotany 1:163–175

Elenga H (1992) Végétation et climat du Zaire depuis 24000 B.P. Analyse palynologique de séquences sédimentaires du Pays Bateke et du littoral. Th. Univ. Aix-Marsaille

Elenga H, Vincens A, Schwartz D (1991) Presence d'elements forestiers montagnards sur le Plateau Bateke (Zaire) au Pleistocène supérieur: Nouvelles donées palynologiques. Palaeoecology of Africa 22:239–252

Elenga H, Schwartz D, Vincens A (1994) Pollen evidence of late Quaternary vegetation and inferred climate changes in Zaire. Palaeogeogr., Palaeoclim., Palaeoecol. 109:345–356

Fischer E, Hinkel H (1992) La Nature du Rwanda. Materialien zur Partnerschaft Rheinland–Pfalz/Ruanda. Mainz

Flohn H (1985) Das Problem der Klimaänderungen in Vergangenheit und Zukunft. Darmstadt

Fredoux A (1994) Pollen analysis of a deep-sea core in the Gulf of Guinea: vegetation and climatic changes during the last 225000 years BP. Palaeogeogr., Palaeoclim., Palaeoecol 109:317–330

Fredoux A, Tastet J-P (1988) Stratigraphie pollinique et paléoclimatologie de la marge septentrionale du Golfe de Guinée depuis 200000 ans. Inst. fr. Pondichéry, trav. sec., sci., techn. 25:175–183

Gasse F, Street FA (1978) Late Quaternary lake-level fluctuations and environments of the Northern Rift Valley and Afar Region (Ethiopia and Djibouti). Palaeogeogr., Palaeoclim., Palaeoecol 24:279–325

Gasse F, Rognon P, Street FA (1980) Quaternary history of the Afar and Ethiopian Rift Lakes. In: Williams MAJ, Faure H (eds) The Sahara and the Nile: Balkema, Rotterdam, pp 361–400

Griffiths JF (1972) Climates of Africa. World Survey of Climatology 10:133–165. Amsterdam

Hamilton AC (1982) Environmental history of East Africa - A study of the Quaternary. Academic Press, London

Hamilton AC (1987) Vegetation and climate of Mt. Elgon during the late Pleistocene and Holocene. Palaeoecology of Africa 18:283–304

Hooghiemstra H (1988) Changes in major wind belts and vegetation zones in NW Africa 20000 yr BP, as deduced from a marine pollen record near Cap Blanc. Review of Palaeobotany and Palynology 55:101–140

Hooghiemstra H, Agwu COC, Beug H-J (1986) Pollen and spore distribution in recent marine sediments: A record of NW-African seasonal wind patterns and vegetation belts. Meteor, Forsch.-Ergebnisse C 40:87–135

Hooghiemstra H, Bechler A, Beug H-J (1987) Isopollen maps for 18000 yr. BP of the Atlantic offshore of northwest Africa: evidence for paleo-wind circulation. Paleoceanography 2:561–582

Hurni H (1982) Klima und Dynamik der Höhenstufung von der letzten Kaltzeit bis zur Gegenwart. Geographica Bernensia G13. Bern

Hutson WH (1980) The Agulhas Current during the Late Pleistocene: analysis of modern faunal analogs. Science 207 4 January 1980:64–66

Jahns S (1996) Vegetation history and climate changes in West Equatorial Africa during the Late Pleistocene and Holocene, based on a marine pollen diagram from the Zaire fan. Vegetation History and Archaeobotany 5:207–213

Jahns S, Hüls M, Sarnthein M (1998) Vegetation and climate history of west equatorial Africa based on a marine pollen record off Liberia (site GIK 16776) covering the last 400000 years. Review of palaeobotany and palynology 102:277–288

Kadomura H (1989) Savannization in tropical Africa. In: Kadomura H (ed) Savannization processes in tropical Africa.. Occasional Papers 17:3–15. Tokyo Metropolitan University

Keay RWJ 1959. Vegetation map of Africa. AETFAT/UNESCO, Oxford

Kendall RL (1969) An ecological history of the Lake Victoria Basin. Ecological Monographs 39/2:121–175

Knapp R (1973) The vegetation of Africa, Gustav Fischer Verlag, Stuttgart

Lauer W (1989) Climate and weather. In: Lieth NR, Werger MJA (eds) Tropical Rain Forests Ecosystems. Amsterdam, Oxford, New York, Tokyo:7–53

Lautenschlager M (1991) Simulation of the ice age atmosphere - January and July means. Geologische Rundschau 80/3:513–534

Lezine A-M (1991) West African paleoclimates during the last climatic cycle inferred from an Atlantic deep-sea pollen record. Quaternary Research 35:456–463

Lezine A-M, Vergnaud–Grazzini C (1993) Evidence of forest extension in West-Africa since 22000 BP. A pollen record from eastern tropical Atlantic. Quaternary Science Reviews 12(3):203–210

Lezine A-M, Le Thomas A (1995) Histoire du massif forestier Ivorien au cours de la dernière déglaciation. 2nd Symposium on African Palynology, 1995, Publ.Occas. CIFEG 1995/31:73–85 Tervuren (Belgium)

Lioubimtseva, E, Faure H, Faure–Denard L, Page N, Wickens GE (1996) Sudan biomass changes since 18000 BP: A test-area for tropical Africa. Palaeoecology of Africa 24:71–85

Livingstone DA (1967) Postglacial vegetation of the Ruwenzori Mountains in Equatorial Afrika. Ecol. Monogr. 37:25–52

Livingstone DA (1971) A 22000-year pollen record from the plateau of Zambia. Limnol. Oceanogr. 16:349–356

Lutze GF et al. (1988) Bericht über die METEOR-Fahrt 6-5 Dakar-Libreville 15.1-16.2.1988. Berichte Geol. Paläont. Inst. 22. Univ. Kiel

Maitima JM (1991) Vegetation response to climatic change in Central Rift Valley, Kenya. Quaternary Research 35:234–245

Maley J (1987) Fragmentation de la forêt dense humide africaine et extension des biotopes montagnards au Quaternaire recent: nouvelles données polliniques et chronologiques - Implications paléoclimatiques et biogéographiques. Paleoecology of Africa 18:307–336

Maley J, Livingstone DA (1983) Extension d'un élément montagnard dans le sud du Ghana au Pleistocène supérieur et à l'Holocène inférieur. Premières données polliniques. C. R. Acad. Sci. 296 II:1287–1292. Paris

Menaut J-C (1983) The vegetation of African savannas. In: Bourlière F (ed) Tropical Savannas. Ecosystems of the World 13. Elsevier, Amsterdam, pp 109–149

Messerli B, Winiger M, Rognon P (1980) The Saharan and East African uplands during the Quaternary. In: Williams MAJ, Faure H (eds): The Sahara and the Nile: 87–132. Balkema, Rotterdam

Morrison MES (1968) Vegetation and climate in the uplands of south-western Uganda during the Pleistocene Period, 1. Muchoya Swamp, Kigezi District. J. Ecol. 56:363–384

Neumann K (1988) Die Bedeutung von Holzkohleuntersuchungen für die Vegetationsgeschichte der Sahara - das Beispiel Fachi/Niger. Würzburger Geographische Arbeiten 69:71–85

Nicholson SE, Nyenzi BS (1990) Temporal and spatial variability of SSTs in the tropical Atlantic and Indic Oceans. Meteroly and Atmospheric Physics 42:1–17

Petit–Maire N et al. (1987) La dépression de Taoudenni (Sahara malien) à l'Holocène. Géodynamique 2/2:154–159

Preuss J (1990) Premières séries d'artefacts lithiques originaires du bassin intérieur du Zaire. In: Lanfranchi R, Schwartz D (eds) Paysages Quaternaires de l'Afrique Centrale Atlantique, Edition ORSTOM: Paris, pp 431–438

Richardson JL, Richardson AE (1972) History of an African Rift lake and its climatic implications. Ecol. Monogr. 42:499–534

Roche E, Bikwemu G (1989) Paleoenvironmental change on the Zaire–Nile ridge in Burundi; the last 20000 years: An interpretation of palynological data from the Kashiru Core, Ijenda, Burundi. In: Mahaney WC (ed) Quaternary and environmental research on East African Mountains, Balkema, Rotterdam, pp 231–242

Rossignol–Strick M, Duzer D (1979) West African vegetation and climate since 22.500 BP from deep sea cores palynology. Pollen Spores 21/1–2:105–134

Runge J (1996) Palaeoenvironmental interpretations of geomorphological and pedological studies in the rain forest "core areas" of eastern Congo (Central Africa). South African Journal 78 2:91–97

Runge J (1998) Rezente und holozäne Vegetations- und Klimadynamik an der Regenwald/Savannengrenze in Nord-Kongo (Zaire) und der Zentralafrikanischen Republik (4°–5°20' N, 23°–25°E). Zbl. Geol. Paläont. Teil I 1/2:91–113

Runge J, Runge F (1995) Late Quaternary paleoenvironmental conditions in eastern Zaire (Kivu) deduced from remote sensing, morpho-pedological and sedimentological studies (phytoliths, pollen, C-14 data). 2nd Symposium on African Palynology, Publ.Occas. CIFEG 1995/31: 109–122. Tervuren (Belgium)

Schneider R (1991) Spätquartäre Produktivitätsänderungen im östlichen Angola–Becken: Reaktion auf Variationen im Passat–Windsystem und in der Advektion des Benguela–Küstenstroms. Berichte, Fachbereich Geowissenschaften; Universität Bremen, 21

Schwartz D, Dechamps R, Elenga H, Lanfranchi R, Mariotti A, Vincens A (1995) Les savanes du Congo: une végétation spécifique de l'Holocene Supérieur. 2nd Symposium on African Palynology, Publ.Occas. CIFEG 1995/31:99–108 Tervuren (Belgium)

Schulz E (1987) Die holozäne Vegetation der zentralen Sahara (N-Mali, N-Niger, SW-Libyen). Palaeoecology of Africa 18:143–161

Schulz E, Joseph A, Baumhauer R, Schultze E, Sponholz B (1990) Upper pleistocene and holocene history of the Bilma region (Kawar, NE-Niger) - Recent data in african earth sciences. Occ. Publ. CIFEG 1990/22:281–284

Schulz E, Pomel S (1992) Die anthropogene Entstehung des Sahel. Würzburger Geographische Arbeiten 84:263–288

Servant M, Servant–Vildary S (1980) L'environment quaternaire du bassin du Tchad. In: Williams MAJ, Faure H (eds) The Sahara and the Nile, Balkema, Rotterdam, pp 133–162

Sirocko F (1996) The evolution of the monsoon climate over the Arabian Sea during the last 24000 years. Palaeoecology of Africa 24:53–69

Sirocko F, Sarnthein M, Lange H, Erlenkeuser H (1991) Atmospheric Summer Circulation and Coastal Upwelling in the Arabian Sea during the Holocene and the last glaciation. Quaternary Research 36:72–93

Sowunmi AM (1981) Nigerian vegetational history from the late quaternary to the present day. Palaeoecology of Africa 13:217–234

Sowunmi A (1991) Late Quaternary environments in equatorial Africa: Palynological evidence. Palaeoecology of Africa 22:213–238

Ssemmanda I, Vincens A (1991) Histoire forestière des plateaux ouest Ougandais depuis 14000 ans. XIIe Symposium de l'Association des Palynologues de Langue Franç aise, 23–27 septembre, Caen

Stager CJ (1988) Environmental changes at Lake Cheshi, Zambia since 40000 years B.P. Quaternary Research 29 :54–65

Talbot MR (1984) Late Pleistocene rainfall and dune building in the Sahel. Palaeoecology of Africa 16:203–214

Talbot MR, Livingstone DA, Palmer PG, Maley J, Melade JM, Delibrias G, Gulliksen S (1984) Preliminary results from sediment cores from Lake Bosumtwi, Ghana. Palaeoecology of Africa 16:173–192

Tiercelin J, Renaut RW, Delibrias G, le Fournier J, Bieda S (1981) Late Pleistocene and Holocene lake level fluctuations in the Lake Bogoria basin, northern Kenya Rift Valley. Palaeoecology of Africa 13:105–120

Van Campo E, Duplessy JC, Rossignol–Strick M (1982) Climatic conditions deduced from 150-Kyr oxygen isotope-pollen record from the Arabian Sea. Nature 296:56–59

Verschuren D (1996) Utilisation de cladocères et chironomides fossiles pour réconstruire l'évelution hydrologique de leur habitat marécageux dans la tourbière de Kashiru (Burundi) depuis 40000 ans BP. Palaeoecology of Africa 24:133–145

Vincens A (1991) Végétation et climat dans le bassin sud-Tanganyika entre 25000 et 9000 BP: Nouvelles données palynologiques. Palaeoecology of Africa 22:253–263

Vincent E (1972) Climatic change at the Pleistocene-Holocene boundary in the southwestern Indian Ocean. Palaeoecology of Africa 6:45–54

Webb RS, Rind DH, Lehman SJ, Healy RJ, Sigman D (1997) Influence of ocean heat transport on the climate of the Last Glacial Maximum. Nature 385, 20 February 1997:689–695

Wefer G (1988) Bericht über die METEOR-Fahrt M6-6, Libreville-Las Palmas, 18.02.1988-23.03.1988.-Berichte, Fachbereich Geowissenschaften. Universität Bremen 3, Bremen

Wickens GE (1982) Paleobotanical speculations and Quaternary environments in the Sudan. In: Williams MAJ, Adamson DAA (eds) A land between two Niles, Balkema, Rotterdam, pp 23–50

White F (1983) The vegetation of Africa. UNESCO/AETFAT/UNSO, Rome

Winter A, Martin K (1990) Late Quaternary history of the Agulhas Current. Paleoceanography 5, 4:479–486

Environmental and Climatic History of the Eastern Kivu Area (D.R. Congo, ex Zaire) from 40 ka to the Present

Jürgen Runge
University of Paderborn, Department of Physical Geography, D-33095 Paderborn, Germany
arung1@hrz.uni-paderborn.de

Abstract: This paper provides a brief review of extensive large-scale terrestrial observations on stratigraphic and pedologic features of soils and sediments in the eastern Kivu region of former Zaire. Along an almost 600 km wide exploration transect from Bukavu in the Western Rift Valley to Kisangani in the Congo basin, Late Quaternary and Holocene environmental changes are discussed. The eastern Kivu has long been believed to be a part of the LGM (Last Glacial Maximum) "core-area" and a refuge area of the central African rain forest, showing more or less environmental stability during drier glacial times. In the course of road building, numerous deep soil openings, especially in the area of Walikale, were subject to palaeoenvironmental interpretation. Yellowish-brown hillwash sediments contained charcoal. These covered stonelines and fossil tree trunks within the mottled and pallid zone of soils. The resulting many radiocarbon dates allowed for the reconstruction of the environmental frame conditions of the eastern Kivu region between 40 ka to the present. There is considerable evidence for a major forest retreat between 36 ka and 13 ka. Several periods with a pre-LGM age from 36–28 ka, a LGM phase between 21–18 ka and a post-LGM period around 13–12 ka were characterized by alternating and contrasting dry and wet seasons with significantly intensified surface morphodynamic processes (sheet wash erosion, gullying) and stronger fluvial dynamics of rivers (alluviation). Annual rainfall in the lowland is suggested to have reached around 1200 mm only. The vegetation cover was mainly dominated by a savanna–woodland mosaic with extended gallery forest systems. Therefore, the conception of a "fluvial refuge" in the eastern Congo is a preferred environmental view for the LGM. Lowland tropical mean temperatures are supposed to have been slightly cooler in the range of 2–4° C lower than today.

15.1. Introduction

For a considerable period of time equatorial rain forests were believed to be one of the earth's most stable ecosystems that had survived over geological time (see Eggert 1992; Kadomura 1995). Over the last 30 years growing progress in absolute isotope dating techniques and an increasing number of palaeoenvironmental research activities in the lower latitudes of the African tropics have shown in many places that there might had been severe climatic turning points during the Quaternary. These resulted in sometimes strong ecological modifications of the environment (Alexandre et al. 1994; Aubreville 1962; Flenley 1979; Hamilton 1982; Runge 1992; Thomas 1994). However, it must be stated that the majority of palaeoenvironmental studies in central Africa deal with pollen records which are normally better preserved at higher altitudes.

Besides some pioneer work carried out by de Heinzelin (1954) and de Ploey (1964) in the former Belgium Congo there are only a few publications on large-scale terrestrial observations using geomorphologic and stratigraphic conceptions to study the Quaternary of Africa in lower regions close to the equator (see Michel 1991; Preuss 1986; Runge 1992, 1996, 1997; Schwartz 1988).

In the present study, extended road cuts, soil cores from swamps and lake level variations along an almost 600 km wide landscape transect between the towns of Kisangani in the Congo basin and Bukavu with the Lake Kivu in the Western Rift Valley (Fig. 15.1) were subject to palaeoenvironmental interpretation. Mainly sedimentological and pedological research on different sediment and soil characteristics helped in the reconstruction of the former climatic and vegetation conditions in the eastern part of the central African Congo basin.

15.1.2 Rain Forest Refugia or "Core Areas" in Central Africa

"Refuge theory holds that forest and non-forest biomes changed continuously in distribution during the geological past, breaking up into isolated blocks and again expanding and coalescing under the varyingly humid to arid climatic conditions of certain geological time intervals, especially during the Quaternary" (Haffer 1982:9). For the eastern Congo/Zaire study area, this basic idea of the "core area" concept has been frequently proposed by Moreau (1966), Livingstone (1979) and Hamilton (1982). They assume that in the course of Pleistocene climatic changes a biological diversification took place that is in parts still recognizable today. It is thought that formerly, i.e. before the glaciations, existing rain-forest areas were separated by a drying out and cooling down of climate into island-like areas of remaining rain forest, so-called "core" or "refuge" areas. Subsequently, after the Last Glacial Maximum (LGM) when the climate reverted again to more humid and warmer conditions, the rain forest spread from the core areas and probably also from isolated sites in the east African mountain regions ("montane refuge") back to the Congo basin. There are numerous thematic maps demonstrating the Late Pleistocene distribution of vegetation in Africa (see Vanzolini 1973; Bowen 1978; Hamilton 1982; Maley 1988; Liedtke 1990, Roche 1991). Most of the maps are based exclusively on biogeographical evidence. The landscape transect of the present study crosses through the postulated Pleistocene refuge area in the eastern Congo. An underlying question of our study was whether geomorphological, pedological and geochemical properties of the area during the last 40 000 years would lead to a verification of the core-area model for this region or not; in other words, will sediments and soils reflect signs of former stable or unstable environmental conditions due to morphodynamic activity that has been affected mostly by modifications in climate and vegetation cover?

15.2. Study Area

Field research was carried out in the environs of Osokari–Walikale (about 1°20' S, 28° E, 600–700 m asl), the Musisi–Karashoma swamp (2°16' S, 28°39' E, 2200–2300 m asl) and in the Lake Kivu basin (2°15' S, 28°50' E, 1460–1500 m asl) in the Kivu province of former eastern Zaire, now the D. R. Congo (Fig. 15.1).

Fig. 15.1. Location map and surveyed landscape transect between Kisangani and Bukavu, Congo.

Precipitation rates in the Osokari–Walikale area range from 1800–2500 mm/yr distributed over 150–200 days. Annual temperature reach about 24.2° C. The area is drained by the major rivers Lowa and Oso that flow into the Lualaba and subsequently the Congo/Zaire river. Geologically it is characterized by Precambrian basement rocks of the Burundien (1800–2100 Ma) that are frequently interspersed by post-tectonic intrusions of granite (950–1000 and 1300–1350 Ma). Overlying sedimentary rocks of silt and sandstones (Série de la Lukuga) date back to the Permian and Carboniferous (République du Zaire 1974). The Osokari–Walikale region occupies a zone in landscape transition situated between the elevated areas of the Western Rift Valley (Monts Mitumba, >3000 m asl.) and the inner Congo/Zaire basin (300–400 m asl). The relief is generally undulating with stronger and gorge-like fluvial incisions in places contrasting with tectonically controlled smaller alluvial basins. Hilly areas are related to weathered granitic domes and to resistant scarps of mica-slate and quartzite. Several uplifted planation levels of a probably Tertiary to Quaternary age can also be recognized.

Landsat-TM satellite images illustrate the clear transition between folded structures of the Kibaran orogenic cycle (Cahen and Snelling 1966) in the north and in the east, and several plain- to basin-shaped structures covered by the Lukuga sediments in the west. Furthermore, the widespread occurrence of alluvial ore and gold deposits indicates that there might be "younger" sediments inside these geomorphological forms that contained palaeoenvironmental data of the landscape history during the LGM.

The easterly situated study areas of the Musisi–Karashoma swamp and Lake Kivu are characterized today by a lower precipitation of about 1200–1700 mm and an average annual temperature of 16.6° to 20.3° C (see Table 15.1). This area of high intensity relief close to Lake Kivu is characterized by Tertiary to Quaternary fracture tectonics and uplift of older basement complexes (Monts Kahuzi and Mont Bièga) in connection with younger volcanism resulting in extended lava plains (see Denaeyer et al. 1965).

15.3. Characteristics of Eastern Kivu Soils and Sediments and Evidence for Modified Paleoenvironments since 40 ka

Most of the soils, pedisediments and alluvial deposits studied in the Osokari–Walikale area show undulating brown to yellow (10YR 5/4–10YR 4/6) hillwash layers, several centimeters to several meters in depth, underlain by similarly undulating stone-lines. These consist in parts of *in-situ* weathered rounded quartz originating from quartz veins. But there are also stone-lines that seem to be of an allochthonous origin. These contain, apart from rounded quartz, pebble-like (fluvial?) detritus mixed up with iron-rich concretions, which seem to be the product of a former morphodynamic process (Runge 1992, 1997). Subsequently, these stony accumulations were buried by fine-grained sandy to silty sediments. Radiocarbon dating of charcoal that was occasionally found in these pedisediments yielded average dates up to 1000–2200 yr BP. Also, some recent radiocarbon ages were measured for charcoal found at depths of 1.0–2.0 m (Fig. 15.5). Recent and former influence of termites (bioturbation) should be kept in mind for such sites. However, regarding the texture of termite mounds and the corresponding hillwash there was no evidence that these hillwash sediments were derived from former termite mounds (Runge and Runge 1995). Interestingly enough, many artifacts exist inside the stone-lines and the covering pedisediments.

Below these hillwash-like sediments and stone-lines at average depths of 2–3 m, soil colors change to orange and strong brown (contrasting of Hematite and Goethite weathering). The content of kaolinic clay increases up to 30% whereas the sandy fractions decrease significantly. Iron rich, nodular pisolites are also frequently found. Probably the pisolites are of a recent pedogenetic origin due to seasonally fluctuations of the ground water table (*in-situ* formation). Values for pH range between 3.9 and 5.0. Deeper developed profiles show typical mottled and pallid soil horizons with a high content of white kaolinic clay and silt which make up 95% of the soil's texture. However, not all the soil profiles are very deep. It is also common that more or less saprolitic rocks (isovolumetric weathering) or even unweathered rocks are at deptha of only 2–3 m, sometimes even less.

The regional distribution of stone-lines in the eastern Congo are such that the commonness of layers of coarse and stony horizons in soils decreases from east to west near Kisangani, whereas they increase if one gets closer to the Walikale area and the Western Rift mountains. This could indicate a regional gradient of formerly changing environments in the sense of strongly modified vegetation cover and a greater dynamic of fluvial surface processes from the western to the eastern part of the study area (Runge 1997).

3.1. Walikale, Osokari

Two extended road cuts, 3 km and 36 km west of Walikale, allowed a deeper insight into the Late Quaternary landscape history (Figs. 15.2 and 15.3). At these locations the allochthonous character of the weathering profiles and of the accumulated sediments respectively becomes evident.

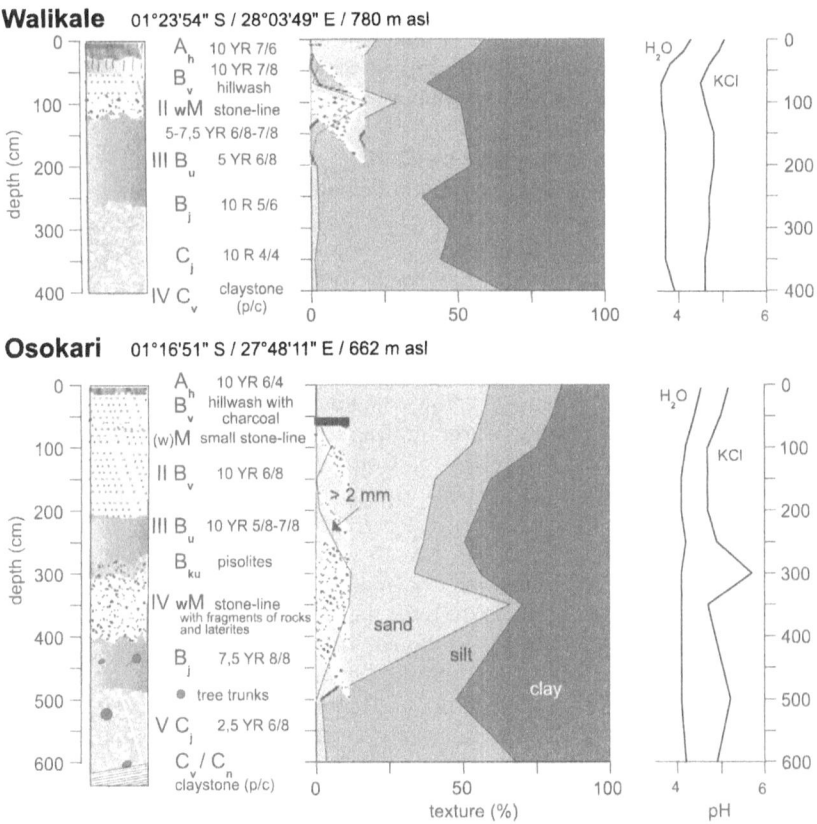

Fig. 15.2. Characteristic soil and sediment features of selected profiles in the eastern Kivu region (soil profile terminology after Arbeitsgruppe Bodenkunde 1982, 1994).

Here, numerous fossil tree trunks were found at depths of 4.5 m to almost 8 m, within the mottled and the pallid zone of the profile. Some trunks were also found in direct contact with the parent rock, which is a less weathered dark bluish-gray to bluish-black colored silty clay to sandstone (Fig. 15.2). AMS Radiocarbon ages of more than 10 fossil wood samples (organic sediment) at different locations of the extended road cut (especially at the Osokari site) yielded dates 12960 ± 330, 13190 ± 390, 31920 ± 250 and 36680 ± 440 yr BP (Fig. 15.5). These dates correspond to a pre-LGM and a high- to late-glacial age. The shape of tree trunks are up to one meter in diameter, indicating that the carbon samples belong to larger tree individuals. The $\delta^{13}C$ figures for the ^{14}C dates come up to –20 to –26‰, supporting the assumption that these trees were certainly species of a tropical environment (C3-plants). An attempt was also made to reconstruct the botanic species by looking at remaining cell structures within thin sections (H. Doutrelepont, Tervuren). However, the plant cells were already strongly modified under the influence of thousands of years of weathering, therefore it was not possible to learn any knowledge on the tree species which had formerly grown in the area. Aside from the richness of organic material within the pallid and mottled zone, the soil texture shows a relatively high amount of silt against the clay fraction. The upper parts of the profiles show comparable features such as stone-lines, pisolites and hillwash sediments as described in many other profiles (see Runge 1992, 1997; Fig. 15.2).

In addition, heavy-mineral analysis (L. Pfeiffer, University of Cologne) was used to get further evidence on the origin of the pedisediments. The heavy-mineral suites of samples of the fraction 0.063–0.200 mm showed more or less similar features with a striking predominance of staurolite, commonly derived from metamorphic rocks, which is not the parent rock material at the Osokari and Walikale study site. Zircon, tourmaline and rutile, which are very resistant against chemical and physical weathering, gave average values up to 25%. Unstable heavy-minerals such as garnet and amphibole were of minor importance. Only garnet occurred in a low percentage below a depth of 600 cm.

The geochemical composition of the profile was analyzed for major and trace elements (W. Rammensee, University of Cologne). In a search for significant geochemical features that could reflect aspects of a former climate controlled landscape dynamic, the investigation focused on the question of the relationship between the parent-rock material and the soil material, examining whether there are relative or absolute accumulations of certain elements within the profile (Fig. 15.3). The content of SiO_2 increases from the bottom to the top whereas one would expect that under stable conditions of humid tropic pedogenesis the content of SiO_2 would decrease in comparison (desilicification). The decrease of quartz is accompanied by a noteworthy increase of sesquioxides between 280–310 cm, especially for the Osokari field site. This zone corresponds to a pisolite layer (30–40 cm wide) that lies close to the present water table (probably *in-situ* formation).

Regarding the trace elements, it becomes obvious that elements with a higher atomic weight such as Barium (Ba), Lanthanum (La), Cerium (Ce) and Neodymium (Nd) are relatively enriched at the border between soil material and parent rock. Lanthanum (atomic weight 138.9) shows striking fluctuations within in the hillwash sediments and the finer grained mottled and pallid zone. As the probable result of a former fluvial to alluvial accumulation process, it reaches up to 232 ppm on the margin to the parent rock, which itself contains only up to 20 ppm of Lanthanum. Zirconium (Zr, atomic weight 91.2) shows a striking enrichment within the profiles from bottom to top (Fig. 15.3). There is equally a close correlation of the content of Zirconium and the amount of sand within the soil/sediment complex (Fig. 15.4). Single sandy layers are regarded to be the result of a former fluvial to alluvial input by sedimentary processes.

A few more observations concerning the geochemical properties of the profiles contradict the relative accumulation of heavier elements due to weathering and pedogenesis. Very low Manganese (Mn) and relatively higher Thorium (Th) concentrations in the study sites are, according to Schellmann (1986) a characteristic feature of granites. Within the weathering product the Mn/Th-ratio is therefore a very low one (< 1.0), whereas the Mn/Th-ratio for sandstone shows a higher value. This seems to be confirmed by the studied profiles where the parent rock material is a silty clay to sandstone whereas the weathered soil material seems to be of a granitic origin (see Runge 1996).

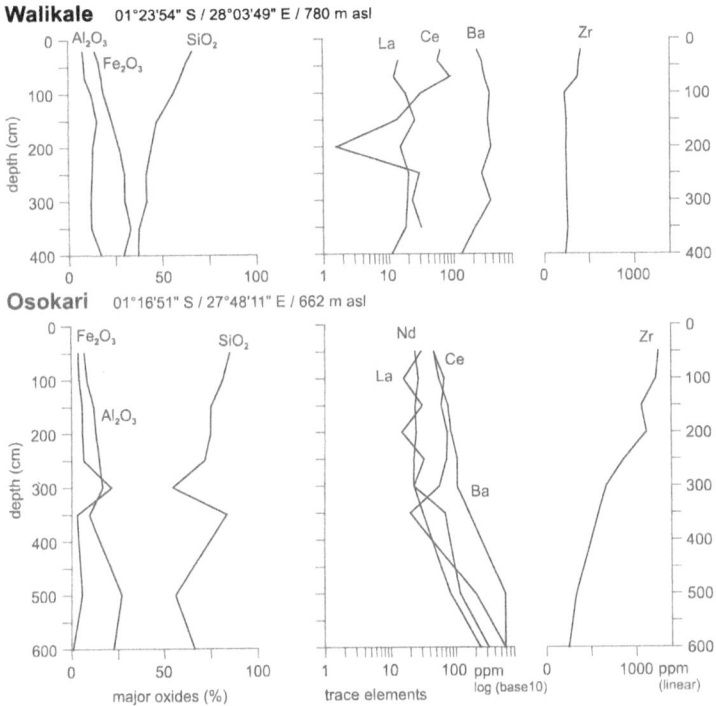

Fig. 15.3. Characteristic soil and sediment features of selected profiles in the eastern Kivu region (major oxides, trace elements).

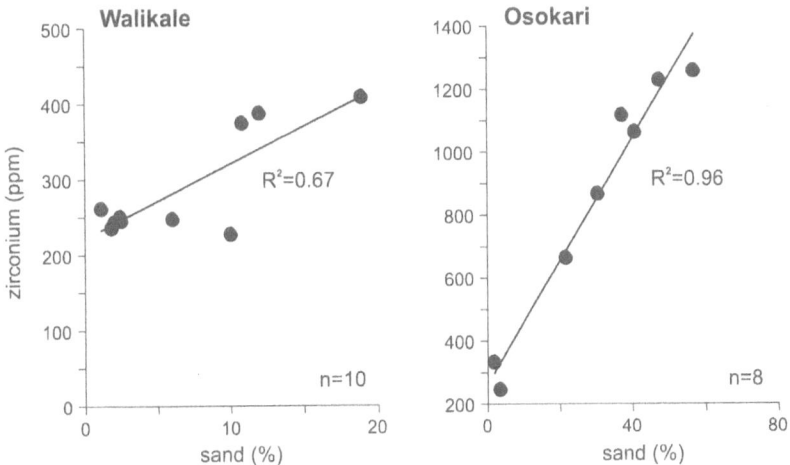

Fig. 15.4. Correlation between the content of Zirconium and sand in soils of the eastern Kivu.

15.3.2. Musisi–Karashoma

Soil coring in the Musisi–Karashoma swamp on the Graben shoulder of the Western Rift (around 2°16' S, 28°39' E) at an altitude of about 2300 m asl, resulted in an almost 5 m long profile containing peat, wood, organic and sandy sediment layers. Radiocarbon dates of 5150 ± 60 yr BP (Vilimumbalo 1993), 4470 ± 135 and 6495 ± 205 yr BP (Runge and Runge 1998) at the base of the profile propose a mid-Holocene initial formation time of the ancient "Lake Musisi" (see Boutakoff 1939). Precipitation values might have been significantly higher during the "Holocene climatic optimum" compared with today (Table 15.1). Concerning the average temperature it is not quite certain at the moment if it has been slightly warmer or cooler at that time. However, there is some pollen evidence that afrosubalpine species as *Hypericum* and *Erica* became significantly less dominant since 6.5 ka, whereas the influence of *Gramineae* and *Cyperaceae* was getting a higher importance. This remarkable environmental change is also expressed by the occurrence of a shorter period of vegetation controlled strong soil erosion on the nearby steep slopes of the Mont Kahuzi, evidenced by distinct sandy layers within the profile (see Vilimumbalo 1993).

Phytolith as well as pollen analyses (F. Runge, University of Paderborn, M. Moscol, Université de Liège) made possible the distinction of two further periods for changing environmental conditions between 4.3–3.5 ka (older *Poaceae* period) and from 1.3–0.3 ka (younger *Poaceae* period) representing slightly cooler and drier climatic frame conditions (Table 15.1, see Runge and Runge 1998). Increasing number of pollen from secondary mountain vegetation species as *Hagenia* proposes that the initial and ongoing influence of humans in the region started as early as 1.3 ka (E. Roche, Tervuren, personal communication).

15.3.3. Lake Kivu

Some peat profiles close to the present Lake Kivu shore-line at Buhandahanda (2°16' S, 28°50' E) indicate younger climate controlled variations of the water level (Table 15.1). Radiocarbon dates of an almost one meter thick peat deposit gave a bottom age of 745 ± 35 yr BP. A subjacent fossil humic horizon, covered by lake debris accumulation (quartz pebbles and other coarse material) was dated up to 970 ± 65 yr BP. This proposes a slight (2–3 meters) decrease of Lake Kivu's water level shortly after 1 ka. More humid climatic conditions inside the Western Rift Valley are shown by the overflow of Lake Kivu at the beginning of the Holocene at 9.5 ka into Lake Tanganyika. Shortly before, at about 12 ka, the lake level decreased more than 300 meters (see Hecky 1978). This extreme reduction in humidity was regarded to represent a climatic sign of the "younger Dryas" event (Table 15.1). At around 15–14 ka, shortly after the peak of the LGM, Lake Kivu's level was only 86 m less than today (Hecky 1978). A comparative view of the regional palaeoenvironmental data from the Lake Kivu area, the Musisi swamp and from the Osokari basin for the period since 40 ka, is shown by a synoptic view in Table 15.1, allowing one to see a regional view of Late Quaternary climate and vegetation changes in the Kivu region.

15.4. Overview of Radiocarbon Dates – Stratigraphy of Pedisediments

An illustrated summary of all the radiocarbon dates from the different study sites in the eastern Kivu (Fig. 15.5) assists coming to some conclusions on the overall stratigraphic features and their time of formation. Relating these dates to the profile depth and the pedological and sedimetological properties, the charcoal findings from the hillwash-like pedisediments in group (a) cover mainly the first 1–2 m of soil profiles. These "soil sediments" had been mainly deposited between 5–2 ka under a more open vegetation and a more seasonal climate than today. Average sedimentation rates were relatively high and are calculated to have reached between 1–40 mm per year (Fig. 15.5).

It is evident that the samples from inside the cores represent very liquid Musisi swamp sediments that are also Holocene in age. Figure 15.5 classifies the Musisi radiocarbon dates separately as group (b) as they cannot be directly compared with the soils and sediments of groups (a) and (c). All other radiocarbon dates gained from below the hillwash sediments and the subjacent stone-line deposits showed a pre-Holocene to LGM and pre-LGM age (c). The fact that fossil tree trunks were found on different profile sections below the stone-lines, supports the assumption that before, during and after the LGM a major retreat, shrinkage and fragmentation of tropical forests might have had occurred. This climate controlled "deforestation" lasted probably for several thousands of years from 36–28 ka, 21–18 ka, and 13–12 ka (Table 15.1). The findings, especially from the Osokari field site, suggest a reduced and opened vegetation cover due to a cooler and drier climate with intensified morphodynamic processes such as alluviation, pedimentation and sheet wash erosion (including episodic gullying).

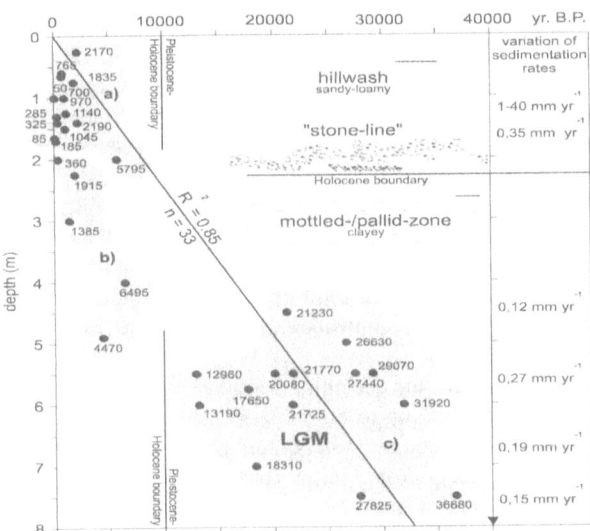

Fig. 15.5. Synoptic view of radiocarbon dates and stratigraphic features of soils and sediments in the eastern Kivu: **a)** charcoal, hillwash pedisediments, **b)** *Cyperaceae* swamp sediments, Musisi–Karashoma, **c)** fossil tree trunks, Walikale–Osokari basin.

Annual rainfall in the lowland of the eastern Congo is suggested to have reached around 1200 mm compared to recent rainfall of about 2400 mm (Table 15.1). At that time, the vegetation cover was mainly dominated by a savanna–woodland mosaic with extended gallery forest systems. Therefore the conception of a "fluvial refuge" (Colyn et al. 1991) in the eastern Congo is a preferred environmental view for the periods before, during and after the Last Glacial Maximum (LGM). Lowland tropical mean temperatures are supposed to have been slightly cooler, about 2–4° C lower than today. Further, the common occurrence of stone-lines in the eastern Kivu leads to the assumption that these pedological features are suited to be useful stratigraphic markers of the Pleistocene/Holocene boundary (Fig. 15.5).

15.5. Results

Different core area conceptions proposed that the rain forest in eastern Kivu remained in larger refuge areas and were characterized by more stable environmental conditions during the Late Pleistocene. We set out to determine if an examination of the soils and sediments in this region would support this assumption or not. In particular, pedological, sedimentological and geochemical examinations as well as radiocarbon dating on several exposures demonstrated that in certain zones of transition between the East African highlands and the inner African lowlands significant environmental changes might, in fact, have occurred.

Obviously pre-weathered sediments containing a high amount of staurolite and a relative enrichment by heavy elements such as Ba, La, Ce, Nd, were found on parent rock (silty clay- and sandstone) which, it would seem, could not be the source of the discussed stratigraphic and pedologic features. The total heavy-mineral content showed striking deviations from the average values that were most probably induced by changing environmental conditions. The existence of garnet, especially in the finer fraction at the bottom of one profile, and significantly increased values of zircon, tourmaline and rutile, both support the assumption that there was formerly an addition of aeolian allochthonous components (Kalahari sands?) into the alluvial to fluvial sediments. In conclusion, it can be proved that many of the soils below and above the stone-lines have an allochthonous character, e.g. they are primarily sediments which have undergone a secondary pedogenetic transformation (pedisediments). Tertiary strongly pre-weathered soils and sediments had been eroded under a less developed vegetation cover from the tectonic active and uplifted topographic areas of the Western Rift mountains into smaller fault controlled alluvial basins neighboring the main Congo basin.

Fossil tree trunks within the ferrallitic section of the profile (below the stone-line) were radiocarbon dated up to 36–12 ka (Last Glacial Maximum, before and after). Hillwash sediments containing charcoal above the stone-line were dated up to 5–2 ka. The estimated age of the initial formation of stone-lines could therefore be the period from the end of the Pleistocene and the beginning of the Holocene. The layers of coarse material are interpreted as former local erosion surfaces with intensive morphodynamic processes (sheetwash, pedimentation) leading to the accumulation of stony material on the ground.

landscape unit transect	Osokari NW Lowa-Oso-Lualaba drainage system transition to the Congo/Zaire basin 1°16'51"S / 27°48'11"E / 662 m asl	Musisi-Karashoma Monts Mitumba chain Graben shoulder of the Western Rift 2°16'16"S / 28°39'37"E / 2278 m asl	Lake Kivu volcanic barrier lake in the Western Rift Valley → SE 2°15'43"S / 28°50'19"E / 1460 m asl
field site / time (ka B.P.)	*2400 mm rainfall, 24,2° C mean temperature, dense tropical rain forest (Gilbertiodendron dewevrei)*	*1700 mm rainfall, 16,6° C mean temperature, mountain forest (Hagenia, Arundinaria, Cyperus spec.)*	*1270 mm rainfall, 20,3° C mean temperature, savanna (Hyparrhenia spec.)*

400 km (transect span)

Osokari:
- drier and slightly cooler at **2 ka** and **5 ka** with episodic **hillwash** formation on local spots with savanna vegetation, alternating dry and wet seasons
- **11-5 ka**: warm and humid, extension of dense rain forest vegetation
- **13-12 ka**: cooler and dry, regression of forest, strong erosion; alluvial accumulation under an alternating semi-humid (to semi-arid?) climate; sedimentation of coarse material, major time of **"stone-line"** formation, degradation of older lateritic crusts (**"younger Dryas"** event?)
- **17-13 ka**: initial recolonisation with rain forest under warmer and more humid climatic frame conditions
- **21-18 ka**: dry and cool, alluviation and pedimentation with possible aeolian input, open savanna vegetation, woodland and gallery forests
- **27-22 ka**: warmer and humid, forest-savanna mosaic
- **36-28 ka**: cool and dry, alluviation and pedimentation

Musisi-Karashoma:
- **0,3-1,3 ka**: slightly drier, increase in *Poaceae* species (phytolith evidence for **"younger Poaceae period"**)
- **2 ka**: warm and dry, mosaic of mountain forest and savanna patches, increasing influence of humans
- **3,5-2 ka**: ongoing regression of mountain forest, slightly cooler, pollen of *Podocarpus* and *Olea*
- **4,3-3,5 ka**: cool and dry, regression of forest, increase in *Poaceae* phytoliths (**"older Poaceae period"**)
- **6,5-5 ka**: very humid, warm(?), initial formation of swamp "lac Musisi", **"Holocene climatic optimum"**
- **9-5 ka**: warm-humid, extension of mountain forest
- ?
- ?

Lake Kivu:
- drier from **1-0,2 ka** and **1,6-1,5 ka**; between **1,5-1 ka** more humid, slightly higher level of Lake Kivu (+2-3 m)
- **3 ka**: more humid, extension of forest
- **5-3 ka**: slightly drier and cooler (?), regression of dense forest, Lake Kivu level -30 m less at **4 ka** than today
- **9-5 ka**: humid and warm, overflow of Lake Kivu at **9,5 ka** into L.Tanganyika
- **12 ka**: cool and dry (**"younger Dryas"** event?), low lake level: -310 m less
- **13-9 ka**: significant trend of warmer and humid climate, extension of dense forest, shortly interrupted around 12 ka
- **15-14 ka**: Lake Kivu level less -86 m compared with today
- **27-15 ka**: drier and cooler, savanna, probably smaller mountain forest refuges
- **32-27 ka**: extension of mountain forest due to warm and humid climate
- **39-32 ka**: cool and dry, mountain forest is replaced by savanna vegetation

Time scale (ka B.P.): present, 5, 10, 15, 20, 25, 30, 35 — Holocene / LGM / Late Quaternary

Table 15.1. Environmental history of the eastern Kivu region from 40 ka to the present.

Therefore, climatic boundary conditions were significantly drier in the past than today (alternating rainy and dry seasons with an open savanna–woodland or even in parts a pure-grass savanna). These new findings from eastern Congo/Zaire are in accordance with the marked dry climatic phase of the *Léopoldvillian* starting around 30 ka in the western Congo. Similar drier conditions can now be supposed for larger parts of the eastern Congo basin before, during and after the LGM.

A Holocene period of strengthened morphodynamic activity was connected with the formation of hillwash sediments (sheet flow erosion, overland flow, seasonally changing vegetation cover), which is a typical feature for a savanna-like environment. This sequence was radiocarbon dated at 2000–1300 yr BP. The ways in which human influence might have played a role in accelerating soil erosion cannot, aside from the Musisi swamp findings, as yet be stated. More archaeological research, in particular concentrating on the former environment and climate, and on human history in this central African area, needs to be carried out.

15.6. Discussion

There are other terrestrial observations on palaeoclimates in adjoining areas that support the concept of a more discontinuous core area extension during the Late Quaternary. Brook et al. (1990) found by studying pollen in cave speleothems at Matupi Cave, only 400 km NNE from Walikale, at about 1100 m asl that 22–12 ka a grass savanna environment with a greater degree of erosion had prevailed in this area. Ladmirant and Roche (1988) discovered on satellite images an ancient erg in the Lusambo area (Kasai province), 500 km SW from Walikale, which is actually overgrown by rain forest and Miombo woodland. In comparison with other East African research sites (for example Kalambo Falls close to Lake Tanganyika) an age of 25–15 ka is assumed by Ladmirant and Roche (1988) for these signs of former arid to semi-arid climate conditions.

However, it is reported from the Ruki-Valley in the central western Congo basin that the environmental change from a semi-arid/semi-humid climate to the humid climate of today took place as early as 17 ka (Preuss 1986), thus at a time when in the adjacent regions the drier climatic conditions were still predominant. Marine palaeoclimatological studies (Jansen et al. 1984) showed that terrigenous sedimentation of the Congo river, indicating arid environmental conditions in the extended drainage basin, lasted until 14.5 ka.

Consequently, a decrease of sediment supply took place during the transition from arid to humid climate around 13 ka. In the Walikale area, this climatic turning point seems to have occurred as early as 17–18 ka when sedimentation rates started to become significantly higher. However, the dates of the stone-lines, with an age of about 13–12 ka, indicates an even younger climate controlled shrinkage of forests shortly before the dawn of the Holocene. The strong accumulation of sediment was affected by morphodynamic processes that occurred under seasonal semi-humid to in parts semi-arid savanna climate.

In conclusion, it now seems that the overall environmental and climate character of the wider central African basin during the Late Quaternary was much drier than it was commonly believed before. This recognition is being increasingly accepted by the majority of the scientific community.

Acknowledgements. The study was financially supported by the Deutsche Forschungsgemeinschaft, DFG, Bonn (Grant Ru 555/2-1, Ru 555/2-2). Logistic support during the field work 1991–1994 was kindly provided by STRABAG International GmbH, Bujumbura and Cologne, and by the Deutsche Gesellschaft für Technische Zusammenarbeit GmbH (GTZ), Bukavu, Congo.

References

Alexandre J, Aloni K , de Dapper M (1994) Géomorphologie et variations climatiques au Quaternaire en Afrique Centrale. Géo-Eco-Trop 16:167–205

Arbeitsgruppe Bodenkunde (1982, 1994) Bodenkundliche Kartieranleitung, 3. und 4. Auflage, Hannover, pp 1–392

Aubreville A (1962) Savanisation tropicales et glaciations Quaternaires. Adansonia, n.s., II,1–2:16–84

Boutakoff N (1939) Géologie des territoires situés à l'ouest et au nord-ouest du fossé tectonique du Kivu. Mémoires de l'Institut Géologique de l'Université de Louvain 9:1–207

Bowen DQ, (1978) Quaternary Geology. A stratigraphic framework for multidisciplinary work. Oxford

Brook GA, Burney DA, Coward JB (1990). Paleoenvironmental data for Ituri, Zaire, from sediments in Matupi Cave, Mont Hoyo. Virginia Mus. Nat. Hist. Memoir 1:49–70

Cahen L, Snelling NJ (1966). The geochronology of Equatorial Africa. North. Holland Publ., Amsterdam

Colyn M, Gautier–Hion A, Verheyen W (1991) A re-appraisal of palaeoenvironmental history in central Africa: evidence for a major fluvial refuge in the Zaire basin. Journal of Biogeography 18:403–407

de Heinzelin J (1954) Sols, paléosols et désertification anciennes dans le secteur Nord-Oriental du Bassin du Congo. Publi. Inst. Nat. Etude Agronom. du Congo Belge, pp 1–168

Denaeyer ME, Schellinck F, Coppez A (1965) Recueil d'analyses des laves du fossé tectonique de l'Afrique Centrale. Geol. Wetenschapen, 45, Tervuren

de Ploey J (1964) Cartographie géomorphologique et morphogènese aux environs du Stanley–Pool (Congo). Acta Geogr. Lovaniensia 3:431–441

Eggert MKH (1992) Über die Flüsse in die Wälder. Zur Besiedlungsgeschichte des äquatorialen Regenwaldes. In: Bollig, M and Bünnagel, D (eds) Der zentralafrikanische Regenwald. Lit-Verlag, Afrikanische Studien 3:53–63

Flenley JR (1979). The Late Quaternary history of the equatorial mountains. Progr. in Physical. Geogr. 3:488–509

Haffer J (1982) General aspects of the refuge theory. In: Prance, GT (ed) Biological diversification in the tropics. Columbia University Press, pp 6–24

Hamilton AC (1982). Environmental history of East Africa: A study of the Quaternary. Academic Press, London, New York

Hecky RE (1978) The Kivu–Tanganyika basin: the last 14 000 years. Polske Archiwum Hydrobiologii 25:159–165

Jansen JHF, van Weering TCE, Gieles R, van Iperen J (1984) Late Quaternary oceanography and climatology of the Zaire–Congo fan and the adjacent eastern Angola basin. Netherlands Journal of Sea Research 17:201–249

Kadomura H (1995) Palaeoecological and palaeohydrological changes in the humid tropics during the last 20.000 years, with reference to equatorial Africa. In: Gregory, KJ, Starkel, L, Baker, VR (eds) Global continental palaeohydrology, pp 177–202

Ladmirant H, Roche E (1988). Important Quarternary climatic changes as evidenced from remote sensing data (Satellite Landsat) and from palaeobotanical studies. Acad. Roy. Sc. Lettr. et des Beaux-Arts de Belgique, 91–96

Liedtke H (1990) Stand und Aufgabe der Eiszeitforschung. In: Liedtke, H (ed) Eiszeitforschung, pp 40–54

Livingstone DA (1979) Quaternary Geography of Africa and the refuge theory. In: Prance, GT (ed) Biological diversification in the tropics. New York, pp 523–536

Maley J (1988) Fragmentation de la forêt dense humide africaine et extension des biotopes montagnards au Quaternaire recent: nouvelles données polliniques et chronologiques. Implications paléoclimatiques et biogéographiques. Palaeoecology of Africa and the surrounding islands 18:307–336

Michel R (1991) Les grottes du Mont Hoyo (NE Zaire). Un paléoenvironnement karstique de plus de 150.000 ans. Géo-Eco-Trop 1–2:1–59

Moreau RE (1966) The bird faunas of Africa and its islands. Acad. Press, New York

Preuss J (1986) Jungpleistozäne Klimaänderungen im Kongo–Zaire–Becken. Geowiss. in unserer Zeit 4:177–187

République du Zaire (1974) Notice explicative de la carte géologique du Zaire au 1/2.000.000. Département des Mines, Direction de la Géologie. Tervuren

Roche E (1991) Evolution des paléoenvironnements en Afrique Centrale et oriental au Pléistocene supérieur et à l'Holocene. Influence climatiques et anthropiques. Bull. Soc. Géogr. de Liège 27:187–208

Runge F, Runge J (1998) Phytolithanalytische und klimageschichtliche Untersuchungen im Musisi–Karashoma–Sumpf, Kahuzi–Bièga Nationalpark, Ost–Kongo (ex Zaire). Paderborner Geogr. Stud. 11:79–104

Runge J (1992). Geomorphological observations concerning palaeoenvironmental conditions in eastern Zaire. Z.Geomorph. N.F., Suppl.-Bd. 91:109–122

Runge J (1996) Palaeoenvironmental interpretation of geomorphological and pedological studies in the rain forest "core areas" of eastern Zaire (Central Africa). South African Geographical Journal 78:91–97

Runge J (1997) Alterstellung und paläoklimatische Interpretation von Decksedimenten, Steinlagen (stone-lines) und Verwitterungsbildungen in Ostzaire (Zentralafrika). Geoökodynamik 18:91–108

Runge J, Runge F (1995) Late Quaternary palaeoenvironmental conditions in eastern Zaire (Kivu) deduced from remote sensing, morpho-pedological and sedimentological studies (Phytoliths, Pollen, 14C-data). 2ᵉ Symposium de Palynologie Africaine, Tervuren (Belgique). Publ. Ocass. CIFEG 31:109–122

Schellmann W (1986) On the geochemistry of laterites. Chem. Erde 45:39–52

Schwartz, D (1988) Histoire d'un paysage: le Loosséké. Paléoenvironnements Quaternaires et podzolisation sur sables Batéké. ORSTOM Collection Etudes et Thèses, 1–285

Thomas MF (1994) Geomorphology of the tropics. J Wiley & Sons, Chicester, New York

Vanzolini PE (1973) Paleoclimates, relief, and species multiplication in equatorial forests. In: Meggers BJ, Ayensu ES, Duckworth WD (eds) Tropical forests ecosystems in Africa and South America: a comparative review, pp 255–258

Vilimumbalo S (1993) Paléoenvironnements et interprétations paléoclimatiques des depôts palustres du Pléistocene superieur et de l'Holocène du Rift centrafricain au sud du Lac Kivu (Zaire). Thèse du Doctorat, Univ. de Liège, Faculté de Sciences, pp 1–212

Quaternary Climate Changes in Southernmost South America Inferred from Lacustrine Sediments Preserved in Volcanic Maars

H. Corbella[1], Ana M. Borromei[2], Mirta E. Quattrocchio[2]

[1]CONICET - Museo Argentino de Ciencias Naturales "B.Rivadavia" Av. A. Gallardo 470, 1405 Buenos Aires, Argentina
hcorbell@ibm.net
[2] Universidad Nacional del Sur, Departemento de Geologia, Catedra de Geologia Historica, Laboratorio de Palinologia, 8000 Bahia Blanca, Argentina
mquattro@criba.edu.ar

Abstract: As part of the Southern Hemisphere Paleo- and Neoclimates Project (UNESCO IGCP-341), sediments deposited in three small basins of volcanic maars which outcrop in the southern Santa Cruz extra-Andean volcanic belt at 52° S latitude were cored up to 59 meters depth. Undisturbed samples were obtained from six coring sites. A preliminary palynological study of sediments from the Magallanes Maar (one sample/meter) seems to indicate several pulses of higher and lower relative humidity. Radiocarbon analyses indicate an Upper Pleistocene age for these sediments.

16.1. The Southernmost South American Maars

In the southernmost part of the South American continent, 300 km East of the Andes and immediately North of the Magellan Strait, there is a tectono-volcanic belt of Quaternary age known as Pali-Aike. In this region, modern volcanics of alkali-basaltic composition had been reported (Caldenius 1932, Altevogt 1969, Codignotto 1975, Mercer 1976, Skewes 1978, Skewes and Stern 1979).

Studies started by our geological team in 1989 (Corbella et al. 1990, 1991; Corbella 1993; Pomposiello et al. 1990, 1991, 1995; Chelotti and Trinchero 1990) have shown that these basaltic bodies outcrop along a belt 50 km wide and 150 or more km long in a general NW direction. Inside this belt the outcrops and hypabyssal bodies are oriented along two principal directions (NW and E-NE) coincident with megalignments and faults recognized on satellite images (see Fig. 16.1).

The volcanic outcrops are plateaus of moderate extension, lava flow cones topped by scoria cones, some pyroclastic cones and more than 50 maars of phreatomagmatic origin. The latter have been surveyed by remote sensing, geological and geophysical methods.

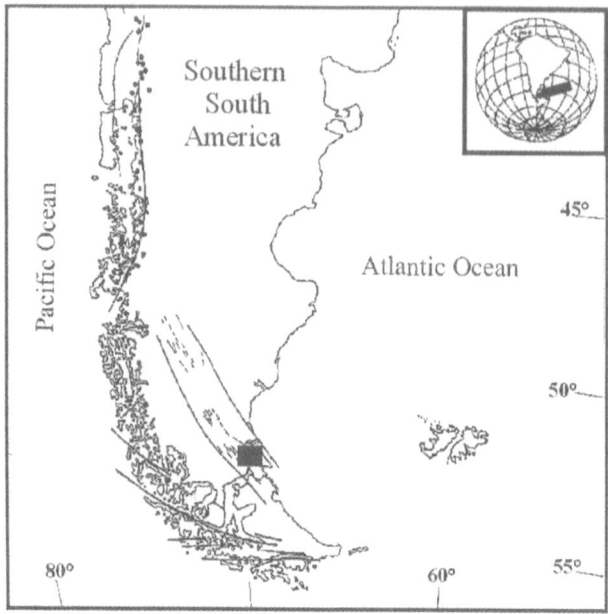

Fig. 16.1. Volcanic outcrops are oriented along a tectonic belt of NW direction. The black solid rectangle indicates where holes were drilled in dry maar depressions to recover paleolacustrine sediments.

16.2. Paleolake Sediments Preserved in Maars

A maar is morphologically a roughly circular depression or crater, 500-1500 m in diameter and 50-100 m deep, with a surrounding ring of finely stratified hyaloclastic sediments. Maars are formed by phreatomagmatic or steam explosions when lava comes into contact with moist soils or surface waters. This phenomenon is able to carve diatremes, i.e. hundreds of meters deep funnel-like cavities in the wall rocks. In Pali-Aike, the existence of diatremes below the bottom of the depressions was verified by gravimetric and seismic surveys (Pomposiello et al. 1991).

In spite of the present dry aspect of the maars – only some of the larger ones have the bottom covered with water during rainy seasons – most of these small endoreic basins could occasionally, in the past, have been occupied by ponds. In that case, they behaved as local depositional basins for periglacial, lake and eolian sediments. In this extremely windy region, where all modern sedimentary deposits have been reworked by the wind, the existence of maar-lake deposits could provide a unique tool for the reconstruction of past environmental changes (e.g.: paleoclimatic fluctuations, volcanic activities, plant successions and history of vegetation). Therefore in 1993, the authors drilled several holes to verify the existence of paleolacustrine sediments below the surface of the dry maar depressions."

16.3. Drilling and Coring

During the austral spring of 1993, YPF (Yacimientos Petroliferos Fiscales) kindly allowed us to employ a seismic drilling equipment and crew. To obtain undisturbed sediments, a core barrel with one sharp edge made of hardened steel was specially made to lodge segments of opaque polyethylene tubes 50 cm long, 60 mm diameter, with a phi steel type, that finally contained the cores. The drilling machine was used, not as a rotary unit, but only to thrust the sampling assembly.

Six holes were drilled at three different dry maars: Magallanes (52° 07' S, 69° 16' W), Flamenco (51° 54' S, 70° 17' W) and Tito (51° 52' S, 69° 24' W). The deepest and most informative of the 3 drilled paleolakes was the one in the Magallanes Maar, 10 km from the Magellan Strait, where three holes (each one 30 m apart) were drilled in soft sediments, reaching a maximum depth of 59 m below the surface before stopping at the basaltic bed rock.

The recovered sediments are now being analyzed with respect to petrography, palynology, diatom content, oxygen isotopic composition and chronology. In this paper we present the preliminary results of the palynological studies and two ^{14}C radiometric ages, performed on the organic matter fraction of samples from 47 and 37 m depth, which mark the time limits of our observations.

16.4. Present Climate

Predominant westerly winds throughout the year cause strong winter precipitations on the western slopes of the Patagonian Andes (3000 to 4000 mm/year). As the distance from the Pacific Ocean increases, the precipitation decreases. The gradient is 800 mm at the base of the eastern Andean slopes, 400 mm East of the great glacial lakes, 200 mm at the central Patagonian region and 180 mm at the Atlantic coast (Centro Editor de America Latina 1981, Mancini 1993).

16.5. Vegetation

The vegetation distribution is a suitable indicator of the precipitation gradient (see Fig. 16.2), with evergreen forests in the West where the precipitation values are the highest, continuing with deciduous beech forests (caducifolios), gramineous steppes bordering the forests, and xeric steppes in the middle and eastern parts of the plateau, followed by small and medium-sized salt fields near the coast (Mancini 1989).

Nothofagus forests [Meridional Forests by Hüeck-Seibert 1981]: Evergreen *Nothofagus betuloides* (guindo/South American cherry) forests are found in small areas in the extreme West of Santa Cruz and South of Tierra del Fuego. *Nothofagus caducifolios* appear on the eastern slopes of the Patagonian Cordillera. Two species predominate: *Nothofagus antarctica* (ñire) on the valleys penetrating the steppes to the East and *Nothofagus pumilio* (lenga) on the mountain-slopes.

Gramineous steppe [Patagonic Steppes and Semideserts, Fuegian and Magellan Sector by Hüeck-Seibert 1981]: The gramineous steppe is a narrow zone along the precordillera that widens near the Magellan Strait and penetrates Tierra del Fuego. *Festuca pallescens* is predominant, although thick bushes and shrubs such as *Berberis, Chiliotrichum diffussum, Empetrum rubrum, Senecio* and *Mulinum spinossum* are also present. There are species favouring the increase of humidity such as *Trifolium repens, Poa pratensis* and *Erigeron myosotis*. In areas where the humid steppe suffers from drier conditions, species such as *Armeria maritima, Azorella fueguiana, Rytidosperma virescens* appear.

Xeric steppe [Patagonic Steppes and Semideserts, Central Sector by Hüeck-Seibert 1981]: The xeric steppe extends East of the cordillera-lakes from sea level up to approximately 500 m above sea level. It looks like a pasture with hard-leaved bushes 30 to 40 cm high, sometimes altered by larger areas of *Verbena tridens* and *Berberis*. We also find different species of *Stipa, Nassauvia ulicina, Polygala darwiniana, Nardophyllum bryoides, Ephedra frustillata*. On the saline soils of the Atlantic littoral *Atriplex, Frankenia patagonica, Lycium ameghinoi, Lepidophyllum cupressiforme* and *Senecio patagonicus* are characteristic. *Prosopis patagonica* and *Schinus polygamus* make up the gallery forests.

16.6. Materials and Analytical Methods

The samples come from 3 holes drilled in the Magellanes Maar where 59 m depth was reached. For the preliminary study one sample per meter was taken.The chemical techniques used for the concentration of the pollen contents were those employed by Heusser and Stock (1984), processing 5 grams of sediment per sample. Before starting the chemical treatment, two tablets with *Lycopodium* spores were added to compute the pollen concentration.

The microscopic identification of the pollen types was done with the aid of reference standards from the Palynotheque of the Palynology Laboratory of the Universidad Nacional del Sur, Argentina. The works of Heusser (1971, 1989), Markgraf and D'Antoni (1978), Hooghiemstra (1984) and Morbelli (1980), among others, were used. To guarantee the presence of significant taxa in the pollen inventory, the concept of minimum area (Bianchi and D'Antoni 1986) was employed, establishing a pollen sum statistically representative for the population under study.

As the samples are sterile (lacking pollen) down to a depth of 28 m, the palynological profile starts at 29 m. Samples 2, 6, 11, 12 and 18 are sterile. Samples 3, 4, 5 and 21 have a very low number of pollen grains. It was not possible to count them in a statistically significant way, so they figure only as present in the diagram.

The percentages (%) and concentrations (grains/gram = gn/g) are shown in Fig. 16.3. As the pollen-concentration depends on many factors, from vegetational communities over preservation to sedimentation-rate, the significance of the pollen concentration should be interpreted with care. The fossil pollen associations are compared with present associations observed in surface samples (Mancini 1989, 1993). The fossil vegetation communities (see Fig. 16.2) were compared to present-day communities following Hüeck-Seibert (1981)

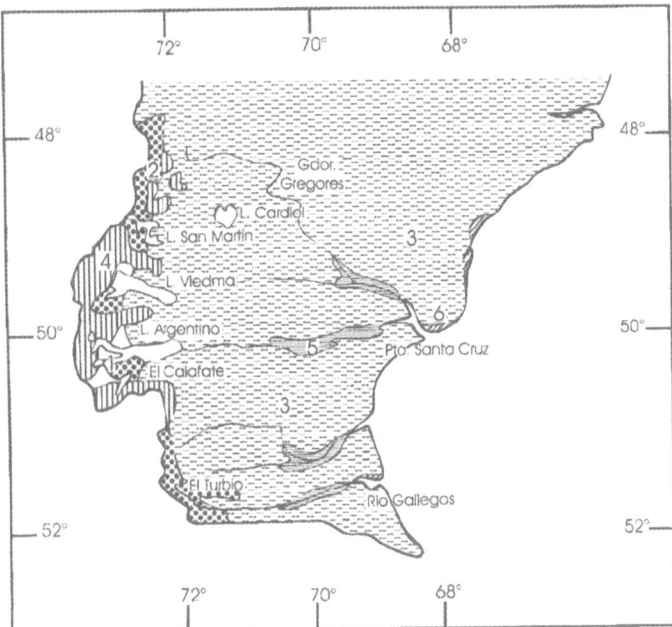

Fig. 16.2. Phytogeographic map of Southern Santa Cruz Province after Hüeck and Seibert (1981). 1: Evergreen forest. 2: Deciduous beech forest. 3: Steppes and Patagonian semideserts. 4: High mountain Andean vegetation. 5: Littoral dunes.

16.7. Observed Pollen

In the palynologic profile (Fig. 16.3) several pollen-associations can be observed. These are from bottom to top:

Sample 20 (56 m): Fine sand with coarse sand.
Flora: Gramineae (40%) predominant, Compositae Tubuliflorae (19%) and *Empetrum* (15%).

Sample 19 (54 m): Silt with coarse sand.
Flora: Compositae Tubuliflorae (34%) and Gramineae (33%), Chenopodiaceae-Amaranthaceae (5%). In both samples the concentration values are minimal. The maximum values correspond to Gramineae with 1200 grains/g in sample 20 and Compositae Tubuliflorae showing 800 grains/g in sample 19. In relation to the rest of the profile, the diversity is low. A small amount of *Nothofagus* (7%, Sample 19) and the presence of *Podocarpus*, Myrtaceae and *Schinus* with values below 1% can be seen.

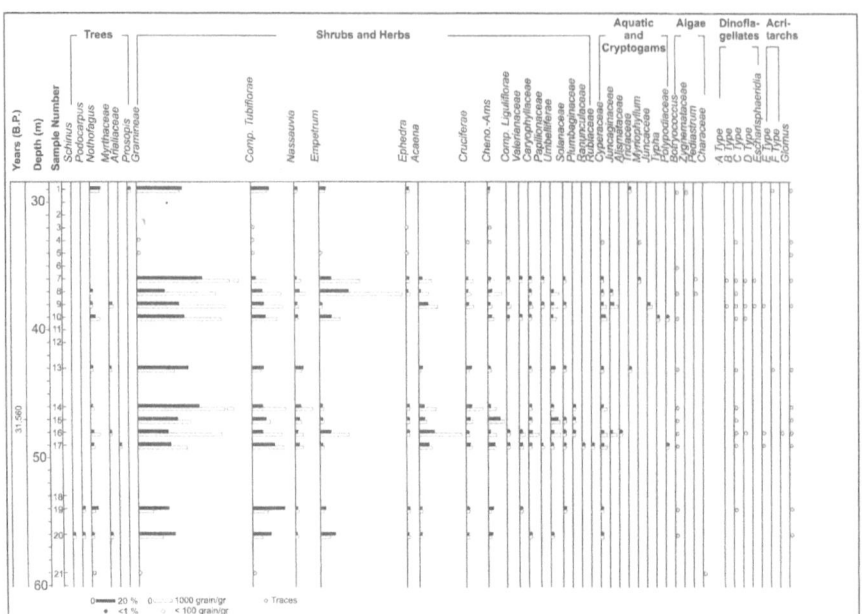

Fig. 16.3. Integrated diagram of relative frequencies (%) and pollen concentration (grains/g) observed in the Magellanes Maar (SW Santa Cruz Province, Argentina).

Sample 17 (49 m): Silt.
Flora: Gramineae (35%), Compositae Tubuliflorae (24%) and *Acaena* (10%). Also Chenopodiaceae-Amaranthaceae (6%), *Nassauvia* (5%) and Valerianaceae (4%). An increase of the concentration values of Gramineae to 2600 grains/g and Compositae Tubuliflorae (1750 grains/g) can be seen.

Sample 16 (48 m): Pelite.
Flora: Gramineae (33%), Acaena (16%), Compositae Tubuliflorae (13%) and *Empetrum* (11%). The concentration values increase for all taxa, reaching the highest values for Gramineae (4500 grains/g) and *Acaena* (2200 grains/g), followed by *Nothofagus* (500 grains/g).

Sample 15 (47 m): Silt.
Flora: Gramineae (42%), Compositae Tubuliflorae (15%) and Chenopodiaceae-Amaranthaceae (12%). Also observed: Umbelliferae (cf. *Azorella* 8%) which reach in this sample the maximum value of the whole profile and *Acaena* (6%). Lower concentration values with respect to the former level are recorded for Gramineae (2800 grains/g), Compositae Tubuliflorae (990 grains/g) and Chenopodiaceae-Amaranthaceae (800 grains/g).

Sample 14 (46 m): Silty sand.
Flora: Gramineae (62%) and Compositae Tubuliflorae (12%). Also *Nassauvia* (6%), Cruciferae (6%) and *Acaena* (6%) can be seen. Gramineae reach here their

maximum value in the whole profile with 9680 grains/g. Compositae Tubuliflorae (1850 grains/g), *Nassauvia* (960 grains/g) and Cruciferae (960 grains/g). *Acaena* shows with 890 grains/g a notable increase with respect to the preceding sample.

Sample 13 (43 m): Silt.
Flora: Gramineae (54%), Compositae Tubuliflorae (13%) and *Nassauvia* (8%). At this level the concentration values for Gramineae fall to 1970 grains/g. The concentration values of the other elements are comparable to those of samples 19 and 20.

Sample 10 (39 m): Pelite.
Flora: Gramineae (50%), Compositae Tubuliflorae (15%) and *Empetrum* (13%). At this level Cyperaceae (4%) and *Nothofagus* (5%) reach their highest value in the profile. The following concentrations are observed: Gramineae (4500 grains/g), Compositae Tubuliflorae (1400 grains/g), *Empetrum* (1140 grains/g) and *Nothofagus* (500 grains/g).

Sample 9 (38 m): Siltstone.
Flora: Gramineae (45%), Compositae Tubuliflorae (16%) and *Acaena* (10%) are the most important elements. Also Umbelliferae (4%), Cruciferae (4%) and Juncaginaceae (4%) are observed. The concentration values are similar to those of sample 10.

Sample 8 (37 m): Sandy silt.
Empetrum (31%), Gramineae (30%) and Compositae Tubuliflorae (12%) are important. Also *Nassauvia* (5%), Chenopodiaceae-Amaranthaceae (5%) and Umbelliferae (4%) can be seen. *Empetrum* reaches its highest concentration value in the profile with 4270 grains/g.

Sample 7 (36 m): Silt.
Flora: Gramineae (70%) and *Empetrum* (12%) are dominant. Also Compositae Tubuliflorae (5%) reach their minimum value in the profile; Rosaceae (4%) are also observed. Gramineae reache 12000 grains/g, the observed maximum concentration.

Sample 1 (29 m): Silt.
Flora: Gramineae (50%), Compositae Tubuliflorae (20%) and *Nothofagus* (10%) are the dominant elements. *Empetrum* (7%), *Nassauvia* (4%) and *Ephedra* (4%) are observed as well. The diversity expressed by the number of taxa is low in this sample. In addition, the concentration values are low as well: For Gramineae, 2300 grains/g and for Compositae Tubuliflorae, 920 grains/g are observed.

16.8. Discussion of the Vegetational History

It was stated above that the upper part of the profile, the first 28 meters, are either barren of palynomorphs or show extremely low contents of palynomorphs. One probable cause could be the chemical oxidation by aerial exposition under drought conditions of the alkaline sediments (Dimbleby 1985). This observation could also be related to a marked continental environment. Horowitz (1992) suggests that in an extremely arid climate the resulting landscape does not permit any kind of vegetational coverage. On the other hand, a maar-depositional environment must be considered. Intensive bacterial activity at the bottom of the water-body could also affect adversely the preservation of the pollen (Horowitz 1992). Beyond palynomorphs the only evidence for Characeae algae is found in sample 21. This might indicate a slightly increased carbonate concentration in the water-body at the time documented by that level (Fernandez 1993).

In the following section the inferred vegetational paleocommunitiesare discussed from bottom to top. The discussion is based on those samples of the profile that contain pollen.

Samples 20 and 19 (56 and 54 m) show a very small variety of taxa. *Nassauvia* and *Ephedra*, representatives of the xeric steppe, appear with quantities less than 1%. So, an environment of gramineous steppe can be deduced by the presence of Gramineae and Compositae Tubuliflorae associated to dwarf shrubby communities of *Empetrum*. The lowest values of pollen concentration and varieties of taxa are recorded at these depths as well as in samples 13 and 1 (43 and 29 m respectively). According to Birks and Birks (1980), the pollen influx decreases under arid or dry climatic conditions. Moreover, with a lowering of the water level, the sediment supply which is caused by erosion of sediments adjacent to the lake increases. Elements transported over long distances are: *Nothofagus*, *Podocarpus*, Myrtaceae, *Schinus*, *Prosopis* and Araliaceae. In steppe surface samples, more than 125 km from the forests (Heusser 1989), the *Nothofagus* values vary between 6 and 9%. In the present profile only samples 19, 10 and 1 (54, 39 and 29 m) are inside that range of values; the rest does not exceed 2%. For those levels we can infer an increase of the forest growth by hydric effectiveness or an intensification of westerly winds. With respect to the planktonic (transported) elements, for this section of the record, the presence of the alga *Botryococcus* would be an indication of lentic conditions such as small ponds or marshes (Fernandez 1993) with little supply of drainage water (Guy-Ohlson 1992). Moreover, the presence of "non marine" dinoflagellates could be interpreted as indicators of "low salinity, nearly fresh, fresh-brackish, limnic-brackish, lacustrine or water strongly influenced by fresh water" (Batten 1989).

Samples 17 and 16 (49 and 48 m) document similar environmental conditions. We see a greater variety of taxa and a notable increase of the concentration values with respect to the base of the profile. After Birks and Birks (1980), this could indicate an increase of the vegetational cover. The gramineous steppe continues with Gramineae and Compositae Tubuliflorae accompanied by various pollen types such as *Acaena*, Valerianaceae, Umbelliferae, Caryophyllaceae, Ranunculaceae, Papilionaceae, Rubiaceae, Solanaceae, Compositae Liguliflorae; taxa that are found nowadays in zones with more humidity than the semideserts (Mancini and Trivi 1992; Pisano 1972).

The *Acaena* peaks accompany the increase of aquatic plants such as Cyperaceae, Juncaginaceae together with the retreat of the halophytes such as Chenopodiaceae-Amaranthaceae and vice versa. *Acaena* is mentioned in meadows together with plants of palustrine and littoral habits (D'Antoni 1978). This behaviour reflects a positive variation of the local hydric balance with a greater extension of the palustrine environments at the expense of the saline ones (D'Antoni 1978) and vice versa.

Empetrum continues to represent the shrubs within the humid steppe. *Nothofagus* shows a significant increase in the concentration values. In relation to the planktonic elements, other types of dinoflagellates and acritarchs appear; the presence of *Botryococcus* continues.

Samples 15, 14 and 13 (47, 46 and 43 m) record a gramineous steppe environment but with elements and indicators that point to a greater aridity as: *Nassauvia*, a reduced amount of *Empetrum* and *Acaena*, the presence of Umbelliferae (cf. *Azorella*) and the disappearance of the pollen types indicative of greater humidity observed in former levels. The halophytic plants increased, while aquatic elements decreased compared to the underlying samples. *Botryococcus* is still present. Dinoflagellates are not observed.

In samples 10, 9, 8 and 7 (39 - 36 m) the highest values of *Empetrum* appear, except in sample 9 where the value decreases to less than 1%. A gramineous steppe-environment accompanied by dwarf shrubby specimens (*Empetrum*) under conditions of increased relative-humidity can be inferred.

The elements of the humid steppe such as: *Acaena*, Umbelliferae, Valerianaceae, Papilionaceae, Caryophyllaceae and Solanaceae continue to be present.

Nassauvia and *Ephedra*, characteristic for the xeric steppes, are not important.

If we consider the planktonic elements, samples 9 and 7 present the greatest variety of "non marine" dinoflagellates in this profile. The presence of *Pediastrum* could indicate pluvial conditions and greater eutrophication (Fernandez 1993).

In sample 1 (29 m) we can no longer observe the pollen types that characterize the humid environments of the gramineous steppe. Gramineae and Compositae Tubuliflorae together with *Nassauvia* and *Ephedra*, coexisting with small amounts of *Empetrum* could indicate less humid conditions than observed in the previous samples. The presence of *Nothofagus*, with the highest value in the profile, may indicate an increase of westerly winds more than an increase of precipitations.

The observation of *Zygnemataceaen* algae could be a sign of stagnant, shallow, mesotrophic, oxygen-rich waters (Van Geel 1979). The development of these algae is induced by an increase of the temperature when the habitat is characterized by an advanced process of desiccation (Ellis and Van Geel 1978). *Botryococcus* could also indicate a shallow water body in low superficial-draining areas. We observe also the presence of acritarchs.

16.9. Conclusions

The section under study could indicate pulses of higher and lower relative humidity.

We can deduce marked drought conditions for the deeper samples located between 59 and 53 m depth (samples 21, 20, 19, 18), compared to higher relative humidity found at 49 and 48 m depth (samples 17, 16). Between 47 and 40 m depth (samples 15, 14, 13, 12, 11) a lower relative humidity is observed. The interval between 39 and 36 m reflects again higher relative humidity (samples 10, 9, 8, 7). Finally, between the depths 35 to 29 m another pulse of drought is observed (samples 6, 5, 4, 3, 2, 1).

References

Altevogt G (1969) Der postglaziale Vulkanismus südlich von Rio Gallegos, Provinz Santa Cruz, Süd-Argentinien. Münster Forsch. Geol. Paläont. 12:3-15

Batten D (1989) Cretaceous freshwater dinoflagellates. Cretaceous Research 10:271-273

Bianchi M, D'Antoni H (1986) Depositación del polen actual en los alrededores de Sierra de los Padres (Prov. de Buenos Aires). VI Congreso Argentino de Paleontología y Bioestratigrafía. Actas 16-27. Mendoza, Argentina

Birks B, Birks H (1980) Quaternary Palaeoecology. Arnold Pub. Ltd., London

Caldenius CC (1932) Las glaciaciones cuaternarias de la Patagonia y Tierra del Fuego. Ministerio de Agricultura, Pub. 95, Buenos Aires, Argentina

Centro Editor de America Latina (1981) Atlas Total de la Republica Argentina. Vientos. Tomo 1 Fasc. 14:216-221. Buenos Aires, Argentina

Chelotti L, Trinchero E (1990) Cuerpos intrusivos subvolcánicos en la Cuenca Austral. Boletín de Informaciones Petroleras 7(3):2-13, Argentina

Codignotto JO (1975) Geología y rasgos geomorfológicos de la Patagonia Austral extraandina, entre el Río Chico de Gallegos (Santa Cruz) y la Bahía de San Sebastián (Tierra del Fuego). Dissertation, University of Buenos Aires, Argentina

Corbella H, Pomposiello MC, Malagnino E, Trinchero E, Alonso MS, Chelotti L, Firpo L (1990) Volcanismo lávico y freatomagmático post-glacial asociado al campo de fracturación Austral, Provincia de Santa Cruz, Argentina. Patagonia Extrandina. 10° Congreso Geológico Argentino, San Juan, Argentina. Actas I:39-42

Corbella H, Pomposiello MC, Chelotti L, Trinchero E, Alonso MS (1991) Cuerpos hipoabisales asociados al volcanismo efusivo cuaternario de la Patagonia Extrandina Austral, Santa Cruz, Argentina. 6° Congreso Geológico Chileno Resumenes 1:510-514

Corbella H (1993) Determinación de las condiciones paleoambientales del Cuaternario del extremo meridional del Continente Sudamericano mediante el estudio de los sedimentos fini-postglaciares preservados en diatremas volcánicas. Workshop Project 341 IGCP/IUGS/UNESCO: Southern Hemisphere Paleo and Neoclimates: A review of the state of the art. Mendoza, Argentina

D'Antoni H (1978) Palinología del perfil del Alero del Cañadón de las Manos Pintadas (Las Pulgas, Provincia de Chubut). Relaciones de la Sociedad Argentina de Antropología. XII:249-262, Buenos Aires, Argentina

Dimbleby GW (1985) The Palynology of archaeological sites. New Phytology 56:12-28

Ellis A, Van Geel B (1978) Fossil Zygospores of Debarya Glyptosperma (De Bary) Wittr. (Zygnemataceae) in Holocene Sandy Soils. Acta Botanica Neerlandica, 27(5-6):389-396.

Fernandez C (1993) Fungal palynomorph and algae from Holocene bottom sediments of Chascomús Lake, Buenos Aires Province, Argentina. Palynology 17:187-200, Texas, USA

Guy-Ohlson D (1992) Botryococcus as an aid in the interpretation of palaeoenvironment and depositional processes. Review of Palaeobotany and Palynology 71:1-15, Amsterdam

Heusser C (1971) Pollen and Spores of Chile. University Arizona Press, Tucson, USA

Heusser C (1989) Late Quaternary vegetation and climate of southern Tierra del Fuego. Quaternary Research 31:396-406

Heusser L, Stock C (1984) Preparation techniques for concentrating pollen from marine sediments and other sediments with low pollen density. Palynology 8:225-227

Hooghiemstra H (1984) Vegetational and climatic history of the High Plain of Bogotá, Colombia: a continuous record of the last 3.5 millions years. Ed. Cramer, Germany

Horowitz A (1992) Palynology of the arid lands. Elsevier, Amsterdam

Hüeck-Seibert P (1981) Mapa de la vegetación de América del Sur. Gustav Fischer, Stuttgart, Germany

Mancini M (1989) Depositación del polen actual en el sur de Santa Cruz. Dissertation, Universidad Nacional de Mar del Plata, Argentina

Mancini M (1993) Recent pollen spectra from forest and steppe of South Argentina: A comparison with vegetation and climate data. Review of Palaeobotany and Palynology 77:129-142

Mancini M, Trivi M (1992) Búsqueda de análogos modernos en el sistema polen del Alero Cárdenas (Provincia de Santa Cruz), VIII Simposio Argentino de Paleobotánica y Palinologia: 81-84, Buenos Aires, Argentina

Markgraf V, D'Antoni H (1978) Pollen flora of Argentina. University of Arizona Press, Tucson

Mercer JH (1976) Glacial History of Southernmost South America. Quaternary Research 6:125-166

Morbelli M (1980) Morfología de las esporas de Pteridophyta presentes en la región Fuego-Patagónica, República Argentina. Opera Lilloana, Tucuman, Argentina

Pisano E (1972) Algunos resultados botánicos de la II Expedición Neozelandesa al Hielo Nor-Patagónico 1971/72. Anales del Instituto de la Patagonia. III(1-2)131-160. Punta Arenas, Chile

Pomposiello MC, Corbella H, Alonso MS, Diaz MT, Trinchero E, Chelotti L (1990) Estudio gravimétrico del volcanismo basáltico posglacial, Patagonia Extrandina Austral. Provincia de Santa Cruz, Argentina. 16a Reunión Científica de Geofísica y Geodesia, Bahía Blanca Argentina

Pomposiello MC, Corbella H, Gonzalez M, Chelotti L, Trinchero E (1991) Gravimetric and Seismic Studies of Magallanes Maar, Southern Extrandean Patagonia, Argentina. 2nd International Congress of the Brazilian Geophysical Society. Salvador, Bahia, Brazil. Actas 413-417

Pomposiello C, Corbella H, Chelotti L, Alonso S (1995) Geophysical and Geological Study of the Austral Fracture Belt at the Atlantic Passive Margin, Southern Patagonia, Argentina. International Union of Geodesy and Geophysics. XXI General Assembly. Boulder, Colorado

Skewes MA (1978) Geología, petrología, quimismo y origen de los volcanes del área de Pali-Aike, Magallanes, Chile. Anales del Instituto de la Patagonia 9:95-106, Punta Arenas, Chile

Skewes MA, Stern CR (1979) Petrology and Geochemistry of alkali basalts and ultramafic inclusions from Pali-Aike volcanic field in southern Chile and the origin of the Patagonian plateau lavas. Journal of Volc. and Geoth. Res. 6:3-25

Van Geel B (1979) Preliminary report on the history of Zygnemataceae and the use of their spores as ecological markers. Proc. IV Inter. Palynological Confer. I:467-469. Lucknow, India

Sedimentary Cores from Mascardi Lake, Argentina: A Key Site to Study Elpalafquen Paleolake

R.A. del Valle[1], J.M. Lirio[1], H.J. Nuñez[1], A. Tatur[2], C.A. Rinaldi[1]
(1) Instituto Antártico Argentino, Cerrito 1248, 1010 Buenos Aires, Argentina.
Liriojm@yahoo.com
(2) Institute of Ecology, Polish Academy of Sciences. Dziekanow Lesny k. Warszawy, 05–092 Lomianky, Poland.

Abstract: Sedimentary cores from Mascardi Lake, as well as outcrops, have been studied to reconstruct late Pleistocene and Holocene environmental conditions in northern Patagonia, Argentina. The Mascardi Lake sequence is a key-site for understanding such conditions, providing evidences of ice retreat, volcanic activity and important sedimentation changes. A significant environmental change occurred around 13 ky BP when the great lake named Elpalafquen became several small basins and some of the present lakes from northern Patagonia started to get their current features. The Mascardi Lake occupied a marginal position, close to the ice front at the western side of the paleolake.

17.1. Introduction

During the Late Pleistocene, a large amount of meltwater accumulated in a proglacial basin located in the southern Manitoba region, Canada, at the edge of the receding North American ice cap, forming a great water-body named "Agassiz Lake" (Broecker *et al.* 1985, 1989). At nearly the same time, a great South American lake named "Elpalafquen" (del Valle *et al.* 1993a, 1993c) covered the area of most of the lakes currently existing in the major part of Nahuel Huapi and Lanín National Parks of Argentina (71° W, 41° S). Both lacustrine and peat bog sediments slowly accumulated within this large basin during the last 18 ky (thousand years). This area, located in the northwestern Patagonian region are subject to detailed studies, the Lagos–Comahue Project, focusing on Quaternary paleo-environmental and paleoclimatic changes; several relevant geological and biological methods have been integrated (Amos et al. 1993, Galloway et al. 1988).

The study area was covered with ice during the late Pleistocene with terminal moraines about 40 km north of San Carlos de Bariloche in Argentina (Flint and Fidalgo 1969). Elpalafquen Lake was formed during the ice retreat and its southwestern marginal side became the Mascardi Lake, in the northwest of the Río Negro Province, Argentina. The 200 m deep Mascardi Lake covers an area of about 25 km^2 and is at 760 m asl (above sea level) within a glacial valley that descends from Mt. Tronador, the highest (3554 m asl) and still glaciated mountain in the region. The mean annual precipitation in the region is about 1500 mm with a mean annual temperature of about 7^0 C.

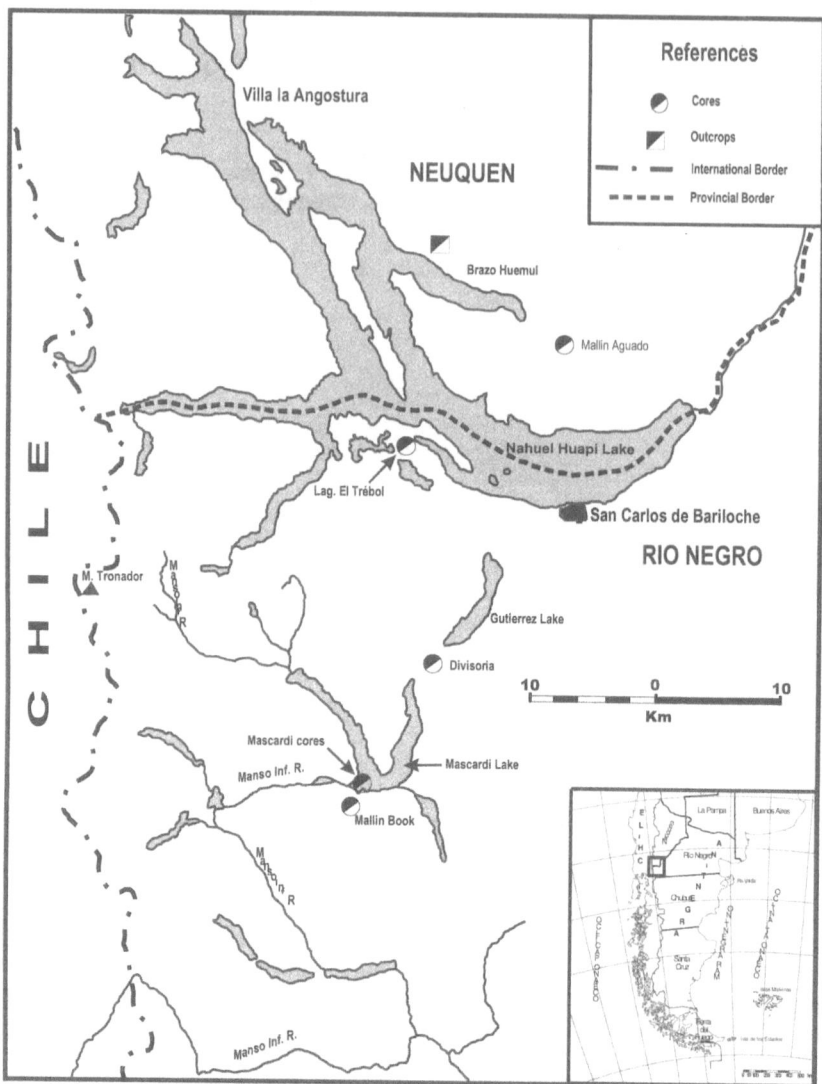

Fig. 17.1. Mascardi Lake and adjacent outcrops.

The vegetation in the narrow valley bottom is largely composed of mixed *Nothofagus dombeyi / N. pumilio* and *N. dombeyi / Austrocedrus chilensis* forest, while the slopes above 1000 m asl are unmixed *N. pumilio* forests up to the treeline at about 1,500 m asl, where also dwarf *N. antarcticus* occurs.

During the 1992–1998 period, the data recorded at Mascardi Lake Weather Station (Fig. 17.1) show strong seasonal variations in temperatures and precipitation, peaking in summer and autumn–winter respectively, even during "El Niño" episodes during 1994–1995 and 1997–1998 (Villarosa et al. 1999).

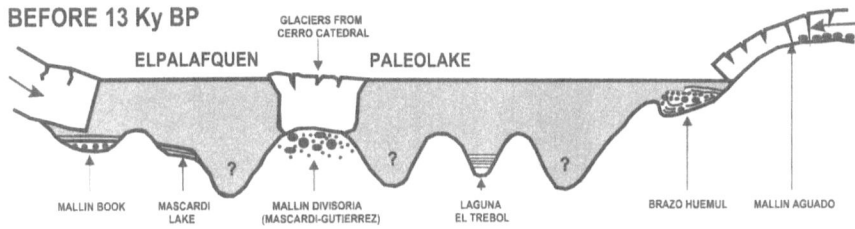

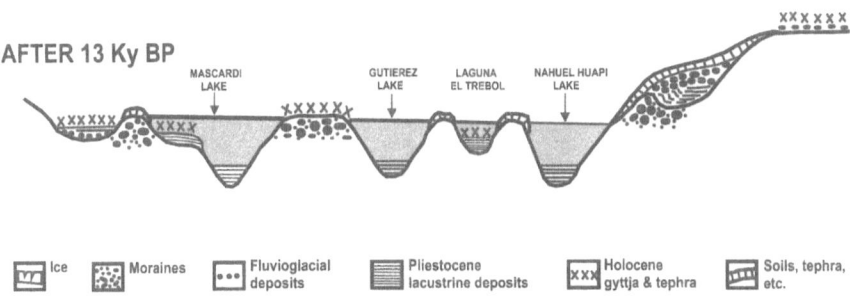

Fig. 17.2. Sedimentary environments in the study area before and after 13 ky BP.

To study current sedimentary processes in detail, a five-year experiment (1993–1998) was carried out. Trypton samplers were installed and checked monthly close to the bottom and near the surface in Mascardi Lake at different distances from the main input of glacial sediments (Manso Superior River, Fig. 17.1). The results show a direct correlation between the accumulation rate with temperature peaks in samplers located near the lake surface, and with precipitations peaks (autumn–winter) in samplers near the bottom (Villarosa et al. 1999). Additionally, a clear inverse relation exists between the accumulation rate and the distance to the mouth of Manso Superior River (Vallverdú et al. 1994). The high rates of clastic sediment input in the lake correlate with periods of high water runoff instead of melt glacier waters. Normally the amount of sediments collected during the precipitation period is about 400% higher than the rest of the year, but during "El Niño" episodes this relationship riches up to 800% (Villarosa et al. 1999).

Sedimentary cores were obtained in 1994 from Mascardi Lake, near the mouth of the outflowing stream Manso Inferior River (Fig. 17.1). Two piston cores with lengths 475 and 455 cm were photographed and visually correlated. Photographic close-ups of cores were enlarged into 13.5 x 23.5 cm prints and used to count varves and measure their thickness using clay layers to define varve boundaries. Additional comparative measurements using X-ray images of cores were done.

17.2. Description of Sediments

There are three different lithological units recorded in Mascardi Lake cores labeled units 1–3 from the top downwards (Figs. 17.3 and 17.4).

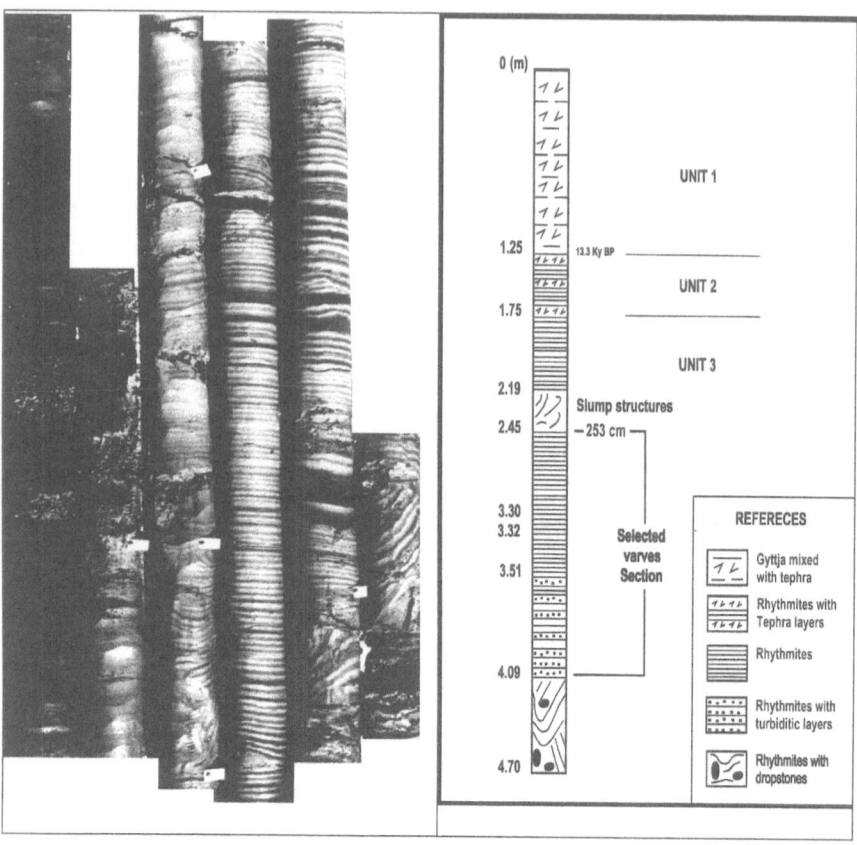

Fig. 17.3. (left) and 17.4. (right): A core of Mascardi Lake. Photograph (left) and geological interpretation (right).

Unit 1 (0–125 cm): Pyroclastic layers ("tephra") interbedded with lacustrine organic-rich sediments ("gyttja" with an organic matter content of about 7%) were deposited during the uppermost Pleistocene and the Holocene. The radiocarbon age of the lower limit of unit 1 is 13 340±160 yr BP (Lund University, Sweden). Lacustrine sediments are mainly formed by brown- to greenish-gray silty clay mixed with reworked diatoms and pyroclastic fragments.

Unit 2 (125–175 cm): This unit consists mainly of greenish-gray clastic varves, poor in organic matter (< 2%). The varve sequence is composed of about 500 couplets formed by very thin (about 1 mm/couplet) alternating silt/clay laminae with an estimated accumulation rate of 1 mm/yr. Only normal grading is evident; wave and current structures are completely absent and no significant internal hiatuses along the sequence were recognized.

Unit 3 (175–470 cm): This unit consists of thin and regularly bedded lacustrine clastic rhythmites (Fig. 17.5) of pinkish–brown to light-gray color. No current or wave structures are present. Sediments are formed by regularly alternating thin silty/clay (light/dark) beds of clastic varved, their mean thickness range from 3.5 (dark layers) to 4.5 mm (light layers). Couplets consist of silty layers overlain by clay layers. Very often the entire couplet appears graded, a sharp contact normally separates successive couplets and occasionally even inner sub-lamination is present inside some isolated couplets (Fig. 17.6). A preliminary estimate of the accumulation rate using varve counting is about 7 mm/yr.

Grain size is dominated by the clay fraction in both light and dark layers, although the silt amount is higher in light layers where variable amounts (±1.00–1.25%) of sand grains are also present. Quartz and lithic fragments dominate the mineralogical composition of coarse-grained clasts (sandy grains to pebbles). Very-fine sand grains are scarce in dark layers where mica flakes dominate.

Grain-size analyses show, from bottom to top of unit 3 (Fig.17.7), a nearly continuous increase of the clay fraction from 20 to 65%, together with a decrease of silt and sand fractions (80 to 35%). The high amount of sand grains and granules (30–20%) in the lower part are closely related to ice-rafting transport and/or turbidite processes, which occasionally dominate the light-gray layers. Turbidite processes were identified at the lower part of unit 3, where fining-upward thin (4.5 mm) light-gray layers contain large amounts (20–30%) of angular coarse sand grains and granules.

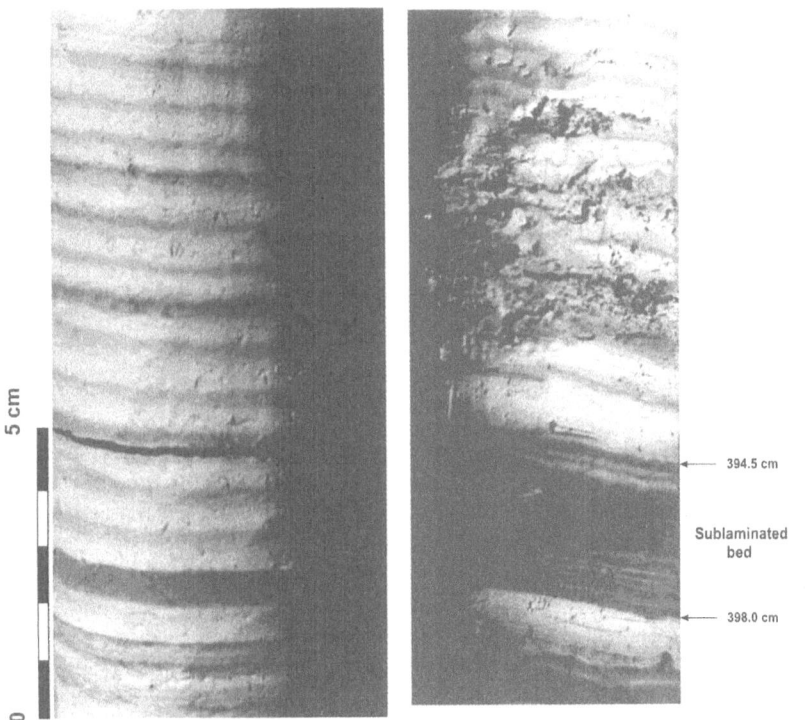

Fig. 17.5. Typical varves in unit 3.   **Fig. 17.6.** Typical sublaminae in unit 3.

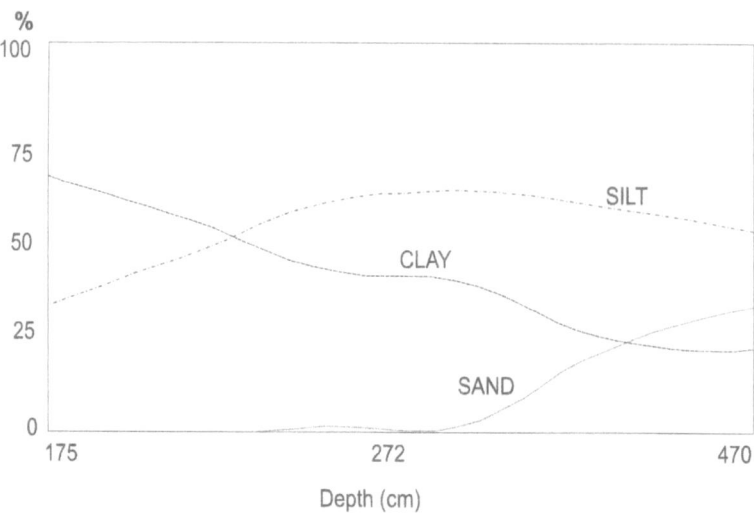

Fig. 17.7. Grain-size distribution in unit 3.

One additional process of sediments transported in cold waters could explain the presence of fine gravely levels within varve sequences without involving ice-rafting transport. The floating, sometime for several kilometers, of granules and small pebble-size sediments in the water surface suspended by superficial tension, has been recently studied in some Patagonian lakes and Antarctic marine waters (del Valle et al. 1993b, Lirio et al. 1994, Möller and Ingólfsson 1994).

Bioturbation is abundant in both light- and dark-gray layers. Irregular and inclined small (0.3 mm in diameter) empty burrows meander in all directions within the silty layers, mainly horizontal in the upper part of clay layers. The very low content of organic matter (<1%) prevented accurate radiocarbon analyses. In spite of this fact, unit 3 is assigned to the Late Pleistocene based on radiocarbon dating, varve counting and correlation with an equivalent sequence (Fig. 17.8) obtained at the Mallín Book site (del Valle et al. 1993c).

17.3. Interpretation of Varves

Following the ideas of several authors (Bogen 1988; Church and Slaymaker 1989; Anderson 1978), the clastic varves may allow a high-resolution reconstruction of glacier changes due to good correlation existing between the glacial activity in the catchment area and the sedimentation rates in the lacustrine basin. Coring in a recent artificial lake located in Muenster, NW Germany (6° W, 52° N) showed about 150 varves, documenting the last 60 years (Smolka, pers. comm.). In such cases one calendar-year might under certain circumstances (open-water sedimentation versus ice-covered sedimentation alternating more than one time per year) show up as two varves per year. The classical varve-theory might be used for dating for such times and in such environments that are characterized by one clearly defined summer and one clearly defined winter: we assume this situation for Mascardi Lake.

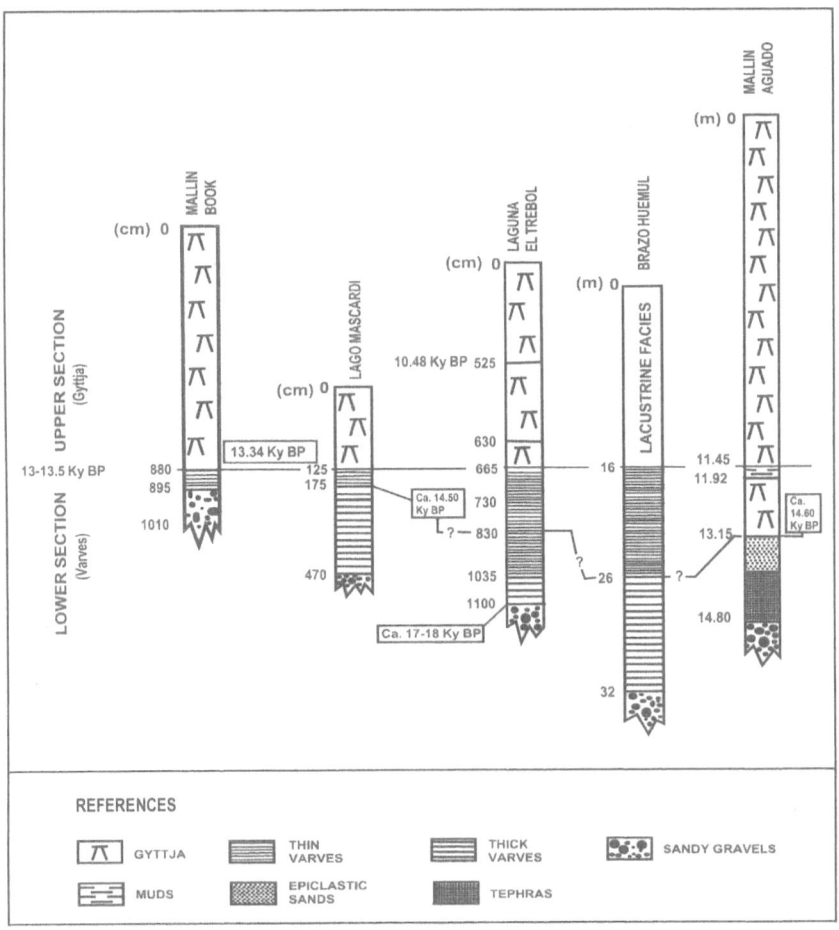

Fig. 17.8. Correlation of wells and sections.

Under present weather conditions, summer periods in Mascardi Lake region are mainly characterized by significant ice melting and lack of precipitation. The high precipitation and runoff rates are prime features of autumn–winter periods, when major floods occur after large rainfalls on fresh snow (S. Rubulis pers. comm.). These climatic and hydrologic features are applied to varve interpretation; the authors use the term winter as a cold-and-rainy period represented by light colored layers, and summer as a warm-and-dry period during which dark layers form.

The silt/clay contact is usually sharp (Fig. 17.5); tiny silty laminae occasionally occur in dark (summer) clay layers. The same applies to isolated clay laminae in the light (winter) silt layers (Fig. 17.6). This tiny internal sublamination is sporadically present in unit 3 at 275.0–280.5 cm, 355.0–382.0 cm and 394.5–400.5 cm. The tiny sublaminae's small size hinders the normal determination of couplet limits. Therefore, traditional varve-counting methods cannot be easily applied to the generation of a time series describing these sediments.

A preliminary attempt to convert a part of the late Pleistocene sequence into a time series is proposed by Compagnucci et al. (1993). For statistical studies, a series of well-preserved rhythmites within unit 3 was selected. This series, located between 253–409 cm (Fig. 17.4), is formed of 191 couplets consisting of alternating dark and light layers and ranging in thickness between 0.7–37.4 mm (mean: 3.6 mm) and 1.4–14.1 mm (mean: 4.7 mm) respectively (Table 17.1).

The standard deviation, skewness and variance in the series of light (winter) layers are lower than in the series of dark (summer) layers. Following the ideas proposed by Smith and Ashley (1985, 1988), usually, at a given site, the thickness of silty layers varies but the thickness of clay layers is relatively constant throughout the basin; the thickness of silty layers depends on the effectiveness of sediment dispersal mechanisms, particularly underflows, whereas the clay thickness is basically a function of settling time and basin depth. In the sequence of Mascardi Lake, an inverse relationship is shown; the thickness variations of winter layers are low, whereas those of the summer layers are high.

Table 17.1. Statistics of varve thickness.

Parameter	Light layers	Dark layers	Couple
Mean (mm)	4.66	3.66	8.29
Median (mm)	4.31	2.87	7.21
Mode (mm)	3.50	0.70	7.00
Standard deviation	1.77	3.58	4.50
Variance	3.15	12.82	20.28
Kurtosis	8.12	43.70	42.44
Skewness	2.05	5.48	5.20
Range	12.73	36.75	48.75
Minimum (mm)	1.40	0.70	2.10
Maximum (mm)	14.13	37.45	50.85
Number of layers	191	191	191

The variable thickness of the clay layers could mainly be caused by different settling times under variable summer conditions. In addition, there are no evidences of rapid oscillations of basin depth during the time represented by varve accumulation (about 200 yr). This shows that the concept of constant background clay sedimentation cannot be applied easily to Mascardi Lake.

According Kuenen (1951), the varve coarser units are formed by turbidity currents originating from ice melting in the summer (without producing erosion in the underlying layers), but in Mascardi Lake they are formed in autumn–winter, when precipitation peaks occur.

Agterberg and Banerjee (1969) established a semi-quantitative model for varve deposition in relation to the distance to ice front. The strong variations in the thickness of varves in units 2 and 3 (Figs. 17.3 and 17.4), and especially within the selected section, could suggest rapid oscillations of the ice front.

Dropstones occur only in the lower part of section 3 (Fig. 17.3), where varves are thicker. Additionally, the amount of clay increases upward (Fig. 17.7), at the same time sand decreases, reflecting the glacial retreat.

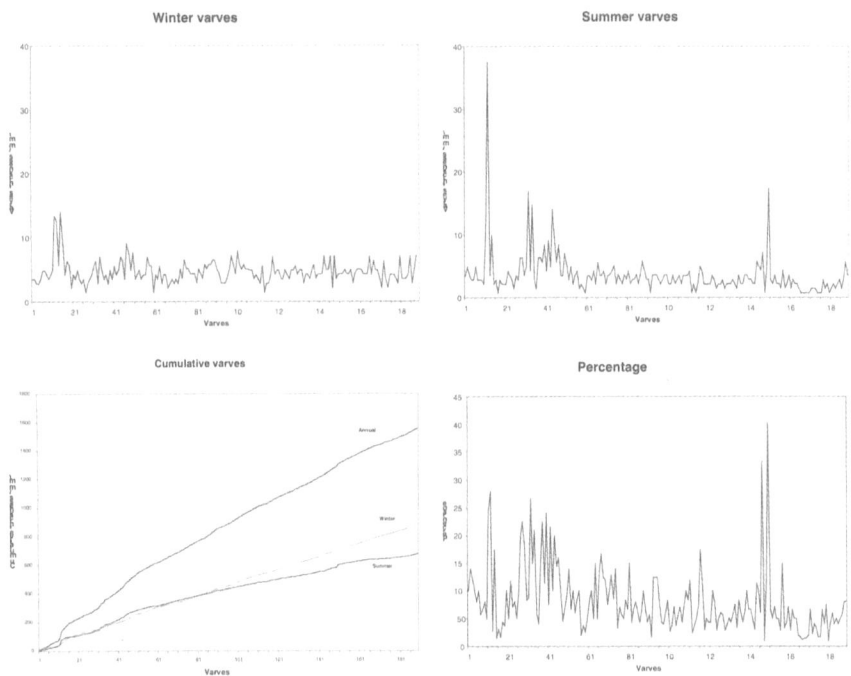

Fig. 17.9. Varve statistics. Explanation in the text.

The cumulative curves of the growth of varves (summer, winter and annual) analyzed between 2.53–4.09 cm (Fig. 17.9) show that the "winter" unit is more regular resembling a linear function and the "summer" unit presents two slopes with an inflection point at about varve number 50. The annual cumulative curve shows some significant similarities with cumulative curves presented by Sylwan (1988) for the Corintos River and Buenos Aires Lake, located at the eastern side of the Andes. The curves reflect a typical ice retreat that took place 13.5–12.5 ky BP (Sylwan 1988). The Mascardi Lake sequence at 2.53–4.09 cm (13.8–14 ky BP) could represent slow ice retreat accompanied by moderate deglaciation conditions.

In this context it should be noted that conspicuous tephra layers are virtually absent in unit 3. This is one of the important differences between layers formed at the marginal zone of Elpalafquen paleolake (i.e. Mascardi Lake area) and those formed at the central part of it (Fig. 17.8; Laguna El Trébol and Brazo Huemul), where tephra layers are commonly interbeded with thinner varves. It is assumed that the ice-cover precluded the local accumulation of tephra layers in the studied area. The lack of significant tephra layers and the presence of thick varve couplets suggest that Mascardi Lake occupied a marginal position, close to the ice front at the western side of the paleolake (Fig. 17.2).

17.4. Results and Conclusions

The presence of varves, dropstones and a rather high accumulation rate of the clay-sized clastic fraction in unit 3, and geomorphologic features found in the region are taken as evidence for glacial activity in the catchment area. The thickness of varve-couplets accumulated in the Elpalafquen Lake reflects the annual accumulation rate determined by interaction of climatic, glacial, fluvial and lacustrine processes (see also Hallet 1979).

The main lithological change (rhythmites/gyttja) in Mascardi Lake, dated at about 13 ky BP is located at the boundary between unit 2 and unit 1 (125 cm) and records a significant late Pleistocene environmental change. This dating correlates with the peat/lacustrine varves boundary (880 cm) in Mallín Book (Figs. 17.8) and confirms early dating at 12.5 ky by Markgraf (1983). She suggested that at this time the glacier had already receded by about half of its total length and reduced its volume considerably.

The continuity of the record in unit 1 (Figs. 17.3 and 17.4; 0–125 cm) implies that there was no significant glacier advance after 13 ky BP and correlates well with the so-called "early deglaciation" found at several places in Argentina and Chile; this event never occurred later than 12.5 ky BP (Mercer 1976, 1982). According to previous results obtained at Mallín Book (Markgraf 1983) the steppe environment expanded considerably westward into the Andean valleys during the lastglacial times, and an arid period occurred (annual precipitation < 500 mm). Before 13 ky BP clastic rhythmites were deposited irregularly in Mascardi Lake throughout the year during a time-span of at least 1000 years, as deduced from the varves of units 2 and 3. During this time no significant breaks in the lacustrine sedimentation occurred.

The strong reworking shown by most of the clasts forming unit 1 together with the practical absence of lamination in gyttja layers strongly contrasts with the delicate rhythmic lamination preserved in the other units. This difference indicates that the water level of the Elpalafquen Lake was tens of meters higher than the Mascardi Lake before 13 ky BP; this is due to the fact that the fine laminated rhythmites were protected from waves action. The authors on Elpalafquen's past shorelines verified this high stand.

The great paleolake broke up around 13 ky BP and the present lakes at northern Patagonia started to get its current shape. Intensive waves action prevented the accumulation of clay in the shallow areas during most of the Holocene and latest Pleistocene. In Mascardi Lake the complete reworking of sand/silt grains and the removal of the clay fraction resulted in a 125 cm thick condensed-section (unit 1).

Variations of varve thickness recorded in the Mascardi Lake sequence reflect variations in the glacier front and the influence of precipitations on northwestern Patagonia during late Pleistocene. The Mascardi Lake sequence provides valuable evidence on the climatic variations to be correlated with past and present climate changes at regional scale, for example those originating El Niño events.

Overall the Mascardi Lake sequence, in connection with other localities, is one of the key-sites for understanding Late Pleistocene and Holocene environmental conditions in northern Patagonia.

References

Agterberg FP, Banerjee Y (1969) Stochastic model for the deposition of varves in glacial Lake Barlow–Ojibway, Ontario, Canada. Can. J. Earth Sci. 6:625–652

Amos AJ, Bianchi MM, Cusminsky GC, Masaferro JI, Roman Ross G, del Valle RA (1993) Nuevas evidencias paleoambientales post-Pleistocenas en lagos y mallines del área de San Carlos de Bariloche (Lat. 41°00'Sur), República Argentina. Submitted to: Symposium "El Cuaternario de Chile y Paleoclimas Cuaternarios de América del Sur", Santiago de Chile

Anderson LW (1978) Cirque glacier erosion rates and characteristics of Neoglacial tills, Pangnirtung Fjord area, Baffin Island. N.W.T., Canada Arctic Alp. Res. 10:749–60

Bogen J (1988) A monitoring programme of sediment transport in Norwegian rivers. IAHS Publ. 174:149–59

Broecker WS, Rind D, Peteet D (1985) Does the ocean–atmosphere system have more than one stable mode of operation? Nature 315:21–25

Broecker WS, Kennet JP, Flower BP, Teller JT, Trumbore S, Bogani G, Wolf W (1989) Routing of meltwater from the Lawrentide Ice Sheet during the Younger Dryas cold episode. Nature 341:318–321

Church M, Slaymaker O (1989) Desequilibrium of Holocene sediment yield in glaciated British Columbia. Nature 337:452–54

Compagnucci RH, Giraldez AE, del Valle RA (1993) Climatic interpretation from Mascardi Lake cyclic deposits of the Late Pleistocene. Submitted to: Workshop Project 341 IGCP/IUGS/UNESCO, Southern Hemisphere Paleo- and Neoclimates: Time series analysis and Paleoclimate, Mendoza

del Valle RA, Tatur A, Amos A, Ariztegui D, Bianchi MM, Cusminsky G, Hsü K, Lirio JM, Martinez Macchiavello JC, Massaferro J, Nuñez HJ, Rinaldi CA, Vallverdú R, Vigna S, Vobis G, Whatley RC (1993a) Elpalafquen, un paleolago de la Patagonia septentrional Andina durante el Pleistoceno tardio. Proyecto PANGEA GLOPALS (IGCP Project 324) Comunicaciones, San Juan:12–15

del Valle RA, Tatur A, Amos A, Ariztegui D, Bianchi MM, Cusminsky G, Hsü K, Lirio JM, Martinez Macchiavello JC, Massaferro J, Nuñez HJ, Rinaldi CA, Vallverdú R, Vigna S, Vobis G, Whatley RC (1993b) Laguna Cari Laufquen Grande: Registro de una fase climática húmeda del Pleistoceno tardío en la Patagonia septentrional. Proyecto PANGEA GLOPALS (IGCP Project 324) Comunicaciones, San Juan:16–19

del Valle RA, Tatur A, Amos A, Ariztegui D, Bianchi MM, Cusminsky G, Hsü K, Lirio JM, Martinez Macchiavello JC, Massaferro J, Nuñez HJ, Rinaldi CA, Vallverdú R, Vigna S, Vobis G, Whatley RC (1993c) Late Pleistocene–Holocene sedimentary core from Mascardi Lake, Nahuel Huapi National Park, Argentina. Workshop Project 341 IGCP/IUGS/UNESCO, Southern Hemisphere Paleo- and Neoclimates: Quaternary Climates, Mendoza

Flint RF, Fidalgo F (1969) Glacial drift in the eastern Argentina Andes between 41 degrees 10' S. and latitude 43 degrees 10 ' S. Geol. Soc. Am. Bull 80(6), 1043-1052

Galloway RW, Markgraf V, Bradbury JP (1988) Dating shorelines of lakes in Patagonia, Argentina. Journal of South American Earth Sciences, Vol.1, 2:195–198

Hallet B (1979) A theoretical model of glacial abrasion. J. Glaciology 31:108–14

Kuenen PH (1951) Mechanics of varve formation and the action of turbidity currents. Geol. Foren. Forhandl. (Stockholm), H1:69–84

Lirio JM, del Valle RA, Nuñez HJ and Lusky JC (1994) Transporte de grava fina por tensión superficial en el mar de Weddell y lagos de Tierra del Fuego. In 3ras Jornadas de Comunicaciones sobre Investigaciones Antárticas, Dirección Nacional del Antártico–Instituto Antártico Argentino, Buenos Aires: 111–112

Markgraf V (1983) Late and postglacial vegetational and paleoclimatic changes in sub-Antarctic, temperate and arid environments in Argentina. Palynology 7:43–70

Mercer JH (1976) Glacial history of southernmost South America. Quaternary Research 6:125–166

Mercer JH (1982) Holocene glacier variations in southern South America. Striae, 18:35–40

Möller P and Ingólfsson Ó. (1994) Gravel and sand flotation: a sediment dispersal process important in certain nearshore marine environments. Journal of Sedimentary Research, Vol. A64, No. 4: 894–898

Smith ND, Ashley G (1985) Proglacial lacustrine environment. In: Ashley G, Show J, Smith ND (eds) Glacial sedimentary environments. Soc. of Econ. Paleont. and Mineralogists Short Course No.16:165–216

Sylwan CA (1988) Patagonian pleistocene glacial varves: an analysis using variation of their thickness. Quaternary of South America and Antarctic Peninsula 6:147–265

Vallverdú RA, Lusky JC and del Valle RA (1994) Presentación preliminar de datos de sedimentación actual en lago Mascardi. *In* Terceras Jornadas de Comunicaciones sobre investigaciones Antárticas (CA Rinaldi, Ed.), Dirección Nacional del Antártico, Buenos Aires: 117–125.

Villarosa G, Massaferro J, Amos AJ and del Valle RA (1999) Variaciones estacionales en las tasas de sedimentación como indicadores de episodios El Niño en el lago Mascardi, Río Negro. *In:* I Congreso Argentino del Cuaternario y Geomorfología, Santa Rosa, La Pampa, Argentina, 27–29 March 1999 (Submitted)

Quaternary Glacial Sequence in the Río Mendoza Valley, Argentina

L. E. Espizua
Instituto Argentino de Nivología, Glaciología y Ciencias Ambientales (IANIGLA–CRICYT), Casilla de Correo 330, 5500 Mendoza, Argentina
lespizua@lab.cricyt.edu.ar

Abstract: As a contribution to data for reconstructing the climatic changes on a S–N transect from Antarctica to Bolivia, a glacial–geological study is presented that focuses on the Río Mendoza valley of the central Andes of Argentina. This valley contains the Cerro Aconcagua (6959 m), the highest peak in the Western Hemisphere. In the Río Mendoza valley, five Pleistocene drifts and one Holocene drift were distinguished by multiple relative-age criteria. U-series ages of travertine layers and fission-track on glass ages of three tephra layers permit a preliminary chronology. Also, an estimate of the magnitude of the Pleistocene snowline depression was obtained, contributing paleoclimatic information. The moraine sequence in the Argentine flank was compared with that studied by Caviedes (1972) along the Río Aconcagua on the Chilean side of the Andes. In addition, the Pleistocene glacial sequence in the Río Mendoza valley was correlated preliminarily with that studied by Espizua (1998) along the Río Grande valley in the southwest of the Mendoza Province.

18.1. Introduction

The aim of this study was to obtain a Quaternary glacial sequence and to constrain drift ages in the upper valley of the Río Mendoza, 32° 45' S. Evidence of repeated alpine glaciation in this valley is well preserved but the record has not been well studied. This sector of the Andes range, culminating in Cerro Aconcagua (6959 m), contains peaks up to 5500 m. Drifts were differentiated and correlated on the basis of relative-age criteria. The most useful were the degree of boulder weathering, development of rock varnish, moraine and terrace relationships, soil-profile development, topographic position of end moraines, distal slope angles of ablation moraines, stratigraphic relations and presence of glacially abraded rock surfaces. Some absolute ages permit the inference of a preliminary chronology (Bengochea et al. 1987; Espizua 1993; Espizua and Bigazzi 1998).

18.2. Morphology, Stratigraphy, and Chronology of the Río Mendoza Glacial Drifts

In the Río Mendoza valley five Pleistocene drifts and one Holocene drift were distinguished in previous studies, each less extensive than its predecessor (Fig. 18.1). From oldest to youngest these are Uspallata, Punta de Vacas, Penitentes, Horcones, Almacenes, and Confluencia drifts (Espizua 1989a, b; 1993).

18.2.1. Uspallata Drift

During the oldest (Uspallata) glaciation, a valley–glacier system flowed along the Río Mendoza valley for 110 km from the Andean drainage divide and 80 km from Cerro Aconcagua, terminating at 1850 m above sea level. The drift is found in few places, it crops out for about 500 m along the northern slope of the main valley, 1.5 km west of Polvaredas where it lies at about 120 m above the Río Mendoza. In the Uspallata valley, the drift crops out for 6 km and unconformably overlies Tertiary sandstones and siltstones, forming a moraine consisting of subparallel ridges immediately north of the Río Mendoza. Polished and striated rhyolite outcrops are found between Polvaredas and Arroyo Picheuta (Fig. 18.1). The moraine has a mean distal slope angle of 6°. Clast lithologies correspond with formations cropping out along the Río Mendoza and include rock types found only west of Puente del Inca. Faceted, polished, and striated clasts of dark, fine-grained volcanic rocks are common. The dacite boulder weathering and the degree of varnish development contrast markedly with those of the next-younger Punta de Vacas drift, implying that a lengthy time interval separated the retreat of the Uspallata glacier from the Punta de Vacas glaciation. Uspallata outwash is preserved in terrace remnants on both sides of the Uspallata valley. The terrace lies 54 m above the Río Mendoza and 29 m above the Punta de Vacas outwash terrace (Espizua 1989a, b, 1993). The Uspallata moraine, in the Uspallata valley, is surrounded and cut by alluvial fans. At Arroyo Chacay a white tephra layer about 1.2 m thick is interstratified within post-Uspallata fan alluvium (Fig. 18.1). This tephra layer (Fig. 18.1, site 1) has been dated by the plateau fission-track method on glass as 170,000 ± 50,000 yr BP. In view of this relationship, and on the basis of the degree of boulder weathering and varnish development, the Uspallata moraine must be at least as old as early Middle Pleistocene (Espizua 1993; Espizua and Bigazzi 1998).

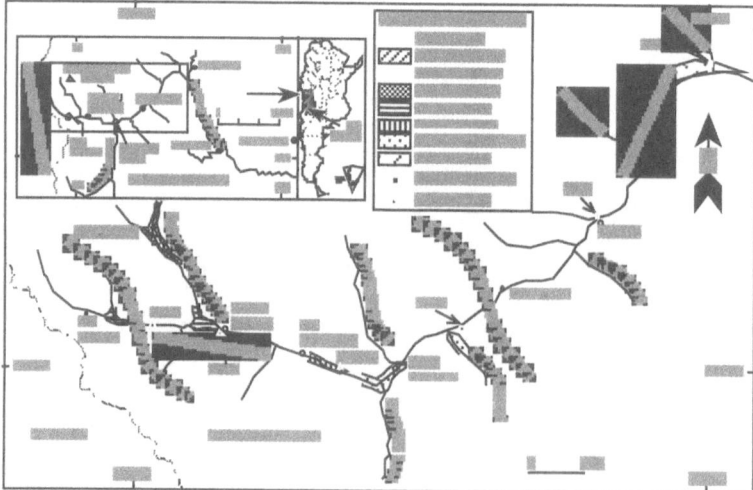

Fig. 18.1. Map showing the Quaternary glaciations in the Río Mendoza valley.

18.2.2. Punta de Vacas Drift

The distribution of the Punta de Vacas drift indicates that glacier ice flowed along Las Cuevas and Los Horcones valleys for 58 and 38 km, respectively, terminating at 2350 m above sea level (Fig. 18.1). The percentage of weathered dacite boulders and the intensity of varnish development on clasts are intermediate between Uspallata and Penitentes drifts. The Punta de Vacas moraine has a mean distal slope angle of 10°, and soil profiles developed on this drift have a reddish-brown (2.5–5.0 YR, chroma 3–4) B2t horizon. Polished and striated fine-grained volcanic boulders are common. Although a terminal moraine has not been recognized, a concentration of large boulders at 2350 m on the valley floor may mark the ice limit. Punta de Vacas outwash forms isolated terrace remnants in the Río Mendoza valley where it is poorly preserved and covered by fan alluvium. The terrace lies 70 m above the Río Mendoza (Espizua 1989a, 1993).

The Río Colorado drift, named after the Río Colorado tributary valley (Fig. 18.1), extends up to the mouth of this valley at an altitude of 2370 m. Along the Río Mendoza valley, the Río Colorado till rests on Punta de Vacas outwash for about 1 km. Desert varnish on clasts, implies that the Río Colorado moraine probably correlates with the Punta de Vacas moraine. Glaciers of Punta de Vacas age, in the Las Vacas and Tupungato, and Río Colorado valleys, apparently did not become confluent, on the basis of stratigraphic and topographic relationships (Espizua 1989a, 1993). At site 3 (Fig. 18.1) Punta de Vacas outwash overlies a tephra layer that dates to 260,000 ± 150,000 yr BP and at site 2 the Punta de Vacas outwash overlies another tephra layer that has a plateau glass fission-track age of 134,000 ± 32,000 yr BP. There are large error limits in these ages; however, the stratigraphic relations of the tephra units related to Uspallata and Punta de Vacas drifts at sites 1, 2 and 3, lead one to infer that the tephra layers may have been deposited during an interval prior to the maximum advance of the Punta de Vacas glaciation during marine oxygen stage (OIS) 6 or maybe in the stage 7 interglacial period (Espizua and Bigazzi (1998). The Punta de Vacas drift probably corresponds to the penultimate glaciation (Bengochea et al. 1987; Espizua 1989a, 1993; Espizua and Bigazzi 1998).

18.2.3. Penitentes Drift

Previous studies have shown that during the Penitentes ice advance, a glacier system flowed down Las Cuevas and Los Horcones valleys to an altitude of 2500 m (Fig. 18.1). Dacite boulders on moraine crests are less weathered than those on the Punta de Vacas till. The varnish is less well-developed on exposed drift clasts. Soils have a reddish-brown (2.5–5.0 YR, chroma 4, but ranges from 3–6) B2t horizon. The moraine has a mean distal slope angle of 10°. Outwash immediately east of the Penitentes moraine in Las Cuevas valley forms a valley train having a braided surface pattern. The Penitentes terrace lies 42 m above the Río Mendoza (Espizua 1989a; 1993).

A travertine layer overlying Penitentes drift has a ^{230}Th/^{232}Th age of 24,200 ± 2000 yr BP and 22,800 ± 3100 yr BP (Fig. 18.1, sites 4 and 5) and 38,300 ± 5300 yr BP (Fig. 18.1, site 6) which is a minimum age for the underlying Penitentes till (Bengochea et al. 1987). These dates imply that this ice advance occurred prior to the last glacial maximum and sometime before about 40 ka (Bengochea et al. 1987; Espizua 1989a, 1993). Pollen analysis in the Rincón del Atuel (34° 45´S) in the Mendoza Province (D'Antoni 1980) permit one to infer a temperature increase between 27,000 and 24,500 yr BP. Markgraf et al. (1986) deduce on the basis of pollen and diatom analyses of radiocarbon-dated lacustrine sediments, that interglacial-type climatic conditions existed 33,000–27,000 yr BP in the temperate Andean region at 40° S latitude.

18.2.4. Horcones Drift

The Horcones glacier advance flowed down Horcones valley to 2750 m near Puente del Inca. The distribution of the drift shows that an independent ice stream from Los Horcones Inferíor and Superíor valleys occupied Los Horcones valley and extended 22 km from the Cerro Aconcagua cirque head. The weathering of dacite boulders and the degree of varnish development on surface clasts are substantially less than for the Penitentes drift. The soil developed on Horcones till lacks a B2t horizon and the moraine has a mean distal slope angle of 21°. To the east of the Horcones terminal moraine lies a well-developed outwash terrace that lies 10–20 m above the Río de las Cuevas and the Río Mendoza (Espizua 1989a, 1993). A travertine layer capping the Horcones outwash (Fig. 18.1, site 7) has a U-series age of 9700 ± 5000 (Bengochea et al. 1987) which is a minimum age for the Horcones outwash. Although the age of Horcones drift is poorly constrained, the available date, weathering characteristics, and morphology imply that it likely represents the culminating advance of the last glaciation (Espizua 1993).

18.2.5. Almacenes Drift

A later readvance or still-stand (Almacenes drift) reached 3250 m. The similarity in relative-age characteristics between this drift and the Horcones drift permits one to infer that this readvance correlates broadly with late-glacial time (about 14,000 or 11,000–10,000 yr BP) (Espizua 1989a, 1993). At Confluencia, where the Río de los Horcones Inferíor joins the Río de los Horcones Superíor, the Horcones till underlies the Almacenes till and rests, in turn, on Penitentes till. Here, in some places, sediments of nonglacial origin separate the drifts. These sediments are interpreted as representing the Penitentes–Horcones and the Horcones–Almacenes nonglacial intervals (Espizua in preparation).

18.2.6. Confluencia Drift

The Confluencia drift terminates at an altitude of 3300 m. The terminal moraine grades into an outwash terrace. The freshness of the drift, the poor development of varnish, together with the proximity of the terminal moraines to active valley glaciers, allow one to deduce that these deposits are of Neoglacial age (Espizua 1989a, 1993). To the west of Horcones valley, in the small Paramillos de las Cuevas valley, a tributary of the Río de las Cuevas valley, two drifts were differentiated on the basis of morphologic characteristics and relative-age criteria; they correlate with Horcones and Confluencia drifts.

18.3. Pleistocene Drifts in the Mendoza Andes

In the Río Grande valley at 35° S latitude in southwest Mendoza in the Andes, Espizua (1998) studied a similar Pleistocene glacial sequence. Two Pleistocene glaciations were distinguished on the basis of relative-age criteria, morphology and a ^{14}C date. These are the Seguro glaciation (pre-last glaciation) and the Valle Hermoso glaciation (last glaciation), the latter consisting of three stadial deposits named from oldest to youngest, the Valle Hermoso I, II and III drifts. The Seguro Drift is inferred to correlate with the Punta de Vacas Drift. The Valle Hermoso I, II and III drifts could be correlated tentatively with Penitentes, Horcones and Almacenes drifts.

The Pleistocene moraine sequence in the Río Mendoza valley was correlated with that studied by Caviedes (1972) and Caviedes and Paskoff (1975). The authors identified moraines at 33° S, in the Aconcagua valley of Chile. These were related to three major advances, which they named from oldest to youngest, Salto del Soldado, Guardia Vieja and Portillo. They inferred that the deposits probably correspond to the prepenultimate, penultimate, and last glaciations. They also inferred that the Ojos de Agua drift represents a stadial event within the penultimate glaciation. A very brief survey of these drifts during this study was made to compare the stratigraphy on the Chilean and Argentinian slopes. Based on morphology, topographic position, and weathering characteristics of the drifts described by Caviedes, the Salto del Soldado drift likely corresponds to the Uspallata drift, the Guardia Vieja drift to the Punta de Vacas drift, and the Ojos de Agua drift to the Penitentes drift.

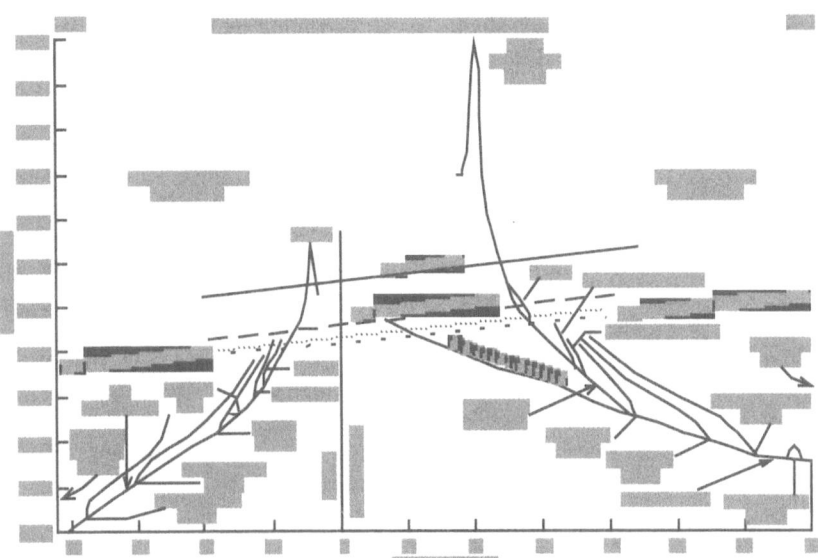

Fig. 18.2. Transect across the Andes near 33° S latitude showing mapped ice limits along the Río Mendoza drainage in the Argentine flank of the Andes (Espizua 1993), and according to Caviedes (1972) in the Río Aconcagua drainage in Chile.

The Portillo moraine system, which includes two distinct phases, compares closely with the Horcones and Almacenes drifts, respectively (Fig. 18.2). Moraines lying just down valley from existing glaciers on both sides of the Andes likely are of Neoglacial age (Espizua 1993).

On the Argentine flank of the Andes at 33° S, an estimate of the magnitude of snowline depression was obtained by comparing the altitude of the present snowline with that of the reconstructed snowline for Pleistocene glaciers (Espizua 1993). The present-day equilibrium-line (ELA) lies at 4500 m. The altitude of the ELA for existing glaciers is 4500 m. Assuming an accumulation-area ratio of 0.65 (Porter 1975) for Pleistocene glaciers, then the depression of the steady-state ELA for the Horcones, Penitentes and Punta de Vacas glaciations were 1000, 1200, and 1250 m respectively (Fig.18.2).

18.4. Summary and Conclusions

The central Andes of Argentina has been repeatedly glaciated during the Quaternary and has been characterized by an Alpine valley-type glaciation. At 33° S latitude, in the Río Mendoza valley, the Quaternary glacial sequence has been defined and ages were constrained. The oldest Uspallata glaciation predates marine oxygen isotopes stage 6; the Punta de Vacas glaciation is inferred to equate with isotope stage 6, the Penitentes moraine may have been deposited during isotope stage 4; the Horcones ice advance and the Almacenes readvance could be correlated with isotope stage 2 (Bengochea et al. 1987; Espizua 1989a, b, 1993; Espizua and Bigazzi 1998). The Confluencia drift is considered to be of Neoglacial age.

The glacial sequence was correlated with that studied by Caviedes (1972) in the Río Aconcagua on the drainage divide of the Chilean range. On the Argentinean side of the Andes, the Pleistocene glacial record of the Río Mendoza valley was correlated tentatively with a similar Pleistocene glacial sequence (Espizua 1998) in the Río Grande valley, at 35° S latitude, 275 km south of the Mendoza valley.

According to Caviedes and Paskoff (1975) and Lliboutry (1956), the modern ELA lies at 4200–4300 m on the Chilean flank of the range. Along a transect across the Andes at 33° S, the present ELA rises eastward from this altitude to 4500 m on the Argentinean side. The apparent parallelism of Pleistocene and present ELA gradients lead to an asymmetry of Pleistocene glaciers. At this latitude the Pleistocene glaciers on the Argentine side of the Andes were larger, had gentler gradients, and terminated at higher altitudes than their Chilean counterparts. The asymmetry was related, in part, to a westward sloping snowline that reflected a predominantly westerly source of moisture and it is also related to topographic asymmetry of the continental drainage divide (Espizua 1993). Streams draining to the Pacific are short and have steep gradients, whereas streams flowing to the Atlantic have low-gradient courses.

In the Argentine Andes, at 35° S latitude, the present snowline drops southward from 4500 m to 3700 m in the Río Grande valley as the mean elevation of the Andean gradually decreases and allows the penetration of moist westerly winds. The Pleistocene glaciers along the Río Grande valley were larger and the terminal moraine positions reached less altitudes than those in the Mendoza valley.

Acknowledgements: I thank C. J. Aguado and R. Bottero for their help in the field. R. Bottero also drew the figures. Funding was provided by the Consejo Nacional de Investigaciones Científicas y Técnicas (CONICET).

References

Bengochea LE, Porter SC, Schwarcz HP (1987) Pleistocene glaciation across the High Andes of Chile and Argentina. In "Abstracts International Union of Quaternary Research INQUA, XIIth International Congress", Ottawa, Canada

Caviedes CL (1972) Geomorfología del Cuaternarío del valle del Aconcagua, Chile Central. Freiburger Geographische Hefte. Selbstverlag der Geographischen Institute Albert–Ludwigs Universitat. Freiburg i. BR (Germany) pp. 9–153

Caviedes CL, Paskoff R (1975) Quaternary glaciations in the Andes of north-central Chile. Journal of Glaciology 14:155–170

D'Antoni HL (1980) Los últimos 30 mil años en el sur de Mendoza, Argentina. III Coloquio sobre Paleobotánica y Palinología. Memorias. Instituto Nacional de Antropología e Historia, 83–102

Espizua LE (1989a) Glaciaciones Pleistocénicas en la Quebrada de los Horcones y Río de las Cuevas, Mendoza, República Argentina. Dissertation, Universidad Nacional de San Juan (Argentina)

Espizua LE (1989b) Secuencia Glacial Pleistocénica en el Río Mendoza (Prov. de Mendoza, Argentina) Geocryology of the Americas, UNESCO–IUGS–IGCP Project 297, First Meeting, Mendoza, Argentina, 50–51

Espizua LE (1993) Quaternary Glaciations in the Río Mendoza Valley, Argentine Andes. Quaternary Research 40: 150–162

Espizua LE (1998) Secuencia glacial del Pleistoceno tardio en el Valle del Río Grande, Mendoza, Argentina. Bamberger Geographische Schriften. Bd. 15, S:85–99, Bamberg, Alemania

Espizua LE Bigazzi G (1998) Fission-track Dating of Punta de Vacas Glaciation in the Río Mendoza Valley. Argentina. Quaternary Science Reviews 17:755–760

Lliboutry L (1956) Nieves y glaciares de Chile. Fundamentos de glaciología: Ediciones de la Universidad de Chile, p 471. Santiago, Chile

Markgraf V, Bradbury JP, Fernandez J (1986). Bajada de Rahue, Province of Neuquén, Argentina: An interstadial deposit in northern Patagonia. Palaeogeography, Palaeoclimatology, Palaeoecology 56:251–258

Porter SC (1975) Equilibrium-line altitudes of Late Quaternary glaciers in the Southern Alps, New Zealand. Quaternary Research 5:27–47

Chapter 4: Prequaternary Climates

The Neogene is a time with increased CO_2 content. As the orography is in many (not in all) aspects similar to present-day conditions, the Neogene is a key-time for understanding the impact of increased CO_2 contents on climate, ocean currents and the ecosystems. The major questions that Neogene studies can answer are:

1) Do the sometimes tremendous climate changes that are observed during the Quaternary also occur in warmer Neogene times?
2) The ocean currents, especially the "global conveyor belt" (the deep water circulation) depends highly on subtle equilibria in the key areas for deep-water formation such as the area south of Iceland and the Weddell polynya (off Antarctica). As these equilibria are mainly dominated by density (temperature dependent) and salinity of the waters, at least in principle a break-down of the global conveyor belt could not be ruled out in times of higher atmospheric temperatures. This would of course show up (mass balances) in the global circulation pattern, including the patterns of the surface currents.
3) Are there patterns of climate change? And if yes, what do they look like?

The above tasks require a global coverage of data. This means that temperature algorithms have been developed that permit, complementary to isotope studies, the quantitative assessment of sea-surface temperatures from faunal communities. These are *not* transfer functions. This task has been performed successfully, as a database of some 600–800 time series of paleotemperatures already exist. Many samples however have not been processible due to evolutive phenomena. These time-series show the following major features:

There are areas that have been throughout the entire Neogene fairly stable. In such sites only minor temperature fluctuations, such as a baseline-shift around 4 million years occur.

Sites that are located near water-mass boundaries, such as between Australia and New Zealand show considerable temperature fluctuations that are regarding their magnitude comparable to Quaternary temperature fluctuations.

Overall the present day current system existed, with some modifications, throughout the entire Neogene. The Oyashio-Kuroshio system was in times prior to about 4 my either very weak or did not exist. This means that those watermasses that move today northward off Japan move in times prior to about 4 my northward further in the east (= in the central Pacific). This means that the Pacific gyre was "smaller".

The closure of the Isthmus of Panama affected the global circulation drastically. A baseline shift of temperature fluctuations as well as a general shift of temperatures towards higher values could be observed factually throughout the world: In the Caribbean, the eastern coast of North America and even between Australia and New Zealand. The area immediately west of Panama experienced a minor decrease of temperatures. This means that the closure of the Isthmus of Panama affected also indirectly (through deep water currents) the global circulation.

Although further details can be found in the respective contributions above lines clearly show that throughout the entire Neogene marine ocean currents in the present-day sense existed. In other words: Even in times with generally warmer conditions ("greenhouse scenarios") a breakdown of the global circulation could *not* be observed.

For understanding *dynamics* of a process observations have to be hanged into a framework that is independent of the observations. Such a framework is called time. In order to study hemisphere- or worldwide processes, such as climate change, a worldwide uniform dating scheme (stratigraphy) is needed. Thus another important task of IGCP-341 was the establishment of a worldwide uniform high resolution stratigraphic standard, at least for the Neogene and also partially for the Paleogene.

On the one hand evolution provides such a worldwide uniform "background clock". On the other hand newly formed species need time to spread.

Classical paleontological (formulated in a simplified manner) assays utilize the evolution by assigning regionally the same age to beds where certain species occur for the first time. This assay contains the background assumption that for certain kinds of species the time to spread is small compared to the time (time-interval) that is dated. Thus it is generally accepted that therefore classical biozones cannot be applied straightforward for worldwide studies because of the time index fossils need to spread.

Therefore within IGCP-341 a new concept was introduced: The evolutionary stratigraphy: Conceptually there is *no* difference between a small basin with limited extension and the world ocean. *Each* fossil has an evolutionary first appearance datum (wherever it is) and *each* fossil has an evolutionary last appearance datum, including "still living" (wherever it is).

These datums can at least in principle be found by establishing a relative order of all fossils observed. These have been in the context of IGCP-341 all planktonic foraminifera, coccolithes, diatoms and radiolaria of the Deep Sea Drilling Project / Ocean Drilling Program (after some screening, excluding disturbed wells etc).

This showed very clearly that by considering the whole ocean the life-spans of most fossils are much greater than previously thought.

Once such an evolutionary sequence, also expressed in million years has been established (data are now available) in the *dating* step new approaches could be utilized:

(1) Concepts like: "Globorotalia truncatulinoides appears for the first time", thus "there the observed bed is xxx.x million years old" have been abandoned completely.

(2) They have been replaced by the concept: "Globototalia truncatulinoides occurres". "Therefore the observed bed is xxx.x million years old OR younger".

(3) By including *all* fossils of a well the time-span of each sample reduces considerably. Thus *each* fossil is an index fossil. *No* fossil can be called undiagnostic.

Therefore now a database with life-spans of about 80 000 fossils, expressed in million years on a worldwide uniform basis exist. In addition now worldwide uniform age-models (age-depth curves) for about 700 drillsites from the DSDP and ODP are available.

Further back in time (see for example Beckmann et al., this volume) the CO_2 content of the atmosphere was higher. Consequently a study from the Salta Basin, focusing mainly on terrestrial data addresses the respective paleoenvironmental questions.

A Worldwide Uniform High–Resolution Stratigraphic Standard with Data for the Neogene and Paleogene

Peter Smolka
Geological Institute, University Muenster, Corrensstr. 24, D-48149 Muenster, Germany
smolka@uni-muenster.de

Abstract: The understanding of process-*dynamics* requires fixing observations into a framework that is independent of the observations themselves. This framework, independent of the characteristics of the process studied, is time. Thus, in the sense of dating, stratigraphy is the basis of studies focusing on the understanding of paleooceanography, paleoclimatology and ecology. Up to now classical biostratigraphy was applicable only locally or regionally due to the diachroneity of zones. However recently, various assays have been introduced for both relative and "absolute" dating. These include cyclostratigraphy, which especially in the Quaternary reaches unprecedented resolution in suitable areas. Evolution as a phenomenon is not diachronous; therefore, a worldwide uniform evolution-based high-resolution stratigraphy for pre-Quaternary times is introduced and tested with both new and old data from the Deep Sea Drilling Project (DSDP) and Ocean Drilling Program (ODP). The results yielded both new evolutive age-ranges (generally larger than previously known) for thousands of planktonic species as well as stable and consistent high-resolution age-models for a large number of drillsites from the DSDP and ODP. Although applied to the Neogene and Paleogene, the new method, called BDLOG-Lingula-Search, can be applied to any fossil class worldwide. The age-ranges and age-models are available on the accompanying CD-ROM.

19.1. Introduction

One of the oldest applications of paleontology is stratigraphy. Therefore, it could be assumed, that after about 200 years of paleontological research, stratigraphy is a standard-tool that is "just" applied for dating as part of studies focusing on other questions, such as paleoclimatology or paleooceanography. The commissions of the IUGS and the International Geological Correlation Program of UNESCO however show that worldwide uniform high-resolution stratigraphy is a current question. This applies especially, if basin-, hemisphere- or worldwide problems are addressed.

Stratigraphy as a tool is needed for two fundamentally different questions:

1) Stratigraphy is needed to *correlate* beds. That is: A horizon of interest, such as a reservoir in petroleum geology must be traced to locate the next drillhole. For this question, *correlation* of parameters is needed, either through well-log analysis, lithologically, by quantitative ecological studies (assemblage zones) and, sometimes across large areas, by tracing sequences of parameters and

beds in the sense of sequence stratigraphy. To be provocative: One must be able to recognize the same lithologic bed at different places, and in the (theoretically) most extreme case, independent of its age. Examples of assays focusing on the *correlation* of beds, also through ecological methods, can be found in the works of Agterberg 1985a,b, Barron et al. 1985, Brower 1985, Brower and Bussey 1985, Cowgill 1972, Hodson et al. 1971, Gradstein 1985, Guex 1991, Kauffmann and Hazel 1977, Marquardt 1978, Shaw 1964, and Wilkinson 1974.

(2) Stratigraphy is needed to *date* beds. This means: The observations have to be hung into a worldwide uniform system called "time" which is independent of the observations themselves. An analog of this approach is the usage of the term "12:00 GMT". In this term, (GMT, not local zonal-time), worldwide has the same meaning. At 12:00 GMT, it is in some place light, in other places, it is at the same time dark, but the meaning of 12:00 *GMT* is the same all over the world.

Classical biostratigraphic assays therefore focused on evolution as a "clock". In order to read the clock at the highest possible resolution, species with the shortest lifespan had been regarded as the most useful ones for dating. These fossils have been called index fossils, implying that the other fossils, having a long evolutive lifespan, are not index fossils and thus "undiagnostic". To clarify, at a certain place, the appearance of a certain species was observed. The time was named after this fossil implying that (ideally) whenever that fossil is observed, the corresponding beds are of the respective age. In addition, those fossils and fossil groups had received most of the attention that occur frequently, evolve rapidly, are easy to recognize, and are widespread.

This approach, using the lifespan of fast-evolving species for both *defining* times AND dating, causes problems because even under open marine conditions the spreading of fossils needs time. Theoretically, planktonic fossils can spread very rapidly, but ecological preferences and land-barriers may prevent the rapid spreading of index fossils into a large number of habitats.

This is explained using an analog from the postglacial history of mankind and the documented (fossilized) artifacts: Compared to the duration of the postglacial history of mankind the so-called epoch of "renaissance" is short. Paintings of "renaissance style" could be used for dating in the sense that "If in Italy a renaissance-style painting is found, it most likely has an age of, say 300-200 yr BP." The problem is, that if similar paintings are found in Denmark, they might be younger, as the ideas leading to the "renaissance painting style" needed time to spread. Other artifacts with longer cultural lifespans, such as horseshoes, are called "undiagnostic" as they can be observed in the 1st as well as 20th centuries.

In South America, however, horseshoes might be used for dating, as they did not occur before the arrival of the Spaniards. BUT: If in South America a horse-shoe is found, nobody would use the horseshoe as a "marker-species" and say: "very simple, 16th century". Everybody would know, that the observation of a horseshoe should be interpreted in the sense "16th century and younger". In art history, there are places where the cultural succession can be studied well. An example is the painting of northern Italy (Venice) between the 16th and 18th centuries.

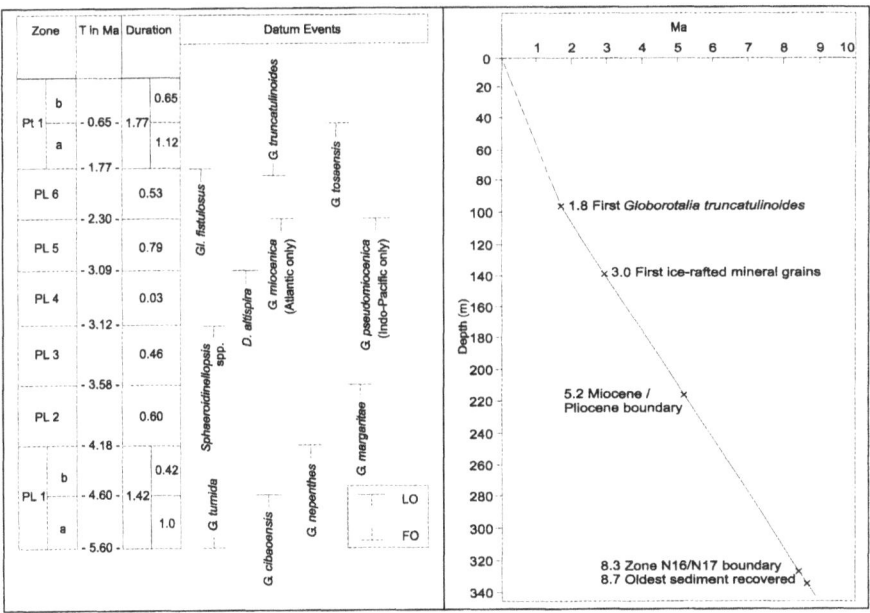

Fig. 19.1. Left: The first (FO) and last (LO) occurrences of planktonic foraminifera are used as indicators for certain ages. This implies that if, while screening a well from bottom to top under normal open-marine conditions, the respective species is found for the first/last time, then the sample has the corresponding age (modified after Berggren et al. 1995). The right figure (modified after Poore 1978) shows the application of this concept for dating DSDP drillsites in the North Atlantic: At a certain depth *Globorotalia truncatulinoides* occurs for the first time. Thus the sample is assigned the age 1.8 Ma.

In geology, similar approaches have been common. Places (outcrops) exist where the evolution of certain fossil-groups can be studied most easily, at least if the attention focuses on certain time intervals. Consequently, such places have been chosen for the definition of time intervals. They are called "stratotypes".

This means that if at the stratotype under discussion a certain fossil or fossil group occurs, the corresponding time is called, for instance, "Serravalian". The same applies to other evolutive phenomena such as the lifespan of certain short-lived species. The lifespan of these short-lived species is used to define time-intervals. They are called biozones implying (often unspoken) that whenever these species occur, the samples have the corresponding age.

It is known that under favorable conditions planktonic species may spread very fast, spreading over thousands of kilometers in less than 100 years. Therefore, another approach that was commonly used, is the definition of time through the first appearance and/or last disappearance of a species. This implies that on a worldwide scale, within a reasonable range of ecological habitats, these species spread fast and appear practically synchronously. Using this concept, a sample is assigned the corresponding age if, analyzing a drillhole from bottom to top, the "marker-species" indicating this age appears for the first time or disappears for the last time. This is demonstrated in Fig. 19.1. The main point of this argument is the word "may" which implies also "may not" if ecological conditions are unfavorable.

In any case the spreading of fossils needs time. This means that in the above-mentioned *classical* sense (GSSP, "global stratotype, section and point") the "Serravalian" is THAT time that corresponds to the first occurrence of a certain fossil in an outcrop in "for example, Italy." Some kilometers away the occurrence of the same fossil may easily have occurred a considerable amount of time later, even if the implications of Walther's facies-law do *not* apply (same ecological conditions). Moore et al. (1993) demonstrated that even under open-marine tropical conditions, the time that planktonic fossils need to spread affects the usability of the concept of marker fossils. In paleontological words (see for example Berggren et al. 1995): If stratigraphic standardization starts with GSSPs, the time reflected by the GSSPs has to be transferred to and from the GSSPs, independent of ecology (in order to avoid diachroneity). This means that the concept of GSSPs is useful for setting up stratotypes. For hemisphere- and worldwide dating (the application step) time is an entity that is in fact abstracted from the GSSPs and its paleoenvironmental conditions. Following the concept of GSSPs certain species have been recognized as marker-species. The appearance of the foraminifer *Globorotalia truncatulinoides* is often used as indicator for the base of the Pleistocene (see Fig. 19.1). The same applies to other marker-species. If larger areas are considered (see the transect in Fig. 19.2) the first appearances of index-fossils (biozones) and/or marker-species vary considerably with space due to ecological preferences.

At this point in the discussion two conclusions are possible:

1) Biostratigraphy is of limited value. It can ideally be used for local or regional *correlation* through the application of local (regional) zonations (reference-tables). For *dating*, other methods such as magnetostratigraphy, isotope-dating and cyclostratigraphy based on stable isotopes (ice-volume changes show up worldwide in the isotope record, practically independent of the local temperature), have to be applied. Although the latter method works well in the Pleistocene and is about to be extended into the Pliocene and Miocene, it has one disadvantage: If in a drillsite large amounts of sediment are missing, especially if there are several unconformities, the cylces are difficult to identify. In addition, many old DSDP drillholes have to be redrilled as the older ones often have large coring gaps.

2) Index fossils/marker species simply do not exist. They are an ideal concept that might apply locally, but not hemisphere- or worldwide. Especially if the dynamics of worldwide processes have to be studied, a clock is needed that is independent of these processes.

Figure 19.2 demonstrates that even generally accepted index fossils and marker species do not fulfill the criterion of synchronous hemisphere- or worldwide appearance. A closer look at Fig. 19.2 not only demonstrates the problem, but also the solution. EACH fossil has an evolutive first appearance datum (wherever it is) and EACH fossil also has an evolutive last appearance datum (including still living), wherever it is. We have a clock evolution, but must reinterpret it; in other words, each fossil is an index fossil, it only needs to be utilized.

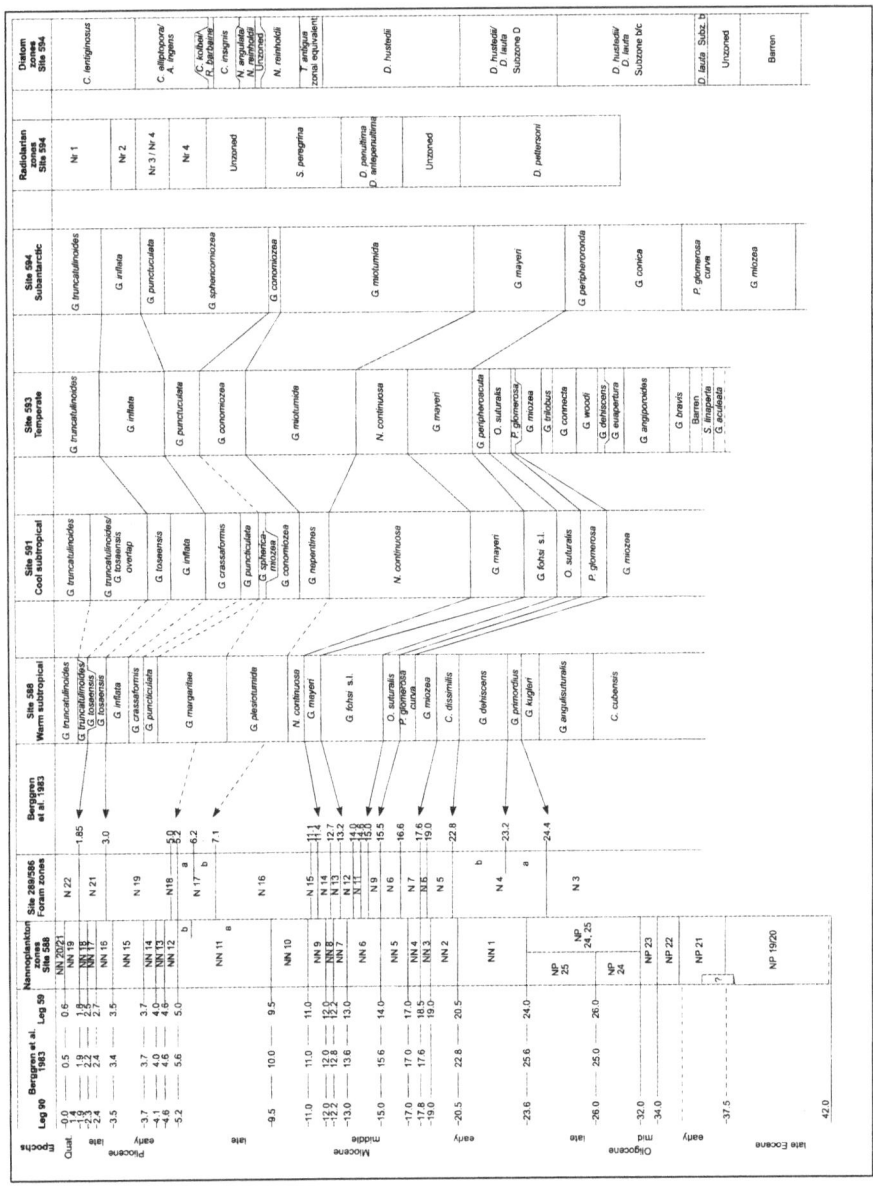

Fig. 19.2. A transect through the South Pacific from cold subantarctic conditions to warm subtropical conditions. The first and last occurrences of zone-defining fossils vary considerably. *Globorotalia truncatulinoides*, which is often used as indicator for the Pleistocene, occurs first between about 3.0–1.8 Ma. *Orbulina suturalis* shows larger differences. It occurs between about 25.5–15.0 Ma. This means that the idea of biozones and/or marker species must be replaced by total evolutive lifespans. If only *this* transect is considered, *G. truncatulinoides* must be regarded as diagnostic for 3.0–0.0 Ma; *O. suturalis* for 25.5–15.0 Ma. This implies that in the future the real evolutive lifespans based on data from the world ocean shall be used (slightly modified after Kennet et al. 1978:14) instead of short interpreted lifespans from stratotypes.

19.2. Evolutive Stratigraphy

Each fossil has a time and a place of origin. From this place it spreads out into the reachable habitats. Depending on conditions a fossil meets (water temperatures, land barriers, competitors in the food chain, etc.) it could either reach another area rather fast (even several thousands of km in 100 years), or very slowly, on the order of millions of years. The latter case applies if, for example, land barriers prevent the immigration of these species into habitats they might principally populate. Examples of such barriers are the Drake Passage, the Bering Straits, the Denmark Straits and Faeroe Shetland Channel, and the Fram Straits (linking the Norwegian Greenland Sea and the North Polar Basin). Even fossils that, due to their ecologically requirements, are principally known to be ubiquitous, can never(!) be used as marker species as (see the Bering Straits); the maximum value of a species is that of a "horseshoe" following above analogs from cultural history.

Before the solution for a worldwide uniform high-resolution stratigraphy is discussed further, some stratigraphy basics have to be recalled, especially as conceptually there is no difference between a small limited basin and the world ocean or even the whole world (if terrestrial and marine biota are considered together). Classic biostratigraphy consists of ten, theoretically simple steps:

1) Paleontologists look into the substratum and find morphotypes that can be distinguished from other morphotypes.
2) They assign them names and call them fossils.
3) They prepare lists of the fossils found in each sample.
4) They consult a handbook to determine when each fossil found has lived.
5) They determine the overlap of all ages of all fossils of a sample. Thus they arrive at a sample age, normally a time-span.
6) If necessary, they separate out groups of fossils that may have lived together.
7) They consider all samples of a well or section together, and arrange the age-ranges in a meaningful order. This is done by assigning new age-ranges for each sample by cutting off contradictory parts of overlaps (for example if a sample in 100 m depth ranges from 10-12 My and a sample in 99 m depth extends from 20–8 My, then the sample in 99 m depth might be adjusted to 10–8 My). In addition, other information such as magnetism may be included. The inclusion of faulted and thrusted sediments requires extension of the previously mentioned principles.
8) They recognize non-documented times and call the interface an unconformity.
9) They draw an age-depth plot or burial graph for the general user who:
10) May apply this for subsequent studies that require time as a worldwide uniform synchronisation plane. This applies for example to paleooceanographical studies.

Since points (4) through (10) can be done manually or by computer programs, high-resolution dating with paleontological methods is theoretically not a problem; the only open question is point (4), a handbook of worldwide age-ranges of all fossils observed. Therefore, the first step for worldwide uniform high-resolution dating consists of the establishment of a reference-table of relative ages.

For establishing such a range-chart no complex mathematics are needed. Conceptually the same methods undergraduate students are taught when they learn to set up a range-chart of "new" fossils (such as Conodonts) found in an hitherto "unknown" basin can be applied:

In the first step, all fossils found are put into a relative vertical order in each section (outcrop/drillhole). In this step, all problematical wells containing overthrusts, reworked horizons, etc., are for reasons of safety not included in the analysis.

In the second step, the relative order of all fossils is reviewed synoptically; thus the oldest, second but oldest, third but oldest fossil, is detected. It is evident that several fossils may have their first occurrence in the same sample. It is also clear that fossils that are present in the oldest sample EVER, may be absent in younger samples and reappear in the youngest samples again. One example is the brachiopod (genus) Lingula, a fossil that evolves in the Cambrian, is in many Mesozoic and Cenozoic samples not found but can be seen in some samples in the youngermost Quaternary. The methodology for establishing a range-chart of fossils for 2, 3 or 5 drillholes is elementary paleontological knowledge established about 200 years ago.

Thus the question arises: What did our *ancestors* do when entering a new basin with unknown outcrops and unknown classes of fossils?

1) They considered each outcrop and the RELATIVE depth-ranges of the fossils.
2) They compared it with the observations of other outcrops.
3) They searched for the relatively oldest species and ticked it off, maybe they called it Lingula.
4) And then, within the basin studied, they searched for the second but oldest, the third but oldest, etc. species. Thus they finally arrived at a range chart of say 150 fossils from 20 outcrops and one basin.

Conceptually there is no difference between twenty outcrops, 150 fossils and a small basin, and about one thousand outcrops (such as the DSDP and ODP drillholes), some 30 000 fossils, millions of individuals and the world ocean. For tracing evolution, not ecology, it is basically the same problem – creating a giant range chart that outlines the evolutive succession. For the creation of this range-chart (calibration-step), a pre-screening process is necessary moving out problematical wells (thrusting, technical mixing, sedimentary mixing, disturbed cores, etc.).

Whenever the database of the ancestors expanded by new expeditions, the range-chart reflecting the "current knowledge" might have been modified. Also regarding this aspect, there is no difference between a regional basin some 200 years old, and the world ocean with drillholes from the DSDP/ODP, as any new drillhole may add new knowledge.

Thus the inclusion of more drillholes is likely to expand the time-span of each fossil considered. What may be regarded as a short-lived index fossil in the classical sense with a lifespan of 1-2 My, may, if the whole world is considered, turn out to be a fossil that actually has a lifespan of 20 My.

This is demonstrated in Fig. 19.2, which shows the diachroneity of biozones. If the evolutionary first and last appearances are sought, then according to Fig. 19.2

the lifespan of *Orbulina suturalis* would extend to a range from its evolutionary "first" occurrence (about 25.5 Ma in hole DSDP-591) to its evolutionary "last" occurrence of about 15.0 Ma in DSDP-558. The same applies to *G. truncatulinoides* that ranges in the Fig. 19.2 example from about 3.0 Ma (DSDP-594) to zero years (still living).

Even using the above species as a "marker" (for lets say 1.8 Ma, see Fig. 19.1) is not possible. The only possible usage that does not cause problems is regarding *Orbulina suturalis,* its use as an indicator for a time-interval ranging (in the example of Fig. 19.2) from 25.5 to 15.0 Ma. This loss of *primary resolution* (which is a gain of safety) might first appear to be painful for some applications.

On the other hand, for dating, *all* fossils can be included: By overlapping several hundreds of fossils in one sample, the common time-span might shrink considerably. Here even long-living fossils contribute to high-precision datings: Assume a species "A" lived 65–26 Ma, and species "B" lived 27–3 Ma. If they occur together in one sample (no mixing), the common time-span for this sample becomes 27–26 Ma.

Consequently, the expected loss of *primary resolution* is (discussed below in more detail) overcompensated by far through the gain of *secondary resolution* as now "each fossil is an index fossil". No species is undiagnostic. Each species can be used for dating.

19.2.1. The BDLOG-Lingula-Search

One method for solving this problem, but not the only one, is a method the author has called the "BDLOG-Lingula-Search". This method begins with a list of all the fossils in a (computer) file and goes through the following steps:

1) Consider all fossils of each outcrop/well together.
2) Consider the first fossil in the file (whichever it is). Keep this species in memory.
3) For each well: Push out all fossils below the first fossil into another file.
4) Consider the file of fossils pushed out, and perform steps 2–4 repeatedly until only one species remains.
5) Tick off the remaining species and assign it the first age (oldest).
6) Create a file containing all species except the one ticked off.
7) Continue with step one.

If steps 2–4 are performed repeatedly, one species is left that has regarding the evolutive succession no species below (before) it. It may however happen that there are species of the same age. Here one question may arise: The "first" species being a candidate for the residue (i.e., not pushed out into another file) is selected in fact randomly (the first in the file). It can be seen immediately that this is not a problem, for the following reason: Assume that the first species of the first well is *Lingula sp.* (which is known to be the oldest, but not by the program). In the run of step two *Lingula sp.* is *not* included in the association of species pushed into the file containing species "below it". This means that other species are now candidates for being the oldest.

Since in steps following (4), fossils that are pushed out are considered again *and* because in SOME well *Lingula sp.* occurs below one of these fossils, *Lingula sp.* comes back again during one of the following steps. Or in other words: While doing so, it is no problem, if for example "Lingula" occurs in some wells for ecological reasons *only* in the Quaternary, in some other in the Mesozoic and in some other not at all. BDLOG (the program) finally cascades down to one remaining oldest fossil, the "Lingula". Then in the second run, another "*Lingula*" is found (the second but oldest) and so forth. Therefore, this method is called the "BDLOG-Lingula-Search".

This first step of the "BDLOG-Lingula-Search" may take some time if all planktonic foraminifera, diatoms, radiolaria and coccolithes of the world ocean are considered synoptically.

Other methods have been programmed and tested as well. They are however more difficult to explain. The basic thought behind these is always however straightforward:

What must students learn when they are taught to establish a range-chart based on paleontological findings (only relative depths and fossils)? What are they taught to do when entering an unknown basin, with 5000 m marls and lots of fossils representing unknown classes, whether these are Bacteria, Conodonts or something else. Whatever students are taught about a stable, paleontological method for establishing range-charts paleontologically: That can be programmed. It can be applied for the world ocean, since it makes no difference whether 10 or 50 outcrops are considered as our ancestors did or whether about 1000 wells with 30000 species and some million individuals are analyzed.

Thus for each fossil the evolutionary first appearance datum (wherever it is) is found. In order to avoid any misunderstandings with other concepts it is abbreviated as eFAD (evolutionary first appearance datum). For searching the evolutionary last appearance datum (eLAD) all wells are reversed upside down and the same procedure is applied. Consequently, at this stage of the discussion, the relative age-numbers of the eLADs have a different meaning than the relative age-numbers of the eFADs.

If only relative age-ranges are sought, the database arrived at up to this step can be used for relative dating.

19.2.2. Calibrating the Relative Ages

The first step yields a set of relative age-ranges that can be used for relative dating. Since at this phase of the analysis nothing is known about the duration of the time-intervals between the numbers of the relative ages, further calibration is useful. It should be kept in mind however that if the "million years arrived at" by any calibration-process change, the relative age numbers based on the evolutionary succession remain. Consequently, this first time-consuming step has to be performed only once.

The most comprehensive dataset of Neogene and Paleogene marine microplankton is the dataset of the Deep Sea Drilling Project/Ocean Drilling Program (DSDP/ODP). This has been critically screened (drilling disturbances, overthrusts, faunal mixing etc). In addition, it has been expanded by new drillholes from the Ocean Drilling Program.

The relative ages of the preceding step can be calibrated against numeric ("absolute") ages in My: For many DSDP/ODP drillholes reliable age-depth-plots that might need some modification due to changed boundary assignments (such as the changed boundary of the Gauss-Chron) do exist. It is obvious that species appear for their first time in different drillholes in different depths (ages). This is *not* a problem because for linking the available relative appearances of the first step with million years only, the oldest and youngest appearance EVER has to be sought. This means that for each species only ONE sufficiently old/young first/last appearance is needed. This means that the approach is extremely resistant against errors: Even if the age-boundaries used by the "DSDP/ODP" workers might contain errors, this is not a problem as the relative ages of step one are merged with the oldest and youngest appearances of the DSDP/ODP.

Step one prescribes the SEQUENCE of species appearance and disappearance. Because (a) about 30 000 species are involved and (b) the calibration must be free of contradictions, the degrees of freedom permitting errors reduce considerably (the primary sequence as prescribed by step one must be maintained).

Each method has limits: It is very seldom in the DSDP/ODP that sea-floor samples exist. In addition, the data-density decreases towards older parts of the Tertiary, especially in the Paleocene. This means that the reliability of the calibration in My is expected to be best between about 2.5 Ma and 40(50) Ma.

On the other hand, even if in the calibration-phase, some biota may have a suboptimal age (in the sense of precision) in some drillholes; it is sufficient that each species has a sufficiently old and/or young age at least in ONE drillhole worldwide. Compared to other assays, where the weakest part of a chain could affect the whole analysis, the BDLOG-Lingula-Search focusing on the *evolutionary* oldest/youngest ages, is inherently safer than other methods. Furthermore, in the *dating-step*, even if some species have a suboptimal age-range, the large amount of species included in the analysis (each fossil is an index fossil) adds to the reliability of the analysis.

This applies also to potential problems caused by low species diversity in boreal waters. Even if some samples of a well have a large time-span caused by low species diversity, the synopsis of *all* species and all samples of a well, including those of other classes contributes to consistent age-models.

19.3. Results

19.3.1. A Worldwide Uniform High-Resolution Range-Chart

For each species an evolutive time-span was determined. This time-span represents the knowledge documented by the available DSDP/ODP drillholes. Thus it can be enhanced by any new drillhole. It turned out that most species have a *larger* age-range than previously known. One extreme example is the above mentioned *G. truncatulinoides* that was previously used as marker for 1.8 Ma. This species has now a time-span of 9.5–0.0 Ma. Crosschecking with DSDP/ODP data-reports and tables have been done (see for this species, the depth and the subsequent resulting(!) age of Poore 1984:432 in connection with Hsü et al. 1984:48).

19.3.2. Evolutive Events

The calibration-step had an unexpected by-product. After having set up the relative age-ranges for all DSDP/ODP drillholes, the frequency of new species per event can be analyzed. A brief analysis shows that on a worldwide scale within the dataset and time interval studied (Neogene and Paleogene) *no* specific evolutionary events can be observed (this is not a statement about events in general). It turned out that "known" events can most likely be attributed to effects of local and/or regional documentation-phenomena such as unconformities. If analyzed on a worldwide scale, species suspected to have evolved (or disappeared) during evolutionary events appeared/disappeared gradually in groups of 3–4 each. It should be noted that these conclusions are drawn on the basis of the *relative* ages of the calibration step. The discussion of evolutionary rates including the quantitative study of movements of evolutionary centers through space and time requires additional work. Therefore, such ideas are left for future studies.

19.3.3. The Reliability of the Generated Age-Depth Curves

Age-models (age-depth-curves) have been set up for all processible DSDP/ODP drillholes. This section discusses selected results. The comprehensive dataset can be found on the accompanying CD-ROM (STRAT-Directory). To judge the reliability of the BDLOG-Lingula-Search the following are questions checked:

1) Are there significant(!) differences between the published "DSDP/ODP-curves" and the "BDLOG-curves"? Does BDLOG generally assign, for example, beds of Miocene age (according to DSDP/ODP) now to the Pliocene or v.v.? *Some* reassignments of this magnitude are no problem; they can be expected due to the significant stratigraphic progress of the last 20 years. If such changes occur in general, this would be a critical point because the reliability of many of the DSDP/ODP stratigraphic assuagements is known.
2) Are in *all* drillholes the curves reproductions of each other? This would be a hint for circular reasoning because classical biozones are diachronous if considered over larger distances (see Fig. 19.2).
3) Do the age-models based on the BDLOG-Lingula-Search curves show more or less details than the "manual" curves? Both should be expected because in cases of varying sedimentation rates, a large number of details are a positive indicator for the reliability of the system; in cases of constant sedimentation rates, few details should be expected in the curves (unless unconformities are crossed).
4) Are the ages assigned by the BDLOG-Lingula-Search generally higher or lower than the classical ones? For some wells or sections, this should be expected, but *overall* reassignments to younger and older ages should average out.

After analyzing age-depth curves for several hundred DSDP/ODP drillholes, it was found that the BDLOG-Lingula-Search fulfills all these criteria. The following section discusses some selected drillholes. In order to demonstrate the wide range of applicability, drillholes from a wide range of ecological environments have been included. In addition, holes drilled with state-of-the-art drilling techniques (hydraulic piston corer) have been included as technically very old wells (rotary coring).

The data are presented in diagrams. The abscissa indicates age in Ma. The ordinate indicates the depth below sea-floor in meters. The solid line curve describes the age-model according to the "BDLOG-Lingula-Search". The dotted line shows the age-model according to the workers of the DSDP/ODP.

The author has attempted to synthesize the DSDP/ODP stage and zone concepts in order to account for the progress of concepts that occurred during the DSDP/ODP history (about 20 years). This means that it is possible that there are some small differences between the "DSDP/ODP-curve" as shown *here* and that one presented in the respective DSDP/ODP-volume. This is because the knowledge about the lower (begin) and upper (end) boundary of the Gauss-Chron changed through time. Therefore, findings of DSDP/ODP-workers referring to the Gauss-Chron have been reinterpreted using current knowledge.

The plot of "DSDP/ODP" versus "new standard" is *not* intended as a plot to calibrate BDLOG against the old. It is only intended as a means of comparison. For ease of reading, the dotted curve (DSDP/ODP-data) contains 3–4 crosses. The solid line (BDLOG-Lingula-Search) contains 3–4 rectangles. The curves themselves consist of all processible data (planktonic foraminifera, diatoms, coccolithes, and radiolaria). This means that each curve consists of tens to several hundreds of datapoints.

19.3.3.1. DSDP-607, Central Subtropical North Atlantic

Site DSDP-607 (Fig. 19.3a) is located in the central subtropical North Atlantic on the mid Atlantic ridge (41° N, 32° W). A 311.3 m long (including DSDP-607A) continuously cored sedimentary section consisting of Pleistocene, Pliocene and Upper Miocene foraminifer and nannoplankton oozes was retrieved with the hydraulic piston-corer and the extended core-barrel (Ruddiman et al. 1987).

The sediments have been studied with different techniques, for example biostratigraphically (Baldauf 1987; Takayama and Sato 1987; Weaver 1987; Weaver and Clement 1986; Westberg-Smith et al. 1987) as well as integrated bio- and magnetostratigraphically (Baldauf et al. 1987; Clement and Kent 1987; Clement and Robinson 1987). Other methods such as stable isotopes have been applied as well.

Comparing the age assignments of the "DSDP-workers" (integrated and corrected for progress such as the age-assignments of magnetic boundaries) with the results of the BDLOG-Lingula-Search (calibrated worldwide across ecological boundaries using evolutive maximum and minimum ages), both curves coincide nearly perfectly. This is especially remarkable because BDLOG does *not* use any regional information. Involved in the dating step is only the evolutionary reference-data (time-spans) and the samples of the well studied.

19.3.3.2. DSDP-77B, Central Tropical Pacific

Site DSDP-77B (Fig. 19.3b) is located in the central tropical Pacific (Hays et al. 1972) near the equator (0° N, 133° W). It was spudded in the pioneer-days of

DSDP using rotary drilling. This means that amongst other items, the depth-assignments of the samples cannot be compared to present-day standards. This site was surveyed paleontologically at a comparably high standard that even today cannot be found everywhere. This applies also to the stratigraphic work that represents the state-of-the-art of that time. Therefore, this well could be submitted to analysis with the BDLOG-Lingula-Search.

Both curves (Fig. 19.3b) coincide well. Note that the dates of the BDLOG-Lingula-Search are slightly younger in the upper Miocene (compared to the "DSDP-workers"). In addition, the age-model of the BDLOG-Lingula-Search shows much more detail than the DSDP-model.

Considering the state-of-the-art applied during DSDP-Leg 9 (Hays et al. 1972), it should be noted that the stability of paleontology in general appears to be remarkably high, especially if all ongoing disputes concerning zones, markers, redatings, synonymies, GSSPs and systematic questions are discussed. In addition, this shows that a data-recycling of older data with new techniques (such as the BDLOG-Lingula-Search) is possible and can produce reliable results.

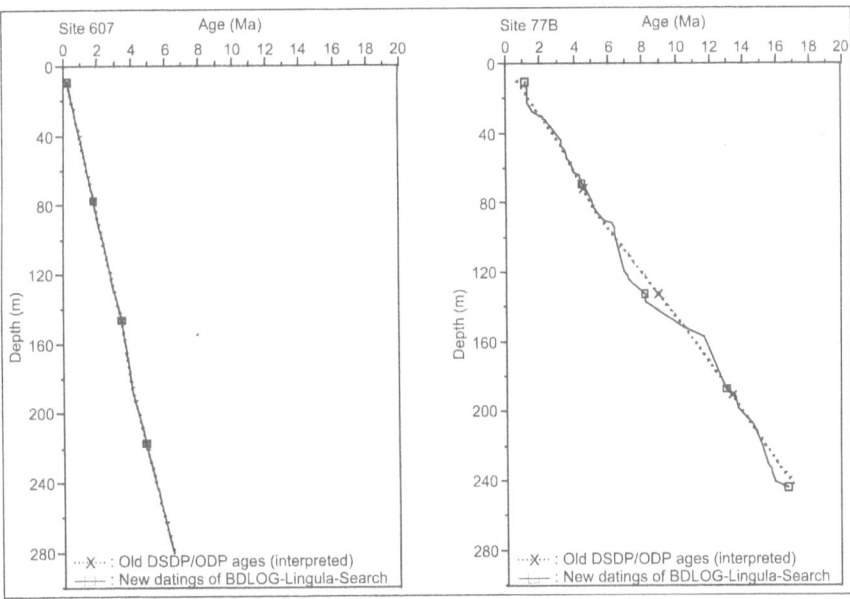

Fig. 19.3. The left figure (a) compares the results of the BDLOG Lingula Search (rectangles) with those of the "DSDP-workers" (crosses) for site DSDP-607 in the central subtropical North Atlantic (Ruddiman et al. 1987). Although the BDLOG-Lingula-Search does *not* get any regional information, both curves coincide nearly perfectly. The right figure (b) shows the results for site DSDP-77 located in the central tropical pacific near the equator. Although this well was cored with old rotary technology some twenty years ago, both curves coincide overall. For the Miocene section BDLOG shows more details than the old curve.

19.3.3.3. DSDP-407, Subartic North Atlantic

DSDP-407, was drilled in the subarctic North Atlantic (64° N, 30° W). It is located on the western flank of the Reykjanes Ridge SW of Iceland (Luyendyk et al. 1978). Today it is influenced by waters from the Norwegian-Greenland-Sea (East Greenland Current) and from the western North Atlantic. This well has been located in cold waters since the Miocene.

In the younger part (0–16 Ma) of this site (Fig. 19.4) the ages based on the BDLOG-Lingula-Search are generally about 1 My older than those according to "DSDP". Note (see above) that it has been attempted to synthesize the curve "according to DSDP" considering improved knowledge on for example magnetic boundaries. This means that the "DSDP-curve" shown here cannot be compared in all details with that of Poore 1978. In addition, the new age-model of BDLOG shows more detail. Furthermore, it must be noted that according to the BDLOG-Lingula-Search, the sedimentation rates are strongly reduced so either condensed horizons or an unconformity (see Poore 1978) can be expected.

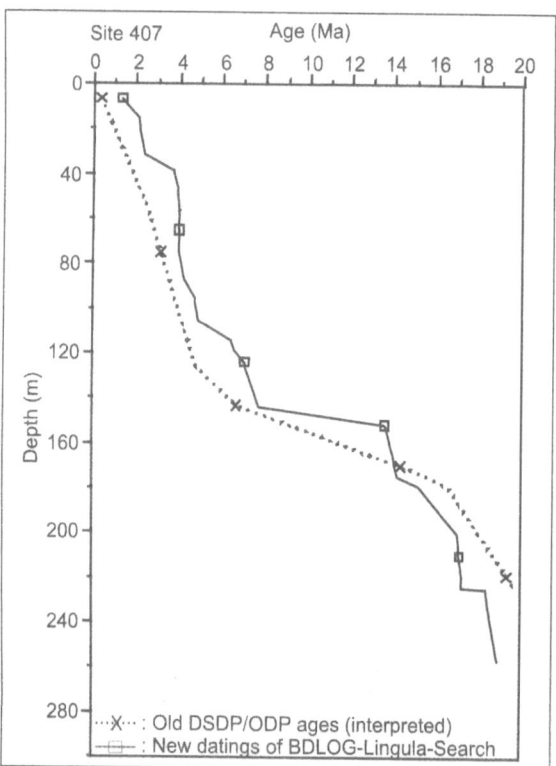

Fig. 19.4. Based on worldwide uniform age-ranges the age-depth curve for the younger part of DSDP-407 is redated towards older times while the lower part is assigned younger ages than those of the "DSDP-workers". Although DSDP-407 is drilled with rotary technology and contains coring gaps, BDLOG allows the possibility for dating wells from boreal low-diversity environments. Around 140 m an unconformity is possible. Further explanations in text.

19.3.3.4. DSDP-284 and DSDP-593, W of New Zealand

Two wells are compared that are located at almost the same position (40° S, 167° E). They are a few nautical miles apart from each other, but drilled in different years with different drilling techniques. DSDP-284 was drilled with rotary technology in 1973 during Leg 29 (Kennett et al. 1975). DSDP-593 was drilled at about the same place in 1982 (Kennett et al. 1987), nine years later, but with a hydraulic piston-corer. The situation is as follows.

A large outcrop is analyzed with one team using rotary drilling in the year 1973. The team takes the samples where it wants. Nine years later another team was sent to the same outcrop. The two teams did not necessarily work at exactly the same place but ("natural variability of a large outcrop") somewhere "in the wall of this outcrop". They took samples at depths of *their* choice. These are not necessarily the same as those of the first team. They analyzed the samples according to the procedures of *their* choice. The results are shown in Fig. 19.5.

Comparing the curves on a depth-to-depth basis and taking the reading-accuracy into account, the smallest difference between two points amounts to about 50 000 yr. Generally the differences are between 300 000 and 500 000 yr.

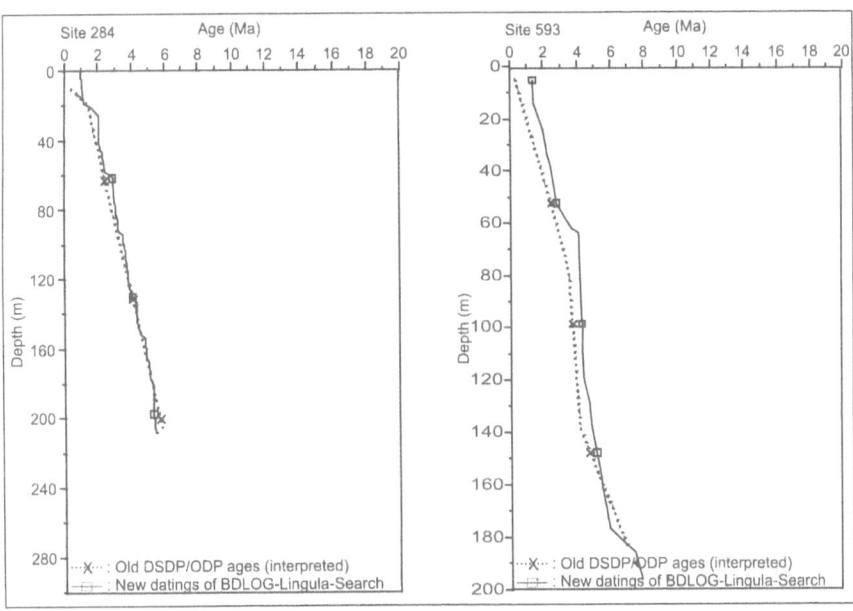

Fig. 19.5. DSDP Sites 284 (rotary drilling) and 593 (hydraulic piston corer) are drilled W of New Zealand only a few nautical miles apart in about the same sediments. The curves DSDP/BDLOG coincide well *between each other*. Both wells are analyzed independently without any regional information for the program. Although methods and workers determining the fossils and sample depths are different and although *some* sedimentological variation could always be expected, the differences between the left curve of DSDP-284 and the right curve of DSDP-593 are minimal: They amount, within the reading accuracy, to some 50 000 to 500 000 yr. According to the BDLOG-Lingula-Search basically the same curve results.

The maximum difference is about 1.2 My. Taking into account that not only drilling technology is different, but also systematics, sampling density and coring gaps, these examples show the high stability of the analysis.

19.3.3.5. DSDP-362 and DSDP-532 off Southwest Africa

The last example demonstrates the stability of the BDLOG-Lingula-Search by showing both the tremendous differences and similarities in the age-depth-curve from the "DSDP-workers" and the BDLOG-Lingula-Search. The sites discussed, DSDP-362 drilled in 1975 (Bolli et al. 1978) and DSDP-532 drilled in 1980 (Hay et al. 1984), are located off SW Africa. Both wells are located some miles apart at basically the same positions (19° S, 10 E). They document the same sedimentary environment, implying (seismics) that the same beds have been cored. Note (Fig. 19.6) the considerable difference of the curves at DSDP-362. Hay et al. (1984), describing DSDP-532, noted also the considerable drilling-disturbance of DSDP-362. In DSDP-532 the "BDLOG/DSDP" curves agree quite well *and* the BDLOG-Lingula-Search-curve of DSDP-362 agrees quite well with that of DSDP-532! Since only fossils and depths are submitted for the analysis, the stability of both paleontology (despite all ongoing disputes) and the BDLOG-Lingula-Search is evident. This is also supported by results achieved at DSDP-284 and DSDP-593 near New Zealand (see above).

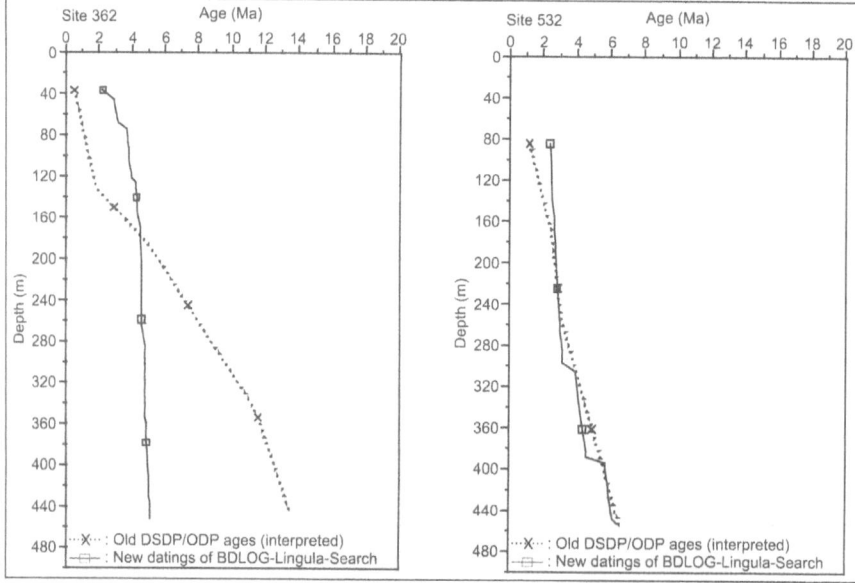

Fig. 19.6. DSDP Site 362 (left) and 532 (right) are drilled at nearly the same positions. For the old site 362 the curve of the DSDP-workers and that of BDLOG differ considerably. For the new site 532 they coincide. In addition, the age-depth-plot elaborated by BDLOG for the old DSDP-Site 362 coincides well with BOTH the BDLOG-curve for 532 AND the new age-depth-plot of the DSDP-workers for DSDP-Site 532. As the program receives only depths and samples (no regional information) stability and reliability of both paleontology and BDLOG is demonstrated.

19.4. Summary and Conclusions

A worldwide uniform range-chart for planktonic foraminifera, diatoms, radiolaria and coccolithes (reference-table) based on evolutive first and last appearances (eFAD/eLAD) was established using the screened and extended dataset of the DSDP/ODP. In the first step, relative ages were established. In the second step, resulting relative ages of about 30 000 fossils were calibrated using reliable, synthesized DSDP/ODP data. This means that by KEEPING the SEQUENCE that was prescribed by step one, the oldest/youngest appearance of each species was sought using a subset of existing and optimized age-depth curves. This yielded a reference table in millions of years. Both steps showed that the evolutive age-ranges of species expands, including those of species that have been formerly considered short-lived. Since a worldwide uniform reference-table for the above-mentioned fossil classes is now available, ALL species can be used for dating. Each species is an index fossil; no species can be called undiagnostic. Even long-living species contribute to high-precision dating by narrowing the overlap of age-spans. If for example, within one sample, species "A" ranges from 65–25 Ma and species "B" from 26 Ma to recent, then, although this is an ideal case, the overlap narrows to 26–25 Ma. The geological task to screen samples for reworking and mixing is still necessary, although communities that numerically could not have lived together (no overlap) can be separated out by the computer program.

There is conceptually no difference between some outcrops in a small basin and the world ocean with about thousand DSDP/ODP drillholes. Therefore, the BDLOG-Lingula-search is a straightforward application of classical paleontological thinking, strictly obeying Walther's facies law *and* evolution. Other methods paleontologists have to apply when entering an unknown basin with unknown fossil classes (for example if in the near future paleontology is introduced on the planet Mars) have been programmed as well.

Thus one of the tasks of IGCP-341, the establishment of a worldwide uniform high-resolution stratigraphy for the Neogene and Paleogene is solved. The data can be found on the accompanying CD-ROM in the STRAT-Directory. For ease of use all files are "pure ASCII" so they can be accessed by any system easily.

Acknowledgement: The patience of the members of UNESCO Project IGCP-341 during the calibration and testing is gratefully acknowledged.

References

Agterberg FP (1985a) Methods of ranking biostratigraphic events. In: Gradstein FM, Agterberg FP, Brower JC, Schwarzacher (eds) Quantitative stratigraphy. Reidel, Dordrecht pp 161–194

Agterberg FP (1985b) Methods of scaling biostratigraphic events. In: Gradstein FM, Agterberg FP, Brower JC, Schwarzacher WS (eds) Quantitative stratigraphy. Reidel, Dordrecht, pp 195–241

Baldauf JG (1987) Diatom biostratigraphy of the middle- and high latitude North Atlantic Ocean, Deep Sea Drilling Project Leg 94. In: Ruddiman WF, Kidd RB, Thomas E (eds) Init. Repts. DSDP, 94:729–762, Washington (U.S. Gov. Print. Off.)

Barron JA, Keller G and Dunn DA (1985) A multiple microfossil biochronology for the Miocene. In: Kennett JP (ed) The Miocene Ocean: paleoceanography and biogeography. Geol. Soc. Amer. Mem. 163:21–36

Berggren WA, Hilgen FJ, Lagenreis CG, Kent DV, Obradovich JD, Raffi I, Raymo ME, Shackleton NJ (1995) Late Neogene chronology: New perspectives in high-resolution stratigraphy. Geol. Soc. Amer. Bull., 107:1272–1287

Bolli H, Ryan WBF (1978) Initial Reports of the Deep Sea Drilling Project, 40, Washington (U.S. Gov. Print. Off.)

Brower JC (1985) Multivariate analysis of assemblage zones. In: Gradstein FM, Agterberg FP, Brower JC, Schwarzacher WS (eds) Quantitative stratigraphy. Reidel, Dordrecht pp 65–94

Brower JC and Bussey DT (1985) A comparison of five quantitative techniques for biostratigraphy. In: Gradstein FM, Agterberg FP, Brower JC, Schwarzacher WS (eds) Quantitative stratigraphy. Reidel, Dordrecht pp 279–306

Clement BM, Kent DV (1987) Geomagnetic polarity transition records from five hydraulic piston core sites in the North Atlantic. In: Ruddiman WF, Kidd RB, Thomas E (eds): Init. Repts. DSDP, 94:831–854, Washington (U.S. Gov. Print. Off.)

Clement BM, Robinson F (1987) The magnetostratigraphy of Leg 94 sediments. In: Ruddiman WF, Kidd RB, Thomas E (eds) Init. Repts. DSDP, 94:635–650, Washington (U.S. Gov. Print. Off.)

Cowgill GL (1972) Models, methods and techniques for seriation. In: Clarke DL (ed) Models in Archaeology. London, Methuen

Gradstein, C (1985) Timescales and burial history. In: Gradstein FM, Agterberg FP, Brower JC, Schwarzacher WS (eds) Quantitative stratigraphy. Reidel, Dordrecht, pp 421–469

Guex J (1991) Biochronological Correlations. Springer, Berlin

Hay WW, Sibuet JC (1984) Init. Repts. DSDP, 75. Washington (U.S. Gov. Print. Off.)

Hays JD (1972) Init. Repts. DSDP, 9, Washington (U.S. Gov. Print. Off.)

Hodson FR, Kendall DG, Tautu P (1971) Mathematics and the archaeological and historical sciences; Edinburgh University Press, Edinburgh

Hsü KJ, LaBreque JL (1984) Init. Repts. DSDP, 73. Washington (U.S. Gov. Print. Off.)

Kauffmann EG, Hazel JE (1977) Concepts and methods of biostratigraphy. Dowden, Hutchinson and Ross, Stroudsburg

Kennett KM, von der Borch CC (1987) Init. Repts. DSDP, 90, Washington (U.S. Gov. Print. Off.)

Kennett JP, Houtz, RE (1975) Init. Repts. DSDP, 29, Washington (U.S. Gov. Print. Off.)

Luyendyk BP, Cann, JR (1978) Init. Repts. DSDP, 49:447–517, Washington (U.S. Gov. Print. Off.)

Marquardt WH (1978) Advances in archaeological seriation. In: Schiffer, M.B. (ed) Advances in archaeological method and theory. Academic Press, London

Moore TC Jr., Shackleton NJ, Pisias NG (1993) Paleooceanography and the diachrony of radiolarian events in the Eastern Equatorial Pacific. Paleooceanography 8(5):567–586

Poore RZ (1984) Middle Eocene through Quaternary planktonic Foraminifers from the southern Angola Basin: Deep Sea Drilling Project Leg 73. In: Hsü KJ, LaBreque JL (eds) Init. Repts. DSDP, 73:429–448. Washington (U.S. Gov. Print. Off.)

Poore RZ (1978) Oligocene through Quaternary planktonic foraminiferal biostratigraphy of the North Atlantic: DSDP Leg 49. In: Luyendyk BP, Cann JR (eds) Init. Repts. DSDP, 49:447–517. Washington (U.S. Gov. Print. Off.)

Ruddiman WF, Kidd RB, Thomas E (1987) Init. Repts. DSDP, 94:763–778. Washington (U.S. Gov. Print. Off.)

Shaw AB (1964) Time in Stratigraphy. New York (McGraw Hill)

Takayama T, Sato T (1987) Coccolith biostratigraphy of the North Atlantic Ocean, Deep Sea Drilling Project Leg 94. In: Ruddiman WF, Kidd RB, Thomas E (eds) Init. Repts. DSDP, 94:651–702. Washington (U.S. Gov. Print. Off.)

Weaver PPE (1987) Late Miocene to recent planktic foraminifers from the North Atlantic: Deep Sea Drilling Project Leg 94. In: Ruddiman WF, Kidd RB, Thomas E (eds) Init. Repts. DSDP, 94:703–727 Washington (U.S. Gov. Print. Off.)

Weaver PPE, Clement BM (1986) Synchroneity of Pliocene planktonic foraminiferal datums in the North Atlantic. Mar. Micropal. 10:295–307

Westberg–Smith MJ, Tway LE, Riedel WR (1987) Radiolarians from the North Atlantic Ocean, Deep Sea Drilling Project Leg 94. In: Ruddiman WF, Kidd RB, Thomas E (eds) Init. Repts. DSDP, 94:763–778 Washington (U.S. Gov. Print. Off.)

Wilkinson EM (1974) Techniques and data analysis-seriation theory. Archaeo-Physika 5:1–142

A new Paleotemperature Transfer-Algorithm and its Application to the Reconstruction of Neogene Oceans

Peter Smolka
Geological Institute, University Muenster, Corrensstr. 24, D-48149 Muenster, Germany
smolka@uni-muenster.de

Abstract: An algorithm has been developed that permits the calculation of fauna-based paleotemperatures for Quaternary and pre-Quaternary times. First, the recent components of fossil faunas are detected for each fossil sample. Second, if these components are sufficiently high, a set of rules and equations is applied that estimates the summer- and winter temperatures of the sample. Because an independent, assumption-free measure to check the explainability of a fossil sample through recent faunas exists, effects of evolution and non-analog ecological conditions (low or zero explainability) can be used to include only explainable samples in the paleotemperature analysis. Based on the processible samples (many samples of the DSDP/ODP are *not* processible) time series of the paleotemperatures have been established. These are discussed at selected sites from the world ocean. A comparison of the time series and subsequent synopses describe water-masses and local gradients. A remarkable shift in paleooceanographic conditions around 4 Ma (million years ago) can be observed in many wells worldwide. This shift is attributed to the closure of the Isthmus of Panama. Additional changes can be observed around 3.2 Ma and 2.6 Ma. These time series have then been used to establish synoptical maps. Overall these maps show that although important and pronounced climatic shifts occurred in many sites of the world, the *main* oceanographic features known from today such as the large ocean currents have prevailed throughout the entire Neogene. This statement does *not* deny the sensitivity of the climate system or the potential impact of climate changes on the hominid population (which can be dramatic if the temperature change observed around 4 Ma happens again). This statement emphasizes that considerable short (less than 500 ky) temperature fluctuations, which are essentially local and regional if they occur again, in the long-range only *modify* the circulation pattern but do not destroy it. A gradient-free world ocean, which would be evidence for a break-down of the circulation system, including the deep-water counterparts, did not occur in the Neogene. The data (samples, time series and maps) are available on the accompanying CD, so they can be linked easily with subsequent model-calculations. In addition, the transfer-algorithms discussed in this paper can be used to process the CLIMAP dataset.

20.1. Introduction

During the younger history of the earth various climatic scenarios occurred. Thus the study of warmer and colder climates that differ from today may serve as a tool for the understanding of potential future climatic equilibria. This is especially important as the relation between the CO_2-concentration and temperature is nearly asymptotic (see Beckmann et al., this volume). For large-scale climatic changes, which have been observed during the younger earth history, either other effects than CO_2 have been responsible for climate change or the CO_2 concentration triggered self-organizing processes of considerable magnitude. In the latter case both magnitudes, time-scales of change and realized climatic equilibria, are tools to set up checklists (not predictions) for phenomena that might be relevant in the future. In addition, the coupling of reconstructed climatic equilibria with climate models helps to assess non-reconstructible parameters such as wind- and precipitation-fields.

One extreme climate that has been studied extensively is the last glacial maximum (LGM). Time intervals with *warmer* extreme climates are the Eem interglacial or the post-glacial Holocene climatic optimum. The study of climates from *these* times shows the differential nature of climate change (see Latrubesse, this volume; Prieto, this volume; Villalba, this volume). Therefore regional studies focusing on these time intervals may result in environmental analogs that might also be relevant to regional decision-making processes, because any reconstructed or potential climatic situation favors some kinds of land-uses while others are inhibited. Thus the knowledge of optimum equilibrium vegetation, either from biome-modeling or paleoclimatic reconstruction, optimizes land-use patterns.

The other question of comparable importance, however, is the overall long-term stability and phenomenology of potential non-analog (warmer?) climates including the location and intensity of ocean Currents.

One time interval that provides ideal conditions to assess this problem is the Neogene. It is a time for which numerous groups and workers have reconstructed parameters worldwide. It is also a time that is covered in the oceans by an overwhelming amount of data from the Deep Sea Drilling Project and Ocean Drilling Program (DSDP/ODP). The Neogene is a time that documents warm climates as well as cold climates, therefore the *patterns* of change can be studied. And finally, it is a time that contains orographic situations that are comparable to present-day scenarios as well as times with orographic modifications (such as the open isthmus of Panama).

On the other hand, the Neogene is also a time with stratigraphic difficulties superceding that of the Quaternary. Consequently, any reconstruction will yield exactly what is necessary: results that describe the *overall* picture which permit statements about nature and timescale of *overall* changes. In addition, the intercomparison of ocean reconstructions, vegetation reconstructions, model-predictions and subsequent biome predictions, helps to optimize the reliability of the "prediction system".

20.2. Methods

Paleontologists have, for a long time, been attracted by the reconstruction of paleoenvironments from fossil census. The classical work of Imbrie and Kipp (1971) showed that quantitative fauna-based paleotemperature analyses are at least in principle possible (see also Cline and Hays 1976; Wolff et al. 1998). Thus hopes exist that quantitative fauna-based paleotemperatures may become a powerful tool not only for the Quaternary but also for Tertiary climate studies.

This is especially important because (1) worldwide databases on Neogene faunas exist and (2) compared to isotope-geology, the testing of new methods using these datasets is only a question of computer time. Analysis of Neogene faunas demonstrated that many phenomena that are now studied in the Quaternary and Holocene, such as rapid climatic shifts and periodic versus chaotic patterns of time series, can also be observed in pre-Quaternary time series of marine microplankton. In addition, we become more and more aware of the fact that the present-day climate is exceptionally stable; natural fluctuations of considerable magnitude are the normal case, not the exception.

Thus one of the objectives of IGCP-341 is the reconstruction of paleotemperatures for pre-Quaternary climates. In a second step, these results are synthesized to time series and maps of sea-surface temperatures (SSTs).

Any method aiming at quantitative reconstructions of Neogene climates must fulfill two criteria:

1)   It must provide quantitative results in degrees centigrade.
2)   The data must cover the whole earth.

For Neogene marine environments, the most comprehensive database covering the whole world is the faunal census of the DSDP/ODP which is available both in printed and partially in digital form. Although different workers provide the microfossil data and the intentions of the research, the documentation, the drilling technology and the data-coverage varies considerably, many drillholes provide well-documented faunal and floral census (limitations are discussed further below). Thus this work utilizes and recycles this data treasure under the aspect of paleotemperature analysis. It is evident that evolutive phenomena prevent a direct application of CLIMAP-type methods. Therefore new methods have been established that permit the assessment of paleotemperatures in times older than the Quaternary.

In addition, technical limitations (manpower) suggest that the necessary worldwide data-coverage cannot be achieved through isotope studies alone (which would require a redrilling of many DSDP holes), but only indirectly through paleotemperature transfer algorithms. In addition, the multiple effects affecting the isotope-ratio, ranging from temperature through ice-volume, salinity and biotic effects, and the existing various correction factors, suggest putting more emphasis on transfer algorithms as an additional powerful tool. Thus the frontier of quantitative fauna-based paleotemperature-reconstruction was pushed well back into Neogene.

The first part of this paper discusses potential methods and thought patterns that can be used to design transfer-algorithms (not transfer functions) for the assessment of paleotemperatures from Neogene faunal communities. A detailed description including equations and reference-data can be found on the accompanying CD in the TEMPER-Directory (see README.TXT in the CD-ROM for more details). Although these principles can be applied to many problems, including terrestrial paleotemperature reconstructions, the second part of this contribution focuses on the reconstruction of Neogene sea-surface temperatures (SSTs) in key-areas of the world. The third part of this paper synthesizes time series describing SSTs to paleotemperature maps of Neogene oceans using the stratigraphic basis discussed elsewhere in this volume. These maps provide the basis for the modeling results presented in the last part of this volume.

20.2.1. Review of Older Methods for Paleotemperature Assessment

This short section briefly reviews older methods to infer paleotemperatures from faunal communities that are *not* used for this contribution.

When the reconstruction of paleotemperatures from biota is discussed, many people think, that for each species the ecological preferences are known so these preferences are more or less overlapped to arrive at a temperature. Often the argument is used that biota may have changed their requirements through time (which is true) or have a very large span of temperatures which they can endure. Both arguments are correct. On the other hand, such a manual assay seems to have the advantage that background information could be included. If such "manual assays" are programmed, they permit also consistency checks comparing measured and estimated data. These assays could operate as follows:

In a first pass the minimum and maximum temperature of each sample element is considered. Generally the most abundant elements are considered first. By overlapping the temperature ranges of *all* present biota (normally species), a temperature window is defined such that the least abundant elements are disregarded if they are outside the window. Experimental studies show (software on the CD) that this window is often very wide, ranging from 8–30° C.

In a second pass all biota are considered again, but their abundance is included. Using the experience of researchers, a temperature may be assigned a value near 8 or near 25 degrees depending on the percentage of "tropical" or "boreal" elements. Indices based on the percentage of "boreal" and "tropical" elements or indices based on the percentage of dextral or sinistral coiling foraminifers such as *Neogloboquadrina pachyderma* may give additional hints.

The procedure that arrives at results similar to this "experience assay" is the weighted mean. The mean of the temperature-range of the first pass is taken as origin. The first component of the weight is the reverse of the squared distance from this mean. Thus very low weights are assigned to temperatures far away from the average of the first pass. The second component of the weight is the abundance (or a power of it). The two weights are multiplied.

Such an assay has some advantages, it can be applied to all biota and it does not need elaborate mathematics. Furthermore such a programmed "manual assay" has the advantage that all samples are treated in the same way. Using the manual assay, it sometimes might happen that the results of some samples may not fit into the overall picture. At this stage, either only results fitting into the overall picture can be considered or (which is more difficult) the causes for deviations are followed up (such as local upwellings permitting the survival of cold-preferent faunas surrounded by warmer environments).

If such an assay is applied computationally, the data from the whole world can be treated several times with different assays. Furthermore, ecological background information that is available in the form of triangle-diagrams and indices may be included in such a program. Since all samples are treated in the same way, potential bias may be reduced. Such bias cannot be reduced completely because the researcher judging the results and designing the program is often the same person.

However, such a "manual assay" also has some disadvantages:

1) It highly depends on the reliability of the reference-data (the absolute and optimum ranges of the biota included in the analysis).

2) Biota may have changed their ecological requirements. If such phenomena occur with biota that have low abundances, this is insignificant (the window of the first pass starts to form with the most abundant biota). The results may be severely affected if abundant biota have changed their ecological requirements. If however such changes of requirements are known, they can be included in such a paleotemperature program if the program knows the sample age.

3) The estimated temperature is an average temperature. The calculation of summer- and winter temperatures is generally not possible.

More about such assays can be found in Feinen (1990), Hutson (1980) and Overpeck et al. (1985).

20.2.2. Transfer Algorithms

Transfer-algorithms are methods that reason from a given set of species and communities to paleotemperatures. In contrast to transfer-functions (see Morley 1989, Imbrie and Kipp 1971; Cline and Hays 1976) transfer-algorithms combine faunal community analysis with equations *and* reasoning. The latter point especially permits the inclusion of background knowledge. The most important point for the applicability of paleotemperature transfer-algorithms to older times is the recognition of "recent components" in "fossil samples", ideally such that an independent, assumption-free quantitative measure exists that describes how good a fossil sample can be described through recent faunas. In general community-assays consist of the following steps:

1) Biota of a set of (ideally) recent samples (often core tops) are counted. These reference samples cover all ecological niches that are relevant to the study.

2) The biota are subject to factor analysis to generalize from a set of species (expressed as abundances of species in the samples) to a set of "derived species" (communities, subfaunas), appearing in the form of percentages in the samples. The factor loadings describe the percentage of each of these communities (subfaunas) in each sample; the factor scores the importance of each species in the communities. Although often treated under statistics, factor analysis is a geometrical method: Geometrically, for example five axes (communities) are used to describe data points in a coordinate system with 27 (41) dimensions (species). The method calculates several statistics that show, how good the locations of the data points can be described by these axes, both with respect to each axis and each sample, each species and each axis, all axes and each sample and some overall performance parameters.

3) Because computationally axes are approximated to samples, additional samples may or may not influence the location of the axes. This means that, *if* a new cluster of observations is added, *and* the cluster contains for example 30% of all samples, *and* the cluster is different from the previous samples,

and the cluster is homogeneous (consists of similar samples in the cluster), the axes may be located in a different position.

In case of a *single* "new" sample and a large "old" data set, the location of the axes is numerically practically unaffected. This means that the new sample is factually explained by the previous data. If the explanation is successful, the measure of success (the communality) will be high for this sample, otherwise it will be low or nil. This has the following useful consequence. If for example, one single Cretaceous sample is factorized together with 115 recent samples, than the location of the axes is factually determined by the recent samples (the Cretaceous sample affects the explanation of the recent samples only on the third right comma digit). Since Cretaceous communities normally cannot be explained by recent communities, the communality will be nil.

A Pliocene or Miocene sample may yield different results. If the biota *and* the relative abundances of a Miocene sample may be explained by the five axes, that is, if the relative abundances of the Miocene sample are comparable to a portion of the recent dataset (for example to the samples coming from recent arctic conditions) then the measure of explanation (the communality) will be high. For this example, the sample can be explained by, for example, the axis one (recent subfauna one) for say 20%, by the recent axis two (recent community (subfauna) two) for 30%, and so forth. The numerical value of the communality demonstrates *independently* of any assumptions(!) whether recent communities (subfaunas) may or may not have existed at the site of the sample in Miocene times. This permits the tracing of the origin of recent communities (subfaunas) and their history through space and time.

It is important to note that communities consist of *both* species *and* the *relative abundances* of species with respect to each other: For example, one community may consist of 10 species with very low-factor scores and 5 species with a high-factor score. The next community may consist of the same 15 species but the first 5 have a low-factor score and the last 10 a high. Thus the submission of one fossil sample *together* with all recent samples permits the recognition independently of any assumptions of recent faunal components in fossil samples. The explainability of this fossil sample is given by the communality (the cumulative factor loadings expressed in percent).

This applies also to samples from the Quaternary as well (18K, CLIMAP, Cline and Hays 1976), so factually a simple factorization of samples from the Pleistocene and subsequent application of recent equations is strictly speaking not possible and may explain the problems that have been noted by Wolff et al. (1998).

4) In the next step the percentages of the subfaunas (the factor loadings expressed in percent) are contoured to get (a) an impression about the spatial distribution of such communities and (b) if necessary, to recognize special ecological conditions.

5) Diagrams of factor loading versus temperature are drawn. Such diagrams may show a tremendous scatter in the beginning. These diagrams may be drawn (a) either for all observed communities without special selection (Fig. 20.1),

(b) for faunal selected communities after recognition of type faunas (see appendix on the CD) or (c) other ecological selection mechanisms (application of indices as selectors).

6)   Figure 20.1 shows the relations between the subfaunas (each cross is a sample as documented in the appendix on the CD) and temperature. With the exception of the diagram for the tropical fauna (subfauna 2) all diagrams show a considerable scatter that prevents any classical straightforward linear or non-linear regression-approach. This scatter is the clue to reliable paleotemperature approximations. The diagram for tropical conditions (subfauna 2) permits approximation of a linear equation (Sa_1). All samples that are explained by this equation appear also in the other diagrams, reflecting other ecological conditions (subfauna 1: boreal; subfauna 3: cool temperate; subfauna 4: warm temperate). Sample 20 (circled), which is well described by equation Sa_2, does *not* need to be described by other equations. It may lie *well off* the segment Sa_4 of the diagram for subfauna 4 (technically it could even be omitted from that diagrams displaying subfauna 1, 2, 4 and 5). Thus the scatter is not a problem, but a contribution to the solution; a sample that is described once by one equation need *not* be considered again. In other words, all samples that have been used to establish *one* equation (such as Sa_2) leave the analysis. This means that each equation (segment) describes *only* the samples in the immediate vicinities. For reasons of printing-space however only one single diagram (instead of 6) is shown here.

A very simple (the algorithm used in this work as outlined in the appendix is more complex) paleotemperature transfer-algorithm that explains only 84 percent of the variance could, based on the diagram of Fig. 20.1 (equations Sa_1 to Sa_9), operates as follows: Note that the following example serves only to explain basic ideas. The transfer-algorithm that is used in this contribution is documented on the CD in the TEMPER-directory.

```
(1) Fauna 2 > 0.05 (tropical): apply Sa₁.
(2) Fauna 2 < 0.05 (extratropical):
(3)     Fauna 3   > 0.8 (moderate cool): apply Sa₃.
              <= 0.8:
(4)           Fauna 4 < -0.1 (moderate warm): apply Sa₅.
(5)           Fauna 4 >= -0.1:
(6)              Fauna 1 > 0.95 (arctic): apply Sa₇.
(7)              Fauna 1 <= 0.95:
(8)                 Fauna 4 >= -0.1: apply Sa₆.
                    else: sample not analyzed.
```

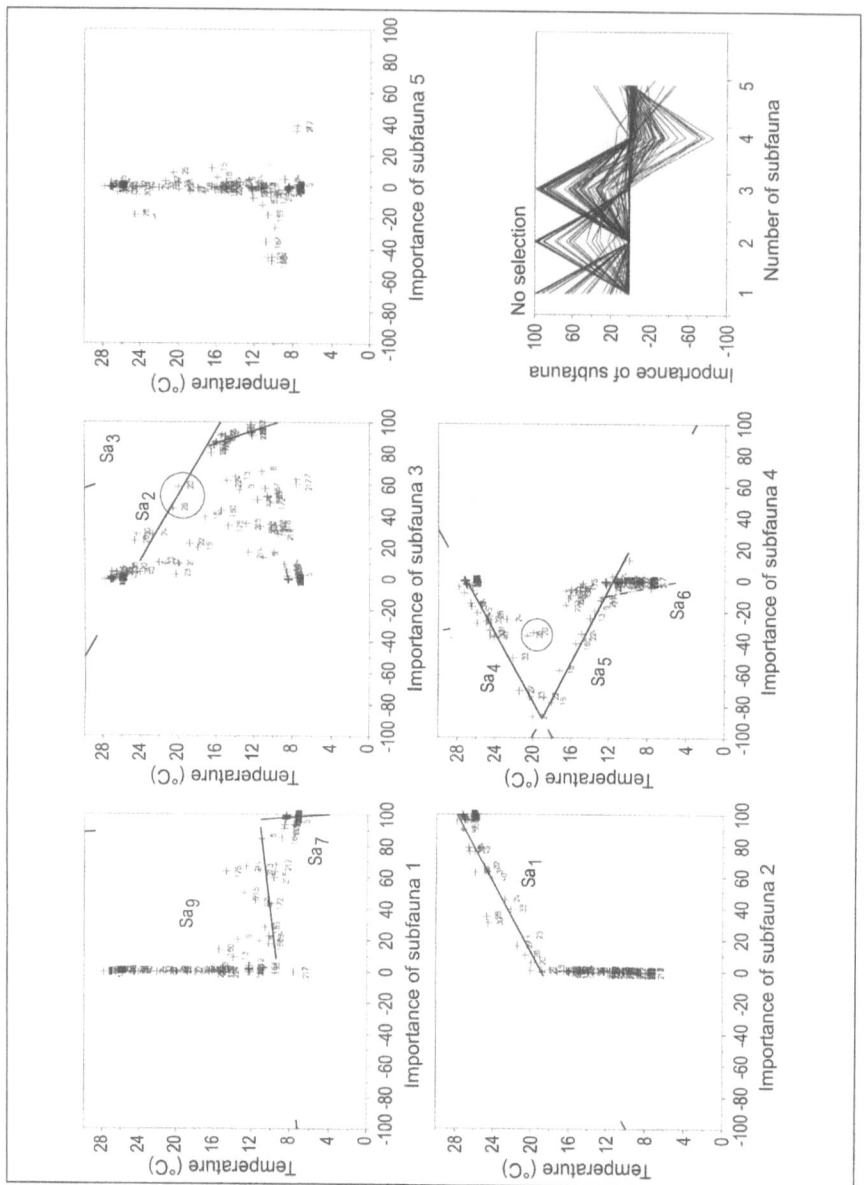

Fig. 20.1. Relations between recent subfaunas resulting from factor analysis and temperature showing the same samples displayed five times. Samples that can be well explained by one equation reflecting specific ecological conditions need *not* be explained by *other* equations. For example, the circled samples that are explained by equation Sa2 need *not* be explained by Sa4 or Sa5 (circle). Each equation focuses only on *those* samples that are not yet explained. So the effective scatter is very low. The top-right diagram showing characteristic curves of ecological conditions that can be used for further selections enabling an optimum flow of reasoning by the program. Further explanations can be found in the text and the appendix on CD-ROM.

The top-right diagram of Fig. 20.1 displays the samples in a different manner: Each curve is a sample. The abscissa displays the number of the subfauna (1 = boreal, 2 = tropical, etc.), the ordinate the signed importance from –100 to 100 percent. This is a diagram similar to the "spider-diagrams" in geochemistry. From this diagram a number of characteristic relative relations (subfauna 1 high *and* subfauna 2 low, etc.) can be seen. Thus this diagram and the above-discussed criteria can be used for *further* step-by-step selections that result in a set of equations. Each equation describes *some* samples well (some ecological conditions). This means that (if applicable) background knowledge from paleontological triangle diagrams or indices (such as the ratio of left- and right-coiling foraminifera of the species *Neogloboquadrina pachyderma*) could be taken.

Furthermore, subfaunas that cannot be used to establish equations, such as subfauna 5, may be used as a selector for specific ecological conditions. One subset of diagrams could contain only those samples that contain subfauna 2 with values greater than 5% and subfauna 5 with values less than –1%. In the next step from the remaining samples, only those are considered that contain subfauna 2 with values greater than 5% and subfauna 5 with values less than or equal to 1%. If in addition, the principles explained above (each equation applies only to *some* samples in the vicinity) are applied, then all ecological conditions could be described by a set of equations and decision paths. An algorithm that operates using these principles (appendix on CD) explains in the control run (measured versus approximated) 96.4 % of the temperature variability.

It should be noted that special attention has to be paid to avoid "critical bifurcations". Critical bifurcations are points where some slight numerical difference may decide whether subsequent steps of the analysis follow, for example, the arctic or boreal decision-path. Therefore the rules have been formulated such that in case of necessary decisions near the decision-point the results of two different equations yield comparable temperatures.

More on the philosophy behind this (inclusion of thoughts from "fuzzy logic") can be found, for example, in Bandemer and Gottwald (1993). These techniques may have a great potential that has not been utilized yet in geology. Assays like these may raise the question of usability and validity, especially as a complex computer program looks strange to those who are expect one simple formula.

The first question is answered very easily: The variability of the summer temperature can be explained for about 96%. Special insight into governing factors (such as genetics) is not provided with such assays. This however applies to some fundamental laws of physics, such as the law of gravitation as well. The law of gravitation is only a possibility to *describe* phenomena that are regarded as the observations of movements of entities in the so-called space-time system in such a way that usable predictions are possible in the easiest way.

The *nature* of gravitation itself is still unknown, although the applicability of the law of gravitation for safety considerations of buildings, just to mention one example, is generally accepted. Thus in the context of usability (see for example the practice of facies analysis from borehole measurements), the description of observations by rules has the same justification as the application of rules in other sciences.

20.2.3. Application of Transfer-Algorithms to Fossil Samples

Algorithms of above-mentioned type are established using recent reference-faunas. Therefore the application of such algorithms to all kinds of fossil samples, including samples from the Quaternary, requires additional measures.

1) The above-mentioned procedures are, at this point of the discussion, only valid for samples that can be explained for 100%. If for instance a fossil sample is explained only for 80% by the factor analysis, a tropical sample might be regarded as outer-tropic because instead of 10% of subfauna two, subfauna two may only have a value of 5%. Therefore the samples should be normalized to the explained fraction of a sample. This procedure contains an implicit assumption: that fraction of the fossil fauna which can be explained by the recent faunas is representative for the ecological conditions studied (in this case the summer temperature).

2) As mentioned above, faunal communities do not only consist of species, but also of the *relative* proportions of the species with respect to each other. If these proportions change, then factorizing all samples of a time interval yields only results that can be interpreted *within* the time interval. Even if qualitatively the species are the same, the communities may be different. This applies also to Pleistocene situations compared to recent reference-faunas. If however equations that are based on recent reference-faunas should be applied, then for the detection of recent components (subfaunas) in fossil communities, each fossil sample has to be submitted together with the whole recent reference-dataset. This submission of one fossil (unknown) sample with the whole recent reference-dataset ensures, that, if this is possible, the fossil sample is explained using recent faunas. In addition, it is ensured that through the communality an independent measure exists for this fossil sample that can be used to decide about the applicability of the paleotemperature equations. It is suggested to apply this procedure also to the CLIMAP data. It is expected that in this case certain inconsistencies might disappear (see Wolff et al. 1998; Cline and Hays 1976).

Since the transfer-algorithm-approach discussed here analyzes whole faunas and not single species, the change of the ecological requirements of *some* species does *not* affect the method. If however whole *communities* changed their requirements, for example the tropical subfauna 2 would have lived under arctic conditions in the Upper Miocene, then methods of this type reach general limits. In addition methods of this type may run into problems if the (sub)faunas are monotypic and the key-species change their ecological requirements. These limitations however apply to paleoecology in general.

The last point to be mentioned is of technical nature. The DSDP/ODP database consists of sites that are documented according to different standards. That is, some sites contain counted fossil census, some only estimated and some contain only information about the presence or absence of biota. Furthermore some species are only recognized by one author, while other species are recognized in general by the research community.

Faunal communities are defined through relative abundances. Therefore faunas that are documented through presence/absence criteria cannot be used for paleotemperature analysis. For instance, assume an arctic community that is dominated by *Neogloboquadrina pachyderma* (S). Leveling this dominance out, would change the relative proportions such that the resultant fauna would erroneously indicate subboreal or cold-temperate waters. This implies that the approach of Sancetta (1978), who treated all samples as if they would only be documented with presence/absence criteria, worked only because she used sites from the tropics (no major differences of relative species abundance).

Because this work includes data from all oceans, a conversion scheme was established that converts estimated results into numerical abundances. During their education paleontologists learn how to estimate abundances. Although there are many subjective components, the results are often remarkably stable (if this would not be the case, faunal lists with estimated abundances would never be accepted for publication).

After interviews with paleontologists, and testing the estimated abundances from DSDP/ODP, records listed as R/F/C/A/D (rare, few, common, abundant, dominant) were assigned the values 3, 10, 50, 150 and 900, respectively. Conversion of counted samples to samples with estimated properties and conversion of these samples to percentages shows that this is a justified compromise. This is also supported by the consistency of the results both with respect to other methods and between each other (temperature maps).

20.3. Discussion of Ocean Currents using Selected Sites and Transects

The above-mentioned transfer-algorithms have been applied to the available and processible DSDP/ODP drillsites from the world ocean. This chapter briefly discusses selected results. The full dataset is included on the CD-ROM in the TEMPER-directory.

This section discusses time series of SSTs following the main current systems, starting with the Oyashio-Kuroshio system off Japan. Although these and other sites lie on the northern hemisphere, it is obvious, that for hydrodynamical reasons they have to be considered as well. The same applies for the necessity to drive atmospheric general circulation models (GCMs) with reconstructed ocean temperatures. Such models also require a worldwide dataset. Discussing the features of the major oceans within the context of a short contribution is a difficult task that normally requires a comprehensive monograph. Because, however, all drillsites are discussed in-depth in the "Initial Reports/Scientific Results" of the DSDP/ODP, this section highlights only the main features of the paleotemperature history of selected key-sites.

20.3.1. Selected Sites from the Pacific

Figure 20.2 shows the location of sites from the Pacific that are referred to in the text. Site DSDP-438A (Figs. 20.2 and 20.3) is located east of Japan, in the waters of the warm Kuroshio Current.

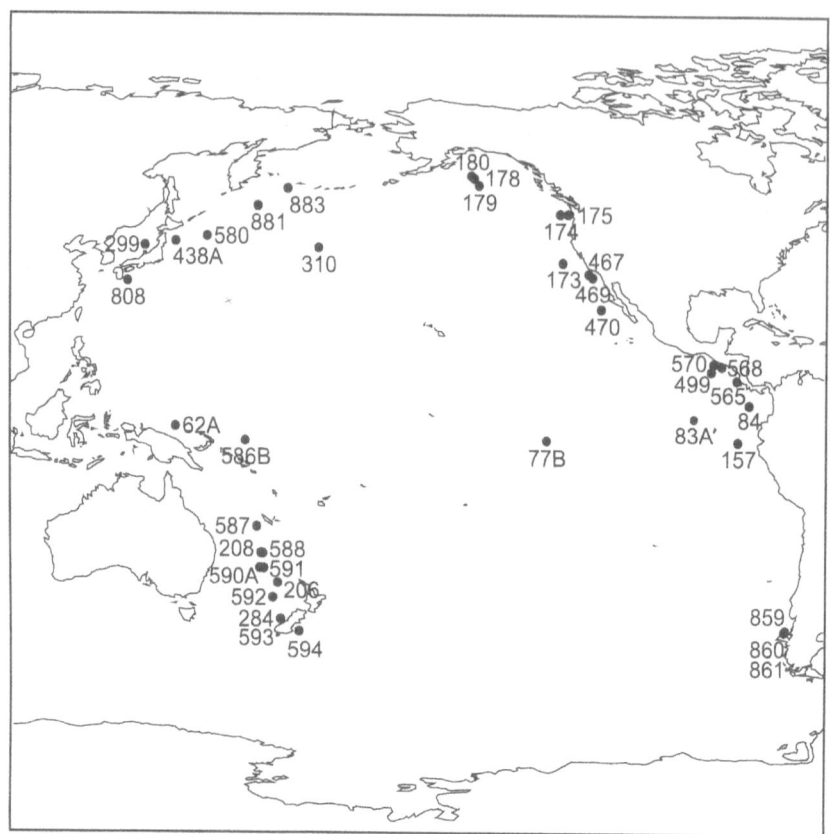

Fig. 20.2. Index map showing discussed drillsites from the Pacific.

In the time interval 8–4 Ma, site 438A (Fig. 20.3a) shows a temperature of about 6° C that changes to 8° C at about 4.2 Ma. Thus between 8–4 Ma, the Kuroshio Current was either less intense (see below) or bypassed Japan at a different position. This interpretation is consistent with evidence from other drillsites that are discussed further below.

Since 4 Ma (see below) pronounced changes between warmer and colder watermasses can be observed. This conforms to observations of Thompson (1980) who discusses, based on the ratio of left- and right-coiling *Neogloboquadrina pachyderma,* shifts of the boundary between Oyashio and Kuroshio watermasses. In addition, Thompson (1980) and Keller (1980) mention oscillating shifts of the Ca-Lysocline that on the one hand demonstrated changes of the water-masses but on the other hand can provide obstacles for paleoclimatic studies based on foraminifers. In addition, Reynolds et al. (1980) studied the presence and absence of Artiscids, selected Spongasters and the presence of thin-walled Collosphaerids to estimate paleotemperatures. They inferred from the varying presence and absence of Collosphaerids considerable temperature fluctuations.

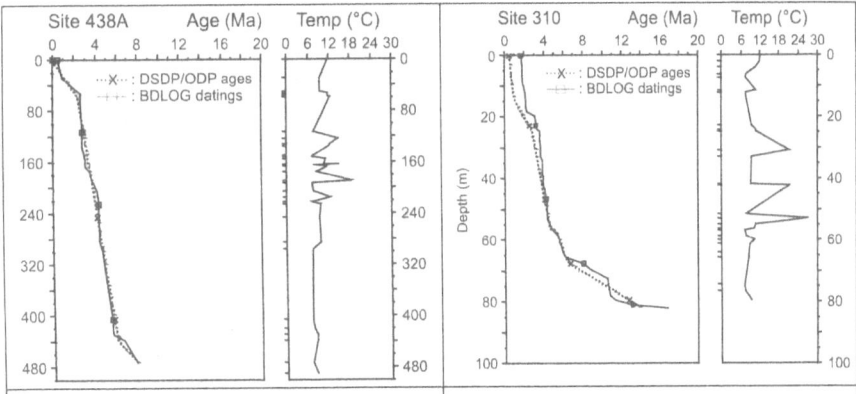

Fig. 20.3a. Site 438A E of Japan shows throughout the Neogene cool conditions that are followed by temperature-fluctuations after about 4 Ma. The upper part (site 438 on CD) shows pronounced temperature-fluctuations documenting both glacial-interglacial cycles *and* the existence of a warm Kuroshio during the Quaternary.

Fig. 20.3b. Site 310, central North Pacific shows during the Neogene cool conditions with some warm-water incursions. Whether these are boundary-eddies of the Kuroshio-System or warm-water incursions from the south must be left open.

Since the number of samples that have been processible with BDLOG (the software that realizes the discussed transfer algorithm and other paleoecological and paleontological methods, such as faunal turnovers, stratigraphy, etc. on a computer) is low, it is concluded that the interval between about 240–160 m describes the conditions most realistically.

Herman (1992) studied the foraminifers of ODP Leg 126 (Bonin Arc, SW of Site 438) and concludes that all samples *there* for the Eocene to Quaternary reflect watermasses of the Kuroshio Current. ODP Site 808 (134° E, 32° N), that is ideally located to study paleooceanographic questions, focuses however on other questions (Hill et al. 1993).

Barron (1995) elaborated a transfer-function (not a transfer-algorithm) for "middle Pliocene" (3.4–3.0 Ma) diatoms. This transfer-function approximates August and February temperatures based on selected diatoms. For site 580 (some degrees north of site 438) he infers temperatures of 22° C (August) and 5–10° C (February). For ODP sites 881 (161° E, 41° N) and 883 (167° E, 51° N), which are located between Kamtchatka and the Aleutes, August temperatures between 15–18° C (significantly warmer than today) and February temperatures between –1 to 4(5)° C are inferred. These temperatures are significantly warmer (ODP-881) resp. slightly colder (ODP-883) than recent temperatures (see discussion of the paleotemperature maps further below). Although this old DSDP-site does not represent state of the art drilling and sampling, it demonstrates that BDLOG is under difficult conditions able to approximate consistent results. Overall however, this site is a key location that should be redrilled for further paleooceanographic studies. DSDP site 299 (Japan Sea, west of Japan) shows for the last 4 My generally cold conditions with temperatures around 8° C (data on the CD-ROM).

Following watermasses further east, the next site is DSDP Site 310 (37° N, 177° E) (see Fig. 20.3b). There are several DSDP/ODP drillholes in the North Pacific which are however not processible by BDLOG for reasons of the faunas (evolution) or due to documentation gaps. Although site 310 is located outside the Oyashio/Kuroshio system, it is a key site for the understanding of the thermal history of the North Pacific (see for example, Romine 1985, Moore and Lombari 1981, Vincent 1981). This applies especially to the extension of cold-watermasses to the south and warm watermasses to the north. DSDP-310 shows, interrupted by short events, from about 16 Ma to about 1.8 Ma a cool North Pacific with temperatures around 8° C. Conclusions about cyclicity cannot be drawn for reasons of sample density. This site should be redrilled with the technology of today, so detailed analyses will be possible.

Following the course of the watermasses (Alaska, California Current), the temperature-baseline increases from the sites off Anchorage (about 7° C) in the Upper Pliocene/Quaternary to about 10° C at about 28° N as shown by site 467 (Fig. 20.4a). In addition, a California Current shows up in the fauna-based temperatures throughout the Neogene (see in addition, from north to south, data from sites 180, 179, 178, 174, 175, 173, 469, 470 and the sites off central America). A southern counterpart off Chile (Humboldt Current) shows up in all processible samples. This means that it existed at least during the Pliocene. Even in this cold current (see site ODP-860), samples documenting slightly warmer conditions can be found. The data from the neighboring sites 859 and 861 are consistent with those from site 860.

The area off Central America is influenced by (1) the northern California Current, (2) the Peru Current (from the south), (3) several threads of the Equatorial Current and (4) coastal upwelling. Consequently, sites with cold-water incursions (site 499, Fig. 20.5a) can be seen as well as sites showing pronounced temperature fluctuations attributed to upwelling or influences of cold waters from the California Current/Peru Current (site 568, Fig. 20.5b).

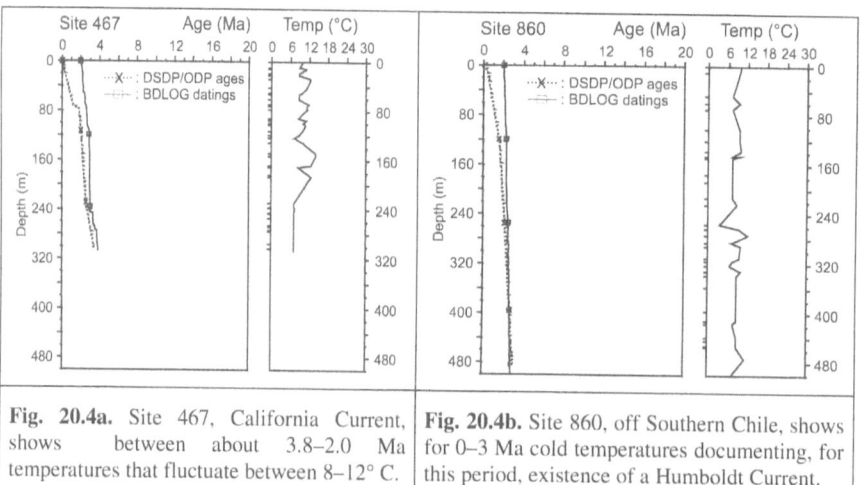

Fig. 20.4a. Site 467, California Current, shows between about 3.8–2.0 Ma temperatures that fluctuate between 8–12° C.

Fig. 20.4b. Site 860, off Southern Chile, shows for 0–3 Ma cold temperatures documenting, for this period, existence of a Humboldt Current.

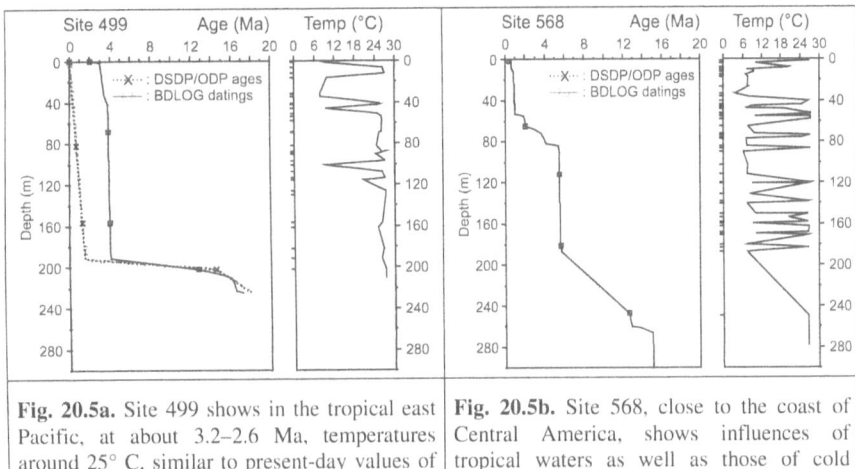

Fig. 20.5a. Site 499 shows in the tropical east Pacific, at about 3.2–2.6 Ma, temperatures around 25° C, similar to present-day values of about 27° C. These temperatures are interrupted by cold-water incursions (upwelling, California Current, Peru current).

Fig. 20.5b. Site 568, close to the coast of Central America, shows influences of tropical waters as well as those of cold waters (upwelling, California Current, Peru current). These fluctuations can be seen throughout the whole documented time.

The tropical central Pacific is characterized by warm conditions through the Pliocene, and very likely also through older parts of the Neogene. Site DSDP-83A shows for the tropical east Pacific, west of central America (5–2 Ma) temperatures that are comparable to present-day values of 27° C (data on the CD). In the time interval 5.0–4.5 Ma, that is in the lower Pliocene, three minor fluctuations to cooler conditions can be observed. This applies basically also to sites DSDP-77B, central tropical Pacific and 586B, eastern tropical Pacific (Figs. 20.6a and b).

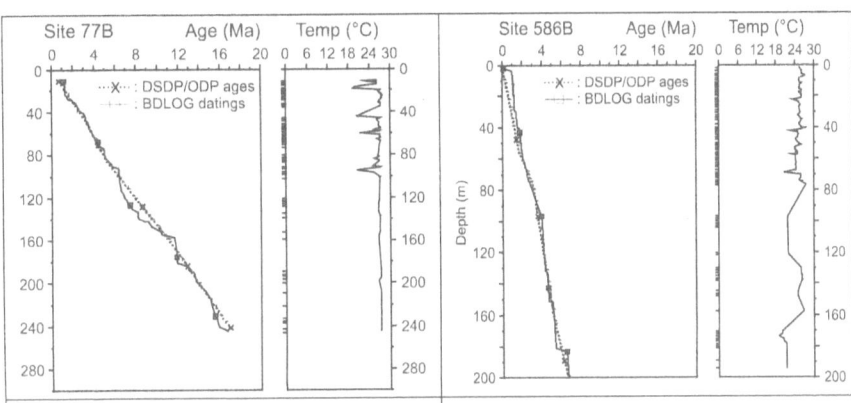

Fig. 20.6a. Site 77B shows for the central tropical Pacific in the time interval between 16 –1 Ma temperatures of about 26° C that are interrupted by some minor incursions of slightly colder waters between 6–0 Ma.

Fig. 20.6b. Site 586B shows for the tropical west Pacific east of New Guinea from about 6–1 Ma temperatures around 26° C Especially between 2.6–0 Ma some short slightly colder intervals can be observed.

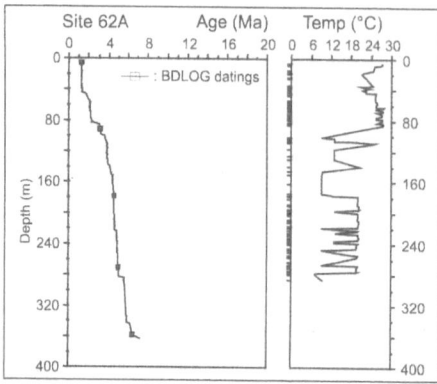

Fig. 20.7. Site 62A, NNE off New Guinea, documents for the time interval between about 6–4 Ma pronounced fluctuations of colder and warmer waters. From about 4 Ma on, generally tropical conditions can be observed. This base-line shift at about 4 Ma is found in many drillsites both in the Pacific and the Atlantic, and is thus caused by a major reorganization of the global circulation, including the deep-water components which cause upwelling (further explanations, also referring to other drillsites, in the text).

Contrasting these findings further west, off New Guinea, pronounced temperature-fluctuations can be observed at site DSDP-62A (Fig. 20.7). Note the shift of the baseline at about 4 Ma, which will be discussed further below. It should be noted at this point that all sites discussed represent different states of drilling-technology, rotary-cores (DSDP-77B, from the pioneer-days of the DSDP) as well as sites drilled with the hydraulic piston corer or other advanced coring techniques. Since intention and technique of the fossil census also changed considerably during the last 20 years, the consistent results of the presented sites are not only a hint to the stability of the temperature transfer-algorithm but also of paleontology in general. This applies especially to the paleooceanographic conclusions that are drawn later from maps that are based on more sites than discussed here.

We now address the area between Australia and New Zealand. This area is today characterized by boundaries of different oceanographic and atmospheric regimes that in a similar or modified form could also have existed in the Neogene. In addition this is an area that is characterized by tropical influences from the north as well as Antarctic fingerprints. Due to the density of both drillholes and samples this is a key area for the understanding of the paleooceanography of the southwestern Pacific. The following section describes from north to south, DSDP sites 587, 588, 208, 590, 591, 592, 284 and SE of New Zealand site 594.

Site 587, the northernmost site of this transect, documents the time interval between 8–1 Ma. Between about 8–4 Ma, summer temperatures (temperatures of the warmer season) fluctuate (with some exceptions) around 20° C. From 4–1 Ma warmer temperatures can be observed. The comparison of site 587 with sites 586B and 62A (off New Guinea) shows an interesting result: Because the waters of site 587, which is located *south* of site 586B show comparably warm values, the fairly cool conditions that can be found at site 62A before 4 Ma can *not* be caused by surface waters flowing from the south. The causes of the cool temperatures found at site 62A are discussed further below. Site 208 (data on the CD) documents for the times younger than 2.6 Ma, tropical environments with temperatures around 25° C. This is consistent with site 587 (Fig. 20.8a) further north and site 588 (Fig. 20.8b) in the immediate vicinity.

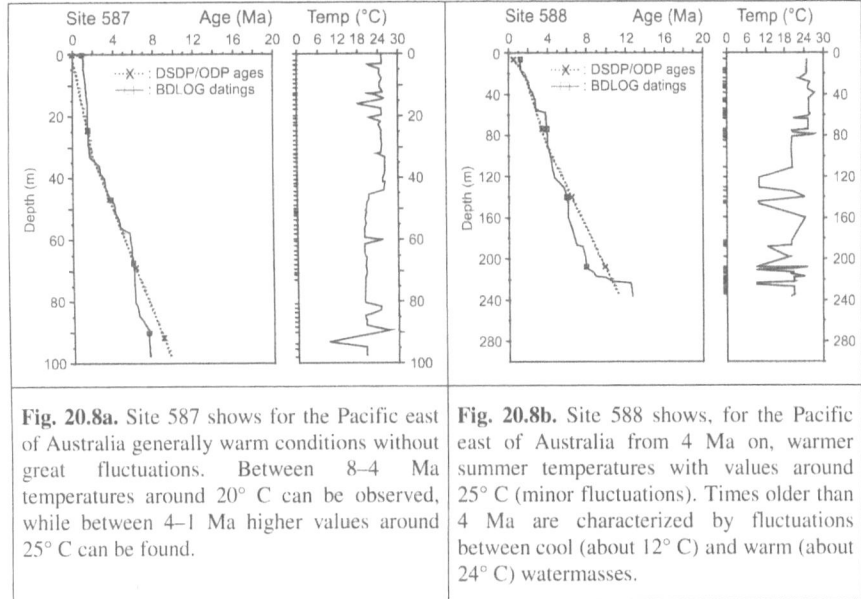

Fig. 20.8a. Site 587 shows for the Pacific east of Australia generally warm conditions without great fluctuations. Between 8–4 Ma temperatures around 20° C can be observed, while between 4–1 Ma higher values around 25° C can be found.

Fig. 20.8b. Site 588 shows, for the Pacific east of Australia from 4 Ma on, warmer summer temperatures with values around 25° C (minor fluctuations). Times older than 4 Ma are characterized by fluctuations between cool (about 12° C) and warm (about 24° C) watermasses.

This demonstrates the stability of BDLOG since workers and drilling technology (rotary drilling of site 208) versus hydraulic piston coring represent totally different standards. The few samples which are older than 2.6 Ma show temperatures around 10° C. This is consistent with the results of site 588 (Fig. 20.8b), where the time interval younger than 2.6 Ma is characterized by tropical conditions (25° C) while before 4 Ma, colder values around 20° C can be observed. Furthermore, the interval before 4 Ma shows fluctuations between cool (12° C) and warm (24° C) conditions. This phenomenon, the temperature shift towards warmer temperatures around 4 Ma, can also be observed in sites 587, 206 (data on the CD), 590A (Fig. 20.9a), 591 (data on the CD) and 592 (Fig. 20.9b).

It could be expected that the change of paleooceanographic conditions affected one of the largest reefs of the world, the Great Barrier Reef. In this context Davies et al. (1991:485) state "A very young age for the Great Barrier Reef (i.e. less than 1 My) is supported by two lines of evidence: (1) dated seismic reflectors traceable beneath the reef and (2) the presence of reef-derived sediments only in the upper part of the section at site 821. The significance of this conclusion is substantial and transcends disciplinary boundaries." On the other hand, Davies et al. (1991:464), describing drillsite ODP-819 state "the entire hole can be assigned to Zone N22-N23, because *Globorotalia truncatulinoides* occurs throughout."

In this context (age of *G. truncatulinoides*) it should be noted that its evolutive maximum age (see Smolka, this volume) is found to be around 9.5 Ma, which conforms with its occurrence at a depth of 150 m in DSDP site 519 (Poore 1984:432). Thus, at least based on above-mentioned fossil, the origin of the Great Barrier Reef cannot be reassigned to a young age. A redating of the Great-Barrier reef may have far-reaching consequences. Since both this work and IGCP-341 focus on other aspects, the discussion is left to other works.

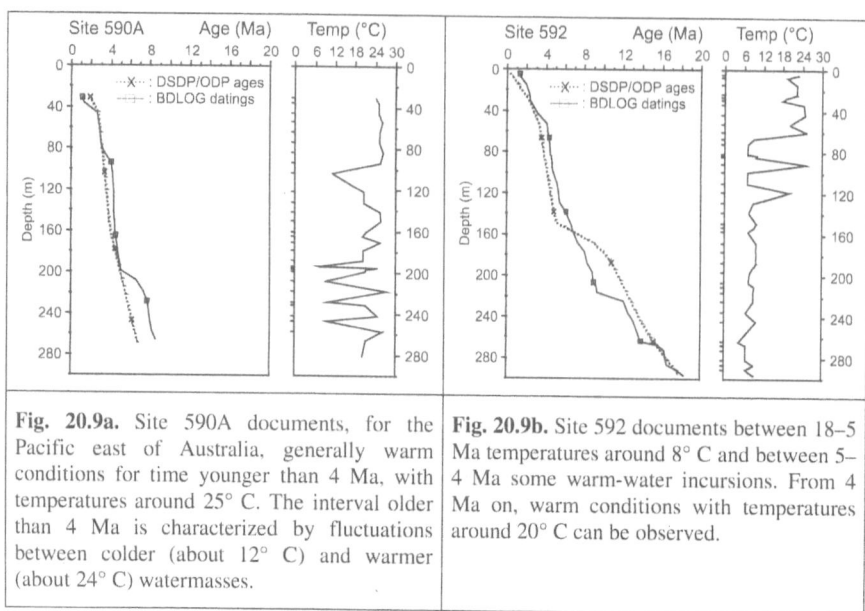

Fig. 20.9a. Site 590A documents, for the Pacific east of Australia, generally warm conditions for time younger than 4 Ma, with temperatures around 25° C. The interval older than 4 Ma is characterized by fluctuations between colder (about 12° C) and warmer (about 24° C) watermasses.

Fig. 20.9b. Site 592 documents between 18–5 Ma temperatures around 8° C and between 5–4 Ma some warm-water incursions. From 4 Ma on, warm conditions with temperatures around 20° C can be observed.

Site 592, which is located further south at about the northern end of New Zealand, documents between 18–5 Ma cool conditions (about 8° C), between 5 and about 4 Ma some warm water incursions, and from 4 Ma on, generally warmer conditions with summer temperatures around 22° C. Note that the fluctuations at site 592 occur *around* 4 Ma.

Sites 593 and 284 are located at the southern end of the Lord Howe Rise off the Southern Island of New Zealand. Both sites coincide stratigraphically well, not only according to the "DSDP-workers" but also according to BDLOG (Smolka this volume). Since most of the samples of site 593 permit stratigraphical analysis by BDLOG, but only a few permit paleotemperature analyses, the older site 284 is discussed here. The few processible samples of site 593 show generally cool conditions (around 10° C) that are interrupted by four "warmer samples".

Site 284 is overall characterized in the younger parts (0.0–3.2 Ma) by generally warmer conditions (about 20° C) that are interrupted by incursions of cold water. The interval from 3.2–6.0 Ma is characterized by numerous oscillations that cover the range of 8, 14 and sometimes even 24° C. Even if the drilling technique and some other limitations are considered, it could be stated that the older interval is generally cooler than the interval younger than 3.2 Ma. This is consistent overall with sites further north, where warmer conditions start at about 4 Ma.

Site 594 is located SE of the Southern Island of New Zealand. The time interval between 14 and about 1.8 Ma is generally characterized by cool conditions with temperatures around 8° C. Only in the youngest part of the site warmer temperatures can be observed (about 14° C).

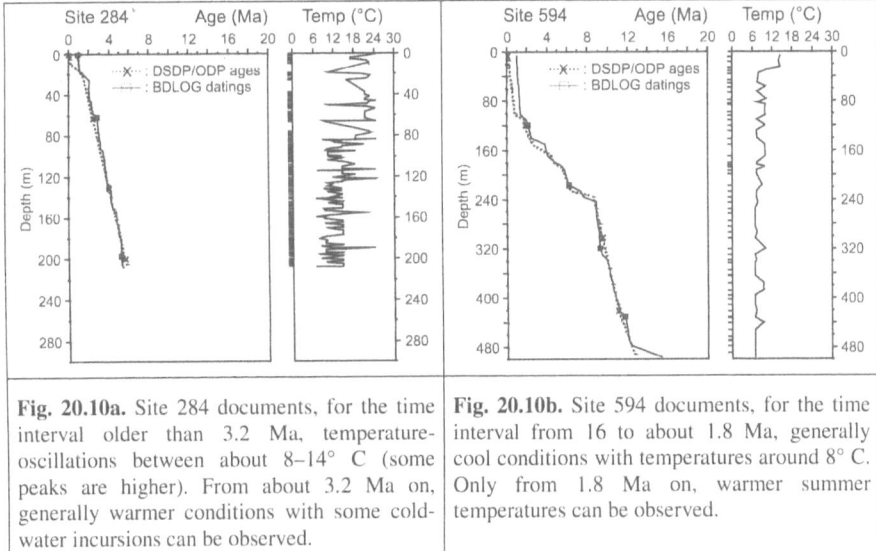

Fig. 20.10a. Site 284 documents, for the time interval older than 3.2 Ma, temperature-oscillations between about 8–14° C (some peaks are higher). From about 3.2 Ma on, generally warmer conditions with some cold-water incursions can be observed.

Fig. 20.10b. Site 594 documents, for the time interval from 16 to about 1.8 Ma, generally cool conditions with temperatures around 8° C. Only from 1.8 Ma on, warmer summer temperatures can be observed.

Overall this transect of drillsites that reaches from site 586B in the north to site 594 in the south, documents the history of watermasses that are influenced by both Antarctic/subboreal conditions on the one hand and tropical conditions on the other. Except in site 284, that might be located exactly near the position of a fluctuating watermass boundary, the northern sites show a shift towards tropical conditions at about 4 Ma. Further south, this shift could be observed at about 3.2 Ma (site 593B documents permanent tropical conditions). The time immediately preceding a change to new stable tropical conditions was characterized by fluctuations between cooler and warmer conditions.

It is remarkable that cool watermasses reached relatively far north in Pliocene times and affected site 588. They did *not* however reach site 587. Therefore, the temperature oscillations that can be observed in the tropical drillsite 62A can *not* be attributed to incursion of cold surface waters from the south, because the more southern site 587 documents tropical conditions throughout. Other causes for the cold-water incursions at site 62A have to be sought (equatorial current, coastal upwelling off New Guinea?). Since site 62A was drilled in the pioneer-days of the DSDP, all statements should be treated with respective care.

In this context the work of Kennett (1986) are mentioned. He studied oxygen and carbon isotopes of Leg 90 at selected sites (588, 590 and 591). There the curve of the oxygen-isotopes of 590 shows a much higher variability than the temperature curve from BDLOG for the same site. Because the isotope-curves contain not only a temperature signal but also an ice-volume and salinity signal, this finding is not a contradiction to the results discussed here. Younger works focusing on the data of ODP Leg 130 and 138, however, use isotope curves for stratigraphy and do not convert isotope-ratios into degrees centigrade.

Before selected sites from other oceans are discussed, the most important aspects, that are also relevant for understanding other oceans, are summarized.

Tropical conditions prevail in the equatorial Pacific for at least the last 16 My. Although this sounds trivial, the title of this contribution, which has also a strong methodological side, should be remembered. The last 16 My are characterized by pronounced evolutionary changes in faunas from all parts of the ocean. The stable results of BDLOG show that evolutionary effects either have no effect or, if they have an effect, the respective samples are not analyzed. This resulted in the low sample density of many wells and many unprocessible wells. BDLOG can thus be regarded as a safe method.

In contrast to the temperatures of the central Pacific in the western equatorial Pacific, colder conditions can be observed before 4 Ma. A discussion of potential factors can be found further below.

The area of the recent Kuroshio Current, east of Japan, is characterized by colder conditions since about 6 Ma, then interrupted since 4 Ma by warmer times.

The North Pacific is characterized as far south as a latitude of 37° N by colder temperatures around 10° C, which are interrupted by incursions of warm waters. The temperature data themselves do *not* permit the deduction of a *permanent* tropical current (Kuroshio) east of Japan during the Neogene. The temperatures are about 12° C higher than off Alaska, but they are far below "tropical/subtropical" conditions. Since a consistent Kuroshio Current in the sense of *permanent* warm water-masses cannot be deduced for most of the Neogene, it is assumed that the warm water-masses reaching site 310 came directly from the south (implying: *not* from the area SW of Japan).

The area SW of Alaska is characterized by cold temperatures throughout the entire Pliocene (which is not a new finding, but consistent with already known data). These cold waters can be followed further south along the west coast of America. In addition, the number of warm-water peaks increases towards the south. An overall increase of the general temperature in the sense of a base-line change *cannot* be observed.

West of Central America two areas can be distinguished: (A) Near the coast tropical and cool conditions change frequently. This is likely to be attributed to varying intensities of upwelling. (B) Sites further west, such as DSDP site 83A do not show this phenomenon.

If an ocean current is reconstructed for one part of an ocean then for reasons of mass-balance a corresponding movement of water can be expected in other parts of the ocean as well (as long as it is not a vertical convection of up- and downwelling). Since for long parts of the Neogene both a California Current and tropical conditions can be deduced, a corresponding Kuroshio-system may have existed, although the respective temperatures are fairly low. This would indicate, (modified slightly later) that during the entire Neogene a current system similar to the present-day system could be expected (with varying intensities, warm water incursions, upwellings, etc.).

In the South Pacific the data density is lower than in the North Pacific. The drillholes off Antarctica could not be included either for stratigraphical reasons or because the available samples had been rejected by BDLOG. Processible sites could be found in the tropics, off Peru, between Australia and New Zealand and in the Humboldt Current off Chile (ODP Leg 141 "Chile Triple Junction" and 112 "Peru Margin"). Therefore, the area between Australia and New Zealand is regarded as a key transect. Here in the northern sites, tropical conditions can be found since 4 Ma, while in the southern sites warm conditions can be found only since 3.2 Ma. Around the southern end of New Zealand, during the entire Neogene cool conditions can be observed.

Originating from the circum-Antarctic current, cool watermasses extended relatively far north between Australia and New Zealand. The boundary between tropical and cool watermasses seldom or never reached site 587. It can be excluded that these watermasses (as deduced from temperatures) reached DSDP site 62A off New Guinea as *cool* surface waters. Statements on the existence of an east Australian countercurrent based on temperature data are not possible (too few sites).

At least for the Pliocene, off Chile a cold current, comparable to the present-day Humboldt Current can be observed. For reasons of the mass budget, the entire Pacific was characterized during the Neogene by a current system that is comparable to the present day. Despite that, at least in the South Pacific, especially in the western part, cold watermasses extended relatively far northward before 4 Ma. This will be discussed further below based on the maps.

20.3.2. Selected Sites from the Atlantic

The main question regarding the South Atlantic for Neogene times are the existence of an Agulhas–Equatorial–Brazil current system on the one hand and the extension of cold Antarctic watermasses to the north on the other.

Site 519 (11° W, 26° S), located in the middle of the South Atlantic, is characterized by tropical conditions (26° C) after 6 Ma. The time interval before 6 Ma documents some colder phases that are interpreted as a boundary between subtropical and Antarctic watermasses.

Sites 357 (35° W, 30° S), 516A (35° W, 30° S) and 518 (38° W, 29° S) are located in the western South Atlantic generally off Rio de Janeiro. Site 518 (Fig. 20.11b) documents, for times since 4 Ma, tropical conditions around 25° C (some samples show pronounced cold-water incursions). In the lower part of this site, both cold and warm temperatures can be observed. Site 516A (Fig. 20.11a), which is located further east, shows pronounced temperature-fluctuations between 8–0 Ma. This is consistent with Barash et al. (1984), who conclude that site 516A was permanently located within the subtropical gyre during the Quaternary. A comparison of site 518 (nearer to the Brazilian coast) with site 516A, especially for the time interval after 4 Ma, shows that a Brazil Current in the present-day sense, which requires oceanographic gradients, came into existence only since 4 Ma. During older times, the data density of DSDP 518 and 516A is insufficient. Thus the current-system can be inferred only through mass-balances and indirect evidence from other locations (see below).

In the *eastern* South Atlantic, site 532B (Fig. 20.12a, 10° E, 19° S) is located on the Walfis Ridge relatively near to the African coast. It shows that generally cold conditions prevail since about 7 Ma (only some warm-water incursions). In this context, studies on the Meteor Rise (ODP sites 703 and 704) should be noted: Here Pujol and Bourrouilh (1991) found cooling events (more polar faunal elements) around 5.2 Ma. Müller et al. (1991) observed cooling-signals for this site around 8.8 Ma.

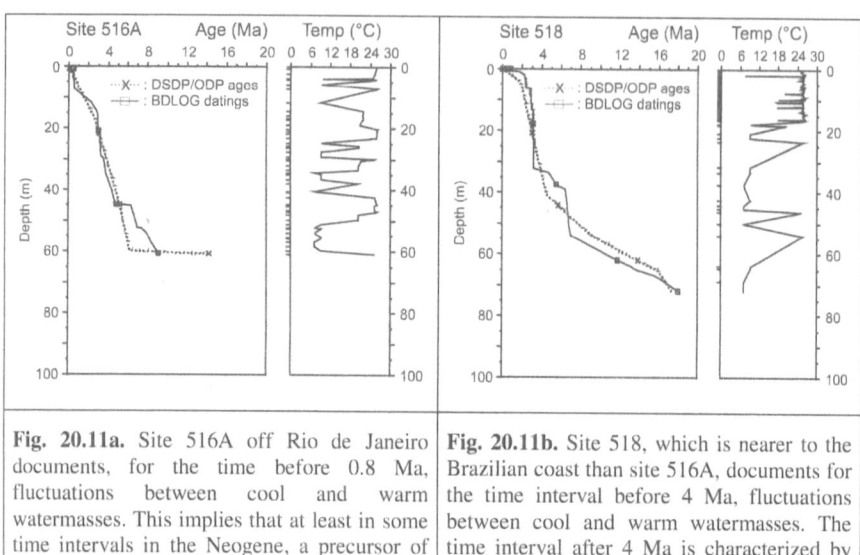

Fig. 20.11a. Site 516A off Rio de Janeiro documents, for the time before 0.8 Ma, fluctuations between cool and warm watermasses. This implies that at least in some time intervals in the Neogene, a precursor of the Brazil current existed.

Fig. 20.11b. Site 518, which is nearer to the Brazilian coast than site 516A, documents for the time interval before 4 Ma, fluctuations between cool and warm watermasses. The time interval after 4 Ma is characterized by generally tropical conditions (Brazil current).

In contrast to these observations, site 525B (2° E, 29° S), also on the Walfis Ridge but further west, documents between about 10–8 Ma warm conditions with temperatures around 24° C. Since 8 Ma, fluctuations of cold and warm intervals can be observed. This supports the above statements regarding the intensity and extension of the subtropical gyre and the Brazil current.

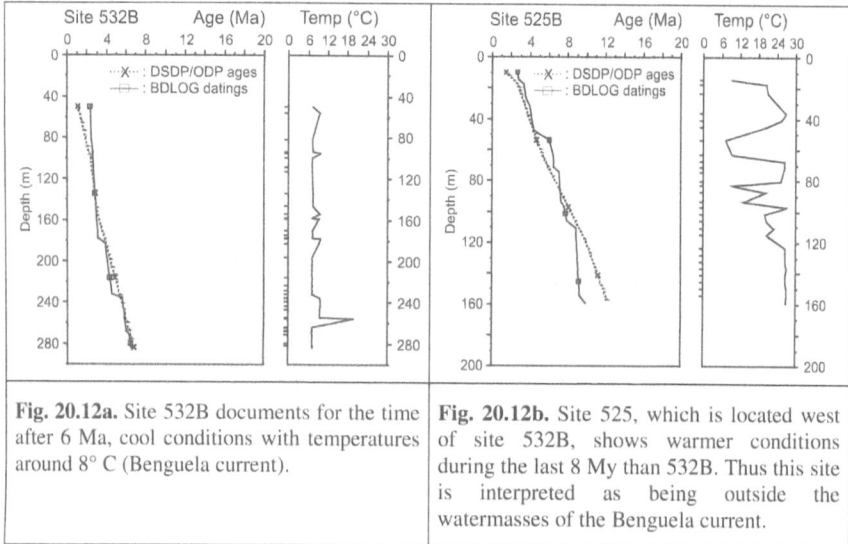

Fig. 20.12a. Site 532B documents for the time after 6 Ma, cool conditions with temperatures around 8° C (Benguela current).

Fig. 20.12b. Site 525, which is located west of site 532B, shows warmer conditions during the last 8 My than 532B. Thus this site is interpreted as being outside the watermasses of the Benguela current.

While site 532 was, during the discussed time interval, permanently inside the "Benguela-system", site 525B was partially inside the subtropical gyre and partially affected by cold Benguela watermasses.

The discussed sites not only document local time series of temperature changes. They also indirectly document, through mass-balances, that at least during the last 10 My in the South Atlantic, a current system existed that was overall comparable to the present. This current consisted of a cold current off the coast of South Africa and a warm current off the coast of Brazil. Both currents had varying intensities and extensions as can be seen in sites close to the system "center" (fluctuations visible in site 516A, Rio Grande Rise) and 525B (Meteor Rise).

Further north, off the coast of eastern North America, only one site is discussed, DSDP-603C (70° W, 35° N). This site (Fig. 20.13) documents in its older part (below 220 m, about 4 Ma), cooler conditions (see also Muza et al. 1987:610).

After some pronounced oscillations these cooler conditions are followed by warm watermasses that are comparable to those of site 533 (data on the CD). From site 533 it can be concluded that a pronounced warm current existed during the last 2 My. Site 603C (Fig. 20.13) shows that this warm current formed in the present-day sense around 4 Ma (or it reached this site at about 4 Ma). This does not exclude that warm watermasses extended northward from the Gulf of Mexico prior to 4 Ma. More about this will be discussed in the last section of this contribution, using maps.

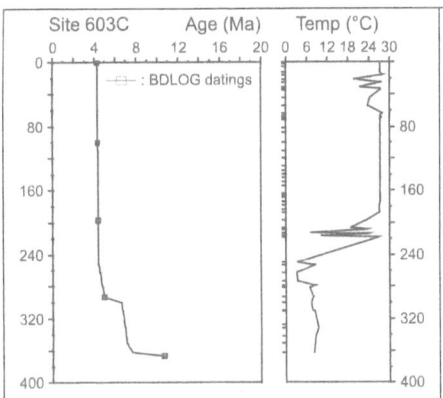

Fig. 20.13. Site 603C, located in the Atlantic off the east-coast of North America, shows at about 4 Ma, a pronounced shift towards warmer conditions. Site 533 further south (data on CD) documents for the last 2 My, comparable summer temperatures. Thus also in this area of the world ocean a pronounced change around 4 Ma could be observed.

The Norwegian–Greenland Sea, the North Atlantic and the Indian Ocean are also covered with analyzed drillholes from the DSDP and ODP. The data can be found on the CD-ROM in the TEMPER-Directory.

20.3.4. Properties of Paleotemperature Maps

For the presentation of sea-surface temperatures (SSTs) three kinds of maps are possible. Each has the following advantages and disadvantages.

1) *Maps based on averaged faunas.* For a given, fairly large time interval, maps of this type compute average faunas (arithmetic means) for each well and each time interval (for example the Upper Pliocene). An advantage is that assignments such as "Upper Pliocene" are readily available (for instance on the DSDP-Tape/DSDP CD-ROM). A disadvantage is that in the resulting averaged samples, species may occur together that never have been observed together. In addition, maps of such large time-slices average over numerous and large climatic fluctuations (see discussion above). The value of such maps is similar to the value of the mean-July temperature from Buenos Aires compared with the mean-July temperature from McMurdo (Antarctica): both are fluctuating considerably but are also very distinct from each other.

2) *Maps documenting a certain horizon such as an isotope stage.* For maps of this type, a unique horizon (see for example Kennett 1985), traceable hemisphere-wide, must exist. In addition maps of this type reflect only very distinct conditions at the time of formation of the horizon. Extrapolations from such a horizon into large times are not valid, due to the high natural variability of climate. Furthermore, depending on the nature of the horizon (isochronous, diachronous), the maps permit different conclusions.

3) *Maps averaging over short time intervals.* Here, for each discussed time interval, temperature values of the samples falling into the respective interval (2–3 Ma, 3–4 Ma, etc.) are averaged. The same applies to the ages of this interval, so (see the data on CD-ROM) it can be checked whether the calculated value is more representative for the top or the bottom of this interval. In addition, the used DSDP/ODP sites have been back-rotated using the hot-spot model of Smith et al. (1994). The same applies to the land–sea masks. Here we note that land–sea masks have been drawn for the Pliocene, Upper, Middle and Lower Miocene, and Oligocene. Coastlines are averages over the respective interval. Therefore in extreme cases, back-rotated drillsites could lie 1–2 mm on a continent. This may happen if the coastline is stored for the center of an interval (such as the middle Middle Miocene) and the drillsite is back-rotated for the lower Middle Miocene. Although this looks unpleasant, the other alternative was to move the drillsite interactively to a wrong position or to reconstruct the coastlines according to a worldwide uniform standard at intervals of 1 My.

Since any reconstruction applies to the whole world, including areas like Greenland and Borneo, and since in many areas of the world people think in terms of members and formations and not in terms of million years, the worldwide synthesis of all Neogene coastlines is regarded as a task for other international initiatives. Further information and references can be found in Zonenshain et al. (1985), Scotese et al. (1988) and Smith et al. (1994). The area of the Bering Straits is not displayed in detail. Here the maps reflect only whether it is closed or open, since this work should not interfere with ongoing seismological studies.

In addition, the land-sea mask is in the step of the map-generation also a paleoceanographical agens. Thus the the mapping-software does *not* interpolate *across* land-boundaries. This means that paleooceanographic gradients across land-bridges, such as the isthmus of Panama, show up correctly. Furthermore, for generating contours, the system uses *only* the data points. The color grid is interpolated *from* the contours (that is, *after* generation of the contours). Thus any kinds of grid-effects, smoothing, oscillations, etc., do *not* occur. Consequently, in areas with a reasonable data density, local gradients have a straightforward interpretation: all gradients that show up are caused by the data. Since the primary data are also shown, the conservative and safety-oriented approach in the map-generating phase is evident.

20.3.5. Average SSTs Between 2 and 3 Ma

This time interval contains data that document paleoclimatic conditions caused by the widespread glaciations after about 2.6 Ma as well as "preglacial" conditions. Overall, the ocean currents appear to be comparable to an average present-day situation (Fig. 20.14). The South Atlantic (off South Africa) is characterized by cold temperatures. For the ocean east of Brazil, warmer conditions are reconstructed. Although on the southern hemisphere the displayed temperatures are winter temperatures, on average the present-day current system with the Benguela and Brazil currents shows up. The North Atlantic is characterized by widespread cold watermasses showing colder temperatures off Greenland than south of the British Isles.

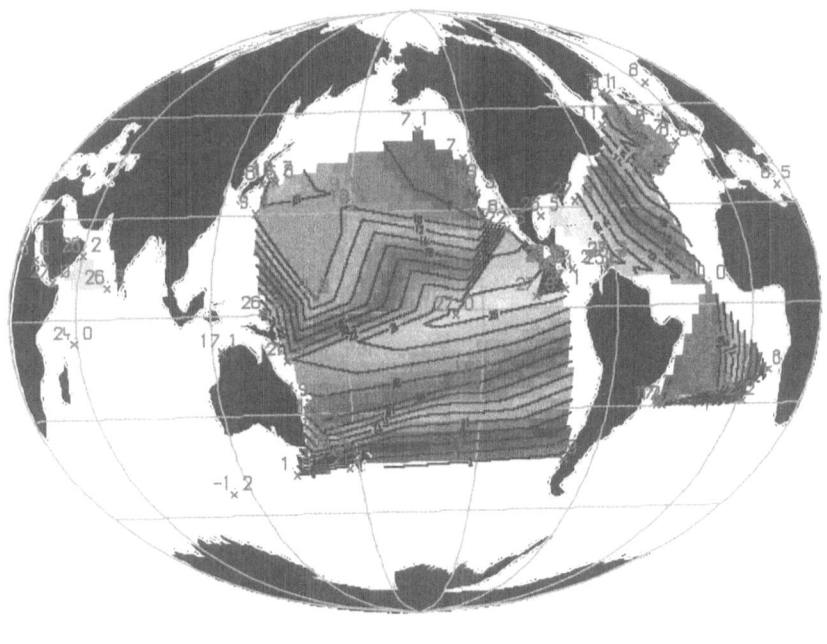

0 0   Celsius   30. 0

Fig. 20.14. Average sea-surface temperatures for the time interval 2–3 Ma. Overall the current-system compares to the present-day situation. Further explanation in the text. A color postscript file of this map can be found on the CD in the TEMPER-Directory.

The Gulf of Mexico is characterized by high temperatures which, together with data off the American east coast, can be used to infer a current-system comparable to the present-day one (Gulf current, West African current, Equatorial current).

West of Panama, in the eastern Pacific, the temperatures are lower than in the Caribbean (closed Isthmus of Panama). Since BDLOG calculates each temperature independently and since the mapping-routines do *not* interpolate across continents, this confirms the stability of the methods. Immediately west of Panama a cold spot can be observed. This is regarded as an effect of coastal upwelling (see the preceding discussions of selected drillsites in this area). A Humboldt Current is visible as well. In the eastern North Pacific large areas of cold waters appear on the map. As the system interpolates between the cold sites of the California Current and warmer sites far away in the central Pacific, this observation shall be interpreted with care until further drillsites are available.

Off Japan in the area of the present-day Kuroshio Current cool temperatures can be observed. Consequently a massive impact of the northern Oyashio Current is deduced. This is *not* problematical, because the time interval 2–3 Ma documents for its larger parts the glacial conditions after 2.6/2.4 Ma.

Between Australia and New Zealand a strong gradient between cold and warm watermasses can be observed. In the Arabian Sea and in parts of the Indian Ocean warm summer temperatures can be found. Overall the current system is in this time interval comparable to that of today.

20.3.5.2. Average SSTs Between 3 and 4 Ma

This time interval is in many cases comparable to that of 2–3 Ma. The South Atlantic is characterized by the "Benguela–Brazil" System. In addition an equatorial temperature gradient between the Caribbean and the Pacific can be observed. Warm watermasses extend relatively far north along the American east coast. In the Pacific, both the California and Kuroshio Currents show up. It is obvious that especially in the North Pacific more drillsites are needed. Because the program has to interpolate between the cold central North Pacific (see Fig. 20.15) and the cold California Current, the North Pacific may appear slightly too cold. In case the maps are used for model-calculations, two alternatives should be run: (1) with a cold Pacific as shown on the map and (2) with a generally warmer Pacific but a local temperature-anomaly off NE Australia.

20.3.5.3. Average SSTs between 4 and 5 Ma

The South Atlantic is characterized by the already known conditions ("Benguela–Brazil" System). Whether the cold watermasses really reached equatorial areas or whether this phenomenon is attributed to interpolation between a data point off South Africa and some equatorial upwelling, must be left open for discussion (see for example, Ruddiman and Sarnthein 1989).

Warm watermasses can be observed in the Caribbean and along the American east coast. The wedge of warm watermasses in the southern North Atlantic is generally possible although the program interpolated across a long distance between a "warm" data point off West Africa and another "warm" data point with 27° C off Venezuela. Further new ODP drillsites would however be useful.

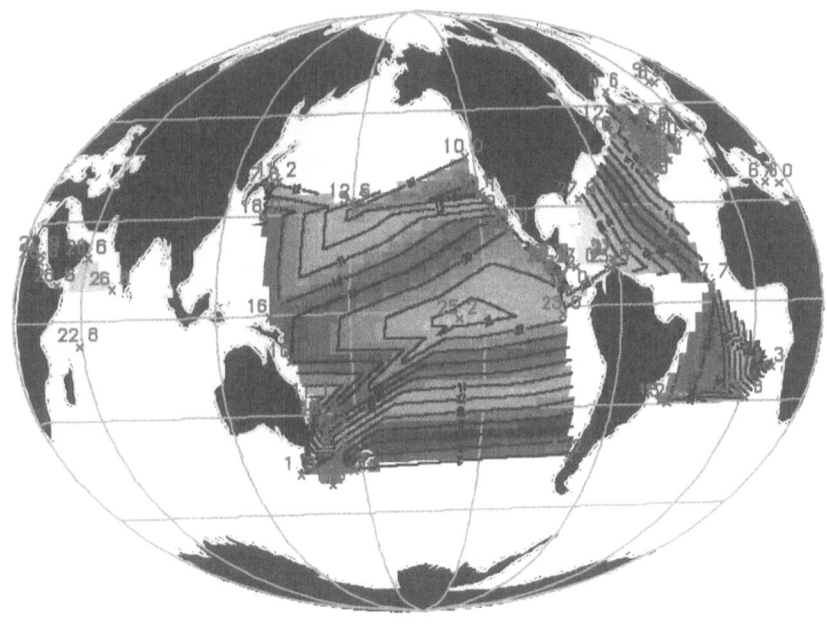

0. 0   Cel si us   30. 0

Fig. 20.15. Average sea-surface temperatures between 3–4 Ma. The South Atlantic and the South Pacific are characterized by currents that have recent equivalents. The same applies to the Pacific. Overall the Pacific and the Indian Ocean appear warmer than the preceding time interval (4–5 Ma). Along the American east coast, warm watermasses reach far northward, which supports a current pattern that differs from that of the preceding time interval. The equatorial east Pacific is characterized by tropical temperatures that are interrupted by coastal upwelling. Between the Caribbean and the equatorial east Pacific, a temperature gradient can be observed. A color postscript file of this map can be found on the CD in the TEMPER-Directory.

In the North Atlantic a small but distinct east-west temperature gradient can be observed. This is a temperature pattern that is similar to the 3–4 Ma time interval but different from the 5–6 Ma time interval (see below).

The equatorial East Pacific is characterized by tropical temperatures. Although it is questionable whether interpolation between the very cold value in the West Pacific and data points further south is possible, it could be stated that overall the West Pacific appears to be colder than the East Pacific. Off Japan a watermass-boundary can be observed. A California Current is also visible in this time. The same applies to the Humboldt Current and the gradient between Australia and New Zealand. The northern Indian Ocean is characterized by warm, but not by tropical conditions. In addition, the wedge of warm waters from the Caribbean into the East Pacific should be noted (open Isthmus of Panama).From the statements above it is evident that oceanographically the maps 4–5 Ma and 3–4 Ma differ fundamentally. Based on the maps alone, this might not be very obvious. In the preceding section in several drillsites a pronounced shift of the temperature-baseline around 4 Ma was observed.

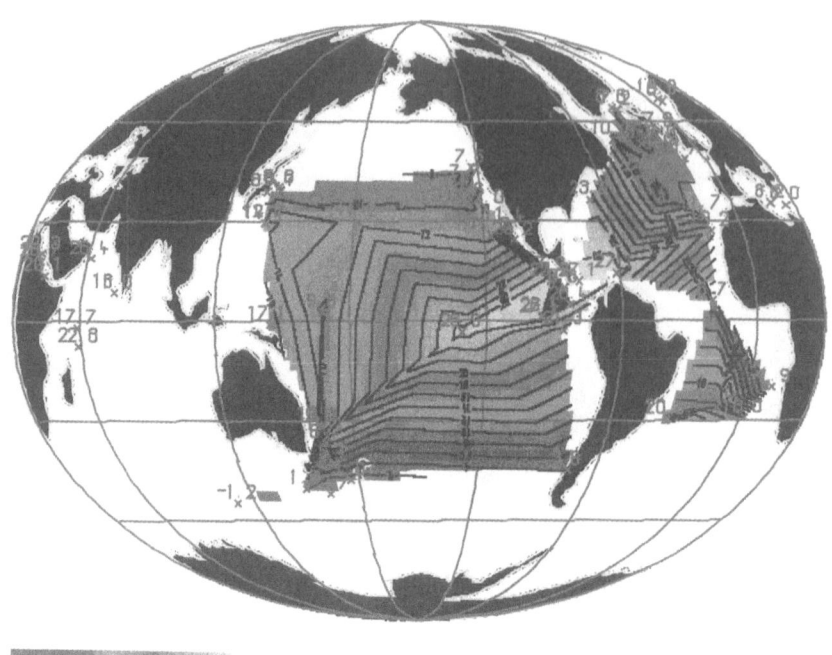

0. 0   Celsius   30. 0

Fig. 20.16. Average sea-surface temperatures between 4–5 Ma. The South Atlantic and the South Pacific are characterized by currents that have recent equivalents. A striking feature is the extension of warm Caribbean waters into the eastern Pacific. In addition, the westernmost Pacific appears to be colder than in the previously discussed time interval (3–4 Ma). See the temperature shift around 4 Ma that is discussed in the text. A color postscript file of this map can be found on the accompanying CD in the TEMPER-Directory.

In the preceding section in several drillsites a pronounced shift of the temperature-baseline around 4 Ma was observed. This shift occurred not only on both sides of the Isthmus of Panama, but in many parts of the world ocean, from the North Atlantic to even the area between Australia and New Zealand. Although the amount of energy that is transferred into the Pacific by the wedge of warm waters is considerable, the effect of these warm surface waters cannot account for the temperature changes off New Guinea, New Zealand and Australia. Thus the closing of the Isthmus of Panama must have affected the whole circulation of the world ocean including the deep-water component. This includes the impact of areas of coastal upwelling as well as the reorganization of the intra-pacific circulation system (California Current, Kuroshio, Oyashio). The properties of the present-day global conveyor-belt are known. The properties of the paleo-global conveyor-belt need however more attention.

As all data are processed independently of each other and as both individual time series and groups of wells and maps show consistent results, this interpretation, which stresses the great impact of orographic changes on the ocean-circulation, is justified. Because the data, which can basically (if all obstacles of the DSDP/ODP are considered) be expected to be very noisy, show consistent patterns, they should be taken seriously. Further additional ODP drillsites located at selected key-positions would however be very useful.

20.3.5.4. Average SSTs between 5 and 6 Ma

The overall picture for the South Atlantic is still the same as discussed. The North Atlantic is characterized generally by cold temperatures. A current system cannot be deduced from the map. In the equatorial North Atlantic the pattern already discussed (warm in the west, colder in the east) seems to form.

East of Asia no hints for a warm Kuroshio can be found. Since on the other hand a pronounced California Current is visible, a potential gyre may not have covered the whole Pacific. The extension of warm watermasses from the area east of Panama into the Pacific should be noted. In this context (California Current, warm watermasses east of Panama, cold waters off Japan) further data, especially high-resolution drillsites in the central North Pacific are essential (see discussion of the key-site 310 further above).

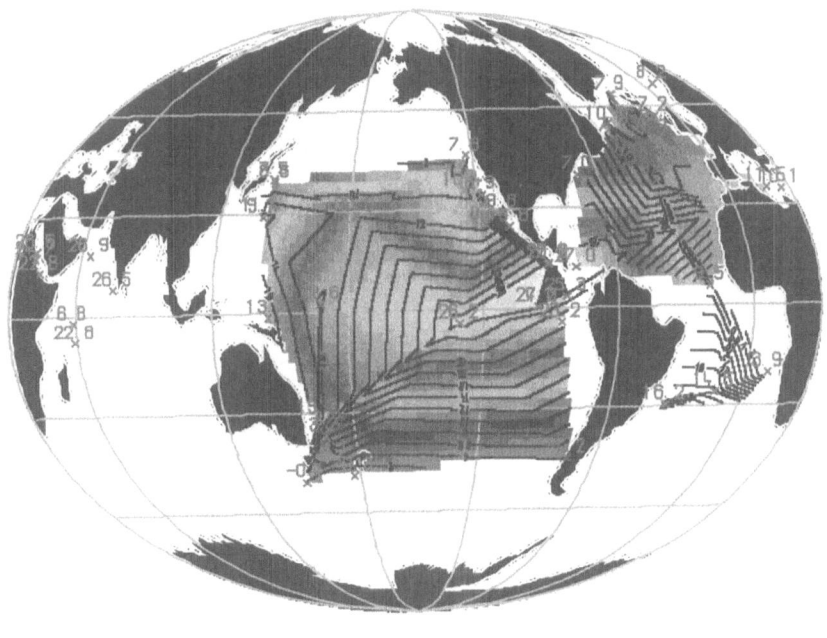

0. 0   Celsius   30. 0

Fig. 20.17. Average sea-surface temperatures between 5–6 Ma. The South Pacific is characterized by currents that have recent equivalents. Warmer waters extend well into the tropical and subtropical Pacific while the extension of warmer waters into the North Atlantic does not yet reach the amounts that can be observed between 4–5 Ma. A color postscript file of this map can be found on the accompanying CD in the TEMPER-Directory.

20.3.5.5. Average SSTs between 6 and 7 My

The South Atlantic documents the known conditions with an asymmetry between a cold eastern and a warm western part. The North Atlantic however appears to differ from the scenarios discussed before: On the one hand it is overall quite cold, on the other hand it appears to be slightly warmer in its central and eastern parts than in the west.

In the Pacific, warm tropical temperatures extend from the area west of Panama further westward into the Pacific. Whether consequently the current system splits into a northern and a southern branch should be left open (data density). Since to the east of Asia no indications for a warm Kuroshio Current can be found but a pronounced California Current does show up, the therefore necessary gyre did not cover the whole North Pacific. If this gyre existed, the northward movements of watermasses occurred in the central North Pacific (see warm water incursions found at site 310). The South Pacific is characterized by the well-known phenomena (Humboldt Current, gradients between Australia and New Zealand). In addition the Indian Ocean documents cooler values in this time interval than in other time intervals.

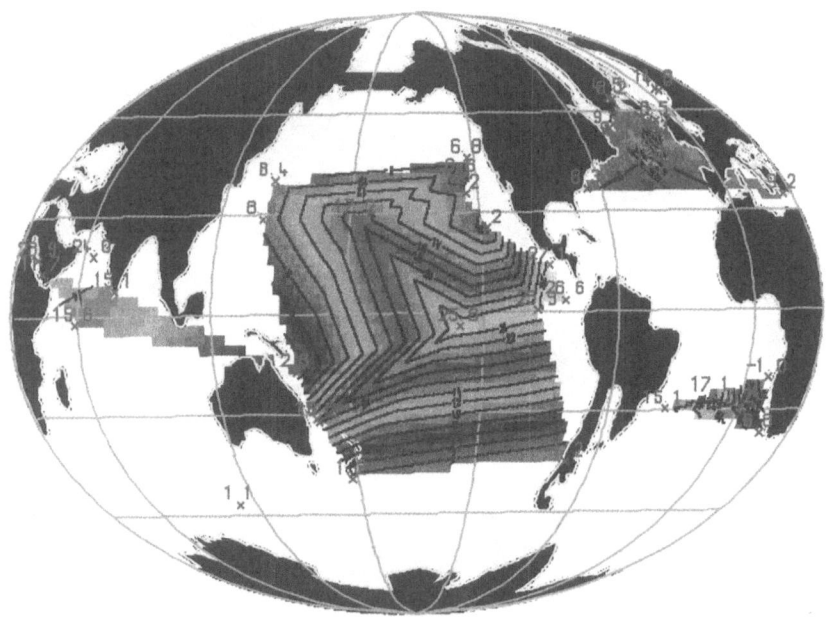

0. 0　　　Celsius　　30. 0

Fig. 20.18. Average sea-surface temperatures between 6–7 Ma. The South Pacific is characterized by currents that have recent equivalents. Overall this time interval appears to be colder than both the younger time intervals and the preceding time interval. A color postscript file of this map can be found on the accompanying CD in the TEMPER-Directory.

In this context the cold values north of Australia are interesting: Although it cannot be said definitely whether this value is typical for the whole interval or whether it is caused by either coastal upwelling or pronounced winter-events, it appears to be consistent with other values from the Indian Ocean. Since all values are calculated independently, consistent results are of a high relevance. On the other hand the data density should be increased by new ODP drillholes.

20.3.5.6. Average SSTs between 7 and 8 Ma

The South Atlantic is characterized by the asymmetry discussed above. In the North Atlantic a differentiation between a cold western part (caused by either a Labrador current or coastal upwelling, note the phosphorite deposits in Florida at this time) and a slightly warmer east Atlantic can be observed. Overall the North Atlantic seems to be characterized during this time by cool conditions and low gradients. Whether this is also a hint for reduced circulation intensities must be left open (see Müller 1992). In the North Pacific warm tropics, a warm Kuroshio, a cold northern North Pacific and a California Current can be observed. The overall picture of the South Pacific (Humboldt Current, cold waters near New Zealand) is comparable to the situations that are already discussed.

0.0   Celsius   30.0

Fig. 20.19. Average sea-surface temperatures between 7–8 Ma. The South Pacific is characterized by currents that have recent equivalents. The same applies to the South Atlantic. A color postscript file of this map can be found on the accompanying CD in the TEMPER-Directory.

20.3.5.7. Average SSTs between 10 and 12 Ma

In the time intervals 8–9 and 9–10 Ma, no general changes take place. Only the data density is decreasing. Therefore, the discussion concludes with the time interval between 10–12 Ma. This time interval shows one data point in the South Atlantic with moderate temperatures, a warm tropical Atlantic and temperate to cool temperatures in the northern Atlantic. If the few data points which are available in the Pacific are interpreted, than a tropical Pacific, a cold California Current and warm waters off Japan (a Kuroshio Current) can be inferred. The same applies to cold waters near New Zealand.

Although a decreasing data density towards older times could be expected, from the data that are available, it could be indirectly inferred that a current system regarding the essential properties comparable to the present-day one *did* exist throughout the Neogene. This is consistent with the observed variations, which locally have been very pronounced (see the sites west of Panama).

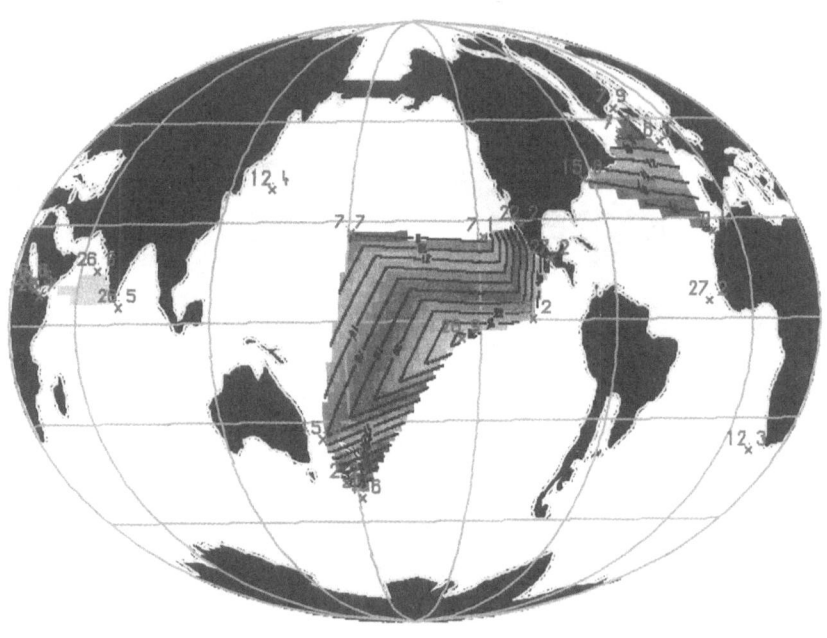

0. 0   Celsius   30. 0

Fig. 20.20. Average sea-surface temperatures between 10–12 Ma. The South Pacific is characterized by currents that have recent equivalents. The same applies to the South Atlantic. From the data points available, a tropical Pacific, a cold California Current, and warm waters off Japan (a Kuroshio Current) can be interpreted as well as cold waters near New Zealand. The same applies for a cold North Atlantic and a warm Arabian Sea. A color postscript file of this map can be found on the accompanying CD in the TEMPER-Directory.

20.4. Summary and Conclusions

An algorithm has been developed that permits the calculation of fauna-based paleotemperatures for Quaternary and pre-Quaternary times.

First, for each sample, the recent components of fossil faunas are detected. Second, if these components are sufficiently high, a set of rules and equations is applied to estimate the summer and winter temperature of the sample. Because an independent, assumption-free measure to check the explainability of a fossil sample through recent faunas exists, effects of evolution and non-analog ecological conditions (low or zero explainability) can be accounted for by exclusion of non-explainable samples from the paleotemperature analysis.

Based on the processible samples (many samples of the DSDP/ODP are *not* processible) time series of the paleo-temperatures have been established. These are discussed at selected sites from the world ocean. It could be demonstrated that the algorithm produces consistent results even if the wells are redrilled after more than ten years (different workers, different drilling technique, etc.). A comparison of the time series and subsequent synopses describe water-masses and local gradients. A remarkable shift in paleooceanographic conditions around 4 Ma can be observed in many wells worldwide. This shift is attributed to the closure of the Isthmus of Panama. Additional changes can be observed around 3.2 and 2.6 Ma.

Overall however it was found that although important and pronounced climatic shifts occurred at many sites on the world, the *main* oceanographic features known from today such as the large ocean currents, seem to have prevailed throughout the entire Neogene. This statement does *not* deny the sensitivity of the climate system or the potential impact of climate changes on the hominid population (which would be dramatic if the temperature change observed around 4 Ma happens again). This statement only emphasizes that considerable short (less than 500 ky) temperature fluctuations, which are dramatic locally and regionally if they occur again, do in the long-range only modify the circulation pattern but do *not* destroy it. A gradient-free world-ocean, which would be proof of a break-down of the circulation system, including the deep-water counterparts, did not occur in the Neogene.

In addition one can reprocess the CLIMAP dataset with the transfer-algorithms discussed here.

IGCP-341 addresses, for climates differing from the present-day climate, similar questions that CLIMAP addressed for the last glacial maximum. Therefore the next step consists of coupling the reconstructed paleotemperatures with atmospheric circulation models. This is discussed both on the CD-ROM and in another contribution (Smolka, this volume).

On the maps a cold eastern part of the South Atlantic and a warm western counterpart appears throughout the Neogene. The same applies to the Humboldt Current and the presence of cold watermasses as far north as New Zealand (influence from Antarctica). Furthermore a California Current can be observed either directly or indirectly throughout the Neogene.

For reasons of mass-balance, a southward-flowing California Current and a northward-flowing Humboldt Current, together with warm tropical waters passing the open straits of Panama into the Pacific, require a corresponding current system, including northward flowing countercurrents. Here, however, warm temperatures *cannot* be observed off Japan in all time intervals. This means that a Kuroshio Current in the present-day sense did not exist through the entire Neogene. Consequently above-mentioned waters must have flown northward not as far west as the present-day Kuroshio, but further east in the central Pacific.

In any case, new drillholes, that should also be sampled and counted as comprehensive as possible, are highly desirable in this important part of the ocean. Especially the coupling of already reconstructed results with physical ocean-models should be used as a tool to optimize the location of the sites.

In the North Atlantic the temperature regime varies. A cold North Atlantic can be observed as well as a North Atlantic that is warmer in the east than in the west. In addition, a North Atlantic showing the present-day configuration shows up on some maps. Although these worldwide maps should be interpreted with care, the transition-time for a situation with an open Isthmus of Panama to a strong gradient across the Isthmus of Panama and warm waters flowing into a northeastern direction, was not less than 1 My. A distinct shift that could be observed in many drillholes needs time to show up as a consistent circulation system.

However, it is known that pronounced climatic fluctuations occurred throughout the Neogene. Therefore it can be concluded that on the long-range the oceanographic system is remarkably stable with respect to atmospheric impact. Changes of physical oceanographic conditions are the most effective mechanism impacting the ocean currents.

These statements do *not* mean that anthropogenic environmental impact should be neglected or is harmless (see even the natural changes of the Amazon rainforest, Latrubesse this volume). It only states that anthropogenic environmental impact may or may not affect those who made it, the hominids through short term-fluctuations or other impact. Especially the non-intended triggering of large-scale natural climate change (see the temperature fluctuations in the drillsites, the natural deforestation of Amazonia that caused the formation of dunes, and the precipitation-changes in South America outlined by Barros, this volume) show that any change from the present-day equilibrium could lead to disastrous consequences under economical aspects. In the long-range (that is in terms averaged over a million years) the geosystem will buffer short-term fluctuations (except orographic changes).

One of the initial questions of IGCP-341 was: Do climatic equilibria departing from the present-day climate significantly affect the geosystem such, that for example the ocean currents would also change significantly, and if so, what do, through modeling, the other parameters such as precipitation look like?

This question can now be answered in general terms: As long as no orographic changes such as the opening of the straits of Panama or obstacles affecting the gulf Current (to be provocative) happen again, no qualitative changes of the ocean Currents may be expected. Short-term fluctuations affecting local areas, such as Europe, New Zealand, etc., may occur. To avoid any misunderstanding, the California Current extended in some times further south than today; in some times it was less intense. If such phenomena happen to other currents, such as the Gulf Current, the consequences for the econoMy of Europe would likely be dramatic. The Gulf Current as a geological phenomenon would however continue to exist.

Consequently, modeling studies dealing with climates departing significantly from the presen can be optimized by focusing on a few key situations. These include: A pre-glacial warmer time interval such as 3–4 Ma; a time interval with an opened Isthmus of Panama such as 4–5 Ma; a generally cooler scenario such as 5–6 Ma; and a scenario with deforested Amazonia/Central Africa.

Overall the temperature reconstructions, the new stratigraphy and the derived maps pushed the limits of paleooceanography considerably forward. This implies also that some interpretations reach intentionally the limits the data provide. Consequently new ODP drillholes and transects should be set such that important paleooceanographic changes are detected better. In addition, data from marine sections on land should be included.

References

Bandemer H, Gottwald S (1993) Einführung in Fuzzy-Methoden. Akademie Verlag, Berlin

Barash MS, Oskina NS, Bylum NS (1984) Quaternary biostratigraphy and surface paleotemperatures based on planktonic foraminifers. In: Barker PF, Carlson RL, Johnson DA (eds): Init. Repts. DSDP, 72:849–870. Washington (U.S. Gov. Print. Off.)

Barron JA (1995) High-resolution diatom paleoclimatology of the middle part of the Pliocene of the Northwest Pacific. In: Rea DK, Basov IA, Scholl DW, Allan JF (eds): Proc. ODP: Sci. Results, 145:179–194. College Station, TX (Texas A & M Univ., Ocean Drilling Program

Cline RM, Hays JD (eds) (1976) Investigation of Late Quaternary Paleooceanography and Paleoclimatology. Geol. Soc. Am. Mem. 145

Davies PJ, McKenzie JA, Palmer–Julson A (1991) Proc. ODP: Init. Repts., 133. College Station, TX (Texas A & M Univ., Ocean Drilling Program)

DSDP (1990): DSDP Paleontology Tape (revised version). National Geophysical Data Center, Boulder, Colorado.

DSDP/ODP (1998): Initial reports / Scientific Results of the Deep Sea Drilling Project / the Ocean Drilling Program Vol. 1–162 (data tables), US Govmt. Printing Office,Washington / Texas A & M University, Ocean Drilling Program, College Station, Tx

Feinen J (1990) Weltweite Entwicklung mariner Planktongesellschaften des jüngeren Känozoikums sowie einige paläoozeanographische und paläoklimatische Implikationen. (Worldwide history of marine plankton-associations of the younger cenozoic and some paleooceanographic and paleoclimatic implications). Dissertation, University of Muenster/Germany

Herman Y (1992) Eocene through Quaternary planktonic Foraminifers from the Northwest Pacific, Leg 126. In: Taylor B, Fujioka K (eds): Proc. ODP: Sci. Results, 126:271–284. College Station, TX (Texas A & M Univ., Ocean Drilling Program)

Hill IA, Taira A, Firth JV (1993): Proc. ODP: Sci. Results, 131. College Station, TX (Texas A & M Univ., Ocean Drilling Program)

Hutson WH (1980) The Agulhas current during the Late Pleistocene: Analysis of modern faunal analogs. Science, 207:64–66

Imbrie J, Kipp NG (1971) A new micropaleontological method for quantitative paleoclimatology: Application to a late Pleistocene Caribean Core. In: Turekian, K.K. (ed): The late Cenozoic glacial ages. Yale Univ. Press, 71–181

Keller G (1980) Planktonic foraminiferal Biostratigraphy and Paleooceanography of the Japan Trench, Leg 57, Deep Sea Drilling Project. In: Thompson PR, Whelan JK (eds): Init. Repts. DSDP, 57:809–834. Washington (U.S. Gov. Print. Off.)

Kennett JP (1985) The Miocene Ocean: Paleoceanography and Biogeography. Geol. Soc. Am. Mem., 163

Kennett JP (1986) Miocene to Early Pliocene oxygen and carbon isotope stratigraphy in the Southwestern Pacific, Deep Sea Drilling Project Leg 90. In: Kennett JP, von der Borch CC (eds): Init. Repts. DSDP, 90, 1383–1412. Washington (U.S. Gov. Print. Off.)

Moore TC Jr., Lombari G (1981) Sea-surface temperature changes in the North Pacific during the late Miocene. Mar. Micropal. 6:581–597

Morley JJ (1989) Radiolarian based transfer functions for estimating paleoceanographic conditions in the south Indian Ocean. Mar. Micropal. 13:293–307

Müller A (1992) Ichthyofaunen aus dem atlantischen Tertiär der USA. Systematik, Palökologie, Biostratigraphie, Evolution und Paläobiogeographie. Habilitation, University of Muenster/Germany.

Müller D, Hodell DA, Ciesielski PF (1991) Late Miocene to earliest Pliocene (9.8–4.5 Ma) paleoceanography of the subantarctic southeast Atlantic: stable istopic, sedimentologic, and microfossil evidence. In: Ciesielski PF, Kristoffersen Y (eds): Proc. ODP: Sci. Results, 114:459–474. College Station, TX (Texas A & M Univ., Ocean Drilling Program)

Muza JP, Wise SW Jr., Covington JM (1987) Neogene calcareous nannofossils from deep sea drilling project site 603, lower continental rise, western North Atlantic: Biostratigraphy and correlations with magnetic and seismic stratigraphy.In: van Hinte JE, Wise SW Jr. (eds) Init. Repts. DSDP, 93:593–660 Washington (U.S. Gov. Print. Off.)

Overpeck JT, Webb T III, Prentice IC (1985) Quantitative interpretation of fossil pollen spectra: Dissimilarity coefficients and the method of modern analogs for pollen data: Quaternary Research 23:87–108

Poore RZ (1984) Middle Eocene through Quaternary planktonic Foraminifers from the southern Angola Basin: Deep Sea Drilling Project Leg 73. In: Hsü KJ, LaBreque JL (eds): Init. Repts. DSDP, 73:429–448. Washington (U.S. Gov. Print. Off.)

Pujol C, Bourrouilh R (1991) Late Miocene to Holocene planktonic foraminifers from the subantarctic South Atlantic. In: Ciesielski PF, Kristoffersen Y (eds) Proc. ODP: Sci. Results, 114:217–233. College Station, TX (Texas A & M Univ., Ocean Drilling Program)

Reynolds RA (1980) Radiolarians from the Western North Pacific, Leg 57, Deep Sea Drilling Project. In: Thompson PR, Whelan JK (eds) Init. Repts. DSDP, 57:735–770. Washington (U.S. Gov. Print. Off.)

Romine K (1985) Radiolarian biogeography and paleoceanography of the North Pacific at 8 Ma. Geol. Soc. Am. Mem. 163

Ruddiman W, Sarnthein M (1989) Proc. ODP: Sci. Results, 108. College Station, TX (Texas A & M Univ., Ocean Drilling Program)

Sancetta C (1978) Neogene Pacific Microfossils and Paleoceanography. Mar. Micropal. 3:347–376

Scotese CR, Gahagan LM, Larson RL (1988) Plate tectonic reconstructions of the Cretaceous and Cenozoic ocean basins. Tectonophysics 155:27–48

Smith AG, Smith DG, Funnel BM (1994) Atlas of Mesozoic and Cenozoic coastlines. Cambridge University Press, Cambridge

Thompson PR (1980) Foraminifers from Deep Sea Drilling Project Sites 434, 435 and 436, Japan Trench. In: Thompson PR, Whelan JK (eds): Init. Repts. DSDP, 57:775–808. Washington (U.S. Gov. Print. Off.)

Vincent E (1981) Neogene planktonic foraminifers from the Central North Pacific, Deep Sea Drilling Project Leg 62. In: Thiede J, Vallier TL (eds): Init. Repts. DSDP, 62:329–354. Washington (U.S. Gov. Print. Off.)

Wolff T, Mulitza S, Arz H, Pätzold J, Wefer G (1998): Oxygen isotopes versus CLIMAP (18ka) temperatures: A comparison from the tropical Atlantic. Geology 26(8), 675–683

Zonenshain LP, Savostin LA, Sedov AP, Volokitina LP (1985) Paleogeodynamic world base maps and paleobathymetry for the last 70 Ma: An explanatory note. Tectonophysics 116:189–207

Paleoclimatic Changes during the Paleocene–Lower Eocene in the Salta Group Basin, NW Argentina

M.E. Quattrocchio[1], W. Volkheimer[2]
(1)Departamento de Geología, Universidad Nacional del Sur Bahía Blanca and CONICET, Argentina
mquattro@criba.edu.ar
(2)IANIGLA–CRICYT, Casilla de Correo 330, 5500 Mendoza, Argentina
Volkheim@lab.cricyt.edu.ar

Abstract: A palynologic study of the Paleocene and Eocene of the Salta Group Basin, NW Argentina, provides climatic information for an area that was, during the Paleogene, located at a geographic latitude similar to the present one (between 23 and 26° S). During the Paleogene the mountain chain of the Andes did not exist. While the Early Paleocene (Danian) is characterized by equivalents of Ulmacean forests (*Phyllostylon*, an endemic species) that today can be found in warm and humid areas, the overlying Late Paleocene (Selandian) Mealla Formation documents somewhat less humid conditions inferred for the "Franja Gris" (= grey band) of the formation, which represents a calcareous pelite plain with shallow lacustrine to palustrine environments and warm climate. An analyzed section from the Late Paleocene (Thanetian) Maíz Gordo Formation is interpreted as dry conditions and a higher altitude montane plant community. The Eocene Lumbrera Formation is palynologically characterized by temperate humid montane paleocommunities.

21.1. Introduction

On a global scale, the Paleocene and Eocene were periods of time with generally much warmer climatic conditions than today. In contrast, we show that during certain parts of the Paleogene, temperatures were similar to those of the present time. We show this by examining the paleoclimatic and paleoenvironmental conditions of Paleocene to Eocene sediments from the Salta Group Basin, NW Argentina.

The Salta Group Basin, currently located between 23–26° S and 63–66° W, was located at a similar paleolatitude during the Paleogene. The Andean Cordillera in the present-day sense did not exist during the Paleogene. In addition, the Antarctic continent still joined Australia and was, in spite of its polar position, largely non-glaciated up to the Early Eocene. Only when Australia separated from Antarctica during the Middle Eocene and began its movement towards its present position, could the cold Circum–Antarctic current form, resulting in a change to cooler conditions in southern South America (for other floristic relations see Boltenhagen 1967).

The geologic history of the Salta Group Basin and adjacent areas (see Fig. 21.1) was partially controlled by a magmatic arc, located in the "Longitudinal Valley" of Chile and the Cordillera Domeyko (Escobar 1980; Jordan 1984), the latter being located approximately between 22° 30´ S and 26° 30´ S, exactly to the west of the Salta Group Basin.

In this paper, we report palynomorphs characteristically from the Paleogene of the Salta Group Basin, in order to provide information for age dating through palynologic analysis. This allows for an interpretation of depositional environments and the reconstruction of vegetation history, prevailing climatic regimes and climatic changes within the basin during the Paleogene. The results are analyzed using the depositional sequences recognized within the Salta Group (Cretaceous–Paleogene) by Gómez Omil et al. (1989).

Samples for palynologic study were collected from the following locations: the Danian Tunal Formation (a lateral equivalent of the top of the Danian Olmedo Formation); localities near the villages of Tilian and Corralito of the Alemanía Sub-Basin; the "Franja Gris" of the Mealla Formation at Sierra de Santa Bárbara in the Olmedo Sub-Basin; the Maíz Gordo Formation at the Corralito locality and the Lumbrera Formation (Faja Verde I and II) at the village "Pampa Grande", both within the Alemanía Sub-Basin (Tab. 21.1, Fig. 21.1). A systematic listing of selected taxa is given in Table 21.2. It includes names of authors and year of publication, following the System of Potonié. Distribution and relative abundances of selected species from Tunal, Mealla, Maíz Gordo and Lumbrera formations are presented as a floral range chart (Table 21.3). Palynomorphs are compared with modern taxa. The terminology for plant communities follows that used by Cabrera (1976), Heusser (1971), and others, for the modern vegetation of Argentina, Chile and worldwide.

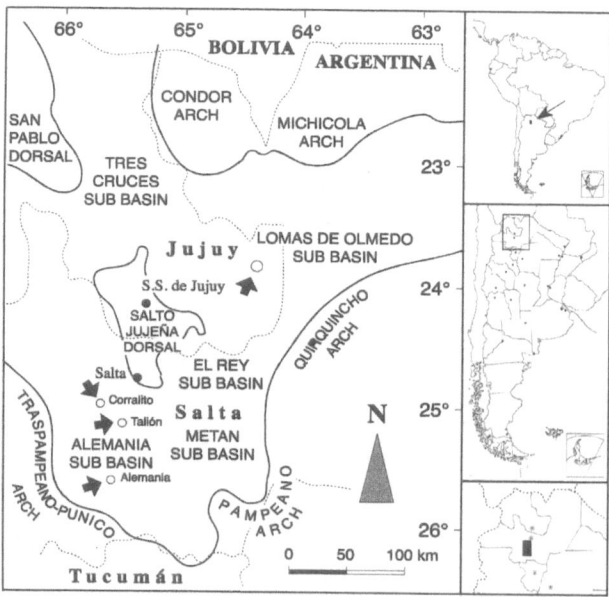

Fig. 21.1. Location map. Salta Group Basin. The basinal outline (solid black line) corresponds to the Balbuena Subgroup (see Table 21.1).

Table 21.1. Stratigraphic setting, Salta Group. The lowermost part of the Salta Group is not including in the Table. This lowermost part of the Salta Group consists of La Yesera Formation, Valanginian to Cenomanian age, and Las Curtiembres Formation, Turonian (?) to Campanian age, both belonging to the Pirgüa Subgroup.

My	Chronostratigraphy		Lithostratigraphy		
	Epoch	Stage	Formation	Subgroup	Group
50.0	E O C E N E		Lumbrera	Santa	S
56.5	P A L E O C E N E	L A T E	Thanetian	Maiz Gordo		A
60.5			Selandian	"Franja Gris" Mealla Fm.	Barbara (post-rift)	
		E A R L Y	Danian	Tunal Olmedo		L
65.0	C R E T A C E O U S		Maastrichtian	Yacoraite ———— Lecho	Balbuena (post-rift)	T
			Campanian Turonian(?)	Los Blanquitos	Pirgüa (syn-rift)	A

21.2. Geologic Setting and Stratigraphic Subdivision of the Salta Group

The continental Cretaceous–Paleogene sequence of NW Argentina (Salta Group) is composed of three major units (Table 21.1). These are, from bottom to top, the Pirgüa Subgroup of Early and Late Cretaceous age, the Balbuena Subgroup of Maastrichtian to earliest Paleocene age, and the Santa Bárbara Subgroup of Paleocene and Eocene age (see also Archangelsky 1976).

There were two main episodes in the basin development. First, a *synrift stage*, represented by the Pirgüa Subgroup redbeds and basalts, and second a *postrift stage*, represented by the Balbuena and Santa Bárbara Subgroups (Salfity and Marquillas 1981; 1994). The positive structural elements (Fig. 21.1) that governed the sedimentation of the Salta Group are the Michicola Arch (Vilela 1965), the Quirquincho Arch (Salfity 1980), the Pampean and Transpampean Arches (Padula and Mingramm 1968) and the San Pablo and Salta–Jujuy highs (Reyes 1972). The grabens or depocenters identified (Fig. 21.1) have been named Tres Cruces, Lomas de Olmedo, Metán, Alemanía, Cerro Hermoso, El Rey and El Charco or Sey (Reyes 1972; Salfity 1980; Schwab 1984, 1985).

The sedimentary history of the Salta Group Basin began early in the Cretaceous with the deposition of the continental La Yesera Formation of Valanginian–Cenomanian–Turonian(?) age (Salfity and Marquillas 1994, Fig. 11). It is characterized by alluvial, fluvial, eolian and lacustrine deposits. This formation is the lowermost unit of the Pirgüa Subgroup of the Salta Group. It is overlain by the fluvial and lacustrine Las Curtiembres Formation (Turonian–Santonian), which is partially a lateral equivalent of the alluvial, eolian, and fluvial Los Blanquitos Formation (Turonian (?) –Campanian). For stratigraphic details, see Gómez Omil et al. (1989), Salfity (1968, 1971, 1982), Salfity and Marquillas (1994) and Volkheimer et al. (1984).

The clastic and carbonate Balbuena Subgroup is composed of the mainly fluvial and eolian Lecho Formation (Maastrichtian) and the carbonate Yacoraite Formation (Maastrichtian to lowermost Danian), deposited in a shallow restricted carbonate basin. The Santa Bárbara Subgroup consists of the Olmedo/Tunal, Mealla, Maíz Gordo and Lumbrera formations (Fig. 21.2). Each of these continental formations has its particular paleogeography and sedimentary history (Gómez Omil 1989). The Olmedo Formation is clastic/evaporitic , with a hypersaline lake in the basinal center and low topography in the surroundings.

The Mealla Formation (Moreno 1970) is characterized by red pelites in the basinal center and coarser clastics near the borders. Sedimentation culminates with a facies of dark calcareous pelites, called "Franja Gris", deposited during flooding of nearly the whole basin. The Maíz Gordo Formation is composed of greenish-grey and reddish pelites and psephites with frequent carbonate intercalations (algal stromatolites amongst them). The subaqueous facies are concentrated in those areas characterized by a low clastic input. The formation reaches more than 500 m of thickness at Lomas de Olmedo. The Lumbrera Formation is mainly composed of fine-grained clastic redbeds. The "Fajas Verdes" are composed of greenish-grey calcareous pelites, and is a widely distributed "guide-bed."

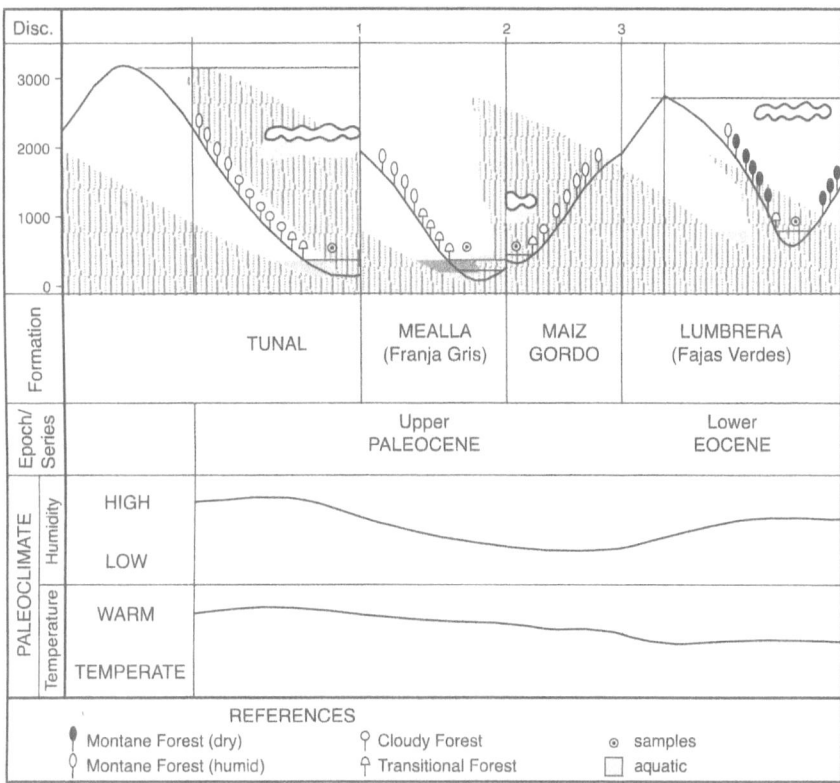

Fig. 21.2. Paleoenvironmental reconstruction for the Paleocene and Eocene of the Alemanía Subbasin (Tunal, Maíz Gordo and Lumbrera Formations) and Olmedo Subbasin (Mealla Formation: "Franja Gris") and inferred overall paleoclimatic development (temperature and humidity) during the Upper Paleocene and Lower Eocene.

21.3. Methods

In terrestrial environments, vegetation analysis is one of the keys for understanding paleoclimates, including both temperature and water availability. Pollen grains and spores are those remnants of plants that have the highest chance of documentation over long time periods. Furthermore, their small size and robustness against various kinds of weathering permits a quantitative assessment of vegetation communities. Pollen grains and spores originate from plants that formed past vegetation in a study area. However, many pollen types can be transported over larger distances. Therefore, careful analysis of the observed communities also permits analysis of local, more distant and regional vegetation growing around the deposition site. Because vegetation is responsive to environmental factors, the study of pollen grains from one site may, if carried out carefully, permit insight into conditions at more remote sites: While palustrine vegetation may grow at the deposition site, additional pollen may help to distinguish between either humid forest environments around that swamp, or desert conditions.

Samples were processed following the techniques of Volkheimer and Melendi (1976). This includes treatment with cold HCl to remove carbonates, subsequent treatment with HF to remove silicates, and then application of oxidation and flotation liquids. This paper includes the study of the taxonomy of pollen and spores, assignment of fossil taxa to modern equivalents, discussion of different plant species, subsequent assessment of vegetation communities and interpretation of pollen assemblages in terms of flora, vegetation and environments. We also characterize the physical environment through the study of lithofacies.

21.4. Results

21.4.1. Paleogene Palynomorphs Identified in the Salta Group Basin

The palynomorphs identified in the Tunal, Mealla, Maíz Gordo and Lumbrera formations are listed in Table 21.2.

21.4.2. Palynostratigraphy: Age of the Microfloristic Assemblages Identified in the Tunal, Mealla, Maíz Gordo and Lumbrera formations.

Considering the pollen assemblage, an Early Paleocene Age (Danian) is assigned to the Tunal Formation. Quattrocchio et al. (in press) defined the *Palynozone* of *Mtchedlishvilia saltenia*, where an interval zone extends from its first appearance to the first appearance of *Rousea patagonica* (Mealla Formation, Salta Group Basin). The first appearance of *Rousea patagonica* is also observed in the Late Danian Salamanca Formation (San Jorge Basin, Patagonia) and the Pedro Luro Formation (Colorado Basin). In this context the work of Pierce (1961) should be noted. The presence of *Simpsonotus* (Mammalia, Henricosborniidae) in the Mealla Formation and the absence of other Casamayoran mammals allowed correlation between the Mealla and the Río Chico Formations of Patagonia (Riochican Age), assigning it a Middle to Late Paleocene Age (Pascual et al. 1981). We note that *Mtchedlishvilia saltenia* (Moroni 1984) is characteristic of the Danian in NW Argentina and the Colorado Basin, but is not present in the Mealla Formation

The Maíz Gordo Formation is Late Paleocene (Riochiquense Mammal Age) (Pascual et al. 1981), with *Corydoras revelatus*, Poeciliidae; *Podocnemis*, Pelomedusidae, Crocodylidae, Notoungulata (Henricosbornidae: *Simpsonotus*). The microflora from the upper part of the Maíz Gordo Formation contains species restricted to the Paleocene, together with Eocene forms (Quattrocchio and Volkheimer 1990).

The Lumbrera Formation is of Eocene Age (Casamayoran Mammal Age) with Siluriformes, Poeciliidae, *Lepidosiren paradoxa*, Pelomedusidae, Phororhacoidea, Bonapartheriidae (*Bonapartherium*), Prepidolopoidae (*Prepidolops*) and others. Based on radiometric analysis ($^{40}Ar/^{39}Ar$), Heizler et al. (1998) reported a Late Eocene age for the Casamayorense of Patagonia. The Lumbrera Formation contains palynomorphs of many biochrons together with endemic species (Quattrocchio and Volkheimer 1990).

Table 21.2. Selected taxa from the Santa Bárbara Subgroup of the Salta Group (Tunal Formation, Mealla Formation: "Franja Gris", Maíz Gordo Formation and Lumbrera Formation.

Psilamonoleti
Laevigatosporites sp. A (of Quattrocchio 1978b)
 	
Laevigati
Azolla sp. (of Quattrocchio et al. 1997)
Biretisportites sp. (of Quattrocchio 1980)
Deltoidospora sp. (of Quattrocchio et al. 1997)
 	
Saccites
Podocarpidites marwickii Couper 1953
 	
Aletes
Inaperturopollenites sp. D (of Quattrocchio 1980)
Inaperturopollenites sp. E (of Quattrocchio 1978b)
Smilacipites saltensis (Quattrocchio 1980)
Smilacipites sp. (of Quattrocchio 1978b)
 	
Plicates
Liliacipites variegatus Couper 1953
Monosulcites minutiscabratus McIntyre 1968
Spinizonocolpites sp. (of Archangelsky 1973)
Ephedripites cf. E. sp. 1 Frederiksen et al. 1983
  (of Quattrocchio and Volkheimer 1988)
Gemmatricolpites subsphaericus Archangelsky
  1973
Clavatricolpites cf. gracilis González Guzmán
  1967
Tricolpites bibaculatus Archangelsky and
  Zamaloa 1966
Tricolpites communis Archangelsky 1973
Tricolpites (Psilatricolpites) lumbrerensis
  Quattrocchio 1980
Tricolpites cf. reticulata Cookson 1947 (of
  Quattrocchio and Volkheimer 1988)
Tricolpites sp. cf. T. reticulata Cookson 1947 (of
  Quattrocchio 1978a)
Rousea patagonica Archangelsky 1973
Psilatricolpites acerbus González Guzmán 1967
Psilatricolpites simplex González Guzmán 1967
Rhoipites minusculus Archangelsky 1973
Rhoipites baculatus Archangelsky 1973
Rhoipites sp. A (of Quattrocchio 1980)
Rhoipites sp. B (of Quattrocchio et al. 1988)
Rhoipites sp. (of Quattrocchio 1978b)
Psilatricolporites salamanquensis Archangelsky
  and Zamaloa 1986
Retitricolporites chubutensis Archangelsky 1973
Retitricolporites medius González Guzmán 1967
Retitricolporites sp. A (of Quattrocchio et al.
  1988)
Ailanthipites sp. (of Quattrocchio et al. 1988)
Nothopollenites sp. (of Quattrocchio 1978b)
	Murornati
	Retitriletes austroclavatidites (Cook) Doring,
	  Krutzsch, Mai and Schultz 1963
	Ischyosporites spp. (in Quattrocchio et al. 1997)
	
	Apiculati
	Baculatisporites sp. (of Quattrocchio 1980)
	Apiculatisporis sp. (of Volkheimer 1972)
	
	Poroses
	Restioniidites spp. (of Quattrocchio et al. 1997)
	Pandaniidites texus Elsik 1968
	Pandaniidites sp. (of Archangelsky 1973)
	Myriophyllumpollenites sp. 1 (of Quattrocchio
	  and Volkheimer 1988)
	Myriophyllumpollenites sp. 2 (of Quattrocchio
	  and Volkheimer 1988)
	Liquidambarpollenites cf. brandonensis Traverse
	  1955 (of Quattrocchio, 1978a)
	Verrustephanoporites simplex Leidelmeyer 1966
	Corsinipollenites menendezii Quattrocchio 1978b
	Echistephanoporites sp. cf. E. alfonsi
	  Leidelmeyer 1966 (of Volkheimer et al. 1984)
	Cricotriporites cf. guianensis Leidelmeyer 1966
	  (of Volkheimer et al. 1984)
	Cricotriporites sp. A (of Volkheimer et al. 1984)
	Triorites sp. (of Quattrocchio et al. 1984)
	
	Incertae sedis
	Mtchedlishvilia saltenia Moroni 1984
	
	Megaspore
	Grapnelispora evansii Stover and Partridge 1984
	
	Fungi
	Inapertisporites sp. A (of Quattrocchio 1978b)
	Inapertisporites ovalis Sheffy and Dilcher 1971
	Multicellaesporites sp. (of Volkheimer et al.
	  1984)
	Diporisporites elongatus V.d. Hammen 1954
	Hypoxylonites (=Diporisporites) sp. (of
	  Quattrocchio 1980)
	Diporicellaesporites sp. (of Quattrocchio 1978b)
	Pluricellaesporites sp. A (of Quattrocchio 1978b)
	
	Algae
	Pediastrum sp. (of Quattrocchio and Volkheimer
	  1988)
	Catinipollis geiseltalensis Krutzsch 1966

Table 21.3. Modern equivalents of some fossil taxa present in the Santa Barbara Subgroup, Paleogene, Salta Group Basin.

Taxa	*Modern Equivalents*
Azolla sp.	Salviniaceae
Dictyophyllidites sp	Dipteridaceae
Baculatisporites sp. 1	Dipteridaceae
Ischyosporites sp.	Schizaeaceae
Retitriletes austroclavatidites	Lycopodiaceae
Laevigatosporites sp. A	Pteridophyta
Gabonisporites vigourouxii	Marsileaceae
Podocarpidites marwickii	Podocarpaceae
Inaperturopollenites sp. D	Araucariaceae
Inaperturopollenites sp. E	Araucariaceae
Ephedripites cf. E. sp. 1	Ephedraceae (?)
Gemmatricolpites subsphaericus	Aquifoliaceae
Rhoipites sp. B	Solanaceae
Retitricolporites sp. A	Anacardiaceae
Ailanthipites sp.	Anacardiaceae
Pandaniidites texus	Pandanaceae
Myriophyllumpollenites sp. 1	Haloragaceae
Myriophyllumpollenites sp. 2	Haloragaceae
Tricolpites cf. reticulata	Gunneraceae
Tricolpites vulgaris	Hamamelidaceae
Rousea patagonica	Adoxaceae
Rhoipites baculatus	Mimosaceae
Rhoipites sp. A	Rutaceae
Retitricolporites chubutensis	Vitaceae/Rutaceae
Restioniidites sp.	Restionaceae
Myriophyllumpollenites sp.	Haloragaceae
Verrustephanoporites cf. simplex	Ulmaceae (Phyllostylon)
Retitricolporites medius	Anacardiaceae
Corsinipollenites menendezii	Oenotheraceae
Nothopollenites sp.	Combretaceae
Cricotriporites cf. guianensis	Ulmaceae (Celtis)
Liquidambarpollenites cf. brandonensis	Hamamelidaceae
Liliacidites variegatus	Iridaceae
Spinizonocolpites sp.	Palmae (Nipa)
Smilacipites saltensis	Liliaceae
Smilacipites sp.	Liliaceae
Catinipollis geiseltalensis	Zygnemataceae (Chlorophyceae)
Pediastrum sp	Chlorophyceae
Inapertisporites sp. A	Fungi
Inapertisporites ovalis	Fungi
Multicellaesporites sp.	Fungi
Diporisporites sp.	Fungi
Diporisporites elongatus	Fungi
Diporicellaesporites sp.	Fungi
Pluricellaesporites, sp. A	Fungi
Hyphae type 1 Elsik 1968	Fungi

21.4.3. Paleocommunities and Paleoclimates

In this study both local and inland assemblages are recognized. The grabens and positive structural elements that governed sedimentation in the Salta Group show characteristic microfloristic assemblages. Since the Paleocene and Eocene, the magmatic arc was located in the "Valle Longitudinal" and "Cordillera Domeyko" in Chile (Escobar 1980, Jordan 1984). But, there is no information about the generation of elevated areas associated with the formation of this arc.

Following De Spirito (1979), the Balbuena Subgroup (Moreno 1970) is reinterpreted: It is composed of the Lecho and Yacoraite formations, but does not include the Olmedo Formation. The Santa Barbara Subgroup is composed of the Olmedo, Mealla, Maíz Gordo and Lumbrera formations. Each one has particular paleogeographic and sedimentary features.

We consider the tectosedimentary study of Gómez Omil et al. (1989) for the Cretaceous–Tertiary Basin of NW Argentina (Salta Group) with the paleofloristic changes observed during the Paleocene–Eocene in this basin. A stratigraphic discontinuity is mentioned in the contact between the Pirgua and Balbuena Subgroups. The upper boundary to the overlying Olmedo Formation is defined by a sedimentological discontinuity, eustatic rupture and a distensive tectonic rupture.

The boundary between the Olmedo Formation and the overlying Mealla Formation corresponds to a tectonic-sedimentary reactivation that developed in different basin positions. This reactivation is documented by a drastic change of the environmental conditions between the Tunal and Mealla Formations. In the "Franja Gris" (the Mealla Formation) a decrease of humid forest representation (Transitional Forest) is observed. We propose that the relative dryness during the Mealla Formation deposition could be correlated with an episode of relative low sea level in the marine Atlantic basins (Quattrocchio et al. 1997).

A new paleogeographic change (see Fig. 21.3) could be inferred by the dominance of dry and higher montane communities in the Maíz Gordo Formation with respect to the Mealla Formation (Franja Gris). The lower member of the Lumbrera Formation also has a soft structural control in relation with distensive faults in the subsurface of the Lomas de Olmedo Subbasin. In addition, it is characterized by a progradation of coarse clastic sediments at the active edges of its Northern flanks and near the village Alemanía. The palynologic assemblages of the members Fajas Verdes I and II of the Lumbrera Formation reflect a lacustrine environment with humid temperate montane paleocommunities at a regional scale. At the same time, elements of the subtropical–tropical forest continue. In the palynologic record the Ulmaceae are poorly represented. This drastic change is also documented by vertebrates since the Early Eocene (Casamayoran Mammal age) in central Chubut (Pascual et al. 1981).

At the same time, in the Colorado Basin there are no Ulmaceae pollen grains observed. This reflects the recession of the Ulmaceae forest to areas further north.

21.4.3.1. The Tunal Formation (Danian)

The presence of Haloragaceae (*Myriophyllumpollenites* sp.) and *Pediastrum* sp. (Chlorophyceae) reflect a lake environment for the Tunal Formation. Due to the assignment of *Verrustephanoporites* cf. *simplex* to *Phyllostylon* (78.0–88.5%)

(Quattrocchio et al. 1988), Ulmaceae, the paleoenvironment of Tunal Formation could be similar to the present Transitional Forest at 350–500 m asl (above sea level) in the Yungas Province, Amazonic Dominion (Cabrera 1976), where this tree is dominant (see also Table 21.4).

The presence of Podocarpaceae (*Podocarpidites marwickii*), Rutaceae (*Rhoipites* sp. A) and Anacardiaceae (*Retitricolporites* sp. A) could indicate an association similar to the Montane Forest District, located in the upper part (1200–2500 m asl) of the Yungas Province.

Today the interval between both mentioned districts is occupied by the Cloudy Forest District (550–1600 m asl). This district could be indicated by the presence of Aquifoliaceae (*Gemmatricolpites subsphaericus*).

The climatic conditions of the Yungas Province (in northwestern Argentina this is known as the subtropical jungle of Tucumán and Oran), are humid and warm with generally rain during summer, some frost in winter and annual pluvial precipitations of around 2500 mm. Thus, these could have been the climatic conditions during the deposition of the Tunal Formation.

Table 21.4. Habitats (meters above sea level) of the modern equivalents of some fossil taxa found in the Tunal, Mealla ("Franja Gris"), Maíz Gordo and Lumbrera formations.

Habitat	Taxa (modern equivalents)
Montane Forest (1200–2500 m):	Podocarpaceae Araucariaceae Gunneraceae Hamamelidaceae Rutaceae Anacardiaceae
Cloudy Forest (550–1500 m):	Aquifoliaceae Lycopodiaceae
Transitional Forest (350–500 m):	Ulmaceae (<u>Phyllostylon</u>, <u>Celtis</u>) Iridaceae
Steppe:	Ephedraceae
Littoral:	Combretaceae
Lacustrine:	Marsileaceae Haloragaceae Clorophyceae (<u>Pediastrum</u>) Salviniaceae <u>Catinipollis geiseltalensis</u>
Palustrine	<u>Mtchedlishvilia saltenia</u> Pandanaceae Oenotheraceae Palmae (<u>Nypa</u>) Restionaceae

21.4.3.2. The Mealla Formation: "Franja Gris" (Selandian)

The Transitional Forest is more impoverished in the "Franja Gris" of the Mealla Formation (Ulmaceae: 10–29%) than in the Tunal Formation (Danian). The first case could be attributed to arid conditions or indicate that these trees lived at some distance from the deposition site. This aridity could also be inferred by the increase of Ephedraceae pollen grains (15.0%) with respect to the Tunal Formation (interpreted with care, as some pollen similar to this genus may have been produced by other plants, for example Araceae) (Frederiksen 1985). This family is tropical and prefers humid and marshy environments at low altitudes.

We note that *Rousea patagonica* (frequently represented in the microfloristic assemblages) is very similar to *Cynomorium coccineum* (Cynomoriaceae), but also similar to Adoxaceae and Salicaceae. Furthermore, it must be pointed out that the *Pediastrum* sp. (green algae), an element present in the Tunal, Maíz Gordo and Lumbrera (Fajas Verdes) formations is absent. These formations are interpreted as deposited in lacustrine basins (Quattrocchio and Volkheimer 1990).

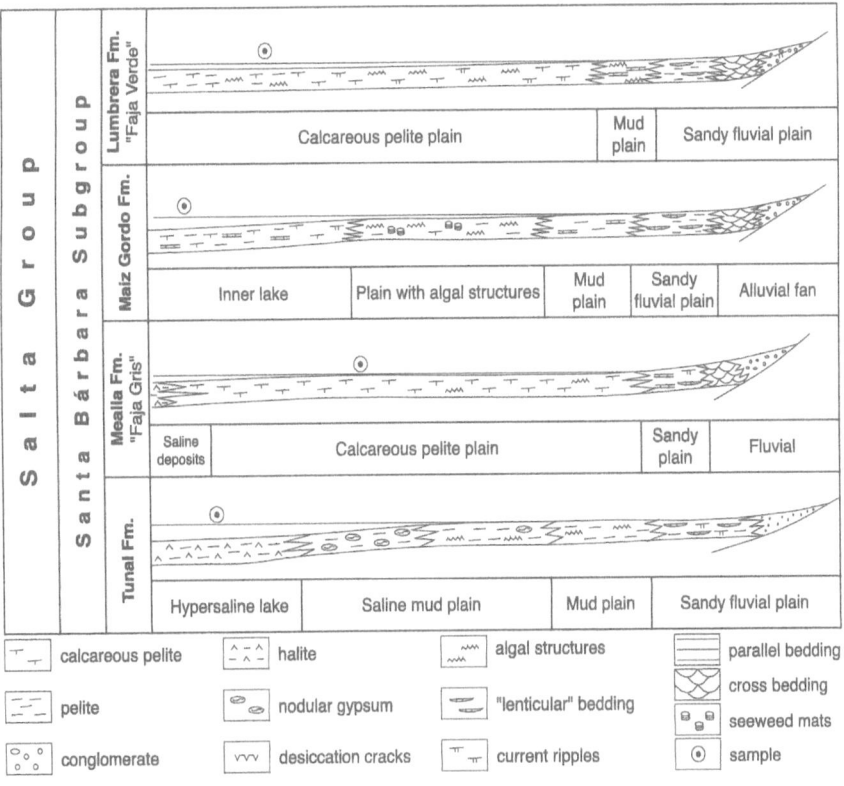

Fig. 21.3. Sedimentary model (without scale) for the Santa Barbara Subgroup, with locations of palynologic samples indicated (after Gómez Omil et al. 1989).

This position of the "Franja Gris" in the Salta Group Basin may imply a deposition in a shallow lacustrine to palustrine environment due to the sedimentologic characteristics (fine sandstone reworked by waves, grey and green pelites, and calcareous levels). There are laminated mudstones that indicate shallow subaquatic conditions. Based on the palynologic and sedimentologic results presented here, the analyzed profile for the "Franja Gris" would be assigned to the "calcareous pelite plain" sub-environment as defined by Gómez Omil et al. (1989). The "Franja Gris" represents a rapid flood and further desiccation in an extremely shallow basin (Gómez Omil et al. 1989).

21.4.3.3. The Maíz Gordo Formation (Thanetian)

The Maíz Gordo Formation at Corralito, southwest of the Salto Jujeña Dorsal, is composed of grey and green fangolites, with intercalations of fine-grained limestone and stromatolitic limestone. This lithofacies characterizes a shallow lake (lithofacies 1 and 2 of Gómez Omil et al. 1989); this is also corroborated by palynologic analysis.

Paleocommunities found in the Maíz Gordo Formation (Volkheimer et al. 1984) correspond to marshes (*Corsinipollenites menendezi*, Oenotheraceae), lakes (*Pediastrum*), sub-tropical moist forest environments (*Verrustephanoporites* cf. *simplex* (23%) occurs in a lower percentage than in the Tunal Formation). The occurrence of *Rhoipites* sp. A (Rutaceae, cf. *Ruta*) in high percentages suggests a higher altitude (montane) paleocommunity and dryness.

21.4.3.4. The Lumbrera Formation (Eocene)

The Lumbrera Formation (Fajas Verdes I and II, Quattrocchio 1978a, b, 1980) is dominated by temperate humid montane paleocommunities documented by the presence of *Tricolpites* cf. *reticulata* (Gunneraceae), *Tricolpites vulgaris* and *Liquidambarpollenites* cf. *brandonensis* (Hamamelidaceae) and *Retitriletes austroclavatidites* (Lycopodiaceae). Swamps and lakes are poorly represented.

21.5. Summary and Conclusions

The Paleocene is one of the key time-slices for the study of green-house climates. We have studied the Salta Group Basin of northwestern Argentina both palynologically and geologically as it represents a key location for questions surrounding climate.

Palynologic analysis is used to reconstruct the environments of deposition and the prevailing climatic regimes during the Paleogene of the Salta Group Basin. The Santa Bárbara Subgroup of the Salta Group is composed of the Olmedo, Mealla, Maíz Gordo and Lumbrera formations (Table 21.1). Based on the pollen assemblage, a Paleocene age is assigned to the Tunal Formation. Due to the absence of *Mtchedlishvilia saltenia* and based on vertebrate data, the Mealla Formation is interpreted to be Middle to Late Paleocene.

The upper part of the Maíz Gordo Formation is Late Paleocene to Early(?) Eocene, based on palynological analysis. The Lumbrera Formation contains palynomorphs of wide stratigraphic range and endemic species. Warm and humid conditions are inferred for the time corresponding to the deposition of the Tunal Formation (a facies of the Olmedo Formation). From palynologic analysis of the "Franja Gris" of the Mealla Formation, we interpret a "calcareous lithofacies." The palynologic record indicates a shallow lake for the Maíz Gordo Formation. The Lumbrera Formation (Fajas Verdes I and II) is dominated by temperate humid montane paleocommunities. The results are analyzed considering the depositional sequences recognized in the Salta Group Basin by other authors.

Acknowledgements: We appreciate the suggestions of Dr. Jose Salfity, Dr. Rosendo Pascual and Dr. Cecilia Del Papa and the critical lecture of the manuscript by Dr. P. P. Smolka and Dr. Bruce Malamud. We also acknowledge bibliography and reference materials from Dr. Juan C. Gamerro and Prof. Alan Graham. The work was supported by grants from CONICET (Consejo Nacional de Investigaciones Científicas y Técnicas) and the National Geographic Society.

References

Archangelsky S (1973) Palinología del Paleoceno de Chubut. I. Descripciones Sistemáticas. Ameghiniana 10(4):339–399

Archangelsky S (1976) Palinología del Paleoceno de Chubut. II. Diagramas polínicos. Ameghiniana 13(1):43–55

Archangelsky S, Zamaloa MC (1986) Nuevas descripciones palinológicas de las formaciones Salamanca y Bororo Paleoceno de Chubut (República Argentina). Ameghiniana 23(1–2):35–46

Boltenhagen E (1967) Spores et pollen du Cretace superieur du Gabon. Pollen et Spores 9:335–353

Cabrera A (1976) Regiones Fitogeográficas Argentinas. Enciclop. Argent. Agricult. y Jardinería 2(1):1–85

Cookson IC (1947) Plant microfossils from the lignites of the Kerguelen Archipielago. B.A.N.Z. Antart. Exp. (A) 2(8):129–142

Couper RA (1953) Upper Mesozoic and Cenozoic Spores and Pollen Grains from New Zealand. New Zealand Geol. Surv. Paleont. Bull. 22:1–77

De Spirito RE (1979) Estudio en detalle y análisis de cuenca de los Subgrupos Balbuena y Santa Bárbara, en el sector centro oriental de la Subcuenca de Lomas de Olmedo y umbral de Cachipunco. Inf. Ined., YPF. Bs. As

Elsik WC (1968) Palynology of a Paleocene Rockdale lignite, Milam County, Texas; Part 2, Morphology and taxonomy. Pollen and Spores 10(3), 599-664

Escobar TF (1980) Mapa Geológico de Chile (escala 1:1.000.000). Serv. Nac. Geol. Min., Santiago

Frederiksen NO, Carr DR, Lowe GD, Wosika EP (1983) Middle Eocene Palynomorphs from San Diego, California, Part. I. AASP. Cont. Ser. 12:1–32

Frederiksen NO (1985) Review of early Tertiary sporomorph paleoecology. Am. Ass. Str. Palyn. Contrib. Series 15:1–92

González Guzmán AE (1967) A palynological study on the Upper Los Cuervos and Mirador Formations (Lower and Middle Eocene; Tibú Area, Colombia). Leiden, 1–67

Gómez Omil RJ, Boll A, Hernandez RM (1989) Cuenca Cretácico–Terciaria del Noroeste Argentino (Grupo Salta). In: Chebli G, Spalletti L (eds) Cuencas Sedimentarias Argentinas. Serie Correlación Geológica 6,:43–64, Universidad Nacional de Tucumán

Heusser CJ (1971) Pollen and Spores of Chile. Modern Types of the Pteridophyta, Gymnospermae and Angiospermae. University of Arizona Press, Tucson 167 p

Heizler M, Kay RF, Madden RH, Mazzoni MM, Re GH, Sandeman H, Vucetich MG (1998) Geochronologic Age of the Casamayoran fauna at Gran Barranca, Chubut Province, Argentina. VII Congreso Argentino de Paleontología y Bioestratigrafía (Bahía Blanca), Octubre, 1988, p. 89

Jordan TE (1984) Cuencas, vulcanismo, y acortamientos cenozoicos, Argentina, Bolivia y Chile, 20–30° Latitud Sur. IX Congreso Geológico Argentino, Actas II: 7–24. Bariloche

Krutzsch W, Lotsch D (1966) Das Oligo-Miozaen-Profil der Lausitz (The Oligo-Miocene cross-section of Lusatia). Zeitschrift fuer Geologische Wissenschaften 3, 158-161

Leidelmeyer P (1966) The Paleocene and Lower Eocene pollen flora of Guyana. Leidese Geologische Mededelingen. 38:49–70, Leiden

McIntyre DJ (1965) Some new pollen species from New Zealand Tertiary deposits. New Zealand J. Bot. 3:204–214

Moreno JA (1970) Estratigrafía y paleogeografía del Cretácico Superior en la cuenca del Noroeste argentino, con especial mención de los Subgrupos Balbuena y Santa Bárbara. Asoc. Geol. Arg. Rev., 24(1):9–44

Moroni AM (1984) Mtchedlishvilia saltenia. sp. en sedimentitas del Grupo Salta, Prov. de Salta. Actas. III Congr. Arg. Paleont. Bioestr. 129–139

Padula EL, Mingramm ARG (1968) Estratigrafía, distribución y cuadro geotectónico-sedimentario del "Triásico" en el subsuelo de la llanura Chaco–Paranense. Jornadas Geológicas Argentinas, 3rd, Comodoro Rivadavia (1966), Actas, 1:291–331

Pascual R, Bond M, Vucetich MG (1981) El Subgrupo Santa Barbara (Grupo Salta) y sus vertebrados. Cronología, paleoambientes y paleobiogeografía. VIII Congr. Geol. Argent., Actas III: 746–758, San Luis

Pierce RS (1961) Lower upper Cretaceous plant microfossils from Minnesota. Bull. Geol. Surv. Univ. Minn. 42:1–86

Quattrocchio M (1978a) Datos paleoecológicos y paleoclimatológicos de la Formación Lumbrera (Grupo Salta). Ameghiniana 15(1–2):173–181. Buenos Aires

Quattrocchio M (1978b) Contribución al conocimiento de la Palinología Estratigráfica de la Formación Lumbrera (Terciario inferior, Grupo Salta). Ameghiniana 15(3–4):285–300

Quattrocchio M (1980) Estudio palinológico preliminar de la Formación Lumbrera (Grupo Salta), loc. Pampa Grande, provincia de Salta. II Congr. Arg. Paleont. Bioestr. y I Congr. Latinoam. Paleont. Actas II:131–149

Quattrocchio M (1984) Sobre el posible significado paleoclimatico de los quistes de dinoflagelados en el Jurasico y Cretacico inferior de la Cuenca Neuquina (Paleoclimatologic significance of dinoflagellate cysts in the Jurassic and Lower Cretaceous of the Neuquen Basin. Actas del III Congreso Argentino de Paleontologia y Bioestratigrafia 3, 107-113

Quattrocchio M, Marquillas R, Volkheimer W (1988) Palinología, Paleoambientes y Edad de la Formación Tunal, Cuenca de Salta (Cretácico–Eoceno), Republica Argentina. IV Congr. Arg. de Paleont. y Bioestr. Actas 3:96–109, Mendoza

Quattrocchio M, Volkheimer W (1988) Microfloras de los estratos limítrofes entre Cretácico y Terciario en las localidades de Tilian y Corralito, Cuenca de Salta. Descripciones sistemáticas. IV Congr. Arg. de Paleont.and Bioestr (1986). Actas 3:109–120, Mendoza

Quattrocchio M, Volkheimer W (1990) Paleogene paleoenvironmental trends as reflected by palynological assemblage types, Salta Basin, NW Argentina. N. Jb. Geol. Palaont. Abh. 181(1–3):377–396

Quattrocchio M, Volkheimer W, Del Papa C (1997) Palynology and paleoenvironment of the "Franja Gris" Mealla Formation (Franja Gris) Salta Group at Garabatal Creek (NW Argentina). Palynology 21:231–247

Quattrocchio M, Ruiz L, Volkheimer W (in press) Palynological zonation of the Paleogene of the Colorado and Grupo Salta basins, Argentina. Revista Española de Micropaleontología

Reyes FC (1972) Correlaciones en el Cretácico de la cuenca andina de Bolivia, Perú y Chile. Revista Técnica de Yacimientos Petrolíferos Fiscales Bolivianos, 1:101–144, La Paz

Salfity JA (1968) Perfil geológico en la quebrada del rio Corralito, Salta. Univ. Nac. Tucuman, Fac. de Cienc. Natur. Sem. I, Salta (Unpublished scientific report)

Salfity JA (1971) Paleogeología de la cuenca del Grupo Salta (Cretácico–Eogénico) del norte de Argentina. VII Congr. Geol. Argent., Actas I:505–515, Buenos Aires

Salfity JA (1980) Estratigrafía de la Formación Lecho (Cretácico) en la Cuenca Andina del Norte Argentino. Universidad Nacional de Salta, Publicación Especial, Tesis 1, 91 pp. Salta

Salfity JA (1982) Evolución paleogeográfica del Grupo Salta (Cretácico–Eogénico), Argentina. V Congr. Latinoamer. Geol., Actas I:11–26; Buenos Aires

Salfity JA, Marquillas RA (1981) Las unidades estratigráficas cretácicas del Norte de la Argentina. In: Volkheimer W, Musacchio EA (eds): Cuencas Sedimentarias del Jurásico y Cretácico de America del Sur, Comité Sudamericano Jurásico y Cretácico, I:303–307, Buenos Aires

Salfity JA, Marquillas RA (1994) Tectonic and Sedimentary Evolution of the Cretaceous–Eocene Salta Group Basin, Argentina. In: Salfity J (ed) Cretaceous tectonics of the Andes, Cambridge University Press, Cambridge, pp 266–315

Schwab K (1984) Contribución al conocimiento del sector occidental de la cuenca sedimentaria del Grupo Salta (Cretácico–Eogénico), en el noroeste argentino. Congreso Geológico Argentino, 9th, Bariloche, Actas, 1:586–604

Schwab K (1985) Basin formation in a thickening crust. The intermontane basins in the Puna and the Eastern Cordillera of NW Argentina (Central Andes). Congreso Geológico Chileno, 4th Antofagasta, Actas 1(2):138–158

Sheffy MV, Dilcher DL (1971) Morphology and taxonomy of fungal spores. Palaeontographica, Abt. B 133(1–3):34–51

Stover LE, Partridge AD (1982) Eocene spore-pollen from the Werillup Formation, Western Australia. Palynology 6, 69-96

Traverse A (1955) Pollen analysis of the Brandon Lignite of Vermont. U.S. Bur. Mines Rep. Invest. 5151

Van der Hammen T (1954) El desarrollo de la flora colombiana en los períodos geológicos. I. Maestrichtiano hasta Terciario mas inferior. Boletín Geológico, II(1):49–106, Bogotá

Vilela CR (1965): El petróleo en las cuencas de Orán y Metán. Jornadas Geológicas Argentinas, 2nd, Salta (1962), in Acta Geológica Lilloana 7:425–438, Tucumán

Volkheimer W (1972) Palinomorfos como fosiles guia (Palynomorphs as guide fossils, Part 1). Revista Minera, Geologia y Mineralogia 30(2) 17-72.

Volkheimer W, Melendi DL (1976) Palinomorfos como fosiles guia; Tercera parte, Tecnicas de laboratorio palinologico (Palynomorphs as guide fossils; Part three, Palynologic laboratory techniques). Revista Minera, Geologia y Mineralogia. Buenos Aires

Volkheimer W, Quattrocchio M, Salfity JA (1984) Datos palinológicos de la Formación Maíz Gordo, Terciario inferior de la Cuenca de Salta. IX Congr. Geol. Arg. Actas IV:523–538, Bariloche

Chapter 5: Modeling

This chapter describes tools that can be used for paleoenvironmental analysis and climate modeling.

These are in particular: The Eximag System on the one hand and the atmospheric general circulation model ccm3.6 from NCAR, ported to Windows NT on the other.

While in modern atmospheric modeling the possible approaches are straightforward, paleoclimate modeling requires additional precautions as the applicability of the results depends highly on the question to be solved, the properties of the model and the nature or the input data, especially the question of "what is prescribed and what should be predicted".

The Eximag System

Gerald Peschel[1], Peter Smolka[2]
(1) Institut für Geologische Wissenschaften, University Greifswald, Germany
(2) Peter Smolka, Geological Institute, University Muenster, Corrensstr. 24, D-48149 Muenster, Germany
Smolka@uni-muenster.de

Abstract: A set of tools is provided that permits, beyond the standard tasks of time-series analysis, the detection of chaos and the simulation of sequences that contain both linear, periodic and chaotic elements. Other tools for modeling self-organizing processes are included.

The Main Features

The Eximag system is a suite of tools used for time-series analysis. Techniques such as moving average, autocorrelation, and Fourier analysis, are common, and although part of the Eximag system, will not be discussed here.

Time series consist of not only linear and periodic components, but also chaotic influences. In addition, these factors may change within a time series. Beyond description, the modeling of time series is an essential part of understanding. Thus, tools are supplied that interactively permit the detection of linear, periodic and chaotic parts of time series. These tools permit one to model respective time series by specifying linear, periodic and chaotic components, including their respective importance. Thus by regional and hemisphere-wide comparison, conditions preceding fundamental oceanographic changes can be traced. Furthermore, after transforming continuous data into discrete intervals, time series can be described *and* modeled utilizing Markov chains. Independent of these paleoclimatic tools, examples (data and the program) are shown to simulate periodic and rhythmic sediments (saline-facies, two-dimensional deltaic sedimentation) through self organizing processes.

Although these tools are primarily intended as a series of "Experiments in Mathematical Geology" their application to the analysis of paleoclimatic time-series shows potential future lines of research which appear to be promising. A more detailed explanation including formulas, algorithmic principles and software can be found on the accompanying CD in the EXIMAG-Directory.

Principles and some Future Perspectives of Paleoclimate Modeling

Peter Smolka
University Muenster, Geological Institute, University Muenster, Corrensstr. 24, D-48149 Muenster, Germany
Smolka@uni-muenster.de

Abstract: The potential result of paleoclimate modeling depends on not only the model, its resolution and the processes included, but also on the question the model should answer.

23.1. Introduction

While many parameters of the environment can be reconstructed, several others such as wind and pressure fields can only be modeled. As this is done generally by atmospheric circulation models, the application of such models to paleoclimatic questions appears to be a routine task. A closer inspection of models, necessary data and potential achievements however shows that, depending on the question, several possible analyses yield different solutions depending on the type of the question. Technically atmospheric general circulation models are weather forecast models that simulate the physical processes of the atmosphere at a reasonable resolution in space (often T42) and time (every 15 minutes). This includes, depending on the type of the model also properties of the land-surface (see the enclosed ccm3.6 from NCAR, Boulder Colorado, USA).

23.2. Modeling Climate

It is obvious that due to both technical limitations ("resolution") and the description of various physical processes, many phenomena have to be formulated in a simplified way. The ability of a model to handle present-day situations as expected is regarded as an indicator for the correctness of both simplified and explicit descriptions of physical processes.

With the advent of questions focusing on global change caused by increased "greenhouse-gases" model simulations have been run that use (formulated in a simplified manner) present-day boundary conditions with increased concentrations of greenhouse gases. According to meteorological definitions, climate is the long-term average of short-term phenomena, called weather. Thus for the detection of potential climate changes, meteorological simulation runs covering many months or years are performed to compare the results statistically. The background of this approach is the meteorological approach behind it: The question is *not*: "(1) Start a run with a synoptic situation of lets say 16th of September 1910. (2) Start another run with the same synoptic situation but increased CO2 contents. (3) Compare differences of the location of the fronts after for example 2 days."

From the meteorological point of view the question is: "(1) Run a weather of the beginning of the century for some years. (2) Compare it with the weather at the end of the century (changed CO2) after a run covering again some years. (3) Are after long simulation runs any differences significant statistically?

Although the first approach is theoretically possible, meteorology applies the second approach following the theoretical definition of climate. This explains in addition in the retrospective why - amongst other reasons - the impact of greenhouse-gases was so difficult to detect: The statistical detection of the change of a certain clearly defined weather phenomenon, such as the "depth" of the lows in a clearly defined region or the "precipitation in a certain region" is quite straightforward as the "measured data" are (formulated simplified) either the same, greater or lower. A theoretical approach testing *global* datasets for warming is due to the differential nature of climate change (areas of warming accompanied by areas of cooling) likely to be misleading as *both* warming *and* cooling occurs in different regions at the *same* time.

The same applies, if geological data are used, to the detection of an anthropogenic climate signal, as both Quaternary and Neogene climate fluctuations (see various contributions in this volume) supercede any anthropogenic impact by far. The essential question to be asked however is: Are we at the moment, maybe triggered by a human experiment, leaving our exceptional field of stability and do we already move on a trajectory leading to one of the many dramatical changes that have been observed in the younger geological history?

The above-mentioned classical meteorological approach ("averaging the weather") means that not only a full annual cycle has to be modeled. In addition, other long-term phenomena, such as snow and ice, oceans, river runoff, etc., have to be considered. This raises the important question of the inclusion of the ocean.

23.3. Introducing the Ocean

The interface of the ocean and the atmosphere, the sea surface, is an equilibrium of long-, medium- and short-term internal ocean dynamics and external influences of the atmosphere. As the formation of sea ice and changing albedo are short-term phenomena, climate models moved from the "data-ocean" (prescribed or externally modeled sea-surface temperatures) to modeling the upper mixed layer of the ocean. Energy fluxes at the bottom of the mixed layer are handled in a simplified way (slab-ocean models).

From a geological perspective the use of such "slab ocean models" is an oversimplified approach. Significant energy fluxes as well as the removal of CO_2 from the atmosphere in areas of deep-water formation (Weddel-Polynya and the area south of Iceland) depend crucially on the inclusion of the deep-ocean circulation. This either, as said, through data (ocean-atmosphere equilibrium documented by sea-surface temperatures) or through realistic modeling of the whole-ocean circulation, including its bathymetry, salinities, various water bodies, etc.

Experience shows that among the "pure-model" experiments (slab-ocean versus fully-coupled models) the fully-coupled models produce the most realist results. Readers interested in the physics of the models or the result of model intercomparison studies are referred to Trenberth (1992) and to the publications of the various climatic research centers such as NCAR (USA) or DKRZ (EU). Examples are Acker et al. (1996), DKRZ, Modellbetreuungsgruppe (1993), Maier-Raimer and Mikolajewicz (1992), Oberhuber (1992), Kluszek et al. (1998).

The effects of ocean circulation can realistically be included either with the help of a reconstructed "data ocean" or with a fully-coupled atmosphere-ocean model. The latter, however, runs into difficulties for older (greenhouse) climatic scenarios as both salinity of the ocean and bathymetry (including that of gateways like the Panama Straits, the Drake Passage, the Faeroe-Shetland-Channel) have to be included.

23.4. Questions, Inputs and Results

The questions and approaches in weather forecast are very straightforward. In paleoclimate modeling, however, the wide range of potential questions permits also a wide range of conceptual experiments, model applications and thus also a wide range of results. In other word, the maximum result to be achieved depends highly on the question.

23.4.1. The Integrated Approach

The most common question is: "How did (does) a climate with increased or decreased CO_2 contents look like" or "How did (does) the paleoenvironment in a certain time-interval look like" (studies of paleoanaloga). For such questions the ocean could be reconstructed. The same applies to known orographic properties like land surfaces and albedos.

Studies following this concept use the model to predict the non-reconstructable parameters such as wind fields, precipitations and, depending on the model, vegetation coverage. As the ocean is "prescribed" (reconstructed) the internal ocean dynamics are included through the distribution of sea-surface temperatures. This has the great advantage, especially in older times, that potentially unknown bathymetric features such as submarine plateaus are, if they have any effect, included indirectly through their climatic impact, namely the sea-surface temperatures (such as areas of upwelling). Thus the potential results of this experiment are: (1) Assessment of the non-reconstructable parameters such as wind-fields. (2) Insight into the reliability of the predictions through intercomparison of predicted vegetation (inferred from the precipitation or modeled) with geological reconstructions. (3) Assessing the ability of the model to handle non-analogue situations, which implies, not only for the climate model but the whole "prediction-system" that intercomparison of model "predictions" (such as precipitations, sea ice distribution etc.) with known geological data helps in addition to verify the prediction-system itself.

In addition to increased insight in non-analogue climates such experiments solve the following important question: "Considering the known instability of climate and the difficulties to predict it: "What would the environment look like if the next climatic equilibrium would, for example, be similar to that of the Upper Pliocene before the great Northern Hemisphere glaciations?" In addition, this integrated approach permits sensitivity studies utilizing known (reconstructed) oceans and various contents of green-house gases. The model on the CD (ccm3.6 from NCAR in the data ocean version with included land-surface model) follows this approach.

23.4.2. Pure Modeling Approaches (Sensitivity Studies)

To understand the ability of a model to handle non-analogue situations, runs with reduced and increased greenhouse gases could be executed. The same applies to tests with various orographies. As long as, for such studies, the ocean is prescribed as data ocean, such studies are sub-studies of those mentioned in the previous chapter. If the ocean is fully modeled with all bathymetric features, a question that is especially difficult if older times (including the Lower Pliocene) are considered, then such studies are sensitivity studies to learn about the *model*. The reason is that such studies would not allow the inclusion of known *and* reconstructed data. The only way to include such data would be to model all relevant processes completely (including bathymetry, landforms, vegetation) and to compare them with reconstructed oceans. In other words, such future studies could be performed using the data of IGCP-341 (included on the CD).

23.5. Modeling Paleoclimate and Paleoanalogs

Above it was said that for meteorological questions in a first step weather is modeled. In a second step the results of such model-runs (that used the same or different boundary conditions, such as changed greenhouse-gases) are compared statistically. While reconstructed time-series of sea-surface temperatures can locally have a very high resolution, the synopsis to maps (oceans) requires, normally some averaging for reasons of dating (stratigraphy). To run a model for reconstructions from the Quaternary, such as the last glacial maximum (18 ka) the averaging might span over some 100 years as synchronous worldwide synopses are even in the Quaternary not possible at a higher resolution. This does *not* mean that the dating methods have deficits. It only means that data from wells reflecting different "state of the art" have to be included in one map.

For older times a worldwide uniform high-resolution stratigraphy covering the Neogene and Paleogene did not exist up to now. This explains why *this* study was carried out under the umbrella of IGCP (International Geological Correlation Program). After this problem had been solved during IGCP-341 (data on CD) it was found that for Neogene times, maps of average temperatures covering time-spans of 1 million years could be drawn. Again, in many drill sites (data on CD) the resolution (spacing of the samples) is much higher.

Thus in both cases (Quaternary and Neogene) any reconstructed ocean reflects an average equilibrium of internal ocean dynamics and external impact of the atmosphere. Metaphorically speaking, such maps are similar to figures that compare average summer temperature of Helsinki (Finland) with that of Ko Samui (Thailand). Although in neither city cold or warm summers can be excluded, the comparison of such average values gives some indication of items such as land-use planning. This means however that the application of climate models to paleo-situations can follow different and more economic principles:

As any reconstructed ocean is already an average equilibrium, the atmosphere driven by this ocean is an average equilibrium as well, even without averaging longer times. This means that computationally the model can be stopped after reaching a numerical-equilibrium state which, depending on the model (with/without land surface model), may be reached after one to three computed months or years (if vegetation is included). Thus, comparing a situation at 4 million years with one at 5 million years, the question to be formulated is: "Are the fronts, lows and highs located at about the same positions or are there differences? Or: Do they exist at all, as a pronounced and strong global circulation should not be taken to for granted?" In other words: Conclusions can *only* be drawn from very *pronounced* differences such as shifts of vegetation-belts, changes of the jet-stream and phenomena of a similar magnitude.. *If* however such phenomena can be observed, their causes must be studied seriously as they are, due to the average nature of the driving oceans, much more significant than a "series of dry years".

23.6. Future Perspectives of Paleoclimate Modeling

By studying already realized climatic equilibria from the past, scenarios may be established that consist of catalogues of "how environments may look like if comparable environmental conditions occur again." In addition, studies which cover a wide range of climatic equilibria may also show that, even if some ruling global parameters (such as the CO_2 content) change, in *some* regions certain key-parameters for land-use planning, such as precipitation, may remain constant while in adjacent regions they change. This means for a land-use-planning authority that if both in the Neogene *and* in the Quaternary *and* in the Holocene (little ice-age, climatic optimum of the 12th century) "nothing changed" or changes have been even more favorite than the present-day situation any potential future climate change is "nothing to fear", but something to adjust to. As paleoclimate studies deal with "real climates" (already realized equilibria) they could in some aspects be a valuable complement to "pure-modeling approaches".

Eyeballing paleoclimatic time-series from the Quaternary and Neogene shown by various authors in this volume, it is obvious that in some regions dramatic shifts occurred while in other over comparable long times the paleotemperatures remained fairly stable. Therefore the following questions arise:

1) How to detect the advent of a fundamental change in advance?
2) How to predict it?
3) How to predict one global climatic equilibrium based on the preceding one or based on a preceding series of climates?

The climate system can be regarded as sufficiently understood if transient model runs covering long time intervals could be performed such that the changing orography (for example during the last 3 my) is superimposed on the model as background-data and the known climate shifts, including the large Northern Hemisphere glaciations, the rapid deglaciation events and related phenomena are reproduced well. This includes ideally the autonomous prediction of the sudden change from Upper Pliocene conditions without large Northern Hemisphere glaciations before 2.6 my to widespread glaciations including an ice-covered polar sea after that time. The following section focuses on a series of potential numerical experiments that might help solving these questions.

23.6.1. Predicting Time-Series with the Help of Neural Networks

Calvo et al. (this volume) showed that time-series based on physical processes could be used for forecasting using neural networks. The same may apply (with some limitations) for the completion of existing geological time-series (bridging "gaps" caused by times of non-deposition / erosion). As paleoclimate time series, from ice cores, isotopes, faunal and floral changes and paleotemperatures (various authors, this volume, data on CD) exist, experiments are suggested to predict a part of a time series from its preceding sequence. If this is done worldwide, the resulting maps could be checked against the existing ones. In case of success, future climatic conditions, could, at least experimentally, be predicted from a set of existing time-series that are distributed in space worldwide.

23.6.2. Detecting and Simulating the Interference of Linear, Periodic and Chaotic Components

Another important question for climate change is not the prediction itself but the assessment of the predictability of climate change. There is no doubt that analytical functions could be written easily as Fourier series. The derivation of Fourier coefficients from existing time-series follows normally a *different* approach. Computationally, the length of the time series is taken as the length of the "first harmonic". The second harmonic has half this length, the third a quarter, and so forth. Although the results can be tested statistically, this means that the results ("harmonics found") depend on the length of the time series. It could be demonstrated that a sine wave of, for instance a one million year wave length, is recognized easily. If, based on the same wavelength another half wave is added (that is, a numerical experiment with *known* wavelength) different harmonics result although the wavelength is the same (tools to verify this can be found in the EXIMAG directory on the accompanying CD).

This means that the *description* of time series through harmonics, even if the results are statistically significant, is not necessarily understanding. This applies also to time series derived from isotope data. The isotope stages are observations. The interpretations about the causes of these fluctuations, are if they are based on Fourier transforms, only expressions of the present state of the art. Eyeballing natural time series (isotope-data, paleotemperatures, faunal communities etc.) some time series appear to be fairly predictable while others appear to document linear, periodic and chaotic properties. Plotting, from old to young, the value of the $n+1^{th}$ point against the value of the nth point of such a time-series shows sequences of points that move either on linear paths, move cyclically or jump "apparently erratically" often creating two or more clusters of points.

Vertical analysis of time series following this approach enables the detection of characteristic patterns (linear, periodic, chaotic) before and during times of climate change. The simulation through the respective equations which permits the experimental simulation of the importance of linear, periodic and chaotic components (tools in the EXIMAG-Directory) is another approach to analyze time series without limitations of classical Fourier transforms.

23.6.3. Neural Networks in Space

It was said previously that neural network applications to time series could be used for predictions, especially if the ruling processes have a physical background. Neural networks can, however, not only be trained with time series but also with data that are distributed *both* in space *and* in time, such as worldwide weather stations. This means that there is the perspective that present-day meteorological and oceanographic data (the routine meteorological and oceanographic measurements including of course the conditions between the ground and the 100 hPa surface) could experimentally be used to train neural networks for weather predictions. As personal computers are now easily available that can contain 700 MB of installed memory, plus Gigabytes of virtual memory on the hard disk, such an experiment can also be carried out in practice.

While the training of large neural networks often needs considerable time, the application to existing problems is extremely fast. Even if only very short time intervals are predicted experimentally, the resulting data-set (weather) could be used as "latest input data-set", while at this moment the "oldest input data-set" is skipped (so the number of "datasets" is kept constant). This leads to the question of transient modeling over long time-intervals, such as the last three million years.

23.6.4. Transient Modeling over Long Time-Periods

It is a generally accepted fact that variation of orbital parameters (Milankovich-Cycles) contribute to climatic fluctuations. Other phenomena such as the Mid-Pleistocene revolution (of cycles) show that the "climate system" itself interacts with orbital forcing differently at different times. We note that physical systems that are characterized by two or more coupled counteracting processes tend to oscillate, even without external forcing.

In this context, some decades ago, F. Strauch brought up a concept about Northern Hemisphere glaciations that might, with considerable modifications and extensions be still worth testing, especially as nowadays with state-of-the-art of modeling, such a test might be technically possible in some years. It is a striking feature that *large* scale Northern Hemisphere glaciations occurred rather suddenly at about 2.6 Ma. It should in this context be noted that of course East Greenland was glaciated much earlier. Furthermore, the Norwegian Greenland Sea was characterized by cold conditions, including minor amounts of ice-rafted debris since the Miocene times. Cold conditions are one precondition of glaciations, moisture (precipitation) is, as the non-glaciated Siberia shows, the other prerequisite. Strauch proposed the concept that the *massive* inflow of the gulf current into the cold Norwegian Greenland Sea across the Iceland Faeroe Ridge (400 m below sea level) and through the Faeroe-Shetland Channel, which was, due to the temperature difference also a massive energy transfer, initiated large-scale precipitation of snow on the neighboring continents. According to that concept (which is now about 25 years old), lowering sea level together with sea ice inhibited that inflow and thus reduced the precipitation during the high glacial events. Melting glaciers reversed that cycle until it could restart again (with the known phenomena of deglaciations including Heinrich events).

Discussions about the subsidence history of the Greenland–Scotland Ridge are beyond the scope of this chapter. It is also evident that this concept is formulated in a highly simplified way (see the large number of isotopic stages). The main idea however is that a set of known physical processes that are in addition modulated by superimposed astronomical forcing could interact such that even the known rapid oscillations at the end of the LGM are caused by the self-organizing interaction of different, counteracting processes. These includes oscillations of the deep-water formulation as well.

During the early 1980's, Strauch proposed to test this hypothesis through climate modeling. This would have required transient modeling of 3 million years starting with a reconstructed Upper Pliocene pre-glacial ocean. In this context it was expected that after the *massive* inflow of warm Atlantic waters into the cold Norwegian Greenland Sea, climate models would have been able to produce the known climate cycles autonomously as only orography (bathymetry) and orbital parameters should be superimposed externally. Although the question is challenging, especially if the modeling starts 5 million years ago with an open Straits of Panama, a lower Himalaya and existing Paratethyan Seas, it is still now (1999) an open question, seeking either improved climate models, new algorithmical experiments and/or faster computers.

The above mentioned ccm3.6 experiment needs, using an NCAR Cray, about 9 minutes per day in single processor operation and 2.5 minutes if running on four processors. Under Windows NT (the version on the CD) it needs 35 minutes on a Pentium II/300 and 18 minutes on a Pentium II/450 IF in addition all other optimizations are performed (the version on CD is optimized for safety). Thus simulating an equilibrium (one month) would take about 35 x 30 minutes = 1050 minutes or about 18 hours. However, simulating one million years would need on the NCAR Cray in the four processor configuration, 2.5 minutes per day, [(2.5 min/dy x 365 dy/yr x 1 million yr) / (60 min/hr x 24 hr/dy)] = 633,680 days = 1736 years computer time. On a standard Pentium this would take about 12 thousand years. In other words, a careful inspection of the code must yield a speed optimization of a factor of twelve thousand (together with faster computers) to turn the vision of Strauch (which was part of the German Paleoclimate Program funded by the Ministry of Research and Technology, BMFT) into reality.

CCM3.6, which is included on the CD, shows the history of system analysis and software development of the last 15 years. This applies especially to the early needs of climate modeling to utilize disk storage and thus perform massive data-transfer. In order to maintain the logic of the model, the data are in the current versions held in core in large buffers and thus transferred to and from these buffers into the working arrays. Thus, a future version might be optimized through the replacement of the large intermediate buffers by the matrices which contain the actual data (reduced internal data transfers). Many equations and equation systems are solved iteratively.

A careful inspection of time spent in the various modules, potentially combined with attempts to solve some equations explicitly, possibly with the help of programs performing symbolic computation, might provide a boost of performance. Whether this yields a factor of 12 thousand is most likely impossible presently, as even a factor of thousand would be a major breakthrough.

The crucial point however with ccm3.6 is the inclusion of the ocean, as the version on CD is driven by a (reconstructed) data ocean (the limitations of slab-ocean models are mentioned above). In this context the transfer and analysis of fully coupled models might be useful. Thus the transient modeling of such long times is still a major challenge. Therefore, paralleling the above approaches, the application of artificial neural networks as sketched above or even the formulation of the meteorological and oceanographical equations in a form suitable for rapid finite-element solver like FastFlow (from CSIRO/Sydney) might show new ways to turn questions into reality (see the documentations of various state of the art models running at NCAR or DKRZ).

The various groups of IGCP-341 contributed not only methods and solutions for the assessment of global change but provided also reference data both from the oceans and the continents (Anhuf, Latrubesse, Runge and others this volume). These permit, beyond any ambitious modeling, calibration of the model results with geological facts, not only for 18 ka, but also for a set of warmer climatic equilibria.

Acknowledgement: The useful hints of E. Kluszek, NCAR during the portation of the model (data formats, etc.) from Cray environment to Windows NT are gratefully acknowledged.

References

Acker TL, Buja LE, Rosinski JM, Truesdale JE (1996) Users Guide to NCAR CCM3. NCAR Technical Note NCAR/TN-421+IA, National Center for Atmospheric Research, Boulder, Colorado

DKRZ Modellbetreuungsgruppe (1992) The ECHAM 3 Atmospheric General Circulation Model. Deutsches Klimarechenzentrum, Technical Report No. 6, Hamburg

Kiehl JT, Hack JJ, Bonan GB, Boville BA, Briegleb BP, Williamson DL, Rasch PJ (1996) Description of the NCAR Community Climate Model (CCM3). NCAR Technical Note NCAR/TN-420+STR, National Center for Atmospheric Research, Boulder, Colorado

Kluzek EB, Olson J, Rosinski JM, Truesdale JE, Vertenstein M (1998) Users Guide to NCAR CCM3.6. National Center for Atmospheric Research, Boulder, Colorado

Maier-Raimer E, Mikolajewicz U (1992) The Hamburg Large Scale Geostrophic Ocean General Circulation Model (Cycle 1). Deutsches Klimarechenzentrum, Technical Report No. 2, Hamburg

Oberhuber JM (1992) The OPYC Ocean General Circulation Model. Deutsches Klimarechenzentrum, Technical Report No. 7, Hamburg.

Trenberth KE (1992) Climate System Modeling. Cambridge University press, 788 pp